T0212905

Lecture Notes in Computer Science 9213

Commenced Publication in 1973
Founding and Former Series Editors:
Gerhard Goos, Juris Hartmanis, and Jan van Leeuwen

More information about this series at http://www.springer.com/series/7407

Jean-Daniel Boissonnat · Albert Cohen
Olivier Gibaru · Christian Gout
Tom Lyche · Marie-Laurence Mazure
Larry L. Schumaker (Eds.)

Curves and Surfaces

8th International Conference
Paris, France, June 12–18, 2014
Revised Selected Papers

 Springer

Editors

Jean-Daniel Boissonnat
INRIA, Sophia Antipolis
Sophia Antipolis
France

Albert Cohen
École Normale Supérieure
Paris
France

Olivier Gibaru
Arts Et Metiers ParisTech Lab de Sciences
 de l'Information
Paris
France

Christian Gout
INSA de Rouen, LMI
Saint-Étienne-du-Rouvray
France

Tom Lyche
Department of Mathematics
University of Oslo
Oslo
Norway

Marie-Laurence Mazure
Université Joseph Fourier
Grenoble
France

Larry L. Schumaker
Department of Mathematics
Vanderbilt University
Nashville, TN
USA

ISSN 0302-9743   ISSN 1611-3349   (electronic)
Lecture Notes in Computer Science
ISBN 978-3-319-22803-7   ISBN 978-3-319-22804-4   (eBook)
DOI 10.1007/978-3-319-22804-4

Library of Congress Control Number: 2015946095

LNCS Sublibrary: SL1 – Theoretical Computer Science and General Issues

Springer Cham Heidelberg New York Dordrecht London

Printed on acid-free paper

Springer International Publishing AG Switzerland is part of Springer Science+Business Media
(www.springer.com)

Preface

The Eighth International Conference "Curves and Surfaces" was held in Paris at the Arts & Métiers ParisTech, Paris Campus, during June 12–18, 2014. The conference was part of an ongoing joint French-Norwegian conference series on curves and surfaces. Previous meetings in France were held in Chamonix in 1990, 1993, and 1996, in Saint-Malo in 1999 and 2002, and in Avignon in 2006 and 2010. The last meeting in Norway was held in Oslo in 2012. Proceedings have been published for all of the meetings.

The 2014 edition of Curves and Surfaces was attended by 280 participants from 35 different countries. The program included 10 invited one-hour survey talks, 10 minisymposia comprising 46 talks, 115 contributed talks, and 2 poster sessions with a total of 36 posters.

The conference was supported financially by the following institutions: Arts et Métiers ParisTech; Institut National de Recherche en Informatique et en Automatique (INRIA); Laboratoire Jean Kuntzmann (LJK, Grenoble); Laboratoire des Sciences de l'Information et des Systèmes (LSIS, Marseille); Maison de la Modélisation et de la Simulation, Nanosciences et Environnement (MaiMoSiNE, Grenoble).

Among the many people who helped with the conference in various ways, we would like to especially thank Yvon Lafranche for preparing the program and abstract booklets. We would also like to thank our webmaster Eric Nyiri for setting up the web site and helping participants with computer issues, and Paul Sablonnière for his help with registration and the book display.

Thanks are also due to all of the invited speakers and to all of the minisymposium organizers whose contributions were critical to making this conference a success. We would also like to thank all other presenters and participants, not forgetting the chair-persons who played a decisive role in ensuring the smooth running of the congress. Finally we would like to express our gratitude to all of the reviewers, who helped select articles for the present volume.

The meeting was held in the very building where Pierre Bézier studied engineering from 1927 to 1930. The meeting included a special session dedicated to him, as well as a party in his honor subsidized by the "Fondation Arts et Métiers". We thank them sincerely.

July 2015

Jean-Daniel Boissonnat
Albert Cohen
Olivier Gibaru
Christian Gout
Tom Lyche
Marie-Laurence Mazure
Larry L. Schumaker

Organization

Organizing Committee and Editors

Jean-Daniel Boissonnat	Inria Sophia Antipolis, France
Albert Cohen	University of Paris 6, France
Olivier Gibaru	Arts et Metiers ParisTech, France
Christian Gout	INSA, Rouen, France
Tom Lyche	University of Oslo, Norway
Marie-Laurence Mazure	University Joseph Fourier, Grenoble, France
Larry L. Schumaker	Vanderbilt University, Nashville, USA

Invited Speakers

Francis Bach	INRIA, France
Robert Ghrist	Pennsylvania University, USA
Lars Grasedyck	RWTH Aachen, Germany
Leonidas Guibas	Stanford University, USA
Kai Hormann	Lugano University, Switzerland
Pencho Petrushev	University of South Carolina, USA
Gabriel Peyre	Ceremade, Paris-Dauphine, France
Konrad Polthier	University of Berlin, Germany
Rebecca Willett	University of Wisconsin-Madison, USA
Grady Wright	Boise State University, USA

Mini-Symposium Organizers

Annalisa Buffa	CNR, Pavia, Italy
Siu-Wing Cheng	Hong-Kong University, China
Michael Floater	University of Oslo, Norway
Tom Grandine	Boeing, Seattle, USA
Remi Gribonval	IRISA, Rennes, France
Philipp Grohs	ETH, Zurich, Switzerland
Anders Hansen	Cambridge University, UK
Mauro Maggioni	Duke University, USA
Mark Pauly	EPFL, Lausanne, Switzerland
Grady Wright	Boise State University, USA

Contents

Finite Element Approximation with Hierarchical B-Splines.   1
  Christian Apprich, Klaus Höllig, Jörg Hörner, Andreas Keller,
  and Esfandiar Nava Yazdani

Non-linear Local Polynomial Regression Multiresolution Methods Using
ℓ^1-norm Minimization with Application to Signal Processing 16
  Francesc Aràndiga, Pep Mulet, and Dionisio F. Yáñez

A New Class of Interpolatory *L*-Splines with Adjoint End Conditions 32
  Aurelian Bejancu and Reyouf S. Al-Sahli

On a New Conformal Functional for Simplicial Surfaces 47
  Alexander I. Bobenko and Martin P. Weidner

Evaluation of Smooth Spline Blending Surfaces Using GPU 60
  Jostein Bratlie, Rune Dalmo, and Børre Bang

Implicit Equations of Non-degenerate Rational Bezier Quadric Triangles 70
  Alicia Cantón, L. Fernández-Jambrina, E. Rosado María,
  and M.J. Vázquez-Gallo

Support Vector Machines for Classification of Geometric Primitives
in Point Clouds. 80
  Manuel Caputo, Klaus Denker, Mathias O. Franz, Pascal Laube,
  and Georg Umlauf

Computing Topology Preservation of RBF Transformations
for Landmark-Based Image Registration. 96
  Roberto Cavoretto, Alessandra De Rossi, Hanli Qiao,
  Bernhard Quatember, Wolfgang Recheis, and Martin Mayr

New Bounds on the Lebesgue Constants of Leja Sequences on the Unit
Disc and on $\Re$-Leja Sequences. 109
  Moulay Abdellah Chkifa

A Curvature Smooth Lofting Scheme for Singular Point Treatments 129
  Elaine Cohen, Robert Haimes, and Richard Riesenfeld

A Consistent Statistical Framework for Current-Based Representations
of Surfaces. 151
  Benjamin Coulaud and Frédéric J.P. Richard

Isotropic Möbius Geometry and i-M Circles on Singular
Isotropic Cyclides   160
    Heidi E.I. Dahl

Symbolic Computation of Equi-affine Evolute for Plane B-Spline Curves . . .   169
    Éric Demers, François Guibault, and Christophe Tribes

On-line CAD Reconstruction with Accumulated Means
of Local Geometric Properties   181
    Klaus Denker, Bernd Hamann, and Georg Umlauf

Analysis of Intrinsic Mode Functions Based on Curvature Motion-Like
PDEs.   202
    El Hadji S. Diop and Radjesvarane Alexandre

Optimality of a Gradient Bound for Polyhedral Wachspress Coordinates   210
    Michael S. Floater

Differential Geometry Revisited by Biquaternion Clifford Algebra   216
    Patrick R. Girard, Patrick Clarysse, Romaric Pujol, Liang Wang,
    and Philippe Delachartre

Ridgelet Methods for Linear Transport Equations   243
    Philipp Grohs and Axel Obermeier

Basis Functions for Scattered Data Quasi-Interpolation   263
    Nira Gruberger and David Levin

Mass Smoothers in Geometric Multigrid for Isogeometric Analysis   272
    Clemens Hofreither and Walter Zulehner

On the Set of Trajectories of the Control Systems with Limited Control
Resources.   280
    Nesir Huseyin, Anar Huseyin, and Khalik G. Guseinov

High Order Reconstruction from Cross-Sections   289
    Yael Kagan and David Levin

Adaptive Atlas of Connectivity Maps   304
    Ali Mahdavi-Amiri and Faramarz Samavati

Matrix Generation in Isogeometric Analysis by Low Rank Tensor
Approximation   321
    Angelos Mantzaflaris, Bert Jüttler, B.N. Khoromskij, and Ulrich Langer

Combination of Piecewise-Geodesic Curves for Interactive Image
Segmentation   341
    Julien Mille, Sébastien Bougleux, and Laurent D. Cohen

A Fully-Nested Interpolatory Quadrature Based on Fejér's Second Rule   357
   Jacques Peter

CINPACT-splines: A Class of C^∞ Curves with Compact Support   384
   Adam Runions and Faramarz Samavati

Error Estimates for Approximate Operator Inversion
via Kernel-Based Methods   399
   Kristof Schröder

Boundary Controlled Iterated Function Systems   414
   Dmitry Sokolov, Gilles Gouaty, Christian Gentil, and Anton Mishkinis

Construction of Smooth Isogeometric Function Spaces on Singularly
Parameterized Domains   433
   Thomas Takacs

Reflexive Symmetry Detection in Single Image   452
   Zhongwei Tang, Pascal Monasse, and Jean-Michel Morel

The Sylvester Resultant Matrix and Image Deblurring   461
   Joab R. Winkler

Author Index   491

Finite Element Approximation
with Hierarchical B-Splines

Christian Apprich[1], Klaus Höllig[1]([✉]), Jörg Hörner[1], Andreas Keller[2],
and Esfandiar Nava Yazdani[1]

[1] IMNG, Fachbereich Mathematik, Universität Stuttgart, Stuttgart, Germany
{apprich,hoellig,hoerner,navayaz}@mathematik.uni-stuttgart.de
[2] Allianz AG, Stuttgart, Germany
andkell@web.de

Abstract. We review the definition of hierarchical spline spaces and
their application to finite element methods. Then we discuss how hierarchical techniques can be implemented using the FEMB program package.
Subdivision algorithms play a crucial role and lead to a very simple program structure. A numerical example illustrates the substantial gains
in accuracy for the adaptive strategy, in particular for higher degree
B-splines.

Keywords: Finite elements · B-splines · Hierarchical bases · Subdivision

1 Introduction

B-splines have become standard tools in approximation, computer aided design,
and graphics. A systematic application to finite element analysis is fairly recent.
Essentially two different strategies have been proposed: weighted [1] and isogeometric [2] methods. The text books [3,4] give a comprehensive description of the
relevant theory for these novel approaches. Which technique is best suited for a
particular problem depends to some extent on the representation and topological
form of the simulation domain D. Isogeometric methods use parametrizations of
subsets of D over rectangles and cuboids which are often provided by NURBS
models for CAD/CAM applications. Weighted methods can handle domains well
which have a natural implicit description, e.g., domains constructed from elementary sets with Boolean operations. There are also problems for which a
combination of both methods might be appropriate [5].

Using B-splines as finite elements bridges the gap between geometry description and numerical simulation. Compared to conventional finite element methods
on unstructured grids, spline-based techniques have several advantages: arbitrary
choice of degree and smoothness, exact representation of boundary conditions,
and simple data structure (one parameter per grid point). As a consequence,
B-splines often yield significantly better results than classical finite element
schemes when highly accurate solutions are sought.

J.-D. Boissonnat et al. (Eds.): Curves and Surfaces 2014, LNCS 9213, pp. 1–15, 2015.
DOI: 10.1007/978-3-319-22804-4_1

Tensor product B-splines span the simplest multivariate spline spaces and can utilize univariate algorithms very efficiently. However, unlike in one variable, knot insertion does not provide truly local flexibility; adding knots has a global effect. The natural remedy is hierarchical refinement. Using nested grids with uniform B-splines of different grid width combines well with adaptive approximation methods.

Given the crucial importance of hierarchical techniques for tensor product spline spaces, it is not surprising that such methods have been subject of intensive research. Building upon the classical articles [6,7], hierarchical splines have been analyzed by a number of authors; the articles [8–11] serve as a few examples which are most relevant to our application. Moreover, several novel concepts, such as T-splines [12] and splines over box-partitions [13], have been introduced.

In this article we consider hierarchical finite element approximation with B-splines, a topic of key importance if B-splines are to compete successfully with classical mesh-based trial functions. For weighted methods, hierarchical B-spline elements were already described in [3], but first implemented in [14] for a special case. For isogeometric elements, adaptive B-spline approximations were recently studied in [15–17] and in the context of T-splines in [18–22]. Our objective is not to improve upon the by now well established theory. Instead, we describe how hierarchical methods can be easily implemented using the FEMB program package [23]. Clearly, incorporating existing software for uniform grids eliminates a great deal of redundant programming effort. As an illustration of our algorithms, we document the substantial gains in accuracy for a typical model problem with a singular solution.

After briefly reviewing some essential components of the FEMB routines in Sect. 2, we define hierarchical splines in Sect. 3. We follow the description in [24,25] where a convenient notation has been suggested, covering many of the concepts introduced so far. Section 4 discusses grid transfer operations based on B-spline subdivision. Then we explain in Sect. 5 how to combine these tools with our programs for assembling and solving Ritz-Galerkin systems for uniform grids. Finally, Sect. 6 illustrates the error behavior of hierarchical B-spline elements at a reentrant corner.

2   FEMB Program Package

The FEMB programs implement finite element algorithms with uniform B-splines for two elliptic model problems: the scalar second order equation

$$- \operatorname{div}(p \operatorname{grad} u) + qu = f, \tag{1}$$

where $p > 0$ and $q \geq 0$ are variable coefficients, and the Lamé-Navier system in linear elasticity

$$- \operatorname{div} \sigma(u) = f, \tag{2}$$

where σ is the stress tensor. For each problem the main steps of a finite element simulation are

- determination of integration parameters,
- assembly of the Ritz-Galerkin system,
- conjugate gradient iteration,
- visualization and computation of the residual.

We briefly discuss these steps for a simple special case of problem (1), a Helmholtz equation with homogeneous Neumann boundary conditions:

$$- \Delta u + u = f \text{ in } D, \quad \partial_\perp u = 0 \text{ on } \partial D, \tag{3}$$

where $D \subset \mathbb{R}^d$ is a bounded domain in two or three dimensions ($d = 2, 3$), and $\partial_\perp$ denotes the normal derivative. This example will be used for illustration purposes throughout the article.

The finite element discretizations of the FEMB programs are based on d-variate uniform tensor product B-splines $b_{k,\xi}$ of a fixed coordinate degree n; see Fig. 1. The index $k = (k_1, \ldots, k_d)$ refers to the lower left grid position, i.e., $b_{k,\xi}$ has the knots $\xi_{\nu,k_\nu}, \ldots, \xi_{\nu,k_\nu+n+1}$ in the ν-th coordinate direction and support

$$(\xi_{1,k_1}, \ldots, \xi_{d,k_d}) + [0, n+1]^d h,$$

where $h = \xi_{\nu,\ell+1} - \xi_{\nu,\ell}$ is the grid width.

The domain is described in implicit form,

$$D: \ w(x) > 0,$$

with a weight function w supplied by the user (cf. [3] and [26] for some principal construction techniques). This description allows to incorporate the geometry information into a B-spline discretization in very simple form, combining the efficiency of regular grids with the flexibility of free-form boundary representations.

Definition 1 (Uniform Splines). *A uniform spline on a domain D is a linear combination*

$$p^\xi = \sum_{k \sim \xi} u_k^\xi \, b_{k,\xi},$$

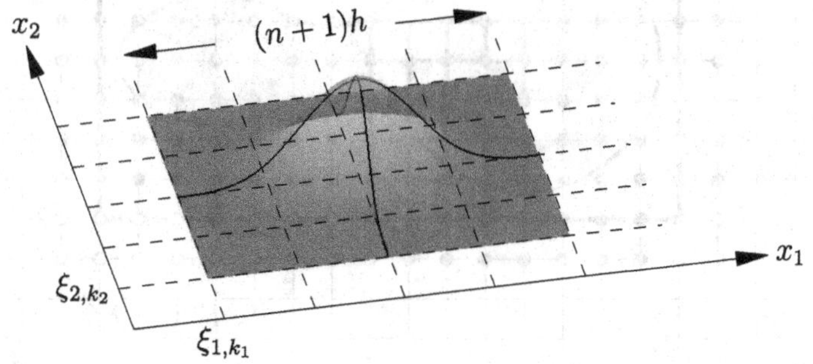

Fig. 1. Uniform tensor product B-spline of coordinate degree $n = 3$ (Color figure online)

where ξ is a d-variate uniform knot sequence,

$$\xi_\nu : \xi_{\nu,0}, \ldots, \xi_{\nu,m_\nu+n}, \quad \xi_{\nu,\ell+1} = \xi_{\nu,\ell} + h, \; 1 \le \nu \le d,$$

with parameter hyperrectangle

$$D_\xi^n = [\xi_{1,n}, \xi_{1,m_1}] \times \cdots \times [\xi_{d,n}, \xi_{d,m_d}] \supset D.$$

The notation $(k_1, \ldots, k_d) \sim \xi$ indicates that the sum is taken over all B-splines corresponding to ξ, i.e., $k_\nu = 0, \ldots, m_\nu - 1$, with the convention that the coefficients u_k^ξ of B-splines with no support in D (irrelevant B-splines) are set to zero.

In Fig. 2, the B-splines $b_{k,\xi}$ corresponding to the knot sequence ξ are marked at the center of their support, using dots for relevant and circles for irrelevant B-splines. Keeping irrelevant B-splines, as described in the above definition, avoids index lists which are necessary for approximations on unstructured grids.

While splines are adequate finite elements for natural boundary conditions, essential or mixed boundary conditions must be incorporated into the finite element subspace. In the FEMB package this is done by using weighted B-splines $wb_{k,\xi}$. Since the hierarchical techniques in this more general case (standard splines correspond to the trivial choice $w = 1$) are completely analogous, we have chosen to focus on the model problem (3), which allows us to explain the algorithms in the simplest setting without any modification of the spline space.

The coefficients u^ξ of a finite element approximation p^ξ satisfy the Ritz-Galerkin system

$$\sum_{j \sim \xi} a(b_{k,\xi}, b_{j,\xi}) \, u_j^\xi = f_k^\xi, \quad k \sim \xi,$$

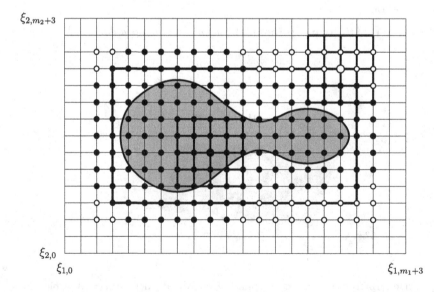

Fig. 2. Relevant and irrelevant cubic B-splines for a bounded domain

where

$$a(\varphi, \psi) = \int_D (\text{grad } \varphi \text{ grad } \psi + \varphi\psi), \quad f_k^\xi = \int_D f b_{k,\xi}$$

for Helmholtz equation (3). We abbreviate this system by

$$G_\xi u^\xi = f^\xi.$$

The matrix G_ξ and the vector f^ξ are assembled by adding the contributions to the integrals from each grid cell using predetermined Gauß parameters (cf. [3,25] for details). Then the coefficients u^ξ are computed with a preconditioned conjugate gradient iteration. An important feature of the final FEMB evaluation and visualization routine is the return of the residual. Unlike for conventional finite element solvers, the spline approximation is continuously differentiable for degree $n > 1$ and can be substituted into the partial differential equation. This provides a very reliable error measure.

In the FEMB package it is assumed that $D \subseteq (0, 1)^d$ and, accordingly, $\xi_{\nu,n} = 0$, $\xi_{\nu,m_\nu} = 1$, $h = 1/(m_\nu - n)$. Changing the relevant program codes slightly, we can relax this requirement. We can provide a MATLAB[1] function which assembles G_ξ and f^ξ for arbitrary uniform knot sequences ξ. In particular, the parameter hyperrectangle D_ξ^n need not contain the simulation region D. This subroutine will be the essential tool for implementing a hierarchical finite element solver.

3   Hierarchical Splines

Roughly speaking, a hierarchical spline is a sum of uniform splines p^ξ with different knot sequences ξ. Typically, such approximations are constructed with an adaptive process. In regions with large error, B-splines $b_{k,\xi}$ are subdivided by halving the grid width and replaced by the resulting B-splines on the finer grid. These refinement steps are repeated until a prescribed tolerance for the error is met. Figure 3 shows an example for bilinear B-splines ($n = 1$).

A knot sequence ξ with grid width $h = 1/4$ and parameter rectangle $D_\xi^1 = [0, 1]^2$ (thick grid lines), covering a heart-shaped domain, is refined near the two corners of the domain boundaries. The knot sequence η near the reentrant corner is further refined leading to a knot sequence ζ with grid width $h = 1/16$. The bilinear B-splines belonging to the hierarchical basis are marked with circles at the center of their support. As mentioned before, B-splines which can be represented on finer grids do not belong to the hierarchical basis as well as B-splines with no support in the domain D. Nevertheless, such B-splines are included with zero coefficients in linear combinations to facilitate programming.

The adaptive construction, leading to a hierarchical approximation, corresponds to a tree structure of knot sequences. To make this more precise, some notation is helpful. We define the extent of a d-variate knot sequence as the smallest rectangle containing all break points:

$$[\xi] = [\xi_{1,0}, \xi_{1,m_1+n}] \times \cdots \times [\xi_{d,0}, \xi_{d,m_d+n}].$$

---

[1] MATLAB® is a registered trademark of The MathWorks, Inc., Natick, MA, U.S.A.

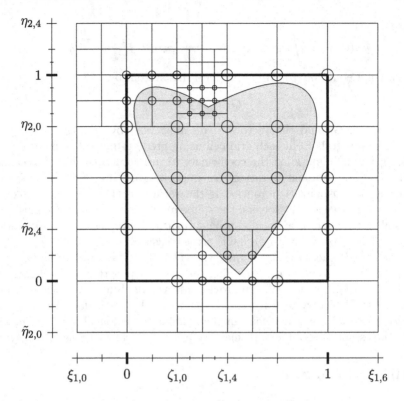

Fig. 3. Grid refinement for bilinear hierarchical B-splines

Moreover, we say that η is the refinement of ξ if η_ν, $\nu = 1, \ldots, d$, is obtained from ξ_ν by adding midpoints in all knot intervals. Finally, we speak of a local refinement, if the knot sequences η_ν are refinements of subsequences of consecutive knots of ξ_ν.

Definition 2 (Hierarchical Spline). *A hierarchical spline p^Ξ is a sum of uniform splines p^ξ corresponding to knot sequences ξ which are nodes of a tree Ξ:*

$$p^\Xi = \sum_{\xi \sim \Xi} p^\xi, \quad p^\xi = \sum_{k \sim \xi} u_k^\xi \, b_{k,\xi} \, .$$

It is required that the children of each node ξ are local refinements of ξ with disjoint extents. Moreover, coefficients u_k^ξ of irrelevant B-splines $b_{k,\xi}$ or B-splines which can be represented on refined grids are set to zero.

In MATLAB, the relevant data for a hierarchical spline p^Ξ can be stored in a cell vector HS of structures, corresponding to the nodes ξ of the tree Ξ, with fields

.parent, .children, .h, .knots, .u .

The first two fields refer to the cell vector indices of the parent and children nodes ξ, .h is the grid width, the field .knots specifies the knot sequence using the lower left and upper right coordinates of $[\xi]$ divided by the grid width,

$$[\xi_{1,0} \; \xi_{1,m_1+n} \; ; \; \cdots \; ; \; \xi_{d,0} \; \xi_{d,m_d+n}]/h,$$

and .u is the d-dimensional array of coefficients with size

$$m_1 \times \cdots \times m_d \, .$$

This natural storage model is illustrated for the example in Fig. 3 above:

```
HS{1}.parent = []; HS{1}.children = [2 3]; HS{1}.knots = [-1 5;-1 5];
HS{1}.h=1/4; HS{1}.u = [0 * * * 0;* * * * 0;0 * * * *;* * * * *;0 * * * *];
HS{2}.parent = 1; HS{2}.children = [4]; HS{2}.knots = [-2 4;6 10];
HS{2}.h = 1/8; HS{2}.u = [0 0 0;* * 0;* * 0;* * 0;0 0 0];
HS{3}.parent = 1; HS{3}.children = []; HS{3}.knots = [2 6;-2 2];
HS{3}.h = 1/8; HS{3}.u = [0 * *;0 * *;0 * *];
HS{4}.parent = 2; HS{4}.children = []; HS{4}.knots = [4 8;12 18];
HS{4}.h = 1/16; HS{4}.u = [* * * 0 0;* * * 0 0;* * * 0 0];
```

We note the forced zeros in the coefficient arrays due to irrelevant or refined B-splines. For example, the B-spline $b_{(4,0),\xi}$ with center at $x = (1,0)$ has support outside of D and hence HS{1}.u(5,1) = 0. The B-spline $b_{(4,0),\eta}$ with center at $x = (3/8, 7/8)$ is a linear combination of the B-splines $b_{k,\zeta}$, $0 \le k_1, k_2 \le 2$. Consequently, HS{2}.u(5,1) = 0. Note that MATLAB indexing starts at 1 while 0 is more convenient for the theory. Moreover, columns in the matrices HS{i}.u correspond to horizontal directions in the figure (analogously as for the commands ndgrid and meshgrid in MATLAB).

In addition to the assumptions on the tree Ξ made in Definition 2, the following property, already proposed in [14], simplifies programming considerably and also has theoretical relevance for stability and the construction of quasi-interpolants.

Definition 3 (Nested Tree). *A tree Ξ is nested if the grid widths of any two B-splines in the hierarchical basis, which are both nonzero at some common point x, are equal or differ by a factor two.*

Requiring a tree to be nested is not a severe restriction. The knot sequences of a tree generated via an adaptive procedure can always be refined to meet the additional requirement. We illustrate this for the example in Fig. 3. Clearly, the tree formed by the knot sequences ξ, η, $\tilde{\eta}$, ζ is not nested since the B-splines $b_{(1,3),\xi}$ (grid width $1/4$, centered at $(1/4, 3/4)$) and $b_{(0,0),\zeta}$ (grid width $1/16$, centered at $(5/16, 13/16)$) are both nonzero for all x in $(2/8, 3/8) \times (6/8, 7/8)$. To conform to the requirement of Definition 3, we add a layer of B-splines with grid width $1/8$ to separate the coarse- and fine-grid B-splines. We replace the knot sequence η by η^* with knots

$$\eta_{1,0}^* = -2/8, \ldots, 6/8 = \eta_{1,8}^*, \quad \eta_{2,0}^* = 4/8, \ldots, 10/8 = \eta_{2,6}^* \, .$$

All B-splines $b_{k,\xi}$ which overlap B-splines $b_{k,\zeta}$ from the finest grid are now replaced by linear combinations of B-splines from η^* with grid width $1/8$. This leads to the following changes in the MATLAB structure listed above.

```
HS{1}.u = [0 * * 0 0;* * * 0 0;0 * * 0 0;* * * * *;0 * * * *];
HS{2}.knots = [-2 6;4 10];
HS{2}.u = [0 0 0 0 0;* * * * 0;* * * * 0;* * * * 0;* * 0 0 0;* * * * 0;* * * * 0];
```

The other structure variables remain unchanged.

4 Grid Transfer Operations

For approximation with hierarchical splines grid transfer operations are essential. Subdivision strategies, systematically introduced by Boehm [27] and Cohen, Lyche, and Riesenfeld [28], provide the canonical mechanism.

Theorem 1 (Subdivision of B-Splines). *A d-variate B-spline $b_{k,\xi}$ with grid width h can be represented as linear combination of B-splines $b_{k,\eta}$ with grid width $h/2$ ($\xi_k = \eta_{2k}$):*

$$b_{k,\xi} = \sum_{\nu_1=0}^{n+1} \cdots \sum_{\nu_d=0}^{n+1} s_\nu\, b_{2k+\nu,\eta}, \quad s_\nu = 2^{-nd} \binom{n+1}{\nu_1} \cdots \binom{n+1}{\nu_d}; \quad (4)$$

see Fig. 4. When manipulating sums it is often convenient to extend the range of summation to all integer vectors $\nu = (\nu_1, \ldots, \nu_d)$. This is possible in view of the convention that the binomial coefficient $\binom{n+1}{\alpha}$ vanishes if $\alpha \notin \{0, \ldots, n+1\}$.

By linearity, the subdivision rule extends to splines as well. It can be used for grid transfer in both directions. We consider each case in turn.

Corollary 1 (Coarse-to-Fine Extension). *If the knot sequence η is the refinement of ξ and $\sum_{k\sim\xi} u_k^\xi b_{k,\xi} = \sum_{j\sim\eta} v_j^\eta b_{j,\eta}$, then the transformation of*

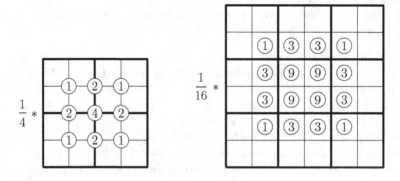

Fig. 4. Subdivision coefficients of a bilinear and biquadratic B-spline

the coefficient vectors has the form

$$v^\eta = S_\xi^\eta\, u^\xi : \quad v_j^\eta = \sum_k s_{j-2k}\, u_k^\xi .$$

In pseudo MATLAB *code, this extension operation can be implemented as*

initialize V with zeros
for $\nu_1,\ldots,\nu_d = 0 : n+1$
  $V(2K + \nu) \leftarrow V(2K + \nu) + s_\nu U$
end

where K denotes the relevant range of indices for the B-splines corresponding to the knot sequence ξ.

The assertions follow from the definitions. Substituting the subdivision formula (4) for B-splines and changing the summation index yields

$$\sum_k u_k^\xi\, b_{k,\xi} = \sum_\nu \sum_k s_\nu\, u_k^\xi\, b_{2k+\nu,\eta} = \sum_j \sum_k s_{j-2k}\, u_k^\xi\, b_{j,\eta} .$$

The middle expression corresponds to the MATLAB pseudo code and the right expression to the extension operator S_ξ^η.

We now consider the reverse operation.

Corollary 2 (Fine-to-Coarse Restriction). *Let the knot sequence η be the refinement of ξ. If, for a bilinear form $a(\cdot,\cdot)$ and a function φ, $v_j^\eta = a(b_{j,\eta},\varphi)$, $j \sim \eta$, then $u_k^\xi = a(b_{k,\xi},\varphi)$, $k \sim \xi$, can be computed with the transformation*

$$u^\xi = S_\eta^\xi v^\eta : \quad u_k^\xi = \sum_j s_{j-2k}\, v_j^\eta .$$

In pseudo MATLAB *code, this restriction operation can be implemented as*

initialize U with zeros
for $\nu_1,\ldots,\nu_d = 0 : n+1$
  $U \leftarrow U + s_\nu V(2K + \nu)$
end

where K denotes the relevant range of indices for the B-splines corresponding to the knot sequence ξ.

Again, the assertions follow by substituting the subdivision formula (4) for B-splines and changing the summation index:

$$u_k^\xi = a(b_{k,\xi},\varphi) = \sum_\nu s_\nu\, a(b_{2k+\nu,\eta},\varphi) = \sum_\nu s_\nu\, v_{2k+\nu}^\eta = \sum_j s_{j-2k}\, v_j^\eta ,$$

with the second to last expression corresponding to the MATLAB code.

The grid transfer operations simplify the assembly and solution of the Ritz-Galerkin system considerably. As will be described in the next section, it suffices to compute integrals involving only B-splines on the same grid level.

5   Solving Hierarchical Systems

We recall from Sect. 3 that a hierarchical spline has the form

$$u^\Xi = \sum_{\xi \sim \Xi} \sum_{k \sim \xi} u_k^\xi \, b_{k,\xi} \, .$$

Accordingly, the Ritz-Galerkin matrix G has block structure with entries

$$G_\eta^\xi : \quad a(b_{k,\xi}, b_{j,\eta}), \; k \sim \xi, \, j \sim \eta \, .$$

We could proceed in a standard fashion, i.e., assemble G using an appropriate storage scheme, and solve the Ritz-Galerkin system

$$GU = F \quad \Leftrightarrow \quad \sum_{\eta \sim \Xi} G_\eta^\xi \, u^\eta = f^\xi, \quad \xi \sim \Xi,$$

by some iterative method. However, since the standard solvers just require the implementation of multiplication by the system matrix, we can avoid the explicit assembly of G and use the routines of the FEMB package for the uniform case in an elegant fashion. To this end we examine the matrix/vector multiplication

$$U \mapsto F = GU, \quad f^\xi = \sum_{\eta \sim \Xi} G_\eta^\xi \, u^\eta \, ,$$

in more detail.

The equation, relating the blocks of the hierarchical vectors U and F, simplifies if we require that Ξ is nested, i.e., if for B-splines in the hierarchical basis only supports of adjacent levels overlap. Since we are also assuming that the extents of knot sequences with the same grid width are disjoint, the matrices G_η^ξ are zero unless $\xi = \eta$, or η is the parent or a child of ξ. Hence, for implementing the multiplication by G_η^ξ, i.e., for computing

$$v^\xi = G_\eta^\xi u^\eta \quad \Leftrightarrow \quad v_k^\xi = \sum_{j \sim \eta} a(b_{k,\xi}, b_{j,\eta}) \, u_j^\eta \, , \tag{5}$$

there are three cases to consider.

(i) $\eta = \xi$:   Since $G_\xi = G_\xi^\xi$ is a Ritz-Galerkin matrix of B-splines of the same grid level, the FEMB routines are applicable with minor modifications (ξ corresponds in general to a grid covering only a subset of the simulation domain D).

(ii) η is the parent of ξ: In Eq. (5) we refine the coarse grid B-splines $b_{j,\eta}$ and obtain by Corollary 1

$$\sum_{j \sim \eta} u_j^\eta \, b_{j,\eta} = \sum_{\ell \sim \zeta} w_\ell^\zeta \, b_{\ell,\zeta}, \quad w^\zeta = S_\eta^\zeta \, u^\eta \, ,$$

with the knot sequence ζ being the refinement of η. This yields the desired conversion to multiplication with a Ritz-Galerkin matrix for B-splines on the same grid level:

$$v_k^\xi = \sum_{\ell \sim \zeta} a(b_{k,\xi}, b_{\ell,\zeta}) \, w_\ell^\zeta \quad \Leftrightarrow \quad v_k^\xi = (G_\zeta \, w^\zeta)_{\tilde{k}} \, ,$$

where $\tilde{k}$ is the index of the B-spline $b_{k,\xi}$ with respect to ζ ($b_{k,\xi} = b_{\tilde{k},\zeta}$). While ξ_ν are subsequences of ζ_ν, the indices do, in general, not match since the labeling starts from 0 for all knot sequences (cf. Fig. 3).

A final adjustment is necessary. Multiplying w^ζ by G_ζ results in a larger vector than needed. Since ξ is a local refinement of η, $[\xi] \subseteq [\eta] = [\zeta]$. As a consequence, some B-splines of the knot sequence ζ are, in general, irrelevant for ξ. Hence, we have to extract the entries relevant for ξ:

$$v_k^\zeta, \; k \sim \zeta \quad \mapsto \quad v_k^\xi, \; k \sim \xi.$$

We denote this truncation operation, which also incorporates the index adjustment, by $v^\xi = I_\zeta^\xi v^\zeta$. Summarizing, we obtain the following procedure.

Theorem 2 (Coarse-to-Fine Multiplication). *If the knot sequence η is the parent of ξ, then*

$$v^\xi = G_\eta^\xi u^\eta \quad \Leftrightarrow \quad v^\xi = I_\zeta^\xi G_\zeta S_\eta^\zeta u^\eta$$

with ζ the refinement of η.

(iii) η is a child of ξ: As in the previous case, we rewrite Eq. (5) in terms of the Ritz-Galerkin matrix G_ζ, where ζ is the refinement of the coarse knot sequence ξ. To this end let $u^\zeta = I_\eta^\zeta u^\eta$ denote the extension of the coefficient vector u^η with coefficients of the B-splines $b_{j,\zeta}$, $j \sim \zeta$, which do not correspond to the smaller knot sequence η ($[\eta] \subseteq [\xi] = [\zeta]$) set to zero, and define

$$\varphi = \sum_{j \sim \eta} u_j^\eta b_{j,\eta} = \sum_{j \sim \zeta} u_j^\zeta b_{j,\zeta}.$$

Then, according to Corollary 2, $v_k^\xi = a(b_{k,\xi}, \varphi)$ can be computed from $w_\ell^\zeta = a(b_{\ell,\zeta}, \varphi)$ with the restriction operator:

$$v^\xi = S_\zeta^\xi w^\zeta, \quad w^\zeta = G_\zeta u^\zeta.$$

We summarize this procedure as follows.

Theorem 3 (Fine-to-Coarse Multiplication). *If the knot sequence η is a child of ξ, then*

$$v^\xi = G_\eta^\xi u^\eta \quad \Leftrightarrow \quad v^\xi = S_\zeta^\xi G_\zeta I_\eta^\zeta u^\eta$$

with ζ the refinement of ξ.

Summarizing, we have reduced the multiplication with the hierarchical matrix G to multiplications with Ritz-Galerkin matrices G_ξ, involving only B-splines with the same grid width. These matrices can be precomputed using the FEMB package before an iterative solver is started. If we employ, e.g., the conjugate gradient method, we have to provide in addition routines for adding hierarchical vectors, multiplication with scalars, and forming of scalar products. This is straightforward.

As a final comment on the programming details, we note that the hierarchical matrix G contains zero rows and columns since we keep irrelevant B-splines for ease of programming. This does not cause any problem though, since the corresponding sections in the coefficient vectors are set to zero. In effect, the iteration "sees" only the invertible part of G.

6   Accuracy of Adaptive Approximations

As a sample problem for testing the proposed method, we chose Helmholtz equation (3) with right–hand side $f(x) = \sin(\pi x_1 x_2)$ on the B–shaped domain shown in Fig. 5. D is the intersection of the square $(0,1)^2$ with the union of the two ellipses

$$E_1 : \; x_1^2 + \frac{(x_2 - a)^2}{a^2} < 1 \quad \text{and} \quad E_2 : \; \frac{9x_1^2}{4} + \frac{(x_2 - b)^2}{(1 - b)^2} < 1,$$

where $a = \frac{11}{10}\left(2 - \sqrt{3}\right)$ and $b = \frac{1}{5}\left(1 + \sqrt{7}\right)$. The boundary of D has a reentrant corner at $x^* = (0.5, 0.55)$ with an interior angle of width $\alpha \approx 1.8089\,\pi$, causing the solution u of (3) to form a singularity at this point. More precisely, u is of asymptotic order $O(r^{\pi/\alpha}) \approx O(r^{0.5528})$ in the region near the corner, where r denotes the distance to x^*, cf. [29]. No other singularities occur at the corners $(0,0)$ and $(0,1)$ since, due to symmetry, u can be regarded as restriction to D of the solution on the entire domain $E_1 \cup E_2$, which has a smooth boundary near these two points.

The shading in Fig. 5 illustrates the growth of $|\operatorname{grad} u|$ in the vicinity of the reentrant corner, indicating that the gradient of u — yet still being square integrable on D — becomes infinite when approaching x^*.

Corner singularities strongly affect the performance of both conventional finite element schemes and methods based on uniform tensor product B–splines,

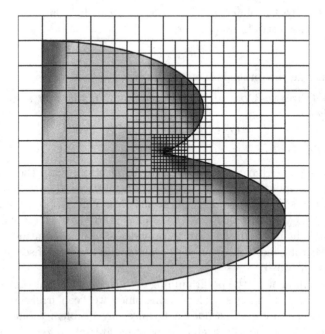

Fig. 5. Singular solution with hierarchical grid (Color figure online)

as they require the solution to belong to some higher order Sobolev space in order to yield optimal rates of convergence. If, as in our example, u has no square integrable second derivatives, the uniform bases cannot release their full approximation power, which results in a slow convergence of numerical solutions. Not even the use of higher spline degrees pays off, as can be seen from Fig. 6 where, regardless of the degree, all uniform spline approximations exhibit about the same poor convergence rate of $O(h^{3/2})$, or $O(N^{-3/4})$ if expressed in terms of the dimension N of the respective spline spaces.

For comparison, we computed approximate solutions to (3) in sequences of hierarchical spline spaces, defined by nested trees with different numbers of levels. Local grid refinement was employed (manually) in regions where the pointwise error of a corresponding uniform approximation was large with respect to the average error. A typical choice of the grid structure for a quadratic spline space is depicted in Fig. 5.

The significant improvement in efficiency achieved through hierarchical refinement near the critical corner is apparent from Fig. 6. The number of parameters necessary to attain similar accuracy is, on an average, smaller by a factor

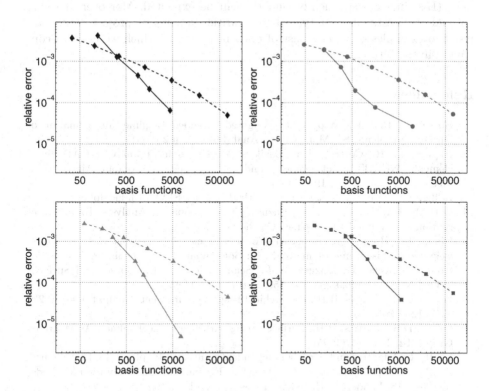

Fig. 6. Relative L^2 error for uniform (dashed) and hierarchical approximations (solid) as a function of the dimension for degrees $n = 1$ ($\blacklozenge$), 2 ($\bullet$), 3 ($\blacktriangle$) and 4 ($\blacksquare$)(Color figure online)

of more than 20, compared with approximations on uniform grids. For example, quadratic elements meet a tolerance of $7.8 \cdot 10^{-5}$ with 1593 B–splines on five hierarchical levels, whereas the error for a finite element space spanned by 18900 quadratic B–splines on a uniform grid of 160×160 cells is still greater than $1.5 \cdot 10^{-4}$.

7 Conclusion

We have described the implementation of hierarchical finite element methods with B-splines using the FEMB package. B-spline subdivision plays a crucial role and leads to a very simple program structure. The numerical results for Helmholtz's equation as a model problem document the substantial gains in accuracy due to adaptive refinement.

For classical mesh-based finite element methods, adaptive techniques have been studied intensively. Hence, for the relatively novel application to B-splines, a number of topics remain to be investigated. An interesting question is, whether the possibility of computing pointwise residuals of the partial differential equations (B-splines are sufficiently smooth) can be exploited. Moreover, deriving optimal error estimates for typical singularities is a major project. These are just two examples of a longer list of open problems to which we hope to contribute in the future.

References

1. Höllig, K., Reif, U., Wipper, J.: Weighted extended B-spline approximation of Dirichlet problems. SIAM J. Numer. Anal. **39**, 442–462 (2001)
2. Hughes, T.J.R., Cottrell, J.A., Bazilevs, Y.: Isogeometric analysis: CAD, finite elements, NURBS, exact geometry and mesh refinement. Comput. Meth. Appl. Mech. Eng. **194**, 4135–4195 (2005)
3. Höllig, K.: Finite Element Methods with B-Splines. SIAM, Philadelphia (2003)
4. Cottrell, J.A., Hughes, T.J.R., Bazilevs, Y.: Isogeometric Analysis: Toward Integration of CAD and FEA. Wiley, Chichester (2009)
5. Höllig, K., Hörner, J., Hoffacker, A.: Finite element analysis with B-splines: weighted and isogeometric methods. In: Boissonnat, J.-D., Chenin, P., Cohen, A., Gout, C., Lyche, T., Mazure, M.-L., Schumaker, L. (eds.) Curves and Surfaces 2011. LNCS, vol. 6920, pp. 330–350. Springer, Heidelberg (2012)
6. Forsey, D.R., Bartels, R.H.: Hierarchical B-spline refinement. Comput. Graph. **22**, 205–212 (1988)
7. Forsey, D.R., Bartels, R.H.: Surface fitting with hierarchical splines. ACM Trans. Graph. **14**, 134–161 (1995)
8. Kraft, R.: Adaptive and linearly independent multilevel B-splines. In: Le Méhauté, A., Rabut, C., Schumaker, L. (eds.) Surface Fitting and Multiresolution Methods, pp. 209–218. Vanderbilt University Press, Nashville (1997)
9. Greiner, G., Hormann, K.: Interpolating and approximating scattered 3D-data with hierarchical tensor product B-splines. In: Le Méhauté, A., Rabut, C., Schumaker, L. (eds.) Surface Fitting and Multiresolution Methods, pp. 163–172. Vanderbilt University Press, Nashville (1997)

10. Rabut, C.: Locally tensor product functions. Numer. Algorithm. **39**, 325–348 (2005)
11. Giannelli, C., Jüttler, B., Speleers, H.: Strongly stable bases for adaptively refined multilevel spline spaces. Adv. Comput. Math. **40**, 459–490 (2014)
12. Sederberg, T.W., Zheng, J., Bakenov, A., Nasri, A.: T-splines and T-NURCCS. ACM Trans. Graph. **22**, 477–484 (2003)
13. Dokken, T., Lyche, T., Petterson, K.F.: Polynomial splines over locally refined box-partitions. Comput. Aided Geom. Des. **30**, 331–356 (2013)
14. Mustahsan, M.: Finite element methods with hierarchical WEB-splines. Dissertation, Universität Stuttgart (2011)
15. Vuong, A.V., Giannelli, C., Jüttler, B., Simeon, B.: A hierarchical approach to adaptive local refinement in isogeometric analysis. Comput. Meth. Appl. Mech. Eng. **200**, 3554–3567 (2011)
16. Schillinger, D., Rank, E.: An unfitted hp-adaptive finite element method based on hierarchical B-splines for interface problems of complex geometry. Comput. Meth. Appl. Mech. Eng. **200**, 3358–3380 (2011)
17. Bornemann, P.B., Cirak, F.: A subdivision-based implementation of the hierarchical b-spline finite element method. Comput. Meth. Appl. Mech. Eng. **253**, 584–598 (2013)
18. Bazilevs, Y., Calo, V.M., Cottrell, J.A., Evans, J.A., Hughes, T.J.R., Lipton, S., Scott, M.A., Sederberg, T.W.: Isogeometric analysis using T-splines. Comput. Meth. Appl. Mech. Eng. **199**, 229–263 (2010)
19. Dörfel, M.R., Jüttler, B., Simeon, B.: Adaptive isogeometric analysis by local h-refinement with T-splines. Comput. Meth. Appl. Mech. Eng. **199**, 264–275 (2010)
20. Scott, M.A., Borden, M.J., Verhoosel, C.V., Sederberg, T.W., Hughes, T.J.R.: Isogeometric finite element data structures based on Bézier extraction of T-splines. Int. J. Numer. Meth. Eng. **88**, 126–156 (2011)
21. Scott, M.A., Li, X., Sederberg, T.W., Hughes, T.J.R.: Local refinement of analysis-suitable T-splines. Comput. Meth. Appl. Mech. Eng. **213–216**, 206–222 (2012)
22. Schillinger, D., Dedè, L., Scott, M.A., Evans, J.A., Borden, M.J., Rank, E., Hughes, T.J.R.: An isogeometric design-through-analysis methodology based on adaptive hierarchical refinement of NURBS, immersed boundary methods, and T-spline CAD surfaces. Comput. Meth. Appl. Mech. Eng. **249–252**, 116–150 (2012)
23. Höllig, K., Hörner, J.: Finite element methods with B-splines: supplementary material (2012). http://www.siam.org/books/fr26/
24. Höllig, K., Hörner, J.: Approximation and Modeling with B-Splines. SIAM, Philadelphia (2013)
25. Höllig, K., Hörner, J.: Programming finite element methods with B-splines. To appear in: Comput. Math. Appl., Special Issue on High-Order Finite Element and Isogeometric Methods (HOFEIM 2014) (2015)
26. Rvachev, V.L., Sheiko, T.I.: R-functions in boundary value problems in mechanics. Appl. Mech. Rev. **48**, 151–188 (1995)
27. Boehm, W.: Inserting new knots into B-spline curves. Comput. Aided Des. **12**, 199–201 (1980)
28. Cohen, E., Lyche, T., Riesenfeld, R.F.: Discrete B-splines and subdivision techniques in computer-aided geometric design and computer graphics. Comput. Graph. Image Proc. **14**, 87–111 (1980)
29. Nazarov, S.A., Plamenevsky, B.A.: Elliptic Problems in Domains with Piecewise Smooth Boundaries. de Gruyter, Berlin (1994)

Non-linear Local Polynomial Regression Multiresolution Methods Using ℓ^1-norm Minimization with Application to Signal Processing

Francesc Aràndiga[1], Pep Mulet[1], and Dionisio F. Yáñez[2,3]([✉])

[1] Department of Matemática Aplicada, Universitat de València,
C/Doctor Moliner, 46100 Burjassot (Valencia), Spain
{arandiga,mulet}@uv.es
http://gata.uv.es/~arandiga, http://gata.uv.es/~mulet
[2] Campus Capacitas, Universidad Católica de Valencia, C/Sagrado Corazón, 5,
46110 Godella (Valencia), Spain
[3] Departamento de Matemáticas, CC. NN. y CC. SS. aplicadas a la educación,
Universidad Católica de Valencia, C/Sagrado Corazón,
46110 Godella (Valencia), Spain
dionisiofelix.yanez@ucv.es
http://www.uv.es/diofeya

Abstract. Harten's Multiresolution has been developed and used for different applications such as fast algorithms for solving linear equations or compression, denoising and inpainting signals. These schemes are based on two principal operators: decimation and prediction. The goal of this paper is to construct an accurate prediction operator that approximates the real values of the signal by a polynomial and estimates the error using ℓ^1-norm in each point. The result is a non-linear multiresolution method. The order of the operator is calculated. The stability of the schemes is ensured by using a special error control technique. Some numerical tests are performed comparing the new method with known linear and non-linear methods.

Keywords: Statistical Multiresolution · Multiscale decomposition · Signal processing · Generalized wavelets

1 Introduction

During the past few years multiresolution analysis (MR) has been widely used in several applications as signal processing (see, e.g., [16–18]). Harten in [24] designed a MR framework based on two operators: decimation and prediction defining them as the composition of two functions: discretization and reconstruction. The discretization represents the nature of the data. The study of

This research was partially supported by Spanish MCINN MTM 2011-22741 and MTM 2014-54388.

J.-D. Boissonnat et al. (Eds.): Curves and Surfaces 2014, LNCS 9213, pp. 16–31, 2015.
DOI: 10.1007/978-3-319-22804-4_2

this function can be found in, for example, [9,10,20]. However, the extensive literature is dedicated to the construction of an accurate reconstruction operator using different techniques, linear (see, e.g., [11,12]) and non-linear. In particular, Aràndiga and Donat in [7] use the Essential Non-Oscillatory (ENO) technique in order to obtain a more adapted prediction operator. Belda et al. extend it introducing Weighted ENO techniques (see [5]). Also, non-linear techniques based on Piecewise Polynomial Harmonic (PPH) are used obtaining interesting results for signal compression and adaptation to edges (see [2]). Recently, in order to design a new prediction operator statistical tools (see, e.g., [15,19,25,26,28]) have been considered as learning processes (see [6]) and approximation by local polynomial regression (LPR) (see, e.g., [12]). In this context, three variables determine the problem. First, the degree, r, of the chosen polynomials. Afterwards, the weight function which assigns a value to each point depending on the distance to the approximated point. Finally, the loss function, $L(x, y)$. Typically, the ℓ^2-minimization norm is used (see [11]). In this context, we extend the method substituting $L(x, y)$ by the ℓ^1-norm which is introduced in the formulation of the problem. Then, a non-linear problem is obtained. Therefore, the order and the stability have to be studied. The order is a measure of the accuracy of the prediction operator. The stability is crucial for MR compression processes since in order to reduce the number of elements to store, we have to truncate several values. Thus, a control on the final error obtained *a priori* between the original signal and the approximate signal is necessary, several papers in the literature have already dealt with this (see, e.g., [3,4,8,18]).

It is not difficult to prove that the order does not depend on the loss function. However, the stability, in this case, is not ensured. In this paper, the error control strategy presented by Harten et al. (see, e.g., [1,7,24]) is used.

The paper is organized as follows: we review Harten's MR and local polynomial regression in Sects. 2 and 3. Subsequently, the error control method is presented in Sect. 4. Finally, some numerical results are displayed, some conclusions are presented and future research is described.

2   Harten's Interpolatory Framework

Harten in [23,24] developed a MR scheme based on four operators: discretization, reconstruction, decimation and prediction. Let $\mathcal{F}$ be a locally integrable function space and V^k a discrete vector space, being k the level of the resolution, then the discretization function, $\mathcal{D}_k : \mathcal{F} \to V^k$, represents the nature of data. Typically, in signal processing is used a point-value method which consists in considering a signal as the values of a function in a grid. Therefore, let us indicate a nested set of uniform dyadic grids on the interval $[0,1]$, as $X^k = \{x_j^k\}$, $x_j^k = j \cdot h_k$, $j = 0, \ldots, J_k$, $J_k \cdot h_k = 1$, with $J_k = 2^k J_0$ and where J_0 is some integer. Then $f^L = \{f_j^L\}_{j=0}^{J_L}$ with $f_j^k := (\mathcal{D}_k(f))_j = f(x_j^k)$. We need to take into account that, as $x_j^{k-1} = x_{2j}^k$, then $f_{2j}^k = f_j^{k-1}$. Thus, in this case, we define the decimation operator which combines two levels of the MR, $\mathcal{D}_k^{k-1} : V^k \to V^{k-1}$ as

$$(\mathcal{D}_k^{k-1} f^k)_j = f_j^{k-1} = f(x_j^{k-1}) = f(x_{2j}^k) = f_{2j}^k, \quad 0 \le j \le J_{k-1}.$$

It must be linear. The reconstruction operator, $\mathcal{R}_k : V^k \to \mathcal{F}$ can be viewed as an interpolating or approximating function. In the literature (see [7,24]), this operator is usually defined using the following sequence: Let r be the degree of the polynomial, s the number of points chosen for the interpolation and

$$z(x) \in \Pi_1^r(\mathbb{R}) = \{g \mid g(x) = \sum_{m=0}^{r} c_m x^m, c_m \in \mathbb{R}, \forall m\}$$

such that $z(x_{j+m}^{k-1}) = f_{j+m}^{k-1}$ for $m = -s, \ldots, s - 1$; with $r + 1 = 2s$. Then, we define:

$$(\mathcal{R}_{k-1}f^{k-1})(x) = z(x).$$

Therefore, we define the prediction operator to obtain the values on the level k as $\mathcal{P}_{k-1}^k := \mathcal{D}_k \mathcal{R}_{k-1}$, then

$$(\mathcal{P}_{k-1}^k f^{k-1})_j = \mathcal{D}_k(\mathcal{R}_{k-1}f^{k-1})_j = (\mathcal{D}_k(z))_j = z(x_j^k). \tag{1}$$

These operators should be consistent following the equation: $\mathcal{D}_k^{k-1}\mathcal{P}_{k-1}^k = \mathcal{I}_{k-1}$, where $\mathcal{I}_k$ is the identity function defined in the space V^k. When the discretization based on point-value is chosen then it is easy to prove that this property is satisfied replicating the odd-values from the low level, as a consequence

$$(\mathcal{P}_{k-1}^k f^{k-1})_{2j} = f_j^{k-1}. \tag{2}$$

From standard 1D interpolation results we obtain that:

$$z(x_{2j-1}^k) = \sum_{l=1}^{s} \beta_l(f_{j+l-1}^{k-1} + f_{j-l}^{k-1})$$

where the coefficients β_l are

$$\begin{cases} s = 1 \Rightarrow \beta_1 = \frac{1}{2}, \\ s = 2 \Rightarrow \beta_1 = \frac{9}{16}, \beta_2 = -\frac{1}{16}, \\ s = 3 \Rightarrow \beta_1 = \frac{150}{256}, \beta_2 = -\frac{25}{256}, \beta_3 = \frac{3}{256}. \end{cases} \tag{3}$$

Observe that $e_{2j}^k = f_{2j}^k - (\mathcal{P}_{k-1}^k f^{k-1})_{2j} = f_j^{k-1} - f_j^{k-1} = 0$. Then, if we define $d_j^k = f_{2j-1}^k - (\mathcal{P}_{k-1}^k f^{k-1})_{2j-1}$ we have that the two sets $\{f^k\}$ and $\{f^{k-1}, d^k\}$ which are equivalent: From $\{f^k\}$ we obtain $f_j^{k-1} = f_{2j}^k$ and $d_j^k = f_{2j-1}^k - (\mathcal{P}_{k-1}^k f^{k-1})_{2j-1}$. Also, from $\{f^{k-1}, d^k\}$ we obtain $f_{2j-1}^k = (\mathcal{P}_{k-1}^k f^{k-1})_{2j-1} + d_j^k$ and $f_{2j}^k = f_j^{k-1}$.

Then, the multiscale decomposition algorithm is:

Algorithm 1 $(f^L \to Mf^L = (f^0, d^1, \ldots, d^L))$

$$\begin{cases} \textbf{for} \quad k = L : 1 \\ \quad f_j^{k-1} = f_{2j}^k, \qquad\qquad\qquad\quad j = 0 : J_{k-1} \\ \quad d_j^k = f_{2j-1}^k - (\mathcal{P}_{k-1}^k f^{k-1})_{2j-1} \quad j = 1 : J_{k-1} \\ \textbf{end} \end{cases} \tag{4}$$

And the multiscale reconstruction algorithm is:

Algorithm 2 $(Mf^L = (f^0, d^1, \ldots, d^L) \longrightarrow M^{-1}Mf^L)$

$$\begin{cases} \text{for} \quad k = 1 : L \\ \quad f^k_{2j-1} = (\mathcal{P}^k_{k-1} f^{k-1})_{2j-1} + d^k_j, j = 1 : J_{k-1} \\ \quad f^k_{2j} = f^{k-1}_j, \qquad\qquad\qquad j = 0 : J_{k-1} \\ \text{end} \end{cases} \tag{5}$$

Thus, $\{f^L\} \leftrightarrow \{\{f^0\}, \{d^1\}, \ldots, \{d^L\}\}$. The compression is obtained when the values $d^0, \ldots, d^L$ are close to zero or zero. The key of these schemes is to construct an accurate prediction operator. Then, when the decomposition algorithm is used, a truncation or quantization technique is applied to the result. It is important to control the *a priori* error (see Sect. 4). In the next section we extend the method developed in [11] using ℓ^1-norm minimization.

3   Design of LPR MR Prediction Operator: Brief Review

In this section, a family of prediction operators is constructed using statistical tools. In particular, we will review the principal components marked in [11].

Therefore, we consider the values of a signal in a determined level of MR $f^{k-1} = \{f^{k-1}_j\}^{J_{k-1}}_{j=0}$ with $f^{k-1}_j = f(x^{k-1}_j)$ and $x^{k-1}_j = jh_{k-1}$.

For a fitting point on level k, x^k_{2j-1}, we approximate the value f^k_{2j-1} with a curve based on the points included in an interval centered in x^k_{2j-1} with length fixed, $(2s-1)h_{k-1}$, with $s > 1$ constant, non-necessary integer. In order to assign a weight to each value f^{k-1}_l such that $x^{k-1}_l \in [x^k_{2j-1} - (2s-1)h_k, x^k_{2j-1} + (2s-1)h_k]$ we define a kern function that will be denoted by K_s, as:

$$K_s(x^k_{2j-1}, x) = \omega\left(\frac{x^k_{2j-1} - x}{(2s-1)\tilde{h}_k}\right) \tag{6}$$

where $\tilde{h}_k$ is a slight perturbation of the value h_k. In [11] the value $\tilde{h}_k = (1 + \epsilon)h_k$, with $\epsilon = 2 \cdot 10^{-3}$ is taken to unify the notation with the interpolation method. And $\omega(x)$ is a non-negative symmetric function (see [26]). We show these functions in Table 1.

Table 1. Kernel functions

rect	$\omega(u) = 1, |u| \leq 1$
tria	$\omega(u) = 1 - |u|, |u| \leq 1$
epan	$\omega(u) = 1 - u^2, |u| \leq 1$
bisq	$\omega(u) = (1 - u^2)^2, |u| \leq 1$
tcub	$\omega(u) = (1 - |u|^3)^3, |u| \leq 1$
trwt	$\omega(u) = (1 - |u|^2)^3, |u| \leq 1$

Let $z(x)$ be a polynomial of degree r, $z(x) \in \Pi_1^r(\mathbb{R})$. We choose a loss-function $L(x, y)$ which measures the distance between the real values on the level $k-1$ and the approximation. In [11] the chosen function is $L(x, y) = (x - y)^2$. Then, the classical least-squares method is obtained. For this paper, we propose a robust LPR with

$$L(x, y) = |x - y|.$$

Hence, the problem is the following:

$$\hat{z}_{2j-1}^k(x) = \underset{z(x) \in \Pi_1^r(\mathbb{R})}{\arg\min} \sum_{l=0}^{J_{k-1}} K_s(x_{2j-1}^k, x_l^{k-1}) L(f_l^{k-1}, z(x_l^{k-1}))$$

$$\mathbf{c}_{2j-1}^k = \underset{c_m \in \mathbb{R}, m=0,\dots,r}{\arg\min} \sum_{l=0}^{J_{k-1}} K_s(x_{2j-1}^k, x_l^{k-1}) \left| f_l^{k-1} - \sum_{m=0}^{r} c_m (x_l^{k-1})^m \right| \tag{7}$$

and we define $\mathcal{R}_{k-1}(x) = \hat{z}_{2j-1}^k(x)$. Consequently, by Eqs. (1) and (2) we have that:

$$\begin{cases} (\mathcal{P}_{k-1}^k f^{k-1})_{2j-1} = \mathcal{D}_k(\mathcal{R}_{k-1}f^{k-1})_{2j-1} = (\mathcal{D}_k(\hat{z}_{2j-1}^k))_{2j-1} = \hat{z}_{2j-1}^k(x_{2j-1}^k), \\ (\mathcal{P}_{k-1}^k f^{k-1})_{2j} = f_j^{k-1}. \end{cases}$$

Remark 1. We consider as $\hat{z}_{2j-1}^k$ the polynomial used to calculate the approximation to f_{2j-1}^k. Notice that we need a different polynomial for each estimation. It is proved in [11] that the prediction operator is independent of the level k and of the point j. In Sect. 3.1 more details are explained.

Remark 2. If $r + 1 = 2\lfloor s \rfloor$ (where $\lfloor \cdot \rfloor$ is the function that rounds a number to the nearest integer less than or equal to it) then the obtained method is the method based on piecewise polynomial interpolation, Eq. (3), independently of the weight and the loss functions that have been chosen.

3.1 Optimization Problem and Non-linearity of the Operator

The choice of the loss function is crucial for the applications. The distance between the real values and the approximation is calculated with this operator. It is easy to prove that if we choose the ℓ^2-norm, then the resulting problem is the classical least-square minimization. In this section, we explain the method using ℓ^1-norm minimization.

By using weight functions such that $w(u) = 0, |u| > 1$, we have that

$$K_s(x_{2j-1}^k, x_{j+l}^{k-1}) \neq 0 \text{ if } -\lfloor s \rfloor \le l \le \lfloor s \rfloor - 1,$$

and our problem can be rewritten as:

$$\hat{z}_{2j-1}^k(x_{2j-1}^k) = \underset{c_m \in \mathbb{R}, m=0,\dots,r}{\arg\min} \sum_{l=-\lfloor s \rfloor}^{\lfloor s \rfloor - 1} K_s(x_{2j-1}^k, x_{j+l}^{k-1}) \left| f_{j+l}^{k-1} - \sum_{m=0}^{r} c_m (x_{j+l}^{k-1})^m \right|. \tag{8}$$

For fixed r, j and s, problem (8) can be written as follows:
Let $\bar{f}_j^{k-1} = (f_{j-\lfloor s \rfloor}^{k-1}, \ldots, f_{j+\lfloor s \rfloor -1}^{k-1})^T$, the $2\lfloor s \rfloor \times (r+1)$ matrix

$$X = \begin{pmatrix} 1 & x_{j-\lfloor s \rfloor}^{k-1} & \cdots & (x_{j-\lfloor s \rfloor}^{k-1})^r \\ 1 & x_{j-\lfloor s \rfloor +1}^{k-1} & \cdots & (x_{j-\lfloor s \rfloor +1}^{k-1})^r \\ \vdots & \vdots & \ddots & \vdots \\ 1 & x_{j+\lfloor s \rfloor -1}^{k-1} & \cdots & (x_{j+\lfloor s \rfloor -1}^{k-1})^r \end{pmatrix} ; \tag{9}$$

and the matrix $2\lfloor s \rfloor \times 2\lfloor s \rfloor$

$$W_{2j-1}^k = \begin{pmatrix} \omega\big(\frac{x_{2j-1}^k - x_{j-\lfloor s \rfloor}^{k-1}}{(2s-1)\tilde{h}_k}\big) & 0 & \cdots & 0 \\ 0 & \omega\big(\frac{x_{2j-1}^k - x_{j-\lfloor s \rfloor +1}^{k-1}}{(2s-1)\tilde{h}_k}\big) & \cdots & 0 \\ \vdots & \vdots & \ddots & \vdots \\ 0 & 0 & \cdots & \omega\big(\frac{x_{2j-1}^k - x_{j+\lfloor s \rfloor -1}^{k-1}}{(2s-1)\tilde{h}_k}\big) \end{pmatrix}. \tag{10}$$

Then, our problem is to calculate

$$\mathbf{c}_{2j-1}^k = \underset{(c_0,\ldots,c_r)\in\mathbb{R}^{r+1}}{\arg\min} \|W_{2j-1}^k(X(c_0,\ldots,c_r)^T - \bar{f}_j^{k-1})\|_1. \tag{11}$$

Therefore, we have that $\hat{z}_{2j-1}^k(x_{2j-1}^k) = A_r(x_{2j-1}^k)^T \mathbf{c}_{2j-1}^k$, where $A_r(x) = (1,x,\ldots,x^r)$ and consequently

$$\begin{cases} (\mathcal{P}_{k-1}^k f^{k-1})_{2j-1} = \mathbf{c}_{2j-1}^k A_r(x_{2j-1}^k)^T, \\ (\mathcal{P}_{k-1}^k f^{k-1})_{2j} = f_j^{k-1}. \end{cases}$$

In order to solve the ℓ^1-norm approximation problem, Eq. (11), different methods can be used (see [14]). In this paper, we reformulate it as a linear program in a convex optimization context (see more details in [13]):

$$\begin{aligned} \text{minimize} \qquad & (1,\ldots,1)^T t \\ \text{s.t.} \quad -t \preceq \; & W_{2j-1}^k(X(c_0,\ldots,c_r)^T - \bar{f}_j^{k-1}) \preceq t. \end{aligned} \tag{12}$$

where $\preceq$ is the componentwise inequality between two vectors.

In the numerical experiments (Sect. 5), we obtain the solutions using a Matlab [27] *package*, called cvx, designed by M. Grant et al. in [21,22].

A measure of the prediction operator's accuracy is the order. It is defined in the following section.

3.2   Order of the Scheme Using ℓ^1-LPR

Definition 1. *Let be $p(x) \in \Pi_1^r(\mathbb{R})$ an arbitrary polynomial of degree less than or equal to r. Then, the order of the prediction operator is r if*

$$\mathcal{P}_{k-1}^k(\mathcal{D}_{k-1}p) = \mathcal{D}_k p, \tag{13}$$

i.e., the prediction operator is exact for polynomials of degree less than or equal to r.

In [11] is proved that the order of the scheme for LPR MR using least squares is equal to the order of the polynomial taken to approximate the real values. Analogously, in this case with $L(x, y) = |x - y|$, we have the same consequence. Therefore, we prove the following theorem.

Theorem 1. *The order of the MR scheme using LPR of degree r is, independently of the weight and the loss functions chosen, r.*

The proof is a direct consequence of the Eq. (7). If the discrete signal f^k is the discretization of a polynomial of degree r then the polynomial that minimizes the functional is exactly the same. Also, the distance between the approximation and real values is 0.

4   Error Control Strategy and Data Compression

In applications as data compression, the details, $\{d_j^k\}_{j=1}^{J_k}$, $k = 1, \ldots, L$ are reduced by means of truncating or quantizing to keep a lower number of values. Thus, if we denote as $\hat{d}_j^k = |d_j^k|_\varepsilon$ being

$$|d_j^k|_\varepsilon = \begin{cases} d_j^k, & \text{if } |d_j^k| \geq \varepsilon; \\ 0, & \text{if } |d_j^k| < \varepsilon. \end{cases} \tag{14}$$

with $j = 0, \ldots, J_k$, $k = 0, \ldots, L$ and ε the introduced threshold parameter, then the error between the approximation and the real values has to be controlled, i.e.

$$\|f^L - \hat{f}^L\|_p \leq C\varepsilon, \qquad p = 1, 2, \infty \tag{15}$$

where $\hat{f}^L$ is the signal obtained after applying the Algorithm 2 and after truncating the details, and C is a constant.

In order to satisfy this error, the linearity of the operator is an essential ingredient (see, e.g., [7,9,12]). In our case, a non-linear prediction operator is designed using ℓ^1-norm minimization. Therefore, a modification of the encoding procedure is necessary to ensure stability. We use the strategy showed in [1,7,10] based on a change of the algorithm. The algorithmic description of the modified encoding is as follows:

Algorithm 3

$$\begin{cases} \textbf{for} \quad k = L : 1 \\ \quad f_j^{k-1} = f_{2j}^k, \, j = 1 : J_{k-1} \\ \textbf{end} \\ \text{Set } \hat{f}^0 = f^0 \\ \textbf{for} \quad k = 1 : L \\ \hat{f}_0^k = f_0^L \\ \quad \textbf{for} \quad j = 1 : J_{k-1} \\ \quad d_j^k = f_{2j-1}^k - (\mathcal{P}_{k-1}^k f^{k-1})_{2j-1}, \, \hat{d}_j^k = |d_j^k|_\varepsilon \\ \quad \hat{f}_{2j-1}^k = (\mathcal{P}_{k-1}^k f^{k-1})_{2j-1} + \hat{d}_j^k, \, \hat{f}_{2j}^k = \hat{f}_j^{k-1} \\ \quad \textbf{end} \\ \textbf{end} \end{cases} \tag{16}$$

With this modification it is not difficult to prove the following proposition.

Proposition 1. *[Aràndiga and Donat [7]] Given a discrete sequence f^L and a tolerance ε, if the truncation parameters ε_k in the modified encoding algorithm (Alg. 3) are chosen so that*

$$\varepsilon_k = \varepsilon$$

then the sequence $\hat{f}^L = M^{-1}\{f^0, \hat{d}^1, \ldots, \hat{d}^L\}$, with M^{-1} defined in Eq. (16), satisfies

$$\|f^L - \hat{f}^L\|_p \leq \varepsilon, \qquad p = 1, 2, \infty. \tag{17}$$

5   Numerical Experiments

We apply the new method to compress signals. The MR schemes in the point-value framework using reconstruction operators obtained from Lagrange polynomial interpolatory techniques are considered. Also, we compare our algorithm with the non-linear method designed by Amat et al. (see [2]) and with the scheme using ℓ^2-norm presented in [11]. Each MR scheme is identified by an acronym. The equivalences are as follows:

PV: Point-value using interpolation techniques with four centered points, Eq. (3), $s = 2$. The predictor operator is:

$$(\mathcal{P}^k_{k-1} f^{k-1})_{2j} = f_j^{k-1},$$

$$(\mathcal{P}^k_{k-1} f^{k-1})_{2j-1} = \frac{9}{16}(f_j^{k-1} + f_{j-1}^{k-1}) - \frac{1}{16}(f_{j-2}^{k-1} + f_{j+1}^{k-1}).$$

PPH: This scheme introduced by Amat et al. [2] is a nonlinear stable algorithm which consists in modifying the PV scheme:

$$(\mathcal{P}^k_{k-1} f^{k-1})_{2j} = f_j^{k-1},$$

$$(\mathcal{P}^k_{k-1} f^{k-1})_{2j-1} = \frac{1}{2}(f_j^{k-1} + f_{j-1}^{k-1}) - \frac{1}{8}pph(d^2 f_j^{k-1}, d^2 f_{j-1}^{k-1}).$$

where $d^2 f_j^{k-1} = f_{j+1}^{k-1} - 2f_j^{k-1} + f_{j-1}^{k-1}$ and pph is the function

$$pph(x, y) = \left(\frac{sign(x) + sign(y)}{2}\right)\frac{2|x||y|}{|x| + |y|}, \quad \forall x, y \in \mathbb{R} \setminus \{0\};$$

$$pph(x, 0) = 0, \forall x \in \mathbb{R}; \qquad pph(0, y) = 0, \forall y \in \mathbb{R}.$$

kern$^p_{r,s}$: LPR MR method, where kern is a weight function showed in Table 1; p is the loss function used, $p = 1, 2$; the degree of polynomial is r and the parameter s is the bandwidth of the LPR, i.e., the number of points used to construct the approximate polynomial.

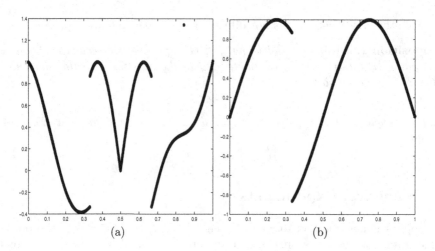

Fig. 1. Functions: (a) f_1, (b) f_2

Table 2. TV of the original function, f_1, and TV of the reconstructions using different weight functions and ℓ^p loss functions, with $p = 1, 2$

Method	f_1	PV	PPH	$\text{tria}^p_{1,3.5}$	$\text{epan}^p_{1,3.5}$	$\text{tcub}^p_{1,3.5}$	$\text{trwt}^p_{1,3.5}$
$p = 1$	9.433	10.157	9.350	9.360	15.288	9.360	9.360
$p = 2$				13.818	15.428	13.648	12.014

We use different functions to analyze the discrete data. It consists of the point-values of the functions: f_1 (Harten function, see [9,10]) and f_2 on the finest grid (see Fig. 1).

$$f_1(x) = \delta(x - x_{865}) + \begin{cases} -x\sin(\frac{3\pi}{2}x^2), & -1 < x \le -\frac{1}{3}, \\ |\sin(2\pi x)|, & |x| < \frac{1}{3}, \\ 2x - 1 - \sin(3\pi x)/6, & \frac{1}{3} \le x < 1. \end{cases}$$

$$f_2(x) = \begin{cases} \sin(2\pi x), & 0 \le x \le 1/3, \\ -\sin(2\pi x), & 1/3 \le x \le 1; \end{cases}$$

Firstly, we perform some experiments to check that the use of the ℓ^1-norm reduces considerably the Gibbs phenomenon at the discontinuities. We take $J_0 = 256$ and $J_L = 1024$, i.e., $L = 3$ and we define a measure of oscillations, called Total Variation (TV) of a function as:

$$TV(f^k) = \sum_{j=1}^{J_k} |f_j^k - f_{j-1}^k|. \tag{18}$$

We compare the reconstructions obtained using different methods with the TV of the original function fixing the bandwidth for all the cases, $r = 1$, $s = 3.5$

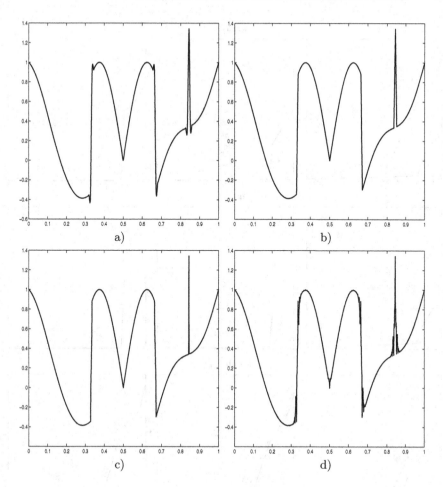

Fig. 2. Reconstructions obtained using different MR schemes: (a) PV, (b) PPH,
(c) $\mathtt{tcub}_{1,3.5}^1$ and (d) $\mathtt{tcub}_{1,3.5}^2$

without keeping any details. In Table 2, we can see that when ℓ^1-norm is used then the oscillations are reduced independently of the weight function chosen. The same rate than in the PPH case is obtained. However, when $p = 2$, the TV ratio is higher because of Gibbs phenomenon as we can observe in Fig. 2.

Afterwards, we present some tests to compress the discrete functions. We use the error control algorithm (Sect. 4) for each method.

In order to reduce the signals, the details are truncated following the strategy presented in the last section, i.e., $\tilde{d}_j^k = |d_j^k|_\varepsilon$ with $j = 0, \dots, J_k$ and ε the constant introduced in Eq. (14). We measure the error in the ℓ^p discrete norm with $p = 1, 2, \infty$ defined by:

$$E_1 = \frac{1}{J_L}\sum_j |f_j^L - \hat{f}_j^L|, \quad E_2 = \sqrt{\frac{1}{J_L}\sum_j |f_j^L - \hat{f}_j^L|^2}, \quad E_\infty = ||f^L - \hat{f}^L||_\infty.$$

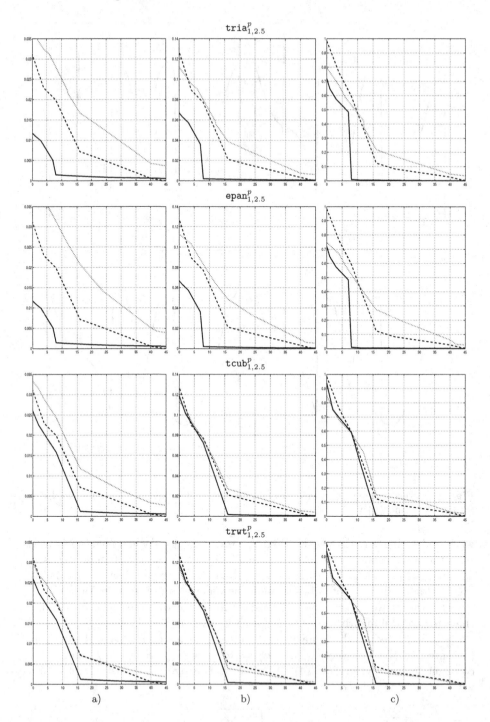

Fig. 3. Errors vs the number of details NNZ using MR schemes to compress the function f_1 with PV (dashed line) and $\text{kern}_{1,2.5}^p$ with $p = 1$ (solid line) and $p = 2$ (dotted line): (a) E_1, (b) E_2 (c) E_∞

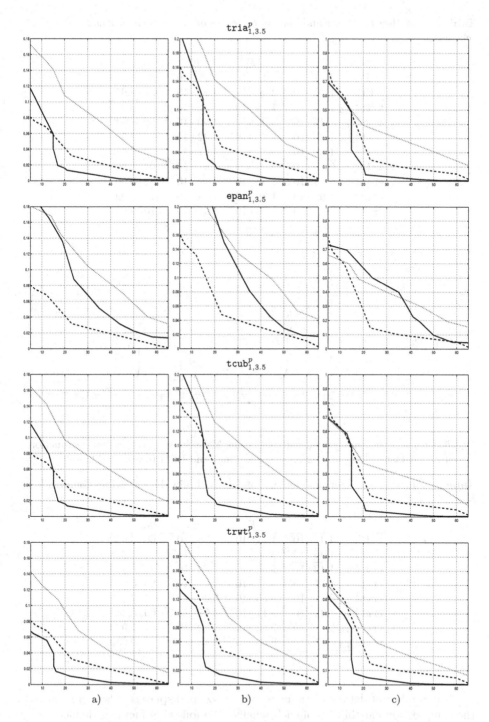

Fig. 4. Errors vs the number of details NNZ using MR schemes to compress the function f_1 with PV (dashed line) and $\mathtt{kern}^p_{1,3.5}$ with $p = 1$ (solid line) and $p = 2$ (dotted line): (a) E_1, (b) E_2 (c) E_∞

Table 3. Function f_1. Errors and number of details obtained with various compression schemes.

	E_1	E_2	E_∞	NNZ	r_c
PV	$4.41 \cdot 10^{-3}$	$1.11 \cdot 10^{-2}$	$4.98 \cdot 10^{-2}$	60	0.940
PPH	$1.35 \cdot 10^{-2}$	$4.56 \cdot 10^{-2}$	$4.99 \cdot 10^{-1}$	22	0.978
$\text{tria}^1_{1,3.5}$	$1.34 \cdot 10^{-2}$	$1.74 \cdot 10^{-2}$	$4.32 \cdot 10^{-2}$	21	0.979
$\text{epan}^1_{1,3.5}$	$1.49 \cdot 10^{-2}$	$1.92 \cdot 10^{-2}$	$4.90 \cdot 10^{-2}$	58	0.942
$\text{tcub}^1_{1,3.5}$	$1.33 \cdot 10^{-2}$	$1.73 \cdot 10^{-2}$	$4.32 \cdot 10^{-2}$	21	0.979
$\text{trwt}^1_{1,3.5}$	$1.07 \cdot 10^{-2}$	$1.44 \cdot 10^{-2}$	$4.97 \cdot 10^{-2}$	22	0.978
$\text{tria}^2_{1,3.5}$	$9.29 \cdot 10^{-3}$	$1.29 \cdot 10^{-2}$	$4.97 \cdot 10^{-2}$	96	0.905
$\text{epan}^2_{1,3.5}$	$9.68 \cdot 10^{-3}$	$1.33 \cdot 10^{-2}$	$4.70 \cdot 10^{-2}$	99	0.902
$\text{tcub}^2_{1,3.5}$	$1.09 \cdot 10^{-2}$	$1.39 \cdot 10^{-2}$	$4.94 \cdot 10^{-2}$	73	0.928
$\text{trwt}^2_{1,3.5}$	$9.87 \cdot 10^{-3}$	$1.26 \cdot 10^{-2}$	$3.97 \cdot 10^{-2}$	70	0.931

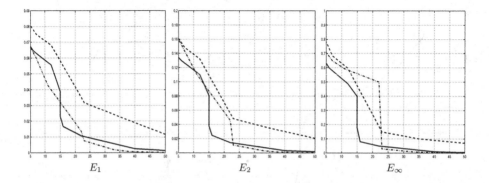

Fig. 5. Errors vs the number of details NNZ using MR schemes to compress the function f_1 with PV (dashed line) and $\text{trwt}^1_{1,3.5}$ (solid line) and PPH (dotdashed line)

Table 4. Function f_1. Errors and number of details obtained with various compression schemes.

	E_1	E_2	E_∞	NNZ	r_c
PV	$4.41 \cdot 10^{-3}$	$1.11 \cdot 10^{-2}$	$4.98 \cdot 10^{-2}$	60	0.940
PPH	$1.35 \cdot 10^{-2}$	$4.56 \cdot 10^{-2}$	$4.99 \cdot 10^{-1}$	22	0.978
$\text{trwt}^1_{0,3.5}$	$1.13 \cdot 10^{-2}$	$1.54 \cdot 10^{-2}$	$4.89 \cdot 10^{-2}$	37	0.963
$\text{trwt}^1_{1,3.5}$	$1.07 \cdot 10^{-2}$	$1.44 \cdot 10^{-2}$	$4.97 \cdot 10^{-2}$	22	0.978
$\text{trwt}^1_{2,3.5}$	$3.51 \cdot 10^{-3}$	$8.05 \cdot 10^{-3}$	$5.00 \cdot 10^{-2}$	55	0.945
$\text{trwt}^1_{3,3.5}$	$4.41 \cdot 10^{-3}$	$1.11 \cdot 10^{-2}$	$4.98 \cdot 10^{-2}$	60	0.940
$\text{trwt}^1_{4,3.5}$	$5.11 \cdot 10^{-3}$	$7.82 \cdot 10^{-3}$	$3.20 \cdot 10^{-2}$	61	0.939
$\text{trwt}^1_{5,3.5}$	$5.25 \cdot 10^{-3}$	$8.00 \cdot 10^{-3}$	$3.15 \cdot 10^{-2}$	61	0.939

The number of details different to zero, NNZ, is displayed. Also, to measure the compression capabilities of each scheme, the following factor is defined:

$$r_c = \frac{J_L - J_0 - |\mathcal{D}_\varepsilon|}{J_L - J_0} \tag{19}$$

Table 5. Function f_2. Errors and number of details obtained with various compression schemes.

	E_1	E_2	E_∞	NNZ	r_c
PV	$2.50 \cdot 10^{-3}$	$4.02 \cdot 10^{-3}$	$9.99 \cdot 10^{-3}$	461	0.543
PPH	$2.65 \cdot 10^{-3}$	$4.10 \cdot 10^{-3}$	$9.97 \cdot 10^{-3}$	425	0.578
$\text{tria}^1_{1,3.5}$	$2.91 \cdot 10^{-3}$	$4.35 \cdot 10^{-3}$	$9.97 \cdot 10^{-3}$	385	0.618
$\text{epan}^1_{1,3.5}$	$2.89 \cdot 10^{-3}$	$4.34 \cdot 10^{-3}$	$9.98 \cdot 10^{-3}$	384	0.619
$\text{tcub}^1_{1,3.5}$	$2.93 \cdot 10^{-3}$	$4.37 \cdot 10^{-3}$	$9.97 \cdot 10^{-3}$	378	0.625
$\text{trwt}^1_{1,3.5}$	$2.81 \cdot 10^{-3}$	$4.27 \cdot 10^{-3}$	$9.97 \cdot 10^{-3}$	398	0.605
$\text{tria}^2_{1,3.5}$	$2.81 \cdot 10^{-3}$	$4.22 \cdot 10^{-3}$	$9.98 \cdot 10^{-3}$	384	0.619
$\text{epan}^2_{1,3.5}$	$2.75 \cdot 10^{-3}$	$4.16 \cdot 10^{-3}$	$9.99 \cdot 10^{-3}$	387	0.616
$\text{tcub}^2_{1,3.5}$	$2.76 \cdot 10^{-3}$	$4.20 \cdot 10^{-3}$	$9.99 \cdot 10^{-3}$	387	0.616
$\text{trwt}^2_{1,3.5}$	$2.76 \cdot 10^{-3}$	$4.18 \cdot 10^{-3}$	$9.99 \cdot 10^{-3}$	393	0.610

where $\mathcal{D}_\varepsilon = \{(j,k) : |\hat{d}^k_j| = |d^k_j|_\varepsilon > 0\}$. If $|\mathcal{D}_\varepsilon| = 0$ then the compression is maximum and $r_c = 1$. When $r_c = 0$ then the algorithm does not produce any compression.

We set the number of points used to approximate the curve and analyze three aspects: the degree of the polynomial, the weight function (Table 1, [26], we only use the most representative one) and the chosen loss function, $p = 1, 2$.

In the next experiment, we take $J_0 = 64$ and $J_L = 1024$, thus $L = 4$. We compare the errors obtained using different weight function using four points (similar to PV), i.e., $s = 2.5$ and the number of details NNZ stored for each MR scheme. The degree chosen for this example is $r = 1$. In this case, Fig. 3, we can observe that the results obtained are similar when **tria** and **epan** kernel functions and **tcub** and **trwt** are used. It is interesting to remark that, when $p = 1$, the errors obtained with our method versus the NNZ are smaller than those obtained with every LPR methods.

In the following experiments, we take $J_0 = 16$ and $J_L = 1024$, thus $L = 6$. In order to analyze the advantages of increasing the number of point without increasing the degree of polynomial used, we fix the bandwidth $s = 3.5$. In Table 3, we can see that when we use the LPR MR with ℓ^1-norm we reduce the number of values different to zero. Indeed, better results are obtained when **tria**, **tcub** and **trwt** are used, being the latter the one that produces a most accurate prediction operator (see Fig. 4). It is significant that PPH considerably reduces the non-zero elements obtaining similar results as LPR methods with ℓ^1-norm. However, the value of E_2 is reduced with $\text{trwt}^1_{1,3.5}$ ($1.44 \cdot 10^{-2}$) in comparison with PPH ($4.56 \cdot 10^{-2}$). It is displayed in Fig. 5. When ℓ^2-norm is introduced, the LPR algorithms do not improve the results of the linear methods, as we can see in Fig. 4.

In Table 4, we show the results when the degree of polynomial change, $r = 1, \ldots, 5$. When the degree is 3 or bigger than 3, the results are similar to using the linear method. It is quite interesting that when a constant ($r = 0$) is used to approximate, the results are better than using PV method.

In the last experiment, with the function f_2 (Table 5), in order to make a more realistic example, we suppose our measuring is not correctly produced. Therefore, we introduce a Gaussian noise with standard deviation $\sigma = 10^{-2}$ in each value of f_2. Taking into account that the signal is quite difficult to compress, with LPR MR we obtain a more accurate prediction operator. Therefore, the number of non-zero elements decreases considerably preserving E_p with $p = 1, 2, \infty$. In this case, the PPH method slightly improves the results obtained using the PV method.

In conclusion, the predictor operator using as loss function the ℓ^1-norm improves the results in comparison with the use of other methods, obtaining a high compression rate. However, we have to calculate the approximation in each point. Therefore the runtime increases.

6   Conclusions and Future Research

In this work, we extend the method developed by Aràndiga et al. in [11] based on Harten's framework [24] replacing the minimization of the ℓ^2- norm by a robust norm ℓ^1. The linearity of the method is lost, therefore, the stability has been ensured by the modification of the direct algorithm. The order of the method has been calculated and it has been used to compress signals with some conclusions:

i. The method obtains a high compression rate in signals with slight noise that are difficult to compress.
ii. The prediction operator adapts point to point.
iii. A bigger number of points can be used to approximate the real value without enlarging the order of the polynomial used.
iv. The weight function is an essential variable in these examples.

Several possibilities can be studied as the stability (without using error control), the use of the different weight functions depending on the signal or the set of functions to approximate the curve. Also, this method can be used in image processing as compression or denoising.

References

1. Amat, S., Aràndiga, F., Cohen, A., Donat, R.: Tensor product multiresolution analysis whith error control for compact representation. Signal Process. **82**, 587–608 (2002)
2. Amat, S., Donat, R., Liandrat, J., Trillo, J.C.: A fully adaptive PPH multiresolution scheme for image processing. Math. Comput. Modelling **46**, 2–11 (2007)
3. Amat, S., Liandrat, J.: On the stability of the pph nonlinear multiresolution. Appl. Comput. Harmon. Anal. **18**(2), 198–206 (2005)
4. Amat, S., Moncayo, M.: Exact error bound for the reconstruction processes using interpolating wavelets. Math. Comput. Simul. **79**, 3547–3555 (2009)
5. Aràndiga, F., Belda, A., Mulet, P.: Point-value WENO multiresolution applications to stable image compression. J. Sci. Comput. **43**(2), 158–182 (2010)

6. Aràndiga, F., Cohen, A., Yáñez, D.F.: Learning-based multiresolution schemes with application to compression of images. Signal Process. **93**, 2474–2484 (2013)
7. Aràndiga, F., Donat, R.: Nonlinear multiscale descompositions: the approach of A. Harten. Numer. Algorithm. **23**, 175–216 (2000)
8. Aràndiga, F., Donat, R.: Stability through synchronization in nonlinear multiscale transformation. SIAM J. Sci. Comput. **29**, 265–289 (2007)
9. Aràndiga, F., Donat, R., Harten, A.: Multiresolution based on weighted averages of the hat function I: linear reconstruction technique. SIAM J. Numer. Anal. **36**, 160–203 (1998)
10. Aràndiga, F., Donat, R., Harten, A.: Multiresolution based on weighted averages of the hat function II: Nonlinear reconstruction technique. SIAM J. Numer. Anal. **20**, 1053–1093 (1999)
11. Aràndiga, F., Yáñez, D.F.: Generalized wavelets design using kernel methods. Application to signal processing. J. Comput. Appl. Math. **250**, 1–15 (2013)
12. Aràndiga, F., Yáñez, D.F.: Cell-average multiresolution based on local polynomial regression. Application to image processing. Appl. Math. Comput. **245**, 1–16 (2014)
13. Boyd, S., Vandenberghe, L.: Convex Optimization. Cambridge University Press, New York (2004)
14. Chan, T.F., Golub, G.H., Mulet, P.: A nonlinear primal-dual method for total variation-based image restoration. SIAM J. Numer. Anal. **20**, 1964–1977 (1997)
15. Cleveland, W.S., Devlin, S.J.: Locally weighted regression: an approach to regression analysis by local fitting. J. Amer. Stat. Assoc. **83**, 596–610 (1988)
16. Cohen, A.: Numerical Analysis of Wavelet Methods. Springer, New York (2003)
17. Cohen, A., Daubechies, I., Feauveau, J.: Biorthogonal bases of compactly supported wavelets. Comm. Pure Appl. Math. **45**, 485–560 (1992)
18. Dahmen, W.: Stability of multiscale transformations. J. Fourier Anal. Appl. **2**(4), 341–361 (1996)
19. Fan, J., Gijbels, I.: Local Polynomials Modelling and its Applications. Chapman and Hall, London (1996)
20. Getreuer, P., Meyer, F.: ENO multiresolution schemes with general discretizations. SIAM J. Numer. Anal. **46**, 2953–2977 (2008)
21. Grant, M., Boyd, S.: Graph implementations for nonsmooth convex programs. In: Blondel, V., Boyd, S., Kimura, H. (eds.) Recent Advances in Learning and Control (a tribute to M. Vidyasagar). LNCS, pp. 95–110. Springer, Heidelberg (2008)
22. Grant, M., Boyd, S.: CVX: Matlab software for disciplined convex programming (web page and software), (2009) http://stanford.edu/~boyd/cvx
23. Harten, A.: Discrete multiresolution analysis and generalized wavelets. J. Appl. Numer. Math. **12**, 153–192 (1993)
24. Harten, A.: Multiresolution representation of data: General framework. SIAM J. Numer. Anal. **33**, 1205–1256 (1996)
25. Hastie, T., Tibshirani, R., Friedman, J.: The Elements of Statistical Learning. Springer, New York (2009)
26. Loader, C.: Local Regression and Likelihook. Springer, New York (1999)
27. MATLAB. version 7.10.0 (R2010a). The MathWorks Inc., Natick (2010)
28. Wand, M.P., Jones, M.C.: Kernel Smoothing. Chapman and Hall, London (1995)

A New Class of Interpolatory L-Splines with Adjoint End Conditions

Aurelian Bejancu$^{(\boxtimes)}$ and Reyouf S. Al-Sahli

Department of Mathematics, Kuwait University, PO Box 5969,
Safat 13060, Kuwait, Kuwait
aurelian@sci.kuniv.edu.kw

Abstract. A thin plate spline for interpolation of smooth transfinite data prescribed along concentric circles was recently proposed by Bejancu, using Kounchev's polyspline method. The construction of the new 'Beppo Levi polyspline' surface reduces, via separation of variables, to that of a countable family of univariate L-splines, indexed by the frequency integer k. This paper establishes the existence, uniqueness and variational properties of the 'Beppo Levi L-spline' schemes corresponding to nonzero frequencies k. In this case, the resulting L-spline end conditions are formulated in terms of *adjoint* differential operators, unlike the usual 'natural' L-spline end conditions, which employ identical operators at both ends. Our L-spline error analysis leads to an L^2-error bound for transfinite surface interpolation with Beppo Levi polysplines.

Keywords: Interpolation · L-spline · Beppo Levi polyspline · Approximation order

1 Introduction

The *thin plate spline* (TPS) interpolant for scattered data was defined by Duchon [11] as the unique minimizer of the squared seminorm

$$\|F\|_{BL}^2 := \iint_{\mathbb{R}^2} \left(|F_{xx}|^2 + 2\,|F_{xy}|^2 + |F_{yy}|^2 \right) \mathrm{d}x\,\mathrm{d}y, \tag{1}$$

subject to F taking prescribed values at a finite number of scattered locations. The minimization takes place in the *Beppo Levi* space of continuous functions F with generalized second-order partial derivatives in $L^2\left(\mathbb{R}^2\right)$.

Recently, Bejancu [5] proposed a new type of TPS surface, passing through several continuous curves prescribed along concentric circles. The new surface minimizes, for $F \in C^2\left(\mathbb{R}^2 \backslash \{0\}\right)$, the polar coordinate version of (1):

$$\|f\|_{BL}^2 := \int_0^\infty \int_{-\pi}^{\pi} \left\{ |f_{rr}|^2 + 2\left|\frac{f_\theta}{r^2} - \frac{f_{\theta r}}{r}\right|^2 + \left|\frac{f_{\theta\theta}}{r^2} + \frac{f_r}{r}\right|^2 \right\} r\,\mathrm{d}\theta\,\mathrm{d}r, \tag{2}$$

© Springer International Publishing Switzerland 2015
J.-D. Boissonnat et al. (Eds.): Curves and Surfaces 2014, LNCS 9213, pp. 32–46, 2015.
DOI: 10.1007/978-3-319-22804-4_3

where $f(r, \theta) := F(r \cos \theta, r \sin \theta)$ denotes the polar form of F. Similar surfaces for *transfinite* interpolation have also been studied in [3, 4] in the case of continuous periodic data prescribed along parallel lines or hyperplanes (see also the survey [6]).

The 'transfinite TPS' surfaces belong to the class of multivariate *polysplines* introduced by Kounchev [14]. In the context of data prescribed on concentric circles $r = r_j$, $j \in \{1, \ldots, n\}$, with $0 < r_1 < \ldots < r_n$, let us denote $\rho := \{r_1, \ldots, r_n\}$ and $\Omega := \{(r, \theta) : r_1 \leq r \leq r_n, -\pi \leq \theta \leq \pi\}$. A function $S : \Omega \to \mathbb{R}$ is termed a *biharmonic polyspline* on annuli determined by ρ if two conditions hold: first, S and its polar form s are piecewise biharmonic, *i.e.*

$$(\partial_{xx} + \partial_{yy})^2 S(x, y) = (\partial_{rr} + r^{-1}\partial_r + r^{-2}\partial_{\theta\theta})^2 s(r, \theta) = 0,$$

on each annulus $r_j < r < r_{j+1}$, $-\pi \leq \theta \leq \pi$, for $1 \leq j \leq n-1$; and second, $S \in C^2(\Omega)$, *i.e.* neighbouring pieces join up C^2-continuously across the interface circles. For sufficiently smooth periodic data functions $u, v, \mu_j : [-\pi, \pi] \to \mathbb{R}, 1 \leq j \leq n$, Kounchev proved that such a polyspline surface is uniquely determined by transfinite interpolation conditions

$$s(r_j, \theta) = \mu_j(\theta), \forall \theta \in [-\pi, \pi], \; \forall j \in \{1, \ldots, n\}, \tag{3}$$

together with boundary conditions $\partial_r s(r_1, \theta) = u(\theta)$ and $\partial_r s(r_n, \theta) = v(\theta)$, $\forall \theta \in [-\pi, \pi]$. He also extended this result to polysplines of higher orders and more general interface configurations in $\mathbb{R}^d$. In the case of *cardinal* interpolation at the bi-infinite set of hyperspheres of radii $r = e^j$, $j \in \mathbb{Z}$, Kounchev and Render [16] constructed polysplines that satisfy growth conditions as $r \to 0$ and $r \to \infty$.

In [5], Bejancu proposed a global polyspline $S : \mathbb{R}^2 \to \mathbb{R}$ for which boundary conditions on the above extreme circles $r = r_1$ and $r = r_n$ are replaced by the requirement that the polar Beppo Levi energy (2) is finite for $f := s$. This *Beppo Levi polyspline* has two additional biharmonic pieces over the extreme annuli $0 < r < r_1$ and $r > r_n$, such that $S \in C^2(\mathbb{R}^2 \setminus \{0\})$. The new surface is automatically continuous at 0, but its partial derivatives can have a singularity at 0.

For sufficiently smooth data, it turns out that there exists a one-parameter family of such Beppo Levi polysplines on annuli determined by ρ, each satisfying the transfinite interpolation conditions (3). Two surfaces S^A and S^B of this family are uniquely determined in [5, Theorem 1] by the following additional conditions: S^A takes an arbitrarily prescribed value at 0, while S^B is biharmonic at 0 (hence, non-singular). Both S^A and S^B are then characterized as genuine TPS surfaces, *i.e.* minimizers of (2), subject to their respective interpolation conditions.

Following the method of separation of variables used by Kounchev [14], the construction of the Beppo Levi polysplines S^A, S^B is obtained in [5, Sect. 4] via the absolutely convergent Fourier representation in polar form

$$s(r, \theta) = \sum_{k \in \mathbb{Z}} \widehat{s}_k(r) e^{ik\theta}, (r, \theta) \in [0, \infty) \times [-\pi, \pi]. \tag{4}$$

For each frequency k, the amplitude coefficient $\widehat{s}_k$ of this representation is a univariate L_k-*spline* for an ordinary differential operator operator L_k, as described in the next section. Moreover, the form of $\widehat{s}_k$ on the extreme intervals $(0, r_1)$ and (r_n, ∞) is determined by the condition that the corresponding Plancherel component of the Beppo Levi energy (2) of s is finite.

The present paper studies the class of such *Beppo Levi L_k-splines* corresponding to non-zero frequencies k (see Sect. 2). In this case, the restrictions satisfied by $\widehat{s}_k$ on the extreme intervals exhibit a twisted symmetry, expressed in terms of adjoint differential operators. Different features appear in the radial case $k = 0$, treated in the companion paper [7], which is connected to Rabut's work on radially symmetric thin plate splines [18].

In Sect. 3, we prove the existence, uniqueness and variational characterization of interpolation schemes with Beppo Levi L_k-splines, as required by the construction of [5]. A part of these results, corresponding to $|k| \geq 2$, has first been obtained in the MSc thesis [2]. For $|k| \geq 2$, we also provide a linear representation of Beppo Levi L_k-splines in terms of dilates of a basis function related to the generalized Whittle-Matérn-Sobolev kernels introduced by Bozzini, Rossini, and Schaback [9]. Further, in Sect. 4, we apply an error analysis of the Beppo Levi L_k-spline schemes to establish the L^2-approximation order $O\left(h^2\right)$ for transfinite surface interpolation with Beppo Levi polysplines on annuli, where h is the maximum distance between successive interface circles. The extension of this work to higher order Beppo Levi polysplines on annuli and their L-spline Fourier coefficients will be addressed in a separate paper.

2 Preliminaries

2.1 Energy Spaces

For each $r \geq 0$, define the Fourier coefficients of $f(r, \theta)$ with respect to θ by

$$\widehat{f}_k(r) := \frac{1}{2\pi} \int_{-\pi}^{\pi} e^{-ik\theta} f(r, \theta)\, d\theta, k \in \mathbb{Z}. \tag{5}$$

The following observation shows the effect on $\widehat{f}_k$ of the condition that the polar Beppo Levi integral (2) is finite. Namely, if f is the polar form of $F \in C^2\left(\mathbb{R}^2 \setminus \{0\}\right)$, then $\widehat{f}_k \in C^2(0, \infty)$ and Plancherel's formula implies the identity [5, (3.2)]:

$$\|f\|_{BL}^2 = 2\pi \sum_{k \in \mathbb{Z}} \left\|\widehat{f}_k\right\|_k^2, \tag{6}$$

where, for each $k \in \mathbb{Z}$, we denote

$$\|\psi\|_k^2 := \int_0^\infty \left\{ \left|\frac{d^2\psi}{dr^2}\right|^2 + 2k^2 \left|\frac{\psi}{r^2} - \frac{1}{r}\frac{d\psi}{dr}\right|^2 + \left|k^2\frac{\psi}{r^2} - \frac{1}{r}\frac{d\psi}{dr}\right|^2 \right\} r\, dr. \tag{7}$$

Let $AC_{\mathrm{loc}}(0, \infty)$ be the vector space of functions $\psi : (0, \infty) \to \mathbb{C}$ that are absolutely continuous on any interval $[a, b]$, $0 < a < b < \infty$. We denote by

Λ_1 the vector space of functions $\psi \in C^1(0, \infty)$ with $\psi' \in AC_{\mathrm{loc}}(0, \infty)$, such that $r^{1/2}\psi''$ and $r^{-1/2}\psi' - r^{-3/2}\psi$ belong to $L^2(0, \infty)$. Also, by Λ_2 we denote the vector space of functions $\psi \in C^1(0, \infty)$ with $\psi' \in AC_{\mathrm{loc}}(0, \infty)$, such that $r^{1/2}\psi''$, $r^{-1/2}\psi'$, and $r^{-3/2}\psi$ all belong to $L^2(0, \infty)$. Note that $\|\cdot\|_k$ is a norm on Λ_2 for $|k| \geq 2$ and a semi-norm on Λ_1 for $k = \pm 1$. The results of Sect. 3 employ the following properties of functions from the spaces Λ_1 and Λ_2.

Lemma 1. (i) *If $\psi \in \Lambda_2$, there exist non-negative constants C_ψ and $\widetilde{C}_\psi$, such that*

$$|\psi(r)| \leq C_\psi \left(r^{3/2} + r|1 - r|^{1/2} \right),$$
$$|\psi'(r)| \leq \widetilde{C}_\psi \left(r^{1/2} + |1 - r|^{1/2} \right), \quad \forall r > 0. \tag{8}$$

(ii) *If $\psi \in \Lambda_1$, there exist non-negative constants C_ψ and $\widetilde{C}_\psi$, such that*

$$|\psi(r)| \leq C_\psi r \left(1 + |\ln r|^{1/2} \right),$$
$$|\psi'(r)| \leq \widetilde{C}_\psi \left(1 + |\ln r|^{1/2} \right), \quad \forall r > 0. \tag{9}$$

Proof. (i) For each $r > 0$, we use the Leibniz-Newton formula

$$r^{-3/2}\psi(r) - \psi(1) = \int_1^r \left[t^{-3/2}\psi(t) \right]' dt$$
$$= \int_1^r \left[t^{-3/2}\psi'(t) - \frac{3}{2}t^{-5/2}\psi(t) \right] dt.$$

Via Cauchy-Schwarz, the last integral is bounded above in modulus by

$$\left| \int_1^r t^{-1} \left[t^{-1/2}\psi'(t) \right] dt \right| + \frac{3}{2}\left| \int_1^r t^{-1} \left[t^{-3/2}\psi(t) \right] dt \right|$$
$$\leq \left| \int_1^r t^{-2}dt \right|^{\frac{1}{2}} \left\{ \left| \int_1^r \left| t^{-1/2}\psi'(t) \right|^2 dt \right|^{\frac{1}{2}} + \frac{3}{2}\left| \int_1^r \left| t^{-3/2}\psi(t) \right|^2 dt \right|^{\frac{1}{2}} \right\}$$
$$\leq |1 - r^{-1}|^{\frac{1}{2}} \left\{ \left\| r^{-1/2}\psi' \right\|_{L^2(0,\infty)} + (3/2)\left\| r^{-3/2}\psi \right\|_{L^2(0,\infty)} \right\}$$

which implies the first of inequalities (8). For the second inequality, we similarly start with

$$r^{-1/2}\psi'(r) - \psi'(1) = \int_1^r \left[t^{-1/2}\psi'(t) \right]' dt$$
$$= \int_1^r \left[t^{-1/2}\psi''(t) - \frac{1}{2}t^{-3/2}\psi'(t) \right] dt,$$

which holds for each $r > 0$, since $\psi' \in AC_{\mathrm{loc}}(0,\infty)$. Hence, we obtain the following upper bound on the modulus of last integral:

$$
\left| \int_1^r t^{-1} \left[t^{1/2} \psi''(t) \right] \mathrm{d}t \right| + \frac{1}{2} \left| \int_1^r t^{-1} \left[t^{-1/2} \psi'(t) \right] \mathrm{d}t \right|
$$

$$
\leq \left| \int_1^r t^{-2} \mathrm{d}t \right|^{\frac{1}{2}} \left\{ \left| \int_1^r \left| t^{1/2} \psi''(t) \right|^2 \mathrm{d}t \right|^{\frac{1}{2}} + \frac{1}{2} \left| \int_1^r \left| t^{-1/2} \psi'(t) \right|^2 \mathrm{d}t \right|^{\frac{1}{2}} \right\}
$$

$$
\leq \left| 1 - r^{-1} \right|^{\frac{1}{2}} \left\{ \left\| r^{1/2} \psi'' \right\|_{L^2(0,\infty)} + (1/2) \left\| r^{-1/2} \psi' \right\|_{L^2(0,\infty)} \right\}.
$$

(ii) For the first inequality, we employ the Leibniz-Newton formula

$$
r^{-1} \psi(r) - \psi(1) = \int_1^r \left[t^{-1} \psi(t) \right]' \mathrm{d}t,
$$

together with the estimate

$$
\left| \int_1^r t^{-1/2} \left(t^{1/2} \left[t^{-1} \psi(t) \right]' \right) \mathrm{d}t \right|
$$

$$
\leq \left| \int_1^r t^{-1} \mathrm{d}t \right|^{\frac{1}{2}} \left| \int_1^r \left| t^{1/2} \left[t^{-1} \psi(t) \right]' \right|^2 \mathrm{d}t \right|^{\frac{1}{2}}
$$

$$
\leq \left| \ln r \right|^{\frac{1}{2}} \left\| r^{1/2} \left(r^{-1} \psi \right)' \right\|_{L^2(0,\infty)},
$$

the last norm being finite due to $r^{1/2} \left(r^{-1} \psi \right)' = r^{-1/2} \psi' - r^{-3/2} \psi$. Since $\psi' \in AC_{\mathrm{loc}}(0,\infty)$, the second inequality is obtained via

$$
\psi'(r) - \psi'(1) = \int_1^r \psi''(t) \, \mathrm{d}t,
$$

followed by a similar estimate, this time in terms of $\left\| r^{1/2} \psi'' \right\|_{L^2(0,\infty)}$. $\square$

2.2   Beppo Levi L_k-splines

As observed in [6], due to the Plancherel-type formula (6), to obtain the variational characterization of the Beppo Levi polyspline s as minimizer of the polar thin plate energy (2), it is sufficient to show that, for each $k \in \mathbb{Z}$, the amplitude coefficient $\widehat{s}_k$ minimizes the corresponding energy component (7). Letting $Q(r, g, g', g'')$ denote the integrand of (7), classical calculus of variations considerations imply that, except at the interpolation locations $r_1, \ldots, r_n$, a minimizer of (7) should satisfy the Euler-Lagrange equation

$$
\partial_g Q - \frac{\mathrm{d}}{\mathrm{d}r} \partial_{g'} Q + \frac{\mathrm{d}^2}{\mathrm{d}r^2} \partial_{g''} Q = 0.
$$

The resulting left-hand side Euler-Lagrange differential operator is given, up to a constant factor, by

$$L_k := r\frac{d^4}{dr^4} + 2\frac{d^3}{dr^3} - \frac{2k^2+1}{r}\frac{d^2}{dr^2} + \frac{2k^2+1}{r^2}\frac{d}{dr} + \frac{k^4-4k^2}{r^3}$$

$$= r\left(\frac{d^2}{dr^2} + \frac{1}{r}\frac{d}{dr} - \frac{k^2}{r^2}\right)^2.$$

Therefore, $\widehat{s}_k$ should necessarily be annihilated by L_k on each subinterval $(0, r_1)$, $(r_1, r_2), \ldots, (r_n, \infty)$.

The null-space $\mathrm{Ker}L_k$ is computed in [14] via the substitution $r = e^t$, $\frac{d}{dt} = r\frac{d}{dr}$, which transforms L_k into a differential operator with constant coefficients in variable t. Standard factorization then implies

$$L_k = \frac{1}{r^3}\left(r\frac{d}{dr} - |k|\right)\left(r\frac{d}{dr} - |k| - 2\right)\left(r\frac{d}{dr} + |k|\right)\left(r\frac{d}{dr} + |k| - 2\right), \quad (10)$$

hence

$$\mathrm{Ker}L_k = \begin{cases} \mathrm{span}\left\{r^2, r^2\ln r, 1, \ln r\right\}, & \text{if } k = 0, \\ \mathrm{span}\left\{r^3, r, r\ln r, r^{-1}\right\}, & \text{if } |k| = 1, \\ \mathrm{span}\left\{r^{|k|+2}, r^{|k|}, r^{-|k|+2}, r^{-|k|}\right\}, & \text{if } |k| \geq 2. \end{cases} \quad (11)$$

Moreover, the condition that the polar Beppo Levi energy component (7) is finite further restricts the form of $\widehat{s}_k$ on the extreme intervals $(0, r_1)$ and (r_n, ∞). Specifically, for $k \neq 0$, evaluating (7) for each of the four generating functions of $\mathrm{Ker}L_k$, we obtain the necessary conditions

$$\widehat{s}_k(r) \in \begin{cases} \mathrm{span}\left\{r^{|k|+2}, r^{|k|}\right\}, & \text{for } r \in (0, r_1), \\ \mathrm{span}\left\{r^{-|k|+2}, r^{-|k|}\right\}, & \text{for } r \in (r_n, \infty). \end{cases}$$

Note that $\mathrm{span}\left\{r^{|k|+2}, r^{|k|}\right\} = \mathrm{Ker}G_k$ and $\mathrm{span}\left\{r^{-|k|+2}, r^{-|k|}\right\} = \mathrm{Ker}R_k$, where

$$G_k := \frac{1}{r}\left[\frac{d^2}{dr^2} - \frac{2|k|+1}{r}\frac{d}{dr} + \frac{|k|(|k|+2)}{r^2}\right]$$

$$= \frac{1}{r^3}\left(r\frac{d}{dr} - |k|\right)\left(r\frac{d}{dr} - |k| - 2\right),$$

$$R_k := \frac{1}{r}\left[\frac{d^2}{dr^2} + \frac{2|k|-1}{r}\frac{d}{dr} + \frac{|k|(|k|-2)}{r^2}\right]$$

$$= \frac{1}{r^3}\left(r\frac{d}{dr} + |k|\right)\left(r\frac{d}{dr} + |k| - 2\right). \quad (12)$$

Remark 1. It can be verified that r^{-3} is the only factor of the form r^α which, when inserted in front of the last two brackets in the right-hand side of the above formulae, turns G_k and R_k into mutually adjoint operators. Indeed, the formal adjoint of G_k is

$$G_k^* = \frac{d^2}{dr^2}\left(\frac{1}{r}\cdot\right) + (2\,|k|+1)\frac{d}{dr}\left(\frac{1}{r^2}\cdot\right) + \frac{|k|\,(|k|+2)}{r^3} = R_k$$

and a similar computation shows $R_k^* = G_k$.

Recall the notation $\rho := \{r_1, \ldots, r_n\}$ used in the Introduction.

Definition 1. *Let $k \neq 0$. A function $\eta : [0,\infty) \to \mathbb{C}$ is called a Beppo Levi L_k-spline on ρ if the following conditions hold:*
  (i) $L_k\eta\,(r) = 0$, $\forall r \in (r_j, r_{j+1})$, $\forall j \in \{1, \ldots, n-1\}$;
  (ii) $G_k\eta\,(r) = 0$, $\forall r \in (0,r_1)$, and $R_k\eta\,(r) = 0$, $\forall r > r_n$.
  (iii) *η is C^2-continuous at each knot $r_1, \ldots, r_n$. The space of all Beppo Levi L_k-splines on ρ will be labelled $\mathcal{S}_k\,(\rho)$.*

Due to conditions (ii), $\mathcal{S}_k\,(\rho)$ is a subspace of Λ_1 if $|k| = 1$, and of Λ_2 if $|k| \geq 2$.

For $k = 0$, the related notion of a Beppo Levi L_0-spline is treated in [7]. In this case, the correct left/right operators G_0 and R_0 on the extreme intervals are not obtained by just letting $k = 0$ in (12). Also, G_0 and R_0 are not anymore mutually adjoint.

The proof of the next result follows from the definition of biharmonic Beppo Levi polysplines on annuli [5].

Proposition 1. *A univariate function $\eta : [0,\infty) \to \mathbb{C}$ is a Beppo Levi L_k-spline on ρ, i.e. $\eta \in \mathcal{S}_k\,(\rho)$, if and only if the polar surface $s\,(r,\theta) := \eta\,(r)\,e^{-ik\theta}$ is a biharmonic Beppo Levi polyspline on annuli determined by ρ.*

We now review some relevant literature. It was pointed out by Kounchev [14, p.91] that, on any interval of positive real numbers, the null space of L_k can be described as an extended complete Chebyshev (ECT) system in the sense of Karlin and Ziegler [13], via the representation

$$r^{|k|}L_k = D_4 D_3 D_2 D_1 = \frac{d}{dr}r^{2|k|+1}\frac{d}{dr}\frac{1}{r^{2|k|+1}}\frac{d}{dr}r^{2|k|+1}\frac{d}{dr}\frac{1}{r^{|k|}},$$

where $D_1 = \frac{d}{dr}\left(\frac{1}{r^{|k|}}\cdot\right)$, $D_2 = D_4 = \frac{d}{dr}\left(r^{2|k|+1}\cdot\right)$, $D_3 = \frac{d}{dr}\left(\frac{1}{r^{2|k|+1}}\cdot\right)$. We also observe, following Schumaker [20, p. 398], that L_k possesses the factorization

$$L_k = M_k^* M_k, \tag{13}$$

where M_k^* denotes the formal adjoint of

$$M_k := \frac{1}{\sqrt{r^{2|k|+1}}}\frac{d}{dr}r^{2|k|+1}\frac{d}{dr}\frac{1}{r^{|k|}} = \sqrt{r}\left(\frac{d^2}{dr^2} + \frac{1}{r}\frac{d}{dr} - \frac{k^2}{r^2}\right)$$

$$= r^{-3/2}\left(r\frac{d}{dr} - |k|\right)\left(r\frac{d}{dr} + |k|\right).$$

Due to (13), a function that satisfies conditions (i) and (iii) of Definition 1 can be characterized as a 'generalized spline' or 'M_k-spline' on $[r_1, r_n]$ in the sense of Ahlberg, Nilson, and Walsh [1], Schultz and Varga [19]. However, our labeling such a function as a 'L_k-spline' agrees with the terminology of Lucas [17] and Jerome and Pierce [12], which is more adequate, in view of the fact that L_k may possess other factorizations of the type (13). Indeed, for $k \neq 0$, our adjoint boundary operators G_k and R_k actually generate, via (10), the factorization

$$L_k = G_k r^3 R_k = \widetilde{L}_k^* \widetilde{L}_k, \text{ where } \widetilde{L}_k := r^{3/2} R_k. \tag{14}$$

This differs from (13) for $|k| \geq 2$, while it coincides with (13) for $|k| = 1$.

On the other hand, the 'natural' end conditions of L-spline literature (see [20]) are always formulated in terms of a single differential operator at both ends of the interpolation domain. It is thus remarkable that adjoint boundary operators as in condition (ii) of our definition have also occured in [4], in the context of exponential L-splines generated as Fourier coefficients of Beppo Levi polyspline surfaces on parallel strips. Such exponential L-splines coincide in fact with Matérn kernels on the full real line (for Matérn kernels on a compact interval, see [10]). As shown in [3], adjoint L-spline end conditions are intimately connected to Wiener-Hopf factorizations for semi-cardinal interpolation.

3   Interpolation with Beppo Levi L_k-splines

3.1   A Fundamental Identity

We employ the notations introduced in the previous section.

Theorem 1. (i) *Let* $k \in \mathbb{Z}$, $|k| \geq 2$, *and an arbitrary Beppo Levi L_k-spline* $\eta \in \mathcal{S}_k(\rho)$. *Also, assume that* $\psi \in \Lambda_2$ *vanishes on the knot-set* ρ:

$$\psi(r_j) = 0, \forall j \in \{1, \ldots, n\}. \tag{15}$$

Then the following orthogonality relation holds:

$$\int_0^\infty r^3 [R_k \eta(r)] [R_k \overline{\psi}(r)] \, \mathrm{d}r = 0. \tag{16}$$

(ii) *The same conclusion holds if* $k = \pm 1$ *and* $\psi \in \Lambda_1$ *satisfies (15).*

Proof. For convenience, let us denote the left-hand side of (16) by $I_k := I_k(\eta, \psi)$. Note that, for any $k \neq 0$, $\eta \in \mathcal{S}_k(\rho)$ implies $R_k \eta(r) = 0$, $\forall r > r_n$, hence we can work with integral I_k on the integration domain $(0, r_n]$. Since

$$r^{3/2} R_k \psi = r^{1/2} \psi'' + (2|k| - 1) r^{-1/2} \psi' + |k| (|k| - 2) r^{-3/2} \psi,$$

the hypotheses imply, via Cauchy-Schwarz inequality, that I_k is an absolutely convergent integral. Using the factorization of the operator R_k and making the notation

$$\eta_1(r) := \left(r \frac{\mathrm{d}}{\mathrm{d}r} + |k| - 2\right) \eta(r),$$

$$\psi_1(r) := \left(r \frac{\mathrm{d}}{\mathrm{d}r} + |k| - 2\right) \psi(r),$$

we have

$$I_k = \int_0^{r_n} r^{-3} \left[\left(r\frac{d}{dr} + |k| \right) \eta_1(r) \right] \left[\left(r\frac{d}{dr} + |k| \right) \overline{\psi_1}(r) \right] dr$$

$$= \sum_{j=1}^n \int_{r_{j-1}}^{r_j} r^{-2} \left[\left(r\frac{d}{dr} + |k| \right) \eta_1(r) \right] \frac{d}{dr} \overline{\psi_1}(r) \, dr$$

$$+ \sum_{j=1}^n \int_{r_{j-1}}^{r_j} r^{-3} \left[\left(r\frac{d}{dr} + |k| \right) \eta_1(r) \right] |k| \overline{\psi_1}(r) \, dr,$$

where all integrals remain absolutely convergent and $r_0 := 0$. Next, we apply integration by parts in each term of the first sum, which is permitted due to the fact that $\psi_1 \in AC_{\text{loc}}(0, \infty)$. Since

$$\frac{d}{dr} \left\{ r^{-2} \left[\left(r\frac{d}{dr} + |k| \right) \eta_1(r) \right] \right\} = r^{-3} \left[\left(r\frac{d}{dr} - 2 \right) \left(r\frac{d}{dr} + |k| \right) \eta_1(r) \right],$$

we obtain

$$I_k = \sum_{j=1}^n \left[\overline{\psi_1}(r) r^{-2} \left(r\frac{d}{dr} + |k| \right) \eta_1(r) \right]_{r_{j-1}}^{r_j}$$

$$- \sum_{j=1}^n \int_{r_{j-1}}^{r_j} r^{-3} \overline{\psi_1}(r) \left(r\frac{d}{dr} - |k| - 2 \right) \left(r\frac{d}{dr} + |k| \right) \eta_1(r) \, dr.$$

Since ψ has continuity C^1 and η has continuity C^2, the first sum of the last display is telescopic, hence we only have to evaluate the boundary terms corresponding to $r := r_n$ and $r := r_0 = 0$. Note that the boundary term at r_n is zero, since the condition $R_k\eta(r) = 0$, $\forall r > r_n$, of a Beppo Levi L_k-spline implies, by continuity, the relation $\left[\left(r\frac{d}{dr} + |k| \right) \eta_1(r) \right]_{r=r_n} = 0$.

For the boundary term at 0, consider first the case $|k| \geq 2$. Then the left end condition $G_k\eta(r) = 0$, i.e. $\eta \in \text{span}\{r^{|k|+2}, r^{|k|}\}$, for $r \in (0, r_1)$, implies

$$r^{-2} \left[\left(r\frac{d}{dr} + |k| \right) \eta_1(r) \right] = O\left(r^{|k|-2} \right), \text{ as } r \to 0.$$

Since, by Lemma 1, $\psi_1(r) = O(r)$, as $r \to 0$, we deduce that the boundary term at 0 vanishes if $|k| \geq 2$. If $|k| = 1$, the left end condition implies $\eta \in \text{span}\{r^3, r^1\}$, for $r \in (0, r_1)$, hence

$$r^{-2} \left[\left(r\frac{d}{dr} + 1 \right) \eta_1(r) \right] = cr, \forall r \in (0, r_1),$$

for some constant c. Since, by Lemma 1, in this case $\psi_1(r) = O\left(r |\ln r|^{1/2} \right)$, as $r \to 0$, it follows that the boundary term at 0 also vanishes if $|k| = 1$.

On the other hand, for each $j \in \{1, \ldots, n\}$, since $\eta \in \mathrm{Ker} L_k$ on the interval (r_{j-1}, r_j), there exists a constant c_j such that

$$\left(r\frac{d}{dr} - |k| - 2 \right) \left(r\frac{d}{dr} + |k| \right) \eta_1(r) = c_j r^{|k|}, \forall r \in (r_{j-1}, r_j).$$

Hence

$$I_k = \sum_{j=1}^{n} c_j \int_{r_{j-1}}^{r_j} r^{|k|-3} \left(r\frac{d}{dr} + |k| - 2 \right) \overline{\psi}(r)\, dr$$

$$= \sum_{j=1}^{n} c_j \int_{r_{j-1}}^{r_j} \frac{d}{dr} \left[r^{|k|-2} \overline{\psi}(r) \right] dr = \sum_{j=1}^{n} c_j \left[r^{|k|-2} \overline{\psi}(r) \right]_{r_{j-1}}^{r_j}.$$

For $|k| \geq 2$, since Lemma 1 implies $r^{|k|-2}\overline{\psi}(r) = O(r)$, as $r \to 0$, and, by hypothesis, $\psi(r_j) = 0, \forall j \in \{1, \ldots, n\}$, we deduce $I_k = 0$, as stated. For $|k| = 1$, we reach the same conclusion without the need to investigate $r^{-1}\overline{\psi}(r)$ as $r \to 0$, since in this case $c_1 = 0$. $\square$

3.2   Existence, Uniqueness, and Optimality

Theorem 2. *Let $\nu_1, \ldots, \nu_n$ be arbitrary real values, where $n \geq 2$. For each $k \neq 0$, there exists a unique Beppo Levi L_k-spline $\sigma \in \mathcal{S}_k(\rho)$, such that*

$$\sigma(r_j) = \nu_j, j \in \{1, \ldots, n\}. \tag{17}$$

Proof. It is sufficient to prove the existence of a unique function $\widetilde{\sigma} \in C^2[r_1, r_n]$ such that $\widetilde{\sigma} \in \mathrm{Ker} L_k$ on each subinterval (r_{j-1}, r_j) with $j \in \{2, \ldots, n\}$, $\widetilde{\sigma}$ satisfies the interpolation conditions (17) in place of σ, and the following endpoint conditions hold:

$$\left[\left(r\frac{d}{dr} - |k| \right) \left(r\frac{d}{dr} - |k| - 2 \right) \widetilde{\sigma}(r) \right]_{r \to r_1^+} = 0,$$

$$\left[\left(r\frac{d}{dr} + |k| \right) \left(r\frac{d}{dr} + |k| - 2 \right) \widetilde{\sigma}(r) \right]_{r \to r_n^-} = 0. \tag{18}$$

Indeed, such a function $\widetilde{\sigma}$ can be uniquely extended to the required Beppo Levi L_k-spline $\sigma \in \mathcal{S}_k(\rho)$ by defining

$$\sigma(r) := \begin{cases} c_1 r^{|k|+2} + c_2 r^{|k|}, & \text{if } 0 < r < r_1, \\[2mm] \widetilde{\sigma}(r), & \text{if } r_1 \leq r \leq r_n, \\[2mm] c_3 r^{-|k|+2} + c_4 r^{-|k|}, & \text{if } r_n < r. \end{cases}$$

To verify this, note that c_1 and c_2 (respectively, c_3 and c_4) are uniquely determined by the conditions that σ and σ' are continuous at r_1 (respectively, at r_n).

The continuity of σ'' at r_1 and r_n then follows automatically from (18) and from the properties $G_k \sigma (r) = 0$, $\forall r \in (0, r_1)$, and $R_k \sigma (r) = 0$, $\forall r > r_n$.

Now, a function $\widetilde{\sigma}$ with the properties stated in the previous paragraph is determined by four coefficients on each of the n subintervals (r_{j-1}, r_j), $j \in \{2, \ldots, n\}$. These coefficients are coupled by three C^2-continuity conditions at each interior knot $r_2, \ldots, r_{n-1}$, the endpoint conditions (18), and the n interpolation conditions (17), which amount to a $4(n-1) \times 4(n-1)$ system of linear equations.

To show that this system has a unique solution, we assume zero interpolation data: $\nu_j = 0$, $j \in \{1, \ldots, n\}$. Then the system becomes homogeneous, since the endpoint conditions and the continuity conditions at the interior knots were already homogeneous linear equations. Let $\widetilde{\sigma}$ be determined by an arbitrary solution of this homogeneous system and let $\sigma \in \mathcal{S}_k (\rho)$ be the unique extension of $\widetilde{\sigma}$ to a Beppo Levi L_k-spline. Taking $\eta = \psi := \sigma$ in (16), we obtain $R_k \sigma (r) = 0$, i.e. $\sigma \in \text{span} \{ r^{-|k|+2}, r^{-|k|} \}$, for $r \in (0, \infty)$. Since $\sigma (r_j) = 0$, $j \in \{1, \ldots, n\}$, and $n \geq 2$, we deduce $\sigma \equiv 0$. Therefore the above homogeneous system admits only the trivial solution, which concludes the proof.   □

Theorem 2 also extends to the case $n = 1$. Indeed, for each integer $k \neq 0$, it is straightforward to verify that there exists a unique function φ_k with the properties: $G_k \varphi_k (r) = 0$ for $0 < r < 1$, $R_k \varphi_k (r) = 0$ for $r > 1$, φ_k is C^2-continuous at $r = 1$, and $\varphi_k (1) = 1$. Its expression

$$\varphi_k (r) = \frac{1}{2} \begin{cases} r^{|k|} \left[(1 + |k|) + (1 - |k|) r^2 \right], 0 \leq r \leq 1, \\ \\ r^{-|k|} \left[(1 - |k|) + (1 + |k|) r^2 \right], 1 < r. \end{cases} \tag{19}$$

was given in [5, (3.10)] for $|k| \geq 2$ and is also seen to hold for $|k| = 1$. Hence, if $\rho = \{r_1\}$, then $\sigma := \nu_1 \varphi_k (\cdot / r_1)$ is the unique Beppo Levi L_k-spline in $S_k (\rho)$, such that $\sigma (r_1) = \nu_1$. As shown by the next result, if $|k| \geq 2$ and $n \geq 2$, the dilates of φ_k also provide a basis for a linear representation of the interpolant of Theorem 2.

Theorem 3. *Assume that $|k| \geq 2$, $n \geq 2$, and let σ be the Beppo Levi L_k-spline satisfying the interpolation conditions (17) of Theorem 2 for given values ν_1, ..., ν_n at the knot-set ρ. Then there exist unique coefficients a_1, ..., a_n, such that*

$$\sigma (r) = \sum_{j=1}^{n} a_j \varphi_k \left(\frac{r}{r_j} \right), \forall r \geq 0. \tag{20}$$

This result was established in [5, Lemma 3] for the special case in which σ satisfies Lagrange interpolation conditions. The proof given there also applies to our general interpolation conditions (17). Note that representation (20) does not hold for $|k| = 1$, but a similar representation for $k = 0$ appears in [7, Theorem 4].

Remark 2. For $|k| \geq 2$, it was observed in [5] that, if we make the notation

$$\psi_k (t) := e^{-t} \varphi_k (e^t) = \frac{1}{2} e^{-|k||t|} \left[(1 - |k|) e^{-|t|} + (1 + |k|) e^{|t|} \right], t \in \mathbb{R},$$

then ψ_k is a positive definite function, due to its positive Fourier transform

$$\widehat{\psi_k}(\tau) = \int_{-\infty}^{\infty} e^{-it\tau} \psi_k(t)\, dt = \frac{4|k|\left(k^2 - 1\right)}{\left[(|k|-1)^2 + \tau^2\right]\left[(|k|+1)^2 + \tau^2\right]}, \quad \tau \in \mathbb{R}.$$

This formula shows that ψ_k belongs to the class of generalized Whittle-Matérn-Sobolev kernels recently studied by Bozzini, Rossini, and Schaback [9].

The next result shows that our Beppo Levi L_k-spline interpolants minimize the functional (7), subject to the interpolation conditions.

Theorem 4. *Given $k \neq 0$ and arbitrary real values $\nu_1, \nu_2, \ldots, \nu_n$, let σ denote the unique Beppo Levi L_k-spline obtained in Theorem 2. Then $\|\sigma\|_k < \|g\|_k$ whenever g satisfies the same interpolation conditions (17) as σ and $g \neq \sigma$, where $g \in \Lambda_1$ if $|k| = 1$, while $g \in \Lambda_2$ if $|k| \geq 2$.*

Proof. Letting $\eta := \sigma$, $\psi := g - \sigma$, the hypotheses imply that ψ satisfies (15), hence (16) holds by Theorem 1. Since $\psi' \in AC_{\text{loc}}(0, \infty)$ and $\psi \in \Lambda_1$ if $|k| = 1$, while $\psi \in \Lambda_2$ if $|k| \geq 2$, we can use the proof of [5, Formula (5.3)] for $k \neq 0$ to show that

$$\int_0^{\infty} r^3 \left[R_k \eta(r)\right]\left[R_k \overline{\psi}(r)\right] dr = \langle \eta, \psi \rangle_k,$$

where

$$\langle \eta, \psi \rangle_k := \int_0^{\infty} \left\{ \eta'' \overline{\psi}'' + 2k^2 \left[\frac{\eta}{r^2} - \frac{\eta'}{r}\right]\left[\frac{\overline{\psi}}{r^2} - \frac{\overline{\psi}'}{r}\right] \right.$$
$$\left. + \left[\frac{k^2 \eta}{r^2} - \frac{\eta'}{r}\right]\left[\frac{k^2 \overline{\psi}}{r^2} - \frac{\overline{\psi}'}{r}\right] \right\} r\, dr.$$

Therefore (16) implies the orthogonality property

$$\langle \sigma, g - \sigma \rangle_k = 0,$$

from which

$$\|g\|_k^2 = \|\sigma\|_k^2 + \|g - \sigma\|_k^2, \tag{21}$$

and $\|g\|_k \geq \|\sigma\|_k$, with equality only if $\|g - \sigma\|_k = 0$. The last relation implies $g \equiv \sigma$ if $|k| \geq 2$, since $\|\cdot\|_k$ is a norm in this case. If $|k| = 1$, the semi-norm $\|g - \sigma\|_k$ vanishes if and only if $g(r) - \sigma(r) = ar$, $\forall r \in (0, \infty)$, for some constant a. Since $g - \sigma$ takes zero values at the knots $r_1, \ldots, r_n$, we deduce again $g \equiv \sigma$, which completes the proof. $\square$

4 Approximation Orders

For each $k \neq 0$, the following result establishes L^{∞} and L^2-error bounds for interpolation with Beppo Levi L_k-splines to data functions from Λ_1 or Λ_2.

Theorem 5. *Let $\rho := \{r_1, \ldots, r_n\}$ be a set of nodes with $0 < r_1 < \ldots < r_n$, $n \geq 2$, and $h := \max_{1 \leq j \leq n-1} (r_{j+1} - r_j)$. For an integer $k \neq 0$, let $g : (0, \infty) \to \mathbb{R}$ be a data function such that $g \in \Lambda_1$ if $|k| = 1$, while $g \in \Lambda_2$ if $|k| \geq 2$. Let $\sigma \in \mathcal{S}_k(\rho)$ be the Beppo Levi L_k-spline of Theorem 2, corresponding to data values $\nu_j := g(r_j)$, $1 \leq j \leq n$. Then, for $m \in \{0, 1\}$, we have the error bounds:*

$$\left\| \frac{d^m}{dr^m}(g - \sigma) \right\|_{L^\infty[r_1, r_n]} \leq \frac{1}{2^{1-m}\sqrt{r_1}} h^{3/2-m} \|g\|_k, \tag{22}$$

$$\left\| \frac{d^m}{dr^m}(g - \sigma) \right\|_{L^2[r_1, r_n]} \leq \frac{1}{2^{1-m}\sqrt{r_1}} h^{2-m} \|g\|_k. \tag{23}$$

Proof. Similar error bounds for $k = 0$ were obtained in [7, Theorems 5 and 6], along the lines of the classical error analysis for generalized splines [1]. The same arguments are also seen to apply to the present case $k \neq 0$, by replacing the seminorm $\|\cdot\|_0$ of [7] with $\|\cdot\|_k$ and using the inequality $\int_0^\infty r |g''(r)|^2 \, dr \leq \|g\|_k^2$, valid for any data function g as in the hypothesis. □

Remark 3. As in [7], the bounds (22) and (23) also imply an L^p-error bound for $p \in (2, \infty)$. Moreover, a similar analysis to that of [7, Theorem 7] shows that the exponents of h in the above error bounds cannot be increased for the classes Λ_1 and Λ_2 of data functions.

The main result of this section applies (23) and the corresponding error bound of [7] for $k = 0$ to obtain a L^2-convergence order for transfinite surface interpolation with biharmonic Beppo Levi polysplines on annuli. To state this result, let W^2 be the Wiener-type algebra of continuous periodic functions $\mu : [-\pi, \pi] \to \mathbb{R}$ with Fourier coefficients $\widehat{\mu}_k$, $k \in \mathbb{Z}$, such that $\sum_{k \in \mathbb{Z}} |\widehat{\mu}_k| (1 + |k|)^2 < \infty$. Note that $W^2 \subset C^2[-\pi, \pi]$ and, as observed in [5, Remark 1], any periodic cubic spline belongs to W^2.

Theorem 6. *Given $F \in C^2(\mathbb{R}^2 - \{0\}) \cap C(\mathbb{R}^2)$ of polar form f such that (2) is finite, assume that $f(r_j, \cdot) \in W^2$ along each domain circle $r = r_j$, $j \in \{1, \ldots, n\}$. Let S be either one of the Beppo Levi polysplines S^A or S^B determined in [5, Theorem 1], satisfying the transfinite interpolation conditions (3) for $\mu_j := f(r_j, \cdot)$, $j \in \{1, \ldots, n\}$, where also $S^A(0) = F(0)$ and S^B is biharmonic at 0. Then, for $m \in \{0, 1\}$, we have the L^2-error bound*

$$\left(\int_{r_1}^{r_n} \int_{-\pi}^{\pi} \left| \frac{\partial^m}{\partial r^m}(f - s)(r, \theta) \right|^2 r \, d\theta \, dr \right)^{1/2} \leq 2^{m-1} \sqrt{\frac{r_n}{r_1}} h^{2-m} \|f\|_{BL}. \tag{24}$$

Proof. For each $r \geq 0$, let $\widehat{f}_k(r)$, $k \in \mathbb{Z}$, be the Fourier coefficients of $f(r, \theta)$ with respect to θ. The smoothness assumptions on F imply that $\widehat{f}_k \in C^2(0, \infty)$, $\widehat{f}_k$ is continuous at $r = 0$, $\forall k \in \mathbb{Z}$, and identity (6) holds. Since $\frac{\partial^m}{\partial r^m}(f - s)(r, \cdot) \in C[-\pi, \pi] \subset L^2[-\pi, \pi]$, the following Plancherel formula is also valid for $m \in$

$\{0, 1\}$ and $r \in [r_1, r_n]$:

$$\frac{1}{2\pi} \int_{-\pi}^{\pi} \left| \frac{\partial^m}{\partial r^m} (f - s)(r, \theta) \right|^2 d\theta = \sum_{k \in \mathbb{Z}} \left| \frac{d^m}{dr^m} \left(\widehat{f}_k - \widehat{s}_k \right)(r) \right|^2.$$

Moreover, since $\sqrt{r} \frac{\partial^m}{\partial r^m} (f - s) \in C([r_1, r_n] \times [-\pi, \pi]) \subset L^2([r_1, r_n] \times [-\pi, \pi])$, we may multiply the above relation by r and integrate both sides to obtain, via Fubini's theorem,

$$\frac{1}{2\pi} \int_{r_1}^{r_n} \int_{-\pi}^{\pi} \left| \frac{\partial^m}{\partial r^m} (f - s)(r, \theta) \right|^2 r \, d\theta \, dr = \sum_{k \in \mathbb{Z}} \int_{r_1}^{r_n} \left| \frac{d^m}{dr^m} \left(\widehat{f}_k - \widehat{s}_k \right)(r) \right|^2 r \, dr.$$

(25)

Note that, for each $j \in \{1, \ldots, n\}$, the transfinite interpolation condition $s(r_j, \theta) = f(r_j, \theta)$, $\forall \theta \in [-\pi, \pi]$, is equivalent to $\widehat{s}_k(r_j) = \widehat{f}_k(r_j)$, $\forall k \in \mathbb{Z}$. Hence, for $k \neq 0$, the error bound (23) implies, for $m \in \{0, 1\}$,

$$\left\| \frac{d^m}{dr^m} \left(\widehat{f}_k - \widehat{s}_k \right) \right\|_{L^2([r_1, r_n])} \leq \frac{1}{2^{1-m} \sqrt{r_1}} h^{2-m} \left\| \widehat{f}_k \right\|_k.$$

(26)

In addition, it follows from [7, Theorem 6] that this error bound also holds for $k = 0$, since $\widehat{s}_0^A(0) = \widehat{f}_0(0) = F(0)$ and $\widehat{s}_0^B \in \text{span} \{r^2, 1\}$ for $r \in (0, r_1)$.

Therefore (25), (26), and (6) imply

$$\frac{1}{2\pi} \int_{r_1}^{r_n} \int_{-\pi}^{\pi} \left| \frac{\partial^m}{\partial r^m} (f - s^{A,B})(r, \theta) \right|^2 r \, d\theta \, dr$$

$$\leq r_n \sum_{k \in \mathbb{Z}} \int_{r_1}^{r_n} \left| \frac{d^m}{dr^m} \left(\widehat{f}_k - \widehat{s}_k \right)(r) \right|^2 dr$$

$$\leq Ch^{2(2-m)} \sum_{k \in \mathbb{Z}} \left\| \widehat{f}_k \right\|_k^2 = \frac{C}{2\pi} h^{2(2-m)} \|f\|_{BL}^2,$$

where $C = 2^{2(m-1)} r_n / r_1$, which establishes (24).   □

A similar approximation order for transfinite interpolation via biharmonic Beppo Levi polysplines on parallel strips has recently been proved in [6]. Related Plancherel representations of the error have been employed before by Kounchev and Render [15] for cardinal polysplines on annuli and by Sharon and Dyn [21] for interpolatory subdivision schemes.

References

1. Ahlberg, J.H., Nilson, E.N., Walsh, J.L.: The Theory of Splines and Their Applications. Academic Press, New York (1967)
2. Al-Sahli, R.S.: L-spline Interpolation and Biharmonic Polysplines on Annuli, M.Sc thesis, Kuwait University (2012)

3. Bejancu, A.: Semi-cardinal polyspline interpolation with Beppo Levi boundary conditions. J. Approx. Theory **155**, 52–73 (2008)
4. Bejancu, A.: Transfinite thin plate spline interpolation. Constr. Approx. **34**, 237–256 (2011)
5. Bejancu, A.: Thin plate splines for transfinite interpolation at concentric circles. Math. Model. Anal. **18**, 446–460 (2013)
6. Bejancu, A.: Beppo Levi polyspline surfaces. In: Cripps, R.J., Mullineux, G., Sabin, M.A (eds.), The Mathematics of Surfaces XIV, The Institute of Mathematics and its Applications, UK, (2013) ISBN 978-0-905091-30-3
7. Bejancu, A.: Radially symmetric thin plate splines interpolating a circular contour map. J. Comput. Appl. Math. (2015). doi:10.1016/j.cam.2015.06.019
8. Bennett, C., Sharpley, R.: Interpolation of Operators. Academic Press, Boston (1988)
9. Bozzini, M., Rossini, M., Schaback, R.: Generalized Whittle-Matérn and polyharmonic kernels. Adv. Comput. Math. **39**, 129–141 (2013)
10. Cavoretto, R., Fasshauer, G.E., McCourt, M.J.: An introduction to the Hilbert-Schmidt SVD using iterated Brownian bridge kernels. Numer. Alg. **68**, 393–422 (2015)
11. Duchon, J.: Interpolation des fonctions de deux variables suivant le principe de la flexion des plaques minces. RAIRO Anal. Numer. **10**, 5–12 (1976)
12. Jerome, J., Pierce, J.: On spline functions determined by singular self-adjoint differential operators. J. Approx. Theory **5**, 15–40 (1972)
13. Karlin, S., Ziegler, Z.: Tchebycheffian spline functions. SIAM J. Numer. Anal. **3**, 514–543 (1966)
14. Kounchev, O.I.: Multivariate Polysplines: Applications to Numerical and Wavelet Analysis. Academic Press, London (2001)
15. Kounchev, O.I., Render, H.: The approximation order of polysplines. Proc. Am. Math. Soc. **132**, 455–461 (2003)
16. Kounchev, O.I., Render, H.: Cardinal interpolation with polysplines on annuli. J. Approx. Theory **137**, 89–107 (2005)
17. Lucas, T.R.: A generalization of L-splines. Numer. Math. **15**, 359–370 (1970)
18. Rabut, C.: Interpolation with radially symmetric thin plate splines. J. Comput. Appl. Math. **73**, 241–256 (1996)
19. Schultz, M.H., Varga, R.S.: L-splines. Numer. Math. **10**, 345–369 (1967)
20. Schumaker, L.L.: Spline Functions: Basic Theory, 3rd edn. Cambridge University Press, Cambridge (2007)
21. Sharon, N., Dyn, N.: Bivariate interpolation based on univariate subdivision schemes. J. Approx. Theory **164**, 709–730 (2012)
22. Wendland, H.: Scattered Data Approximation. Cambridge University Press, Cambridge (2005)

On a New Conformal Functional
for Simplicial Surfaces

Alexander I. Bobenko$^{(\boxtimes)}$ and Martin P. Weidner

Institut Für Mathematik, Technische Universität Berlin,
Straße des 17. Juni 136, 10623 Berlin, Germany
bobenko@math.tu-berlin.de

Abstract. We introduce a smooth quadratic conformal functional and
its weighted version

$$W_2 = \sum_e \beta^2(e) \quad W_{2,w} = \sum_e (n_i + n_j)\beta^2(e),$$

where $\beta(e)$ is the extrinsic intersection angle of the circumcircles of the
triangles of the mesh sharing the edge $e = (ij)$ and n_i is the valence of
vertex i. Besides minimizing the squared local conformal discrete Willmore energy W this functional also minimizes local differences of the
angles β. We investigate the minimizers of this functionals for simplicial
spheres and simplicial surfaces of nontrivial topology. Several remarkable facts are observed. In particular for most of randomly generated
simplicial polyhedra the minimizers of W_2 and $W_{2,w}$ are inscribed polyhedra. We demonstrate also some applications in geometry processing,
for example, a conformal deformation of surfaces to the round sphere.
A partial theoretical explanation through quadratic optimization theory
of some observed phenomena is presented.

1 Introduction: Discrete Conformal Willmore Functional

The Willmore energy of a surface $S \subset \mathbb{R}^3$ is given as

$$\int_S (H^2 - K) = 1/4 \int_S (k_1 - k_2)^2,$$

where k_1 and k_2 denote the principal curvatures, $H = 1/2(k_1 + k_2)$ and $K = k_1 k_2$
the mean and the Gaussian curvatures respectively. For compact surfaces with
fixed boundary a minimizer of the Willmore energy is also a minimizer of total
curvature $\int_S (k_1^2 + k_2^2)$, which is a standard functional in variationally optimal
surface modelling.

In the last years various discretizations of the Willmore functional and of the
corresponding flow were investigated. They are mostly used for surface fairing.

This research was supported by the DFG Collaborative Research Center TRR 109,
"Discretization in Geometry and Dynamics".

© Springer International Publishing Switzerland 2015
J.-D. Boissonnat et al. (Eds.): Curves and Surfaces 2014, LNCS 9213, pp. 47–59, 2015.
DOI: 10.1007/978-3-319-22804-4_4

For surface restoration with smooth boundary condition based on a discrete version of the Willmore energy see [6]. More recently quadratic curvature energy flows were discretized in [14] using a semi-implicit scheme. A two step discretization of the Willmore flow was suggested in [13].

An important feature of the Willmore energy is its conformal invariance, i.e. invariance under Möbius transformations. A conformally invariant discrete analogue of the Willmore functional for simplicial surfaces was introduced in [3] and studied in [5]. Recently there was a big progress in development of conformal geometry processing in general [7] and in particular in investigation of discrete conformal curvature flows [8].

The discrete conformal Willmore energy introduced in [3] is defined in terms of the intersection angles of the circumcircles of neighboring triangles.

Definition 1. *Let S be a simplicial surface in 3-dimensional Euclidean space. Denote by $\mathcal{E}$ and $\mathcal{V}$ its edge set and its vertex set respectively. Let $\beta(e_{ij})$ be the external intersection angle of the circumcircles of the two triangles incident with the edge $e_{ij} \in \mathcal{E}$ as shown in Fig. 1. Then the* discrete conformal Willmore *functional $W(S)$ of S is defined as*

$$W(S) := \sum_{e_{ij} \in \mathcal{E}} \beta(e_{ij}) - \pi|\mathcal{V}|, \tag{1}$$

where $|\mathcal{V}|$ is the number of vertices.

We call the realization of a polyedron *inscribed*, if all its vertices lie on a round sphere. Note that in general we do not require such a realization to be convex. On the other hand we call a polyhedron *inscribable* or *of inscribable type* if there exists a convex, non-degenerate (i.e. without coinciding vertices) inscribed realization. Recall that for inscribed simplicial polyhedra convexity is equivalent to the Delaunay property of the triangulation. The functional W has two important properties that justify its name.

Theorem 1. *Let S be a simplicial closed surface. Then the following properties hold for the functional $W(S)$.*

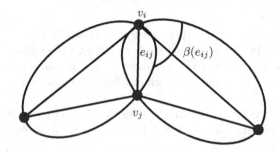

Fig. 1. Definition of the external intersection angle $\beta(e_{ij})$.

(i) $W(S)$ is invariant under conformal transformations of the 3-dimensional Euclidean space (Möbius transformations).

(ii) $W(S)$ is non-negative and it is equal to zero if and only if S is a convex inscribed polyhedron.

The first property follows immediately from the definition since Möbius transformations preserve circles and their intersection angles. Conformal invariance is an important property of the classical Willmore energy [2,15]. The second property is the discrete analogue of the fact that the classic Willmore functional is non-negative and that it is equal to zero if and only if the surface at hand is a (round) sphere. For a proof of (ii) see [4]. Let us note that the minimizer of W for combinatorial spheres is not unique: W vanishes for any inscribed convex polyhedron, i.e. for any Delaunay triangulation of the round sphere.

The functional W can be used in geometry processing to make the surface "as round as possible". In [5] the associated gradient flow is discussed. It works nicely for smoothing surfaces in many cases. However the functional is not smooth for surfaces that have some of the angles $\beta(e_{ij})$ equal to zero. This happens when the circumcircles of two neighboring triangles coincide. To minimize W numerically it works out quite well to simply set the gradient equal to zero as soon as the angle of the corresponding edge attains a value below a certain threshold [5].

In this paper we introduce a smooth conformal energy for simplicial surfaces, which behaves similar to the discrete Willmore energy (1). We have observed several surprising features of the minimizers of this functional. Only very few of them we can explain. The other remain to be challenging problems for future research.

2   Quadratic Circle-Angles Functional

A very natural manner to smoothen W is to consider a quadratic modification of (1).

Definition 2. *Let $\beta(e_{ij})$ be the external intersection angle of the circumcircles as in Definition 1. Then the* quadratic circle-angles (QCA) functional $W_2(S)$ *is given by*

$$W_2(S) := \sum_{e_{ij} \in \mathcal{E}} \beta(e_{ij})^2 - c. \qquad (2)$$

The normalization constant $c = 4\pi^2 \mathbf{1}^t (MM^t)^{-1} \mathbf{1}$ depends only on the combinatorial properties of S. Here M is the incidence matrix $M \in \mathbb{R}^{|\mathcal{V}| \times |\mathcal{E}|}$ of the edge graph of the surface and $\mathbf{1}$ is the vector $(1, \ldots, 1)^t \in \mathbb{R}^{|\mathcal{V}|}$. This choice of c will be justified in Sect. 4. Observe that W_2 is smooth at $\beta = 0$.

A priori it is not clear for which realization (of a given combinatorics) W_2 is minimal. Here an interesting case is the one of inscribable polyhedra because there we can directly compare the result with the minimal realization under the discrete conformal Willmore functional W.

Besides W_2 we have considered some other modifications among which the most promising is a weighted version of W_2.

Definition 3. *Denote by n_i the valence of the vertex $v_i \in \mathcal{V}$. Then the* weighted QCA functional *is given by*

$$W_{2,w}(S) := \sum_{v_i \in \mathcal{V}} \left(\left(\sum_{v_j \sim v_i} \beta(e_{ij}) \right)^2 + \frac{1}{2} \sum_{v_k \sim v_i} \sum_{v_j \sim v_i} (\beta(e_{ij}) - \beta(e_{ik}))^2 \right) - c_w$$

$$= \sum_{e_{ij} \in \mathcal{E}} (n_i + n_j) \beta(e_{ij})^2 - c_w.$$

The constant $c_w = 4\pi^2 \mathbf{1}^t (MN^{-1}M^t)^{-1} \mathbf{1}$ again only depends on the combinatorial structure of S. Here M is the incidence matrix and $N \in \mathbb{R}^{|\mathcal{E}| \times |\mathcal{E}|}$ is the diagonal matrix with the value $n_i + n_j$ in the row (and column) corresponding to the edge e_{ij}. Again the choice of c_w will be motivated in Sect. 4. The motivation for the essential part of the functional is the following. For every vertex of the surface, compute the local discrete Willmore functional, square it and add the squares of all angle differences that occur at the given vertex. Hence besides minimizing the squared local discrete Willmore functional, the functional also minimizes local angle differences. A nice feature is that the functional allows a simple formulation using the valences of the vertices. This also shows that $W_{2,w}$ is nothing but a weighted version of W_2. In fact W_2 and $W_{2,w}$ behave in a similar way, as we shall see in the next section.

3   Minimization of the QCA Functional for Various Types of Discrete Surfaces

All examples have been computed within the VaryLab environment available at http://www.varylab.com using the limited-memory variable metric (LMVM) method from the TAO project. It only requires the implementation of a gradient, which it uses to compute approximations to the Hessian based on previous iterations. See [12] for details. All examples from this article are available as *.obj-files at http://page.math.tu-berlin.de/~bobenko. In this section we only describe the observations made during numerical experiments and the statements are not rigorous. A theoretical analysis is given in the next section.

3.1   Inscribable Simplicial Polyhedra

Consider a polyhedron of inscribable type. By Theorem 1 minimizing W yields a convex inscribed realization. An amazing fact about the minimizers of W_2 and $W_{2,w}$ is the following

Observation 1. *For many randomly generated simplicial polyhedra, the minimizers of W_2 and $W_{2,w}$ are inscribed polyhedra which are convex in many cases. Moreover these minimizers seem to be unique.*

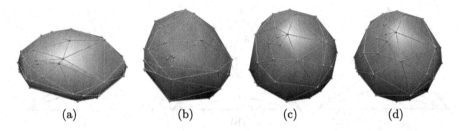

Fig. 2. (a) The original random ellipsoid. (b) The ellipsoid after minimizing W. (c) The ellipsoid after minimizing W_2. (d) The ellipsoid after minimizing $W_{2,w}$. (b), (c) and (d) are convex inscribed polyhedra.

In fact, W_2 and $W_{2,w}$ do not only reproduce the qualitative behavior of W in many cases, but they perform better in a certain sense. As an example consider the ellipsoid in Fig. 2. It has been obtained by placing 50 vertices randomly on the surface of an ellipsoid and computing their convex hull. The fact that minimizers of the functionals W_2 and $W_{2,w}$ are spherical is surprising. The functionals W_2 and $W_{2,w}$ yield considerably more uniform triangulations of the sphere, which is not very surprising. Indeed, we have incorporated this feature explicitly into the definition of $W_{2,w}$ by adding the terms that involve the differences angles at incident edges. The functional W_2 shows the same behavior since values that are close to each other yield a smaller sum of squares. Then the rate of numerical convergence is faster, i.e. it takes considerably less iterations of the numerical solver to obtain a gradient norm below a certain threshold. For the example in Fig. 2 this reflects in the following numbers. After 100 minimization steps for W, its value is still of order 10^{-2}. In contrast, computing 100 minimization steps for W_2 (resp. $W_{2,w}$) yields a realization where the value of W is of order only 10^{-9} (resp. 10^{-10}). Because of our choice of the normalization constants we have W_2 of order 10^{-10} and $W_{2,w}$ of order 10^{-8} after minimizing the respective energy during 100 steps. We have also considered different initial realisations of the same combinatorial structure. This way, minimizing W can lead to different realizations, all of them satisfying $W = 0$. For W_2 and $W_{2,w}$ we have always obtained the same realization up to conformal symmetry.

The next example is given by the first graph in Fig. 3. It is an inscribable polyhedron, i.e. the minimizer for W satisfies $W = 0$. For the minimizer of W_2 we compute $W_2 = 0$ but $W > 0$. A closer investigation reveals that the minimizer of W_2 is a non-Delaunay triangulation of the sphere. In fact, there is one non-Delaunay edge. It is highlighted in the graph by a dotted line. In contrast, the minimizer of $W_{2,w}$ satisfies $W_{2,w} = 0$ and also $W = 0$, i.e. it is a Delaunay triangulation of the sphere. This is an example where W_2 and $W_{2,w}$ yield qualitatively different results. There are also examples for which the minimizer of $W_{2,w}$ is a non-Delaunay triangulation of the sphere. One such example is shown in Fig. 3(b).

There are examples that are not covered by Observation 1. The problem is that there are polyhedra of inscribable type that do not have a realization that

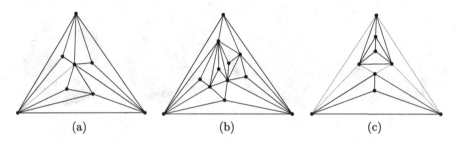

Fig. 3. Three graphs of inscribable type. (a) The graph of a polyhedron for which W_2 is minimized by a non-Delaunay triangulation of the sphere. (b) The graph of a polyhedron for which both W_2 and $W_{2,w}$ are minimized by a non-Delaunay triangulation of the sphere. (c) The graph of a polyhedron that does not converge while minimizing W_2 or $W_{2,w}$.

minimizes W_2 or $W_{2,w}$. Consider the graph in Fig. 3(c). A minimization of W_2 or $W_{2,w}$ leads to a realization where several edges collapse. We postpone an explanation of this behavior to the next section.

3.2   Noninscribable Simplicial Polyhedra

In the case of non-inscribable polyhedra, the investigation of the minimizers for W, W_2 and $W_{2,w}$ is a considerably more difficult task. However, we observe some remarkable phenomena in this case as well.

Consider the example in Fig. 4. It is not inscribable in a strong sense, but there are convex inscribed realizations with several collapsed edges. Thus if we exclude such degenerate realizations, then W does not have a minimum for this polyhedron. The minimizer for W_2 contains a self-intersection but interestingly enough, all its vertices do still lie on a sphere. It is also remarkable that we have $W = 2\pi$ for this realization and that the gradient of W vanishes. It is however not a global minimum for W.

3.3   Surfaces of Higher Genus

An interesting observation can be made for the minimum of W_2 of one particular triangulation of the torus (Fig. 5). The minimum is attained at the triangulation of a torus of revolution and the ratio of the two radii (measured between appropriate vertices) is equal to $\sqrt{2}$ (up to numerical accuracy). The gradient of W also vanishes for this realization, however this critical point of W is unstable. Starting from the realization in Fig. 5 and minimizing W instead of W_2, the numerical solver does not reach the minimal realization.

Recall the famous Willmore conjecture [15] which states that the smooth tori of revolution with a ratio of $\sqrt{2}$ of the two radii (and their Möbius equivalents) minimize the Willmore energy for tori. The conjecture has recently been proven by Marques and Neves [11].

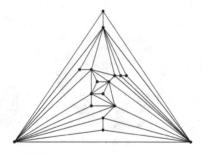

Fig. 4. The graph of a polyhedron of non-inscribable type. Its minimizer for W_2 contains self-intersections. Minimizing $W_{2,w}$ leads to several collapsed edges.

Computing the value of W_2 for the minimal realization gives us $W_2 = 3.998\pi^2$. By refining the triangulation this value seems to converge to $4\pi^2$. In the smooth case the minimal value of the Willmore energy for tori is equal to $2\pi^2$.

3.4   Applications in Geometry Processing

The Willmore energy functional plays an important role in digital geometry processing and geometric modelling. Applications of the discrete Willmore functional (1) for non-shrinking surfaces smoothing, surface restoration and hole filling were demonstrated in [5]. As already mentioned, the main drawback of the functional W is its non-smoothness.

The functionals W_2 and $W_{2,w}$ can be applied to the same problems and have some advantages comparing to W.

An example is shown in Fig. 6. The model is not closed and is treated with fixed boundary conditions, i.e. the boundary curve and tangent planes along it

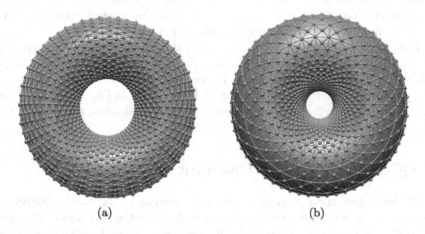

(a)   (b)

Fig. 5. (a) The original triangulation of a torus of revolution and (b) the result after minimizing W_2.

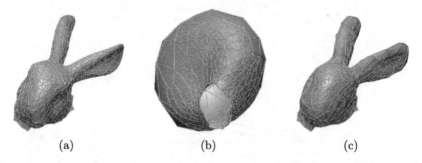

Fig. 6. (a) The original model. (b) The result after 1000 minimization steps for W_2. (c) The result after 1000 minimization steps for W.

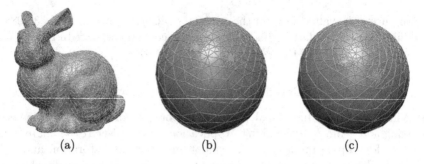

Fig. 7. (a) The Stanford bunny without holes. (b) The minimizer of W_2 after 4000 steps. (c) The minimizer of $W_{2,w}$ after 4000 steps.

are fixed. The ears of the bunny head cause the solver to run into problems when minimizing W. The realization where it gets stuck has many angles β with a value smaller than 10^{-3} with the smallest angle being even of order 10^{-5}. Hence the realization is very close to a critical point. In contrast, minimizing W_2 makes the bunny head already very spherical after 1000 steps.

The complete bunny shown in Fig. 7(a) is the Stanford bunny in which the holes in the bottom have been filled. Minimizing W_2 leads to a spherical shape with a discrete Willmore energy of 2π. The experiments with the weighted energy $W_{2,w}$ yield even better results. Starting with the model in Fig. 7(a), the surface converges to an inscribed convex realization. After 4000 steps, the value of the discrete Willmore energy is of order 10^{-3}.

4   QCA Functional and Quadratic Optimization

For W the minimizers of inscribable polyhedra are convex inscribed realizations. It would be nice to characterize the minimal realizations of these polytopes under W_2. In particular it would be interesting to know in which case they are minimizers of W, i.e. are convex and inscribed. To investigate the problem, we consider a quadratic program corresponding to W_2. At the end of the section we consider also $W_{2,w}$ where similar arguments can be applied.

Suppose we are given the graph G of a simplicial polyhedron of inscribable type. Denote by $\mathcal{V}$ and $\mathcal{E}$ its vertex set and edge set respectively. Now we ignore the geometry and simply consider the intersection angles as arbitrary weights on the edges. The inscribable polyhedra were characterized in [10].

Theorem 2. *Let $\mathcal{P}$ be a convex polyhedron with vertex set $\mathcal{V}$ and edge set $\mathcal{E}$. Let β be a weighting of the edges with $0 < \beta(e_{ij}) < \pi$ for all edges $e_{ij} \in \mathcal{E}$. Then there exists a convex inscribed realization of $\mathcal{P}$ with intersection angles of the circumcircles β if and only if the following conditions are satisfied.*

(i) $\displaystyle\sum_{e_{ij} \sim v_i} \beta(e_{ij}) = 2\pi$ *for every $v_i \in V$. The sum runs over all edges incident with v_i.*

(ii) $\displaystyle\sum_{k} \beta(e_k) > 2\pi$ *for all cycles $e_1^*, \ldots, e_n^*$ in the graph of the dual polyhedron that do not bound a face, where e_k^* is the dual edge that corresponds to e_k.*

Moreover, such a realization is unique up to conformal symmetry if it exists.

Denote by $M \in \mathbb{R}^{|\mathcal{V}| \times |\mathcal{E}|}$ the incidence matrix of the graph. The set of all $x \in \mathbb{R}^{|\mathcal{E}|}$ that satisfy the constraint that the weights sum up to 2π around each vertex is then given by solutions of the linear equation $Mx = 2\pi\mathbf{1}$ where $\mathbf{1} = (1, \ldots, 1)^t \in \mathbb{R}^{|\mathcal{V}|}$. Since we are dealing with the case where G is the graph of a simplicial polyhedron, the matrix M is of full rank $|\mathcal{V}|$ and in particular MM^t is invertible. Thus the following two quadratic programs always have a (unique) solution:

$$\text{minimize } \|x\|^2 = x^t x \text{ subject to } Mx = 2\pi\mathbf{1}, \tag{3}$$

$$\text{minimize } \|x\|^2 = x^t x \text{ subject to } Mx \geq 2\pi\mathbf{1}. \tag{4}$$

By $\| \cdot \|$ we denote the Euclidean norm. Furthermore, all inequalities between vectors are to be understood component-wise. The angle sum $\sum_{e \sim v} \beta(e)$ for any vertex $v \in \mathcal{V}$ is at least equal to 2π for every realization of any surface (see [4]). This means that the solution space of $Mx \geq 2\pi\mathbf{1}$ is a superset of all realizable angle sets. In order to find a sufficient condition for the minimum of W_2 to be inscribed and convex, we state the following

Proposition 1. *Let x and y be the unique solutions of* (3) *and* (4) *respectively. Then x and y are equal if and only if the (unique) solution of $MM^t\lambda = 2\pi\mathbf{1}$ is non-negative in every component.*

Proof. Suppose that the two minima do not coincide, that is $\|x\| > \|y\|$. Let $\delta = y - x$ be the difference of the two solutions. Then we have

$$M\delta = My - Mx \geq 2\pi\mathbf{1} - 2\pi\mathbf{1} = 0. \tag{5}$$

Furthermore we know that $\|x\|^2 > \|y\|^2$ and hence

$$0 > \sum_{i=1}^{|E|} \left((x_i + \delta_i)^2 - x_i^2 \right) = \sum_{i=1}^{|E|} \left(2x_i\delta_i + \delta_i^2 \right) > 2\sum_{i=1}^{|E|} x_i\delta_i.$$

Thus we obtain

$$\delta^t x < 0. \tag{6}$$

On the other hand if there is a vector δ satisfying (5) and (6), we see that $\varepsilon\delta + x$ with some small $\varepsilon > 0$ satisfies $M(\varepsilon\delta + x) \geq 2\pi\mathbf{1}$ and $\|x + \varepsilon\delta\| < \|x\|$. Because of $\|y\| \leq \|x + \varepsilon\delta\|$ this implies that y and x cannot coincide.

Hence the equality $x = y$ is equivalent to the non-existence of $\delta \in \mathbb{R}^{|E|}$ satisfying (5) and (6). By Farkas' lemma (see [16]), such a δ exists if and only if there is no $\lambda \geq 0$ with $M^t\lambda = x$. Since M^t is injective, λ is unique if it exists. It remains to show that it always exists and that it is equal to the unique solution of $MM^t\lambda = 2\pi\mathbf{1}$.

The vector x is the solution of the minimization of $x^t x$ subject to $Mx = 2\pi\mathbf{1}$. The respective Lagrange function is given by

$$L(x, \tilde\lambda) = x^t x - \tilde\lambda^t M x,$$

where $\tilde\lambda$ is the Lagrange multiplier. The critical point is given by

$$2x^t - \tilde\lambda^t M = 0 \Leftrightarrow M^t \tilde\lambda = 2x.$$

Here we see that up to a multiplication by 2, a solution λ of $M^t\lambda = x$ is given by the Lagrange multipliers. Thus the solution always exists and since it is unique, it has to coincide with the solution of $MM^t\lambda = Mx = 2\pi\mathbf{1}$. $\square$

For any incidence matrix M define

$$\beta(M) := 2\pi M^t(MM^t)^{-1}\mathbf{1}. \tag{7}$$

The matrix $MM^t \in \mathbb{R}^{|\mathcal{V}|\times|\mathcal{V}|}$ is the adjacency matrix of the graph with the valences of the vertices on the diagonal. It is called the signless Laplacian of the graph (see [9]). The matrix $M^t(MM^t)^{-1}$ is known as the Moore-Penrose pseudoinverse of M (see [1]). The proposition shows that in the case $2\pi(MM^t)^{-1}\mathbf{1} \geq 0$ it suffices to check whether the vector $\beta(M)$ satisfies $0 < \beta(M) < \pi$ component-wise and condition (ii) from Theorem 2. If this is the case then the minimum of W_2 is inscribed and convex and this minimum is unique up to conformal symmetry. We will derive some sufficient conditions for this.

If we assume $2\pi(MM^t)^{-1}\mathbf{1} > 0$ instead of $2\pi(MM^t)^{-1}\mathbf{1} \geq 0$ then $0 < \beta(M)$ is obviously satisfied. Also $\beta(M) < \pi$ holds. Indeed, let us assume that there exists an edge e with $\beta(e) \geq \pi$. Let us denote the corresponding weight by $\beta_e = \beta(e)$. Consider a perturbation of $\beta(M)$ as in Fig. 8. Around each vertex the β's sum up to 2π and the perturbation sums up to 0. In particular we have $\beta_a + \beta_b \leq \pi \leq \beta_e$ and $\beta_c + \beta_d \leq \pi \leq \beta_e$ and thus for any ε satisfying $0 < 5\varepsilon < \beta_f + \beta_g$,

$$(\beta_e - 2\varepsilon)^2 + (\beta_a + \varepsilon)^2 + (\beta_b + \varepsilon)^2$$
$$+ (\beta_c + \varepsilon)^2 + (\beta_d + \varepsilon)^2 + (\beta_f - \varepsilon)^2 + (\beta_g - \varepsilon)^2$$
$$= \beta_e^2 + \beta_a^2 + \beta_b^2 + \beta_c^2 + \beta_d^2 + \beta_f^2 + \beta_g^2$$
$$+ 2\varepsilon(5\varepsilon + \beta_a + \beta_b + \beta_c + \beta_d - 2\beta_e - \beta_f - \beta_g)$$
$$\leq \beta_e^2 + \beta_a^2 + \beta_b^2 + \beta_c^2 + \beta_d^2 + \beta_f^2 + \beta_g^2 + 2\varepsilon(5\varepsilon - \beta_f - \beta_g)$$
$$< \beta_e^2 + \beta_a^2 + \beta_b^2 + \beta_c^2 + \beta_d^2 + \beta_f^2 + \beta_g^2.$$

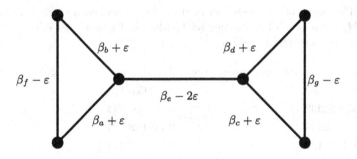

Fig. 8. Perturbing the edge weights on a subgraph.

This contradicts the minimality of $\beta(M)$ and hence we have $\beta(M) < \pi$.

The more complicated question is whether (ii) from Theorem 2 is satisfied. The general answer is no as the example in Fig. 3(c) shows. For the angles $\beta(M)$ there is a cocycle (highlighted in the graph) with the angle sum strictly less than 2π. This reflects in the fact that when we minimize W_2 numerically several edges collapse.

Let us formulate this claim.

Proposition 2. *Let $\mathcal{P}$ be a polyhedron of inscribable type with incidence matrix M. Let λ be given by $\lambda = 2\pi(MM^t)^{-1}\mathbf{1}$ and let $\beta(M)$ be given by (7). Assume that the following two properties are satisfied:*

(i) $\lambda > 0$ component-wise
(ii) $\beta(M)$ satisfies condition (ii) from Theorem 2.

Then the convex inscribed realization given by the angles $\beta(M)$ is a global minimizer of W_2. Furthermore, the minimum is unique up to conformal symmetry.

For the angles $\beta(M)$ we have

$$\beta(M)^t\beta(M) = 4\pi^2\mathbf{1}^t(MM^t)^{-1}\mathbf{1},$$

which motivates the choice of the normalization constant in the definition of W_2.

Empirical data suggests that condition (i) is not necessary and can be weakened to $\beta(M) > 0$. The problem is that we have no tool to characterize realizable angles as soon as they do not correspond to convex inscribed realizations.

We have seen that it can happen that W_2 is minimized by an inscribed but non-Delaunay realization (Fig. 3(a)). The corresponding angles of the minimizer and the abstract angles given by $\beta(M)$ are shown in Table 1. Since the first value in the right-hand column is negative, these values cannot correspond to realizable angles. It is however remarkable that the sign change is the only difference between the two columns (up to numeric accuracy). This phenomenon is still to be clarified.

Finally we briefly mention how to perform a similar treatment for $W_{2,w}$. The main ingredient is the diagonal matrix $N \in \mathbb{R}^{|\mathcal{E}| \times |\mathcal{E}|}$ that has the value $n_i + n_j$ in

Table 1. The angles of the minimizer of W_2 obtained numericaly versus the abstract angles $\beta(M)$ given by (7) for the simplicial surface in Fig. 3(a). All values are divided by π and sorted in ascending order.

Angles after numerical minimization	Abstract angles $\beta(M)$ given by (7)
0.0295374462	-0.0295374466
0.0559262333	0.0559262314
0.1364420123	0.1364420121
0.1587825189	0.1587825174
0.2392983002	0.2392982982
0.2475447917	0.2475447935
0.3247619735	0.3247619762
0.3504010798	0.3504010795
0.5057350595	0.5057350626
0.5142814330	0.5142814304
0.5163805365	0.5163805383
0.5696079164	0.5696079166
0.6085913486	0.6085913487
0.6724641987	0.6724642027
0.7026013894	0.7026013944
0.7579278849	0.7579278807
0.7831171776	0.7831171752
0.8856735920	0.8856735887

the row and column corresponding to the edge $e_{ij} \in \mathcal{E}$. Recall that n_i denotes the valence of the vertex $v_i \in \mathcal{V}$. Thus we now consider the quadratic programs that minimize $x^t N x$ subject to $M x = 2\pi \mathbf{1}$ or $M x \geq 2\pi \mathbf{1}$ respectively. Furthermore, instead of $\beta(M)$ we now consider $\tilde{\beta}(M)$ given by

$$\tilde{\beta}(M) = 2\pi N^{-1} M^t (M N^{-1} M^t)^{-1} \mathbf{1}. \tag{8}$$

An analog of Proposition 2 then reads as follows.

Proposition 3. *Let $\mathcal{P}$ be a polyhedron of inscribable type with incidence matrix M. Let λ be given by $\lambda = 2\pi (M N^{-1} M^t)^{-1} \mathbf{1}$ and let $\tilde{\beta}(M)$ be given by (8). Assume that the following two conditions are satisfied:*

(i) $\lambda > 0$ component-wise
(ii) $\tilde{\beta}(M)$ satisfies condition (ii) from Theorem 2.

Then the convex inscribed realization given by the angles $\tilde{\beta}(M)$ is a global minimizer of $W_{2,w}$. Furthermore, the minimum is unique up to conformal symmetry.

Again, this motivates the choice of the normalization constant

$$\tilde{\beta}(M)^t N \tilde{\beta}(M) = 4\pi^2 \mathbf{1}^t (M N^{-1} M^t)^{-1} \mathbf{1}$$

in Defintion 3.

References

1. Ben-Israel, A., Greville, T.N.E.: Generalized Inverses, 2nd edn. Springer, New York (2003)
2. Blaschke, W.: Vorlesungen über Differentialgeometrie III. Grundlehren der mathematischen Wissenschaften. Springer, Heidelberg (1929)
3. Bobenko, A.I.: A conformal energy for simplicial surfaces. In: Goodman, J.E., Pach, J., Welzl, E. (eds.) Combinatorial and Computational Geometry. Math. Sci. Res. Inst. Publ., vol. 52, pp. 135–145. Cambridge University Press, Cambridge (2005)
4. Bobenko, A.I.: Surfaces from circles. In: Bobenko, A.I., Schröder, P., Sullivan, J.M., Ziegler, G.M. (eds.) Discrete Differential Geometry. Oberwolfach Semin., vol. 38, pp. 3–35. Birkhäuser, Basel (2008)
5. Bobenko, A.I., Schröder, P.: Discrete Willmore flow. In: Desbrun, M., Pottmann, H. (eds.) Eurographics Symposium on Geometry Processing. pp. 101–110. Eurographics Association, Vienna, Austria (2005)
6. Clarenz, U., Diewald, U., Dziuk, G., Rumpf, M., Rusu, R.: A finite element method for surface restoration with smooth boundary conditions. Comput. Aided Geom. Des. 21(5), 427–445 (2004)
7. Crane, K.: Conformal geometry processing. Ph.D thesis, Caltech (2013)
8. Crane, K., Pinkall, U., Schröder, P.: Robust fairing via conformal curvature flow. ACM Trans. Graph. 32(4), 61:1–61:10 (2013)
9. Cvetković, D., Rowlinson, P., Simić, S.K.: Signless laplacians of finite graphs. Linear Algebra Appl. 423(1), 155–171 (2007)
10. Hodgson, C.D., Rivin, I., Smith, W.D.: A characterization of convex hyperbolic polyhedra and of convex polyhedra inscribed in the sphere. Bull. Amer. Math. Soc. (N.S.) 27(2), 246–251 (1992)
11. Marques, F.C., Neves, A.: Min-max theory and the Willmore conjecture. Ann. Math. 179, 683–782 (2014)
12. Munson, T., Sarich, J., Wild, S., Benson, S., Curfman McInnes, L.: TAO 2.0 users manual. Technical Memorandum ANL/MCS-TM-322, Argonne National Laboratory, Argonne, Illinois (2012)
13. Olischläger, N., Rumpf, M.: Two step time discretization of Willmore flow. In: Hancock, E.R., Martin, R.R., Sabin, M.A. (eds.) Mathematics of Surfaces XIII. LNCS, vol. 5654, pp. 278–292. Springer, Heidelberg (2009)
14. Wardetzky, M., Bergou, M., Harmon, D., Zorin, D., Grinspun, E.: Discrete quadratic curvature energies. Comput. Aided Geom. Des. 24(8–9), 499–518 (2007)
15. Willmore, T.J.: Riemannian Geometry. Oxford Science Publications. The Clarendon Press, Oxford University Press, New York (1993)
16. Ziegler, G.M.: Lectures on Polytopes. Graduate Texts in Mathematics, vol. 152. Springer, New York (1995)

Evaluation of Smooth Spline Blending Surfaces Using GPU

Jostein Bratlie[✉], Rune Dalmo, and Børre Bang

R&D Group in Mathematical and Geometrical Modeling,
Numerical Simulations, Programming and Visualization,
Narvik University College, PO Box 385, 8505 Narvik, Norway
{jbr,rda,bb}@hin.no
http://www.hin.no/Simulations

Abstract. Recent development in several aspects of research on blending type spline constructions has opened up new application areas. We propose a method for evaluation and rendering of smooth blending type spline constructions using the tessellation shader steps of modern graphics hardware. In this preliminary study we focus on concepts and terminology rather than implementation details. Our approach could lead to more efficient, dynamic and stable blending-type spline based applications in fields such as interactive modeling, computer games and more.

1 Introduction

The purpose of this article is to introduce a concept for evaluation and rendering of smooth blending type spline surfaces using features available in recent versions of modern rendering pipelines [9], most notably the open graphics library (OpenGL) [10], maintained by the Khronos group, and Microsoft® DirectX® (DirectX) [8]. Since the year 2000, a family of blending-type spline constructions named expo-rational B-splines (ERBS) [6] and, later, generalized expo-rational B-splines (GERBS) [1,2] has been introduced and explored by the research and development (R&D) group Simulations at Narvik University College (NUC). GERBS type splines enjoy some properties, including Hermite interpolation at the knots [5] and minimal support combined with C^k-smooth basis functions, which makes them attractive to interactive geometric modeling and smooth representations of parametric curves and surfaces.

Despite the flexibility of the construction, there is a price to pay, in particular with respect to interactive geometric modeling, due to the cost of the spline basis function evaluator. The performance is constrained by the following limitations:

1. Evaluation of the expo-rational basis function (ERB) requires an integration step.
2. The graphics rendering hardware is designed to support triangle constructions and simple cases of classic splines on Bézier form, i.e. mapped to the interval [0,1].

J.-D. Boissonnat et al. (Eds.): Curves and Surfaces 2014, LNCS 9213, pp. 60–69, 2015.
DOI: 10.1007/978-3-319-22804-4_5

The first issue was addressed by Zanaty in [12], where a relation between the ERBS basis function and Sigmoidal functions was explored. The second issue is addressed in this article.

Recently, in [7], Lakså expressed generic blending functions, including GERBS, in terms of classic B-splines. This is interesting since the rendering pipelines mentioned above were designed to be used with ordinary B-splines.

In this work we consider tessellation techniques which are now standardized across vendor specific application programming interfaces (APIs). Therefore, it is possible to adapt and use such tessellation steps to obtain rendering methods applicable to B-spline type constructions. We seek to describe the relevant technology; blending-type splines and rendering pipelines, as well as the concepts necessary for evaluation and rendering.

In the following sections we describe GERBS as an adjusted recursive definition of classic B-splines, similar to [7], followed by an overview of the relevant steps of the graphics pipeline present in modern graphics processing unit (GPU) hardware. Next, we introduce and define the critical components of the proposed rendering- and evaluation method followed by a description of the method itself. Finally, we give our concluding remarks, where we discuss some theoretical performance results, and suggest topics for future work.

2   Spline Blending Functions

The blending functions of GERBS [1,6] is presented in [7] as an adjusted recursive definition of the B-spline associated with the knots $(t_i)_{i=0}^{k+d}$:

$$B_{d,k}(t) = B \circ \omega_{d,k}(t)B_{d-1,k}(t) + (1 - B \circ \omega_{d,k+1}(t))B_{d-1,k+1}(t), \qquad (1)$$

where $\omega_{d,i}(t) = \frac{t-t_i}{t_{i+d}-t_i}$, $B_{0,i}(t) = \begin{cases} 1; & \text{if } t_i \leq t < t_{i+1}, \\ 0; & \text{otherwise,} \end{cases}$ and, in the case of GERBS, the degree $d = 1$, and B is a C^k-smooth blending function possessing the following set of properties:

1. $B : I \to I$ $(I = [0,1] \subset \mathbb{R})$,
2. $B(0) = 0$,
3. $B(1) = 1$,
4. $B'(t) >= 0, t \in I$.
5. $B(t) + B(1 - t) = 1, t \in I$.

The last property is optional and specifies point symmetry around the point $(0.5, 0.5)$, however, we assume this property in the present study.

B-functions come in a wide range of flavors including trigonometric, polynomial, rational and expo-rational. The perhaps most simple example of a B-function is $B(t) = t$. One example of a C^∞-smooth B-function, which belongs to the family of logistic expo-rational B-splines (LERBS) was presented in [12] and can be expressed, as a logistic expo-rational B-funcion, as follows:

$$B(t) = \frac{1}{1 + e^{\left(\frac{1}{t} - \frac{1}{1-t}\right)}}. \qquad (2)$$

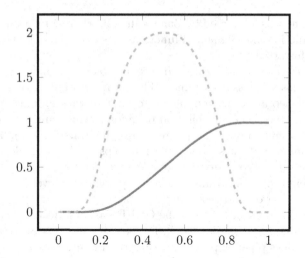

Fig. 1. Plots of a logistic ERB and its first derivative, as solid and dotted lines, respectively.

In contrast to the classic ERBS basis function, it follows from Eq. 2 that evaluation of this particular $B(t)$ does not require an integration step. A plot of $B(t)$ in Eq. 2, where $t \in [0, 1]$, is shown in Fig. 1. An essential remark is that the exponential function is implemented in GPU hardware.

A tensor product B-function spline surface is defined in [5, 7] as follows:

$$S(u, v) = \sum_{i=1}^{n} \sum_{j=1}^{m} \ell_{i,j}(u, v) B_{1,i}(u) B_{1,j}(v), \tag{3}$$

where $\ell_{i,j}(u, v)$ are *local surface patches* which are blended together by the C^k-smooth basis functions B. We note that using local surface patches as coefficients facilitates blending of points (Bézier and B-spline surfaces), points and vectors (Hermite interpolation surfaces) or even scalar- point- or vector valued functions (GERBS). Furthermore, we note the ERBS Hermite interpolation property [5] which states that ERBS type spline blending constructions interpolate the position and all existing derivatives of the local functions at every knot.

3   GPU Tessellation

The most recently added shader component in the GPU rendering pipeline is the tessellation shader. It consists of three sub components, steps two through four, between the vertex shader and the geometry shader, as shown in Fig. 2.

The tessellation shader steps operate on a type of primitive called *patch*. The tessellation patch primitive can be of three different types; isoline, triangle or quad.

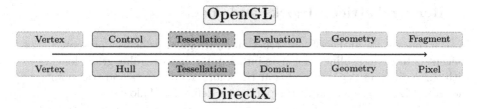

Fig. 2. The three tessellation shader steps shown as parts of the OpenGL and DirectX rendering pipelines. The control- and evaluation shaders, using OpenGLs terminology, which are illustrated as green boxes, are programmable. The primitive generator, or tessellation step, is fixed.

The tessellation step is controlled through three different substeps; control, tessellation and evaluation. The control and evaluation substeps are programmable while the tessellation substep is hardware implementation specific. In the following we provide a brief description of each substep of the tessellator.

  I. *Control* specifies which type of patch primitive to be considered and the amount of tessellation applied to each patch. It provides control of tessellation inside the patch and on the boundary of the patch independently.
 II. *Tessellation*, which is not programmable, only controllable, performs the actual tessellation. It generates primitives. We can think of this step as where the topology of the tessellation is induced.
III. *Evaluation* determines the position of the new tessellated vertex. It is based on affine transformations and performs in a manner similar to what the vertex shader does for a vertex, when combined with a basic primitive, such as point, line, triangle-strip and more.

A vertex generated by the tessellation step has a normalized position within the tessellation patch. This means that the evaluation step works on coordinates on the range $[0, 1]$ and, in the case of triangle type patch primitives, barycentric coordinates.

By considering the blending construction in Eq. 1 we propose building an evaluator based on the tessellation steps. Then, the tessellation patches of type line, triangle and quad, provided by the control step, take the roles as a render block on a blending type spline curve, -triangle surface and -surface, respectively. As we shall see in the following section, this is one layer in a hierarchical blending construction.

We conclude this section by mentioning briefly the roles of the two following steps. The tessellator determines parameter values stating where in the parametric domain a blending type curve or surface is to be evaluated. Finally, the tessellation evaluator is a shader implementation of a blending-type spline evaluator.

4 Render-Lattice, -Blocks and -Loci

In [7] a concept for an ERBS-construction on irregular grids was presented. The concept is to divide spline knot nets into regular and irregular grids, leaving us with three different types of points at the knots; regular-, T- and star points, as shown in Fig. 3. In [7] T- and star-points are defined as follows:

- A *T*-point is defined as a grid (parameter) line ending in an orthogonal grid line.
- A *Star*-point is defined as a point where several grid lines meets in a non-orthogonal way.

We propose the following descriptive names for a few of the grid components, when they are used for a rendering purpose:

- *Render lattice* to describe a grid structure arising from the net of spline knots.
- *Render locus* to describe loci in the render lattice, closely related to spline knots and regular-, T- and star-points.
- *Render block* to describe each line or face in a render lattice, i.e. the subset of the lattice that will be handled by a patch-type primitive.

The render block concept is closely connected to the description of the patch primitives of the GPU hardware. This provides some basic properties shared by all render blocks:

1. The domain of a render block is a parametric domain.
2. The parameter variables take values in the range $[0, 1]$.
3. In order to meet the requirements associated with the ERBS Hermite interpolation properties, the outer tessellation levels of two adjacent patch primitives must be the same.

Below follows a set of definitions which describe the concepts behind the names introduced above.

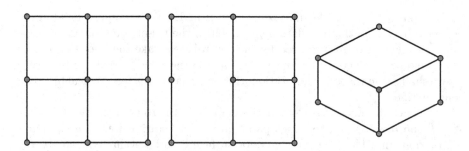

Fig. 3. Three different types of render loci. From left to right: regular, T and star render loci.

Definition 1. *A render locus is an extension to the points defined in [7]. A render locus is defined as a locus in a render lattice associated with a spline knot on a regular or irregular spline net. A render locus can be one of three basic types, depending on the point type of the spline knot, as described in [7]. The three types are regular, T and star.*

Definition 2. *A render block is an extension to the patch-type primitive of modern tessellation based GPU architecture. The parametric domain of the render block is limited by the boundary given by two, three or four render loci. The number of render loci is decided by the type of patch-type primitive, which for line, triangle, or quadratic patch-type primitives are two, three or four, respectively. A point in the domain of a line, triangle or quadratic render locus has a normalized position $((u), (u, v, w), (u, v)$, respectively) in the parametric domain $\Omega \in [0, 1]$.*

Definition 3. *A render lattice is defined in such a way that it coincides with the spline knot nets of the blending-type spline construction. Each locus on the render lattice is called a render locus. In a valid render lattice, render loci is divided and partitioned such that the render lattice consists of adjacently connected render blocks.*

Examples of regular and irregular render lattices with quadratic render blocks are shown in Fig. 4. Throughout this article we shall consider regular render lattices with quadratic render blocks and regular render loci. T- and star-type render loci are subjects for future work.

4.1 Quadratic Render Block with Regular Render Loci

A render block on a regular render lattice where all render loci are regular is defined in the following way:

$$P(u, v) = \sum_{i=1}^{2} \sum_{i=1}^{2} \ell_{i,j} \circ \omega_{i,j}(u, v) \, B_j(v) \, B_i(u),$$

where $u, v \in [0, 1]$ are parameters of the render block surface, and $\omega_{i,j}(u, v)$ are "map-to-local" functions mapping the parametric domain of the render block to the domain of the local surface patch associated with the given render locus.

The parametric domain of the patch-type primitive is normalized and the knot interval is always local, therefore, the "map-to-local" functions can be simplified from how they are described in [5]. For each parametric direction, u, v, the local mapping function, shown here for s, is defined as

$$\omega(s) = \gamma + \kappa \times s,$$

where γ is the parametric offset of the local patch and κ is the scaling factor to the parametric domain of the local patch.

Figure 5 shows an example of a blending spline made up of 3×3 render loci, associated with one local patch each, and rendered over a render lattice consisting of four render blocks.

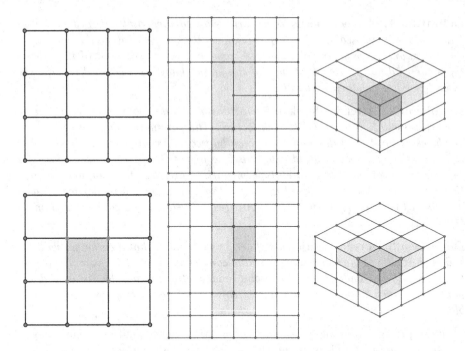

Fig. 4. Three different render lattices. A render block is highlighted in each render lattice on the bottom row. The left column shows regular render loci. Regular and T-type render loci are shown in the middle column, whereas the right column contains regular and star-type render loci. The illustrations are provided to show the equivalence to the illustrations in [[7], Fig. 4].

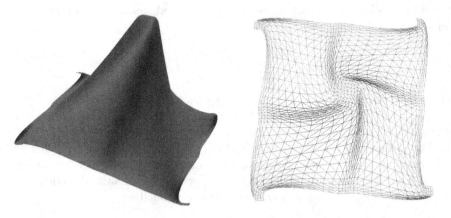

Fig. 5. The "Tower" surface: a blending spline tensor-product surface made up by 3×3 knots with an associated local plane surface patches. The local surfaces in the corners are locally rotated 90 degrees in the xy-plane and 45 degrees in xz-plane. The rendering lattice is here made up by 3×3 render loci (one for each spline knot locus) and four render blocks. On the left is a shaded version of the resulting surface, while on the right is a wireframe version showing the tessellation of the four render blocks.

4.2   Implementation Strategies

The method proposed above provides a rendering strategy for "global" geometric objects which are depending on evaluation of "local" geometry, possibly in several layers. When it comes to implementation, several issues related to optimization and performance, including the following, could be considered:

1. Evaluation of blending functions
2. Evaluation of local geometry
3. Data organization
4. Shader roles

In this article we shall not focus om implementation details, however, we find it appropriate to comment some of the above mentioned issues. Efficient evaluation of the original ERB was one major issue which, after it was addressed in [12], lead to the idea behind this article.

The second and third issues are tightly coupled. For this reason we prefer to see them in connection. We propose two valid, but different, strategies here. The first is based on using a general evaluation scheme of pre-sampled data. It is unnecessary to customize the shaders for each type of local patch, such as, for example, a cap of a sphere vs. a Bézier patch, as each local patch has to be pre-evaluated. One downside of this is that the pre-evaluated data must be stored and managed, another is that the precision of the local surface data is given by the resolution of the pre-sampling approximation step. On the positive side we note that, as a consequence of pre-evaluation, the evaluation time of a local patch would not depend on the patch type. Additionally, since the nature of the blending spline smoothly blends the local geometry, the resolution of the data of the local patches does not need to be as high.

The second strategy for dealing with the second and third issues is using custom built shaders for a given configuration of local render loci of a render block. We mention the following drawbacks of this approach; on-the-fly evaluation of local patches could be more expensive than a look-up in a pre-evaluated table. Furthermore, each shader on a render block must be regenerated whenever there is a change in the configuration of its render loci. Some major upsides with this approach are that it facilitates code modularization and the possibility of pixel-accurate [4,11] resolution. Code modularization is specific to individual GPU architectures and APIs, but as an example, using OpenGL, one could divide each of the tessellation evaluation shader parts (BS evaluator, B-function evaluator, local patch evaluator and others) into shader objects and choose the appropriate ones when used.

The last issue is achieving efficiency by shader design. Using the features provided by the tessellation shader it is possible to generate patch geometry on the fly by circumventing the vertex shader altogether, providing only the coefficients of the local geometry to the shader. The final geometry is generated by the tessellation shader and the local geometry only exists as an evaluation result. A change in position, orientation or coefficient data of the local geometry would directly cause deformation to the rendered geometry. This change would not cause any additional computational costs as far as shader evaluation is concerned and a stable framerate would be maintained. This literally means keeping the spline representation all the way to the graphics hardware. Furthermore, it facilitates affine spatial transformations of the spline coefficients before they are provided to the rendering pipeline.

5   Concluding remarks

We have introduced a method for smooth rendering of blending-type splines where standard features of the tessellation shader architecture are exploited. The presented work has described a method for smooth rendering of blending spline constructions using tessellation based GPU architecture. In addition, fundamental render block types and its terminology has been proposed and described as well as strategies for implementation.

Some notable features of this method include, but are not limited to:

– Predictable and constant rendering time (in the sense of modification of the spline construction)
– The rendering method is local with respect to the render block

The method could be suitable for visualization in computer games or computer-generated imagery (CGI). Furthermore, since modification of the underlying spline construction adds little strain to the rendering method, it supports animation, simulations and interactivity, with little computational overhead.

The rendering method preserves the geometry description and topology until it is discretized in the hardware. Independent of implementation, the sampling is performed by the hardware instead of a defined procedure followed by pushing to the GPU.

Variable level of detail (LoD) per render block could be achieved by using well known methods for setting the inner tessellation levels through the control step of the tessellation shader [4,11]. The outer tessellation levels could be used to adjust and minimize artifacts over the boundary between two adjacent render blocks. This is of interest when the blending spline construction is used in application areas which results in large render patches, such as terrain representation.

The facts that the rendering method is strictly local (limited to a render block), and that it operates directly on the underlying spline construction, makes applications within interactive geometric modeling and sculpting interesting topics for future work.

A preliminary implementation supporting rendering of regular tensor-product surface render latices was created. The surfaces shown in Fig. 5 were rendered using this preliminary implementation, which proves the concept. We propose focusing on efficient design for render blocks containing T- and star-type render loci on irregular render lattices in the next stages of research and development.

Inter-operable features between general purpose GPU (GPGPU) specialized architecture API, such as open computing language (OpenCL) or compute unified device architecture (CUDA), and graphics architecture API, such as OpenGL or DirectX, are available. For this reason, developing appropriate data structures and strategies for efficient data sharing and communication, is desirable.

A more comprehensible study on the efficiency of the solution should be conducted, as part of any specific implementation.

References

1. Dechevsky, L.T., Bang, B., Lakså, A.: Generalized expo-rational B-splines. Int. J. Pure Appl. Math. **57**(6), 833–872 (2009)
2. Dechevsky, L.T., Lakså, A., Bang, B.: Expo-rational B-splines. Int. J. Pure Appl. Math. **27**(3), 319–362 (2006)
3. Kiss, G., Giannelli, C., Jüttler, B.: Algorithms and data structures for truncated hierarchical B-splines. In: Floater, M., Lyche, T., Mazure, M.-L., Mørken, K., Schumaker, L.L. (eds.) MMCS 2012. LNCS, vol. 8177, pp. 304–323. Springer, Heidelberg (2014). http://dx.doi.org/10.1007/978-3-642-54382-1_28
4. Hjelmervik, J.: Direct pixel-accurate rendering of smooth surfaces. In: Floater et al. [3], pp. 238–247
5. Lakså, A.: Basic properties of expo-rational B-splines and practical use in Computer Aided Geometric Design. Ph.D. thesis, University of Oslo (2007). (Dr.philos.)
6. Lakså, A., Bang, B., Dechevsky, L.T.: Exploring expo-rational B-splines for curves and surfaces. In: Dæhlen, M., Mørken, K., Schumaker, L. (eds.) Mathematical methods for Curves and Surfaces, pp. 253–262. Nashboro Press, Brentwood (2005)
7. Lakså, A.: ERBS-surface construction on irregular grids. In: Pasheva, V., Venkov, G. (eds.) 39th International Conference Applications of Mathematics in Engineering and Economics AMEE13. AIP Conference Proceedings, vol. 1570, pp. 113–120. AIP Publishing (2013)
8. Microsoft® corporation: Direct3D 11 features (2009). http://msdn.microsoft.com/en-us/library/ff476342(VS.85).aspx
9. Schäfer, H., Nießner, M., Keinert, B., Stamminger, M., Loop, C.: State of the art report on real-time rendering with hardware tessellation. Eurographics 2014 - State of the Art Reports, pp. 93–117 (2014)
10. Segal, M., Akeley, K.: The OpenGL graphics system: a specification (Version 4.0 (Core Profile)), March 2012. http://www.opengl.org/registry/doc/glspec40.core.20100311.pdf
11. Yeo, Y., Bhandare, S., Peters, J.: Efficient pixel-accurate rendering of animated curved surfaces. In: Floater et al. [3], pp. 491–509
12. Zanaty, P.: Application of Generalized Expo-Rational B-splines in Computer Aided Design and Analysis. Ph.D. thesis, University of Oslo (2014)

Implicit Equations of Non-degenerate Rational Bezier Quadric Triangles

Alicia Cantón, L. Fernández-Jambrina[✉], E. Rosado María,
and M.J. Vázquez-Gallo

Universidad Politécnica de Madrid, 28040 Madrid, Spain
{alicia.canton,leonardo.fernandez,eugenia.rosado,
mariajesus.vazquez}@upm.es
http://dcain.etsin.upm.es/~discreto/

Abstract. In this paper we review the derivation of implicit equations for non-degenerate quadric patches in rational Bézier triangular form. These are the case of Steiner surfaces of degree two. We derive the bilinear forms for such quadrics in a coordinate-free fashion in terms of their control net and their list of weights in a suitable form. Our construction relies on projective geometry and is grounded on the pencil of quadrics circumscribed to a tetrahedron formed by vertices of the control net and an additional point which is required for the Steiner surface to be a non-degenerate quadric.

Keywords: Quadric · Steiner surfaces · Rational Bézier triangles

1 Introduction

Bézier triangles [7] are an alternative to tensor product patches as an extension of the Bézier formalism from curves to surfaces. In fact they were already present in De Casteljau's original work. Though they are not widely used as tensor product patches, they are useful in finite element methods and in gaming and animation, since the triangular geometry is more versatile for building surfaces and avoids the formation of singular points.

Quadrics are extensively used in engineering and therefore a usual requirement for a design formalism is that it may represent quadrics in an exact fashion. Quadric patches can be described as rational quadratic Bézier triangles, though not every rational quadratic Bézier triangle is a quadric patch. A characterisation can be found in [2]. In general rational quadratic Bézier triangles are quartic surfaces known as Steiner surfaces. This family of surfaces includes ruled cubics and quadrics as subcases.

The relation between rational quadratic Bézier triangles and Steiner surfaces has been studied since the very beginning of CAGD. In [11] properties of Steiner surfaces are derived and they are postulated as candidates for surface design. In [9] a control polyhedron is used for representing quadric patches. The authors of [6] define a generalised stereographic projection on the sphere to derive results for

© Springer International Publishing Switzerland 2015
J.-D. Boissonnat et al. (Eds.): Curves and Surfaces 2014, LNCS 9213, pp. 70–79, 2015.
DOI: 10.1007/978-3-319-22804-4_6

quadratic and biquadratic patches. In [4] algebraic geometry is used for studying surfaces that can be parametrised quadratically. In [5] general Bézier triangles are studied as projections of Veronese surfaces and the quadratic case is classified.

In [1] algebraic geometry methods are used to determine whether a rational quadratic Bézier triangle is a quadric patch and an algorithm is provided for classifying them. A tool named Weighted Radial Displacement is proposed for constructing Bézier conics and quadrics in [10].

In this paper we address the calculation of implicit equations for non-degenerate quadrics in rational Bézier triangular form. Our goal is to find coordinate-free expressions that involve just the control net and weights for the patch, using algebraic projective geometry, as we did in [3]. This is useful, for instance, to compute geometric characteristics of the surfaces.

In Sect. 2 we review rational Bézier quadratic patches, introduce notation and define a pencil of quadrics through the corners of the control net of the patch and an additional point where the conics located on the boundary of the patch meet. In order to determine the coefficients of the pencil of quadrics, in Sect. 3 we derive an expression for the bilinear form of a conic circumscribed to a triangle in terms of its control points and weights. In Sect. 4 we show that the data we have from each boundary conic of the Steiner surface is compatible precisely if the surface is a quadric. In Sect. 5 we obtain the bilinear form for the Steiner quadric. Section 6 to several examples.

2   Quadric Steiner Surfaces

We consider rational Bézier quadratic triangles,

$$
c(u, v, w) = \frac{\displaystyle\sum_{i+j+k=2} \frac{2!}{i!j!k!} \omega_{ijk} c_{ijk} u^i v^j w^k}{\displaystyle\sum_{i+j+k=2} \frac{2!}{i!j!k!} \omega_{ijk} u^i v^j w^k}, \qquad \begin{array}{l} u + v + w = 1, \\[4pt] u, v, w \in [0, 1], \end{array} \tag{1}
$$

defined by its control points, $\{c_{002}, c_{011}, c_{020}, c_{101}, c_{110}, c_{200}\}$, and their respective weights, $\{\omega_{002}, \omega_{011}, \omega_{020}, \omega_{101}, \omega_{110}, \omega_{200}\}$, which are real numbers.

Such surface patches are bounded by three curves, defined respectively by the equations $u = 0$, $v = 0$, $w = 0$. For instance, the arc at $u = 0$ is parametrised by

$$
c_u(v) = \frac{\displaystyle\sum_{j=0}^{2} \binom{2}{j} \omega_{0j2-j} c_{0j2-j} v^j (1-v)^{2-j}}{\displaystyle\sum_{j=0}^{2} \binom{2}{j} \omega_{0j2-j} v^j (1-v)^{2-j}}, \qquad v \in [0, 1],
$$

and hence it is a conic arc with control polygon $\{c_{002}, c_{011}, c_{020}\}$ and weights $\{\omega_{002}, \omega_{011}, \omega_{020}\}$.

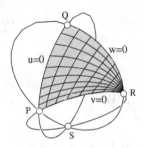

Fig. 1. Characterisation of quadric Steiner surfaces

Similarly, the conic arc at $v = 0$ has control polygon $\{c_{002}, c_{101}, c_{200}\}$ and weights $\{\omega_{002}, \omega_{101}, \omega_{200}\}$, whereas the control polygon of the one at $w = 0$ is $\{c_{020}, c_{110}, c_{200}\}$, with list of weights $\{\omega_{020}, \omega_{110}, \omega_{200}\}$. We assume from now on that these conics are non-degenerate.

The quadratic surface patch in (1) is generically a quartic surface, named Steiner surface [11], but in some particular cases it is a ruled cubic or a quadric. We are interested in the latter case due to the relevance of quadric surfaces. A characterisation of quadric Steiner surfaces is available in [2]:

– If the Steiner surface is a non-degenerate quadric, the three conic sections meet at a point S and their tangents span a plane there (see Fig. 1).
– If the three conic sections meet at a point S and their tangents span a plane there, the Steiner surface is a quadric.

The existence of point S is useful for our purposes. We label the points at the corners of the surface patch as $P = c_{002}$, $Q = c_{020}$, $R = c_{200}$.

The three conic arcs defined by $u = 0$, $v = 0$ and $w = 0$ are respectively located at planes that we denote u, v, w. We consider an additional plane t through P, Q, R (see Fig. 2).

In order to simplify the notation, we also call t, u, v, w the linear forms associated to the respective planes. Since they are defined up to a constant, we fix them by requiring

$$t(S) = u(R) = v(Q) = w(P) = 1. \tag{2}$$

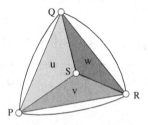

Fig. 2. Tetrahedron inscribed in a quadric Steiner surface

Since a quadric is determined by nine independent conditions, the pencil of quadrics through P, Q, R, S has five independent coefficients [12]. It is easy to check that the bilinear form C for such pencil in a coordinate-free fashion is

$$C = \lambda_{tu}tu + \lambda_{tv}tv + \lambda_{tw}tw + \lambda_{uv}uv + \lambda_{uw}uw + \lambda_{vw}vw, \tag{3}$$

in terms of the linear forms for the planes containing the faces of the tetrahedron.

We have the bilinear form for the quadric except for the unknown coefficients. In the following section we determine the coefficients λ_{ij} by restricting C to the planes u, v, w. Since the intersection of the quadric with such planes are conics with known control polygons and weights, we determine the coefficients up to proportionality factors.

3   Bilinear Forms for Conic Sections

In order to determine the free coefficients of our pencil of quadrics, we need the bilinear forms for the conic sections of each of the faces of the tetrahedron. On Fig. 3 we have the conic on the face u of the tetrahedron.

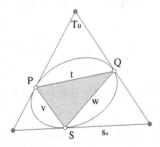

Fig. 3. Conic circumscribing a triangle

This conic has a bilinear form which is proportional to

$$C_u = \lambda_{tv}tv + \lambda_{tw}tw + \lambda_{vw}vw. \tag{4}$$

In this Section t, v, w designate the straight lines which are the intersections of the respective planes with the plane u, as well as their linear forms. That is, for simplicity in the notation in this section, we identify v with its restriction $v|_u$ on the plane u .

The conic on u is determined by noticing that the polar line of the control point $T_u = c_{011}$ is t, since this is the line linking the points P, Q where the tangent lines from T_u meet the conic arc. That is, $C_u(T_u, X)$ is proportional to $t(X)$ for all X on the plane. For simplicity, we follow in most cases the following notation: the polar line of a point A is the line a.

This conic arc from P to Q is parametrised by

$$c_u(t) = \frac{\omega_{002}P(1-t)^2 + 2\omega_{011}T_u t(1-t) + \omega_{020}Qt^2}{\omega_{002}(1-t)^2 + 2\omega_{011}t(1-t) + \omega_{020}t^2}, \qquad t \in [0,1],$$

but the weights are defined up to a Möbius transformation of the interval $[0,1]$ onto itselt [8],

$$t(\tilde{t}) = \frac{\tilde{t}}{(1-\rho)\tilde{t}+\rho}, \quad \tilde{t} \in [0,1],$$

which produces a new list of weights for the same conic arc,

$$\tilde{\omega}_{002} = \rho^2\omega_{002}, \quad \tilde{\omega}_{011} = \rho\omega_{011}, \quad \tilde{\omega}_{020} = \omega_{020}. \tag{5}$$

We may use this degree of freedom to reparametrise the conic arc so that

$$S = \lim_{\tilde{t}\to\infty} c_u\left(t(\tilde{t})\right) = \frac{\tilde{\omega}_{002}P - 2\tilde{\omega}_{011}T_u + \tilde{\omega}_{020}Q}{\tilde{\omega}_{002} - 2\tilde{\omega}_{011} + \tilde{\omega}_{020}}, \tag{6}$$

which has the advantage of writing the barycentric combination for S in terms of the control polygon with coefficients that are simply the weights for the curve. Thus,

$$T_u = \frac{\tilde{\omega}_{002}P + (2\tilde{\omega}_{011} - \tilde{\omega}_{002} - \tilde{\omega}_{020})S + \tilde{\omega}_{020}Q}{2\tilde{\omega}_{011}},$$

we can write the linear form for the polar line for T_u in the frame $\{P, S, Q\}$ with respect to the conic C_u as

$$C_u(T_u, X) \propto (\lambda_{tv}\tilde{\omega}_{020} + \lambda_{tw}\tilde{\omega}_{002})\, t(X) + (\lambda_{tv}(2\tilde{\omega}_{011} - \tilde{\omega}_{002} - \tilde{\omega}_{020}) + \lambda_{vw}\tilde{\omega}_{002})\, v(X)$$
$$+ (\lambda_{tw}(2\tilde{\omega}_{011} - \tilde{\omega}_{002} - \tilde{\omega}_{020}) + \lambda_{vw}\tilde{\omega}_{020})\, w(X),$$

where the usual symbol $\propto$ means "proportional to".

Requiring that t be the polar line of T_u, we get the unknown coefficients of its bilinear form,

$$C_u \propto \tilde{\omega}_{002}tv + \tilde{\omega}_{020}tw + (\tilde{\omega}_{002} - 2\tilde{\omega}_{011} + \tilde{\omega}_{020})vw. \tag{7}$$

If S were a point at infinity, instead of (6) we would need

$$S = \tilde{\omega}_{002}P - 2\tilde{\omega}_{011}T_u + \tilde{\omega}_{020}Q \Rightarrow T_u = \frac{\tilde{\omega}_{002}P - S + \tilde{\omega}_{020}Q}{2\tilde{\omega}_{011}},$$

and we read the coefficients again imposing that t is the polar line of T_u,

$$C_u \propto \tilde{\omega}_{002}tv + \tilde{\omega}_{020}tw + vw. \tag{8}$$

We can summarise this result in the following lemma:

Lemma 1. *The bilinear form for a conic circumscribed to a triangle PQS with sides t, v, w as in Fig. 3 is*

$$\begin{cases} \tilde{\omega}_{002}tv + \tilde{\omega}_{020}tw + (\tilde{\omega}_{002} - 2\tilde{\omega}_{011} + \tilde{\omega}_{020})vw, & \text{if } \tilde{\omega}_{002} - 2\tilde{\omega}_{011} + \tilde{\omega}_{020} \neq 0, \\ \tilde{\omega}_{002}tv + \tilde{\omega}_{020}tw + vw, & \text{if } \tilde{\omega}_{002} - 2\tilde{\omega}_{011} + \tilde{\omega}_{020} = 0, \end{cases}$$

where $S = \tilde{\omega}_{002}P - 2\tilde{\omega}_{011}T_u + \tilde{\omega}_{020}Q$ up to a constant.

The tangent line s to the conic at S is then $\tilde{\omega}_{002}v + \tilde{\omega}_{020}w$, with tangent vector $s_u = \tilde{\omega}_{020}\overrightarrow{SP} - \tilde{\omega}_{002}\overrightarrow{SQ}$.

4   Reparametrising the Quadric

The latter theorem provides some of the unknown coefficients in (3) up to a constant. Since we have made use of a special choice of weights on u to reach this result, we have to check that we can make it on the three boundary conics at a time in order to apply it to the whole quadric. We try to reparametrise the three boundary conics as in (6).

After reparametrising the conic on the plane u, the new list of weights is

$$\{\rho^2\omega_{002}, \rho\omega_{011}, \omega_{020}, \rho\omega_{101}, \omega_{110}, \omega_{200}\},$$

for some constant ρ and we obtain a tangent vector $s_u = \omega_{020}\overrightarrow{SP} - \rho^2\omega_{002}\overrightarrow{SQ}$ to the conic at S.

If we reparametrise the conic arc from R to Q on the plane w, the list of weights changes again,

$$\{\rho^2\omega_{002}, \rho\omega_{011}, \omega_{020}, \sigma\rho\omega_{101}, \sigma\omega_{110}, \sigma^2\omega_{200}\},$$

for some constant σ and we get a new tangent vector $s_w = \omega_{020}\overrightarrow{SR} - \sigma^2\omega_{200}\overrightarrow{SQ}$ to the conic at S.

Finally, if we needed to reparametrise the conic arc from P to R on the plane v, the list of weights would change to

$$\{\tau^2\rho^2\omega_{002}, \tau\rho\omega_{011}, \omega_{020}, \tau\sigma\rho\omega_{101}, \sigma\omega_{110}, \sigma^2\omega_{200}\},$$

for some constant τ and we would obtain another tangent vector $s_v = \sigma^2\omega_{200}\overrightarrow{SP} - \tau^2\rho^2\omega_{002}\overrightarrow{SR}$ to the conic at S.

The last reparametrisation obviously spoils the previous ones, but we may check whether it is necessary or not.

If the Steiner patch is a non-degenerate quadric, the three tangent vectors are to lie on a plane [2]. The determinant of these vectors,

$$\det(s_u, s_v, s_w) = \begin{vmatrix} \omega_{020} & -\rho^2\omega_{002} & 0 \\ \sigma^2\omega_{200} & 0 & -\tau^2\rho^2\omega_{002} \\ 0 & -\sigma^2\omega_{200} & \omega_{020} \end{vmatrix} = \rho^2\sigma^2\omega_{002}\omega_{020}\omega_{200}\left(1 - \tau^2\right),$$

tells us that they form a plane if and only if $\tau = 1$, that is, the reparametrisations to locate S at $t = \infty$ on the three conics are compatible.

From now on we omit the tildes over the weights, assuming that we are using a set of weights with this property,

$$
\begin{aligned}
S &= \frac{\omega_{002}c_{002} - 2\omega_{011}c_{011} + \omega_{020}c_{020}}{\omega_{002} - 2\omega_{011} + \omega_{020}} = \frac{\omega_{002}c_{002} - 2\omega_{101}c_{101} + \omega_{200}c_{200}}{\omega_{002} - 2\omega_{101} + \omega_{200}} \\
&= \frac{\omega_{200}c_{200} - 2\omega_{110}c_{110} + \omega_{020}c_{020}}{\omega_{200} - 2\omega_{110} + \omega_{020}},
\end{aligned}
\tag{9}
$$

if S is a point. If it is a point at infinity, S has in principle three different representatives for each conic,

$$
S_u = \omega_{002}c_{002} - 2\omega_{011}c_{011} + \omega_{020}c_{020}, \quad S_v = \omega_{002}c_{002} - 2\omega_{101}c_{101} + \omega_{200}c_{200},
$$

$$
S_w = \omega_{200}c_{200} - 2\omega_{110}c_{110} + \omega_{020}c_{020},
$$

which are parallel vectors. We write the bilinear form for the conic on u as

$$
\omega_{002}tv + \omega_{020}tw + t(S_u)vw,
$$

to overcome this problem.

5   Bilinear Forms for Steiner Quadrics

If S is a point, we have obtained bilinear forms for the conics on u, v, w as

$$
\begin{aligned}
C_u &\propto \omega_{002}tv + \omega_{020}tw + (\omega_{002} - 2\omega_{011} + \omega_{020})vw, \\
C_v &\propto \omega_{002}tu + \omega_{200}tw + (\omega_{002} - 2\omega_{101} + \omega_{200})uw, \\
C_w &\propto \omega_{020}tu + \omega_{200}tv + (\omega_{020} - 2\omega_{110} + \omega_{200})uv,
\end{aligned}
$$

and we can fit all pieces of information in the bilinear form:

Theorem 1. *The bilinear form for a non-degenerate Steiner quadric patch, bounded by three non-degenerate conic arcs, with vertices of the control net $\{c_{002}, c_{011}, c_{020}, c_{101}, c_{110}, c_{200}\}$ and weights $\{\omega_{002}, \omega_{011}, \omega_{020}, \omega_{101}, \omega_{110}, \omega_{200}\}$, fulfilling that the intersection S of the boundary conics is written as in (9) is*

$$
\begin{aligned}
C ={}& \omega_{020}\omega_{002}tu + \omega_{002}\omega_{200}tv + \omega_{200}\omega_{020}tw + \omega_{002}(\omega_{020} - 2\omega_{110} + \omega_{200})uv \\
&+ \omega_{200}(\omega_{002} - 2\omega_{011} + \omega_{020})vw + \omega_{020}(\omega_{002} - 2\omega_{101} + \omega_{200})uw,
\end{aligned}
$$

where u is the linear form of the plane containing $c_{002}, c_{011}, c_{020}$ which satisfies $u(c_{200}) = 1$, v is the linear form of the plane containing $c_{002}, c_{101}, c_{200}$ which satisfies $v(c_{020}) = 1$, w is the linear form of the plane containing $c_{020}, c_{110}, c_{200}$ which satisfies $w(c_{002}) = 1$ and t is the linear form of the plane containing $c_{002}, c_{020}, c_{200}$ which satisfies $t(S) = 1$.

If S is a point at infinity, the bilinear form is just

$$
\begin{aligned}
C ={}& \omega_{020}\omega_{002}tu + \omega_{002}\omega_{200}tv + \omega_{200}\omega_{020}tw \\
&+ \omega_{002}t(S_w)uv + \omega_{200}t(S_u)vw + \omega_{020}t(S_v)uw.
\end{aligned}
$$

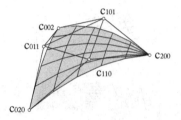

Fig. 4. Elliptic paraboloid

This result provides a procedure for computing a bilinear form for a non-degenerate Steiner quadric patch in a coordinate-free fashion using just the vertices of the control net and their respective weights:

1. Compute the normalised linear forms for the planes t, u, v, w.
2. Obtain S as intersection of the planes u, v, w and check if the patch belongs to a non-degenerate quadric.
3. Obtain an equivalent list of weights fulfilling (9).
4. Use Theorem 1 to obtain the bilinear form for the quadric patch.
5. The implicit equation for the quadric patch is then $C(X, X) = 0$.

6   Examples

We use the previous results to compute implicit equations for several quadric patches:

Example 1. Net: $\begin{bmatrix} (0,0,0) & (1,0,1) & (2,0,0) \\ (0,1,1) & (1,1,1) \\ (0,2,0) \end{bmatrix}$ and weights: $\begin{bmatrix} 1 & 1 & 1 \\ 1 & 1 \\ 1 \end{bmatrix}$ (Fig. 4):

The faces of the tetrahedron are the planes

$$u : \frac{y}{2} = 0, \quad v : \frac{x}{2} = 0, \quad w : 1 - \frac{x+y}{2} = 0, \quad t : -\frac{z}{2} = 0.$$

The planes u, v, w meet at the point at infinity

$$S = (0, 0, -2) = c_{002} - 2c_{011} + c_{020} = c_{002} - 2c_{101} + c_{200} = c_{200} - 2c_{110} + c_{020}.$$

The linear forms for the planes have been normalised according to (2). Hence the bilinear form for this surface is

$$C = tu + tv + tw + uv + vw + uw,$$

and the implicit equation, in cartesian coordinates is

$$0 = \frac{2x + 2y - x^2 - y^2 - xy - 2z}{4},$$

which corresponds to an elliptic paraboloid.

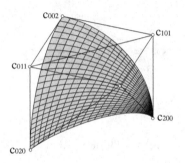

Fig. 5. Sphere

Example 2. Net:
$$\begin{bmatrix} (0,0,1) & (1,0,1) & (1,0,0) \\ & (0,1,1) & (1,1,1) \\ & & (0,1,0) \end{bmatrix}$$
and weights:
$$\begin{bmatrix} 1 & 1 & 2 \\ & 1 & 1 \\ & & 2 \end{bmatrix}$$
(Fig. 5):

The faces of the tetrahedron are the planes

$$u : y = 0, \quad v : x = 0, \quad w : \frac{1-x-y+z}{2} = 0, \quad t : \frac{1-x-y-z}{2} = 0.$$

The planes u, v, w meet at the point $S = (0,0,-1)$. The bilinear form for this surface is

$$C = 2tu + 2tv + 4tw + 2uv + 2vw + 2uw,$$

and the implicit equation in cartesian coordinates is

$$0 = 1 - x^2 - y^2 - z^2,$$

which corresponds to a sphere.

Example 3. Net:
$$\begin{bmatrix} (0,0,0) & (1,0,0) & (2,0,2) \\ & (0,1/2,0) & (1,1/2,0) \\ & & (0,1,-1/2) \end{bmatrix}$$
and weights:
$$\begin{bmatrix} 1 & 1 & 1 \\ & 1 & 1 \\ & & 1 \end{bmatrix}$$
(Fig. 6):

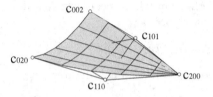

Fig. 6. Hyperbolic paraboloid

The faces of the tetrahedron are the planes

$$u : y = 0, \quad v : \frac{x}{2} = 0, \quad w : 1 - \frac{x}{2} - y = 0, \quad t : \frac{2z - 2x + y}{4} = 0.$$

The planes u, v, w meet at the point at infinity

$$S = (0, 0, 2) = c_{002} - 2c_{011} + c_{020} = -4(c_{002} - 2c_{101} + c_{200}) = \frac{4}{3}(c_{200} - 2c_{110} + c_{020}),$$

and with the choice of bilinear form for t we have

$$t(S_u) = 1, \quad t(S_v) = -\frac{1}{4}, \quad t(S_w) = \frac{3}{4}.$$

Hence the bilinear form for this surface is

$$C = tu + tv + tw + \frac{3}{4}uv + vw - \frac{1}{4}uw,$$

and the implicit equation, in cartesian coordinates is

$$0 = \frac{2z - x^2 + y^2}{4},$$

which corresponds to a hyperbolic paraboloid.

References

1. Albrecht, G.: Determination and classification of triangular quadric patches. Comput. Aided Geom. Des. **15**(7), 675–697 (1998)
2. Boehm, W., Hansford, D.: Bézier patches on quadrics. In: Farin, G. (ed.) NURBS for Curves and Surface Design, pp. 1–14. SIAM (1991)
3. Cantón, A., Fernández-Jambrina, L., Rosado-María, E.: Geometric characteristics of conics in Bézier form. Comput. Aided Des. **43**(11), 1413–1421 (2011)
4. Coffman, A., Schwartz, A.J., Stanton, C.: The algebra and geometry of steiner and other quadratically parametrizable surfaces. Comput. Aided Geom. Des. **13**(3), 257–286 (1996)
5. Degen, W.: The types of triangular Bézier surfaces. In: Mullineux, G. (ed.) The Mathematics of Surfaces VI, pp. 153–170. Clarendon Press, Oxford (1996)
6. Dietz, R., Hoschek, J., Jüttler, B.: An algebraic approach to curves and surfaces on the sphere and on other quadrics. Comput. Aided Geom. Des. **10**(3–4), 211–229 (1993)
7. Farin, G.: Triangular Bernstein-Bézier patches. Comput. Aided Geom. Des. **3**(2), 83–127 (1986)
8. Farin, G.: Curves and Surfaces for CAGD: A Practical Guide, 5th edn. Morgan Kaufmann Publishers Inc., San Francisco (2002)
9. Lodha, S., Warren, J.: Bézier representation for quadric surface patches. Comput. Aided Des. **22**(9), 574–579 (1990)
10. Sánchez-Reyes, J., Paluszny, M.: Weighted radial displacement: a geometric look at Bézier conics and quadrics. Comput. Aided Geom. Des. **17**(3), 267–289 (2000)
11. Sederberg, T., Anderson, D.: Steiner surface patches. IEEE Comput. Graph. Appl. **5**, 23–36 (1985)
12. Semple, J.G., Kneebone, G.T.: Algebraic Projective Geometry. Oxford University Press, London (1952)

Support Vector Machines for Classification of Geometric Primitives in Point Clouds

Manuel Caputo, Klaus Denker, Mathias O. Franz, Pascal Laube$^{(\boxtimes)}$, and Georg Umlauf

Institute for Optical Systems, University of Applied Sciences Constance, Konstanz, Germany
pascal.laube@gmail.com

Abstract. Classification of point clouds by different types of geometric primitives is an essential part in the reconstruction process of CAD geometry. We use support vector machines (SVM) to label patches in point clouds with the class labels tori, ellipsoids, spheres, cones, cylinders or planes. For the classification features based on different geometric properties like point normals, angles, and principal curvatures are used. These geometric features are estimated in the local neighborhood of a point of the point cloud. Computing these geometric features for a random subset of the point cloud yields a feature distribution. Different features are combined for achieving best classification results. To minimize the time consuming training phase of SVMs, the geometric features are first evaluated using linear discriminant analysis (LDA).

LDA and SVM are machine learning approaches that require an initial training phase to allow for a subsequent automatic classification of a new data set. For the training phase point clouds are generated using a simulation of a laser scanning device. Additional noise based on an laser scanner error model is added to the point clouds. The resulting LDA and SVM classifiers are then used to classify geometric primitives in simulated and real laser scanned point clouds.

Compared to other approaches, where all known features are used for classification, we explicitly compare novel against known geometric features to prove their effectiveness.

1 Introduction

Since laser scanners and similar devices are wide spread and have become less expensive, the need for automatic CAD reconstruction methods increases. Originally such devices were used only for quality control in the manufacturing process, but today also for reverse engineering and reconstruction of technical and non-technical components. For the reconstruction of geometric primitives like cones, planes, or spheres, detection is usually implicit by fitting different primitive types and deciding based on error thresholds. In noisy point clouds these often RANSAC-based methods [FB81] may fit the wrong geometric primitive because it yields the smallest error. Explicit classification can solve this

© Springer International Publishing Switzerland 2015
J.-D. Boissonnat et al. (Eds.): Curves and Surfaces 2014, LNCS 9213, pp. 80–95, 2015.
DOI: 10.1007/978-3-319-22804-4_7

phenomenon. Denker et al. [DHR+13] proposed a system for online reconstruction, in which a fixed set of heuristic rules based on estimation of local differential geometric properties is used for primitive detection. Cylinders, spheres, and planes are the only supported primitive types of the system and additional methods for the detection of other primitives such as cones or tori are needed.

Our system classifies the geometry of small patches of a point cloud using machine learning methods. In machine learning input samples classified by learning to discriminate different classes within feature spaces from training samples. Thus, these machine learning algorithms require feature vectors of a fixed length as input. The computation of these feature vectors from raw 3d point data is called feature extraction.

Feature extraction in point clouds and other 3d data is an important topic, that has been addressed from various perspectives. Most of the existing work is either intended for meshed data or for 3d shape recognition in object databases. It can be used to detect complex objects like cars, lamp posts or parking meters in 3d scans of urban environments [GKF09] or to create object recognition for the purpose of robot-object interaction [HHYR12]. Describing such 3D-shapes is possible by a variety of geometric features extracted from point clouds. Simple point features like shape distributions can be used to measure the similarity between different 3d shapes [OFCD02]. The distributions are represented as histograms sampled from angles, distances, areas, and volumes of random point tuples. Considering the local neighborhood of points, features based on point pairs and their normals are introduced in [WHH03]. Additional distributions for identifying shapes contained in point clouds, based on distances, are proposed in [MS09]. Another neighborhood based method for shape description uses curvatures and curvature directions. Hetzel et al. [HLLS01] use histograms based on normals and curvatures to identify objects in range images. In [RMBB08] point neighborhoods are used to describe a 16 dimensional feature histogram for point cloud segmentation.

Recently, geometric features are used for classification based on machine learning methods. Endoh et al. [EYO12] use locally linear embedding to learn object shapes. They explore clustering to reduce the number of required training samples. The impact of supervised, semi-supervised, and unsupervised dimension reduction as a learning method for shape-classification is studied by in [YTSO08] for the surflet pair feature of [WHH03]. Using support vector machines with point features as well as curvature features to classify surfaces is proposed in [AFB+12] for a small set of geometric primitives including edges and corners. The mentioned feature-based methods only use a small subset of possible point could features. It is also assumed that using all features leads to high discriminative power in classification and tests of particular feature and feature-combination performance are missing.

Our approach is to take a small patch of the original point cloud, extract its geometric features and use a pre-trained machine learning algorithm to detect which geometric primitive is most likely represented by the point cloud patch. This approach needs no fixed heuristic rules or thresholds and can be extended to additional primitives, provided enough training data exist. We explicitly test single and combined feature performance for detection.

2   Methods

To apply machine learning algorithms such as linear discriminant analysis (LDA) or support-vector machines (SVM) a training phase is required. In the training phase the LDA and SVM are exposed to a large set of pre-classified point clouds to learn to discriminate between the different classes. The information about these classes is implicitly represented in distributions of geometric features exacted form the local geometry in the point cloud. Acquiring real scanned point clouds for the training is difficult, because scanning a sufficiently large number of training point clouds would require a substantial amount of time and a large number of different real world models. Therefore, the training point clouds are generated using a laser scanning simulation with a built-in error model.

In this section first the geometric features are described, Sect. 2.1. Based on the feature histograms the machine learning approaches are presented in Sect. 2.2. To generate the necessary training point clouds a scan simulation is used (Sect. 2.3), showing the resulting feature spaces in Sect. 2.4.

2.1   Geometric Features and Feature Histograms

The feature histograms used for LDA or SVM are required to have a fixed length. To achieve this, the extracted geometric features are arranged in histograms with 64, 96, 128, 256, or 512 bins. These histograms are usually normalized and might be cropped to eliminate the influence of outliers.

The first four geometric features we used for comparison depend only on the location of the points in the point cloud and are adopted from [OFCD02].

F.1 *Point angles* are computed as the angles between two vectors spanned by three random points.
F.2 *Point distances* δ are computed by the Euclidean distance between two random points.
F.3 *Triangle areas* are computed from the square root of the triangle area of three random points.
F.4 *Tetrahedron volumes* are computed from the cubic root of the tetrahedron volumes V of four random points $\mathbf{p}_1, \ldots, \mathbf{p}_4$, i.e.

$$V = |(\mathbf{p}_1 - \mathbf{p}_4) \cdot ((\mathbf{p}_2 - \mathbf{p}_4) \times (\mathbf{p}_3 - \mathbf{p}_4))|/6.$$

The points required for these features are mutually different, uniformly distributed random points from the point cloud. The resulting feature histograms have 64 bins and are normalized to $[0, 180]$ for F.1 and to $[0, 1]$ for the others.

Geometric features that do not only depend on point locations are normal angles and normal directions.

F.5 *Normal angles* are given by angles between normals at two random points.
F.6 *Normal directions* are coordinates of the normalized normal at all points.

Thus, the feature histogram for F.6 is the concatenation of the three 32 bin histograms for the individual coordinates ranging from -1 to 1.

Geometric features that depend on the curvature are defined as follows:

F.7 *Curvature angles* are computed as the angles between the two corresponding principal curvature directions $\mathbf{v}_1$, $\mathbf{v}_2$ at two random points.

F.8 *Curvature directions* are given by the six coordinates of two normalized principal curvature directions $\mathbf{v}_1$, $\mathbf{v}_2$ at all points. Optionally, each coordinate can be weighted by the absolute value of the principal curvatures κ_1 or κ_2.

F.9 *Curvature differences* are computed from the absolute differences of the principal curvatures κ_1, κ_2 and Gaussian $K = \kappa_1 \kappa_2$ and mean curvature $H = (\kappa_1 + \kappa_2)/2$ at two random points, optionally weighted by distance.

The geometric feature F.7 results in two 64-bin histograms normalized to $[0, 180]$ concatenated to one 128-bin histogram. The feature histogram for F.8 is the concatenation of the six 32-bin histograms for the individual coordinates ranging from -1 to 1. For the feature histogram for F.9 the four 32-curvature-bin are cropped to the 0.05 to 0.95 percentile range, normalized, and concatenated.

In order to combine the classification capabilities of individual geometric features, they can be combined into more general features. In [WHH03] combined feature based on two surflet pairs are proposed. These surflet pairs are defined as point-normal-pairs $(\mathbf{p}_1, \mathbf{n}_1)$ and $(\mathbf{p}_2, \mathbf{n}_2)$ with normalized normals $\mathbf{n}_1, \mathbf{n}_2$. From two surflet pairs a local, right-handed, orthonormal frame is computed

$$\mathbf{u} = \mathbf{n}_1, \qquad \mathbf{v} = ((\mathbf{p}_2 - \mathbf{p}_1) \times \mathbf{u})/\|(\mathbf{p}_2 - \mathbf{p}_1) \times \mathbf{u}\|, \qquad \mathbf{w} = \mathbf{u} \times \mathbf{v}.$$

This frame yields three geometric attributes

$$\alpha = \arctan(\mathbf{w} \cdot \mathbf{n}_2, \mathbf{u} \cdot \mathbf{n}_2), \qquad \beta = \mathbf{v} \cdot \mathbf{n}_2, \qquad \gamma = \mathbf{u}(\mathbf{p}_2 - \mathbf{p}_1)/\|\mathbf{p}_2 - \mathbf{p}_1\|.$$

Together with the point distance δ these attributes define the surflet pair feature:

F.10 *Surflet pairs* are computed as the tuple $(\alpha, \beta, \gamma, \delta)$ for two random points.

Thus, the surflet pair feature yields a 128-bin histogram normalized to $[0, 1]$.

To construct other combined features we combined those features that proved effective as individual features. The combined feature histograms are concatenated from the histograms of the individual features.

F.11 *Triple combinations* are combined from the three best geometric features depending on point, normal, and curvature information: F.4, F.5, F.7.

F.12 *Simple surflet combinations* are combined from F.4, F.5, and F.10.

F.13 *Extended surflet combinations* are combined from F.10 and F.11

Remark 1. For comparison we tested nine additional geometric features, e.g. the shape index [KvD92]. However, all these features proved less effective in the experiments with a true positive classification rate below 0.5. These additional geometric features are given in Appendix A.

Normal and Principal Curvature Estimation. For the computation of the geometric features normals, principal curvatures and principal curvature directions at random points in the point cloud must be estimated.

To estimate a local normal at a point $\mathbf{p}$, the set P of its 100 closest neighbors is determined. Computing a principal component analysis (PCA) for P yields the eigenvector corresponding to the smallest eigenvalue of the covariance matrix. This eigenvector $\mathbf{n_p}$ is used to estimate the normal at point $\mathbf{p}$.

To estimate the principal curvatures and principal curvature directions at $\mathbf{p}$ the polynomial fitting of osculating jets of [CP05] is used. The points of P are approximated by a bi-variate height function $z(x, y)$ over the estimated tangent plane defined by $\mathbf{n_p}$. Then, $z(x, y)$ is computed by the truncated Taylor expansion of order n using the n-jet J_n

$$z(x, y) = J_n(x, y) + o\left(\|(x, y)\|^n\right), \qquad J_n(x, y) = \sum_{k=0}^{n} \sum_{i=0}^{k} \frac{b_{k-i,i}}{(k-i)!i!} x^{k-i} y^i.$$

The n-jet has the same differential geometric properties up to order n as the underlying surface. Since J_n is only approximated, it yields approximated differential geometric quantities. The $(n+1)(n+2)/2$ jet coefficients $b_{k-i,i}$ are computed by least squares approximation. They are used to approximate the Weingarten map of the surface. Its eigenvalues and eigenvectors yield estimates for the principal curvatures κ_1, κ_2 and principal curvature directions $\mathbf{v}_1$, $\mathbf{v}_2$. For the implementation we used the CGAL-library [CGA].

2.2 Machine Learning

Two machine learning algorithms were used: *Linear Discriminant Analysis* (LDA) and a d-class *Support Vector Machine* (SVM) for open and closed classification.

Supervised learning methods are designed to classify new data with respect to a large body of pre-classified training data. The training data are represented as feature vectors which are chosen to separate the different classes from each other. These features live in a high-dimensional feature space F. In supervised learning the objective is to compute hypersurfaces separating the classes in feature space. These hypersurfaces are then used to classify new input data by deciding on which side of the hypersurfaces the data reside. In case of SVMs the additional objective is to maximize the margin between two classes in relation to their separating hypersurface. Figure 1 illustrates this concept in case of separating hyperplanes in a two-dimensional feature space.

Linear Discriminant Analysis. The training data consist of feature data $\mathbf{x}_i \in F$ that are manually assigned to the correct class $y_i \in \{1, \ldots, d\}$ for d-class classification. So, the training data are given by

$$\mathcal{X}_d = \{(\mathbf{x}_1, y_1), \ldots, (\mathbf{x}_n, y_n) \mid \mathbf{x}_i \in F, y_i \in \{1 \ldots d\}\}.$$

For LDA the data are assumed to be separable by a hyperplane which is represented by its normal ω, the *weight vector*, and *bias* b as

$$\omega^t \mathbf{x} + b = 0,$$

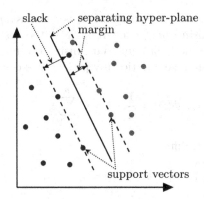

Fig. 1. Separating hyperplane, margin, support vectors, and slack for a two dimensional feature space.

for $\mathbf{x} \in F$. For only two classes this results in a decision function $f_{\omega,b}(\mathbf{x}) = \mathrm{sign}(\omega^t \mathbf{x} + b)$ that assigns the class labels ± 1 to a previously unseen test data point $\mathbf{x}$. Assuming a normal distribution with means μ_1 and μ_2 and common co-variance Σ for the two classes, the weight vector and bias are given by

$$\omega = \Sigma^{-1}(\mu_1 - \mu_2), \qquad b = (\mu_1 + \mu_2)/2.$$

This concept can be extended to d-class classification (see below), [DHS01].

Support Vector Machine. An SVM has the additional objective to maximize the margin between two classes. This results in finding ω and b such that

$$\Phi(\omega) = \omega^t \omega / 2 \tag{1}$$

is minimized with respect to the constraints

$$y_i(\omega^t \mathbf{x}_i + b) \geq 1, i = 1, \ldots, n, \tag{2}$$

see [CST00]. The minimum of (1) subject to (2) is computed by dualization using Lagrange multipliers. The Kuhn-Tucker-conditions imply that the weight vector ω can be represented in terms of the Lagrange multipliers $\alpha = (\alpha_1, \ldots, \alpha_n)$ and the training data as $\omega = \sum_{i=1}^{n} \alpha_i y_i \mathbf{x}_i$. So, instead of minimizing (1) the resulting Lagrangian

$$\Phi^*(\alpha) = \sum_{i=1}^{n} \alpha_i - \frac{1}{2} \sum_{i=1}^{n} \sum_{j=1}^{n} \alpha_i \alpha_j \, y_i y_j \, \mathbf{x}_i^t \mathbf{x}_j. \tag{3}$$

is maximized with respect to the constraints

$$\sum_{i=1}^{n} \alpha_i y_i = 0, \qquad \alpha_i \geq 0, i = 1, \ldots, n.$$

If in the solution $\alpha_i \neq 0$, the corresponding $\mathbf{x}_i$ is called *support vector*.

For noisy data requiring linear separability is too restrictive. Thus, *slack variables* ξ_i are used to allow for some data inside the margin, see Fig. 1. The objective function to minimize additionally penalizes excessive slack variables

$$\Phi(\omega) = \frac{1}{2}\omega^t\omega + C\sum_{i=1}^{n}\xi_i \tag{4}$$

with respect to the constraints

$$y_i(\omega^t\mathbf{x}_i + b) \geq 1 - \xi_i, i = 1, \ldots, n. \tag{5}$$

The minimum of (4) subject to (5) is again computed via dualization and Lagrange multipliers yielding the maximization problem (3) with the additional constraints $\alpha_i \leq C, i = 1, \ldots, n$.

For some problems, the classes cannot be separated by hyperplanes, but only by nonlinear hypersurfaces. So, the scalar product $\mathbf{x}_i^t\mathbf{x}_j$ is replaced by a *kernel* $K(\mathbf{x}_i, \mathbf{x}_j)$. The kernel is chosen such that it corresponds to a scalar product in a high-dimensional feature space. For our tests we used the *Gaussian RBF Kernel*

$$K(\mathbf{x}_i, \mathbf{x}_j) = \exp(-\gamma\|\mathbf{x}_i - \mathbf{x}_j\|^2).$$

With this kernel the classifier for new data $\mathcal{X} \in F$ is given by $f_{\alpha,b}(\mathbf{x}) = \text{sign}(m_{\alpha,b}(\mathbf{x}))$ with the *data margin*

$$m_{\alpha,b}(\mathbf{x}) = \sum_{i=1}^{n}\alpha_i y_i K(\mathbf{x}_i, \mathbf{x}) + b.$$

Multi-class Classification. For d-class classification the SVM concept is extended using the *one-versus-all* approach. Given the d-class training data $\mathcal{X}_d$ the one-versus-all approach uses d binary SVMs. Each binary SVM is trained to separate one class from all others. These d binary SVMs can be used in two different ways to classify new data $\mathbf{x} \in F$. For *closed SVM classification* (cSVM) the class with the largest data margin $m_{\alpha,b}(\mathbf{x})$ is selected to be the class $\mathbf{x}$ belongs to. For *open SVM classification* (oSVM) the class with the largest non-negative data margin $m_{\alpha,b}(\mathbf{x})$ is selected. If all margins are negative, $\mathbf{x}$ is not classified.

For more details on SVM based learning methods refer to [CST00,SS01]. For our implementation we used the SHARK library [IHMG08].

Model Selection. Model selection refers to the process of selecting the best parameters C and γ for the SVM. For this optimization usually a grid search in the two-dimensional (C, γ)-space is used. We used a two-dimensional non-uniform grid on which all pairwise parameter combinations are evaluated. To compare the different models, k-fold cross validation is used. We used $k = 5$. First, the training data $\mathcal{X}$ are divided into k equally sized sub-sets $S_i, i =$

$1, \ldots, k$. Then, for each grid point the SVM is trained k-times on each training set $\mathcal{T}_i = \mathcal{X} \setminus \mathcal{S}_i$. After the training on $\mathcal{T}_i$ the SVM is tested with the data from $\mathcal{S}_i$ yielding a true-positive-rate. For each grid point the k true-positive-rates are averaged. The best (C, γ)-combination is given by the grid point with the largest true-positive-rate. Optimization results are shown in Appendix A.

2.3 Laser Scanning Simulation and Training Data Generation

For the training a large set of training point clouds is required which is generated by a laser scanning simulation. This simulation is based on tracing the rays of a virtual laser line probe onto virtual geometric primitives. It simulates the behavior of real laser scanners by sweeping fans of rays from a scan position over the geometric primitive. To define the position and pose of the laser probe two spheres around the centroid of the geometric primitive are constructed with radii of 8 and 0.02 units. Each scan position and pose is defined by a random point $\mathbf{p}_1$ on the outer sphere, a view direction $\mathbf{s}_1 = \mathbf{p}_2 - \mathbf{p}_1$ by a random point $\mathbf{p}_2$ on the inner sphere, and a random view-up vector $\mathbf{s}_2 \perp \mathbf{s}_1$. The plane perpendicular to $\mathbf{s}_2$ contains the reference fan, that is a cone of rays with aperture angle φ_1 and axis $\mathbf{s}_1$. The reference fan contains φ_1 / φ_3 equidistant rays. To generate additional fans at $\mathbf{p}_1$ the reference fan is rotated around $\mathbf{s}_3 = \mathbf{s}_2 \times \mathbf{s}_1$ by φ_2-angle steps. $\mathbf{p}_1$, $\mathbf{p}_2$ and $\mathbf{s}_2$ are uniformly distributed. This setup is shown in Fig. 3.

Additionally, the laser scanner model includes two types of systematic errors. The first error affects the scanner position $\mathbf{p}_1$ by a random offset adding the same error to all points of one fan. The second error affects the distance measure for each individual scanned point. Both errors are normally distributed with zero mean and standard deviation of 0.0125 and 0.0025, respectively.

There are six classes of geometric primitives: planes, cylinders, cones, spheres, ellipsoids, and tori. The training data were generated by extracting point cloud patches from a simulated scan of a complete primitive. Each point cloud patch was defined to contain all points of the point cloud within a cube of edge length

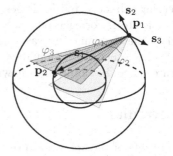

Fig. 2. Point cloud patch of a sphere.

Fig. 3. The setup of the simulation of a laser scanner.

0.7 centered at a random point of the point cloud. Figure 2 shows an example point cloud patch.

We use 9600 point cloud patches with equal distribution from one of the six primitive classes as training data. 7680 patches are used for training, 1920 are used to evaluate the method by the true-positive-rate. The geometric parameters for the primitives are normally distributed with mean and standard deviation as in Table 1. To compute the geometric features from the point cloud patches often random point pairs, triplets, or quadruplets were chosen. In these cases 2^{17} feature values were sufficient to yield stable feature histograms.

Table 1. Properties of normal distribution of primitive parameters.

Primitive parameter	Mean	Standard deviation	Primitive parameter	Mean	Standard deviation
Plane radius	3.0	0.5	Sphere radius	2.0	0.4
Cylinder height	3.0	0.5	Ellipsoid radius	2.0	0.4
Cylinder radius	2.0	0.4	Torus radius	5.0	0.6
Cone height	3.0	0.5	Torus tube radius	2.0	0.4
Cone radius	2.0	0.3			

2.4   Feature Space Visualization

To better understand the classification properties of individual SVMs the separating hyper-surfaces are visualized. Because the feature spaces have dimensions between 64 and 512, PCA is used to reduce the features spaces to three dimensions. The SVMs were then trained on the dimension reduced training data. For visualization the training data are visualized by colored points: blue for planes, green for cylinders, red for cones, cyan for spheres, magenta for ellipsoids, and yellow for tori. For each binary SVMs of the one-versus-all approach the scalar field of the data margin was computed. Using the marching cubes algorithm [LC87] a meshes of the boundary surface was extracted and rendered in the 3d feature space. Examples are shown in Fig. 4.

3   Results

The results of the application of LDA and SVMs for geometric primitive recognition are evaluated with respect to the classification results in terms of true-positive-rates.

In this section we show classification results of the geometric features in Sect. 2.1 used with LDA- and SVM-classification. All results are based on the

Fig. 4. Visualization of the separating hyper-surface of the SVM to classify planes (blue), cones (red), spheres (cyan), cylinders (green), ellipsoids (magenta), and tori (yellow) from different perspectives in the 3d feature space of the simple surflet combinations feature F.12 (Color figure online).

same training data of point cloud patches from the six primitive classes. Table 2 shows the confusion matrix for the closed d-class SVM for the simple surflet combination feature F.12. In the confusion matrix the rows are the primitive class of the test data and the columns are the detected classes, e.g. in the first row there are 311 planar point clouds, of which 304 are classified as planes and 7 are classified as cones. Table 3 shows the true-positive-rates for all features.

The visualization of the reduced feature spaces in Fig. 4 allows, much like the confusion matrix, to identify two or more classes which may not be separated properly. This is the case if two or more colors are mixed inseparable in one region of the feature space.

Figure 5(c) shows real scan data colored according to the detected primitives. For coloring the simple surflet combinations feature F.12 in a closed SVM

4   Discussion

The discussion of the results is split into one section on the classification results and one section on the quality and influence of the training data.

4.1   Classification Results

The geometric features of Sect. 2.1 categorized as point-based, normal-based, curvature-based, and combined features are discussed separately.

With a true-positive-rate of 0.758 the tetrahedron volume feature F.4 has the highest true-positive-rate among point-based features. While performing well for planes it confuses ellipsoids with spheres and tori with cylinders. The reasons

Table 2. Confusion matrices for features F.12, F.10 and F.5 used with a six-class cSVM in heat-map coloring.

	F.12 Simple surflet combi.						F.10 Surflet pairs						F.5 Normal angles					
	Planes	Cyl.	Cones	Spheres	Ellips.	Tori	Planes	Cyl.	Cones	Spheres	Ellips.	Tori	Planes	Cyl.	Cones	Spheres	Ellips.	Tori
Planes	304	0	7	0	0	0	305	0	6	0	0	0	297	0	14	0	0	0
Cyl.	0	298	2	0	0	18	0	299	2	0	0	17	0	291	3	0	0	24
Cones	6	22	294	0	0	3	31	20	266	1	0	7	54	30	230	0	0	11
Spheres	0	0	0	302	16	0	0	0	0	281	37	0	0	0	0	282	36	0
Ellips.	0	0	0	83	240	0	0	0	1	81	238	3	0	0	0	98	208	17
Tori	0	76	1	0	1	247	0	75	4	0	3	243	0	110	3	2	13	197

Table 3. Test results as true positive rate for all learning algorithms and features of our artificial data set.

Feature histogram	LDA	cSVM	oSVM	Feature histogram	LDA	cSVM	oSVM
F.1 Point angles	0.643	0.713	0.680	F.7 Curvature angles	0.628	0.695	0.603
F.2 Point dist	0.336	0.366	0.134	F.8 Curvature dir	0.312	0.443	0.302
F.3 Triangle areas	0.594	0.644	0.558	F.9 Curvature differences	0.356	0.420	0.185
F.4 Tet. vol	0.612	0.758	0.733	F.10 Surflet pairs	0.720	0.850	0.841
F.5 Normal angles	0.678	0.784	0.755	F.11 Triple combi	0.765	0.840	0.820
F.6 Normal dir	0.432	0.582	0.533	F.12 Simple surflet combi	0.779	0.878	0.866
				F.13 Ext. surflet combi	0.803	0.865	0.848

for this seems to be that the training data for those classes are very similar, see Table 1, and point-based features do not capture curvature information sufficiently.

The normal angles feature F.5 has a classification rate of 0.784 and by that the highest classification rate among normal-based features. It best performs for planes and cylinders but is weak for classifying ellipsoids and tori, see Table 2.

With 0.695 the curvature angle feature F.7 has the highest classification rate of the curvature-based features. It best classifies planes and has the weakest performance for ellipsoids and tori. This is due to the fact that the training data for curved primitives are relatively uniform, see Table 1. This effect can be observed in all confusion matrices, where basically the same primitives are confused by all features.

The combined features perform almost equally well. The surflet pair feature F.10 performs best for planes, cylinders, and spheres, and has better performance for cones, ellipsoids, and tori than F.4 or F.7. However, it confuses cylinders with tori, see Table 2. The simple surflet combination feature F.12 performs best

(a) Wooden toy firetruck. (b) Point cloud of scanned (c) Point cloud of firetruck
firetruck.   colored by primitive class.

Fig. 5. A real scan showing a part of a wooden toy firetruck colored by primitive class using the simple surflet combinations feature F.12: planes (blue), cylinders (green), cones (red), spheres (cyan), ellipsoids (magenta), tori (yellow) (Color figure online).

among all described features. It also separates spheres and ellipsoids as well as cylinders and tori best, see Table 2.

For all features cSVM results were superior to LDA and oSVM results. The reason for this is that oSVM only labels results with non-negative margin and LDA can only handle linear separation.

Using cSVM with the simple surflet combination feature F.12 we colored a real scan of a wooden toy firetruck, see Fig. 5. Because of static patch size in the training process the point cloud was scaled to 10 % of its original size. The three main colors are red, blue, and yellow which correspond to cones, planes, and tori, respectively. While planes and tori are often labeled right, transitions between different primitives are classified as cones. Taking into account the cone hypersurface from Fig. 4 one can see that the cone class border is very close to all other feature classes. This results in frequent miss-classification as cones.

4.2 Training Data

One disadvantage of using machine learning for primitive classification is the need for a large set of training data. Since there is no simple way to collect and produce enough real scans for training we uased simulated scans. However, our scan simulator generates scans that have different characteristics compared to real scans. They are different in density and uniformity of the scanned point clouds. Distribution of the scan lines across the surface of the scanned object are different for each human operator. Usually real laser scanners are manually guided and therefore distances between the scan lines vary. The scan simulator rotates the fan by a defined angle and generates uniform distances between different scan lines.

The error model of our simulator only covers errors in the measured distance and in the position of the laser or the camera. Error caused by specular reflections are not simulated, but can occur for polished objects. This could be solved by using real scans for the training or extending the scan simulator. Improving the scan simulator could be done by using real random scan paths instead of simple random points for sweeping and including an error model for surface reflections.

Patches extracted from point clouds typically contain a few hundred points. For training, those patches were extracted from point clouds containing only one primitive. When classifying real scenes, patches will often contain not only a part of one, but multiple or no primitives at all. In open classification those patches might remain unclassified. In closed classification they would be classified to the wrong class. Decreasing the size of the input patch would make classification of smaller patches possible. The smallest possible patch would be the local neighborhood of one point.

We estimate curvatures and normals using the 100-nearest neighbors of a point. For our simulated point clouds this value was sufficient to yield stable values for normals and curvatures. In case the point cloud is very dense, more points are needed to cover a sufficiently large patch for estimation. The same applies for sparse point clouds were too many neighbors result in insufficient neighborhoods. Using small neighborhoods in very noisy point clouds would lead to high approximation errors. Because features based on normals and curvatures heavily rely on point neighborhood, they are not scale invariant.

5   Conclusion

We present a method for primitive recognition based on point cloud classification using support vector machines. Our set of features is normal-, point-, and curvature-based. Based on simulated scans a large set of training data was generated to train and compare results of LDA and open/closed SVM classification. The geometric features were compared with respect to true-positive-rates and optimal SVM parameters. Resulting classifiers were applied to a real scan.

We have evaluated the discriminative power of different features and feature combinations for primitive classification. Especially closed SVM classification used with a combined feature is showing promising results in classifying primitives. Results of curvature based features did not meet our expectations. This might be because of high similarity between some classes within the training set. By using our method as a pre-processing step in fitting, deciding the geometric primitive for noisy point clouds can be improved. This might also lead to less iterations for iterative fitting methods.

For future work we intend to detail the pros and cons of various geometric features and their dependence on the distribution of the training data for SVM classification. To generate simulated scans that match real scans as close as possible is another aspect we plan to investigate. Additionally we plan to extend the machine learning methods to multi-output SVMs and other methods from machine learning to allow for a lager set of geometries to recognize.

A   Additional Geometric Features and Model Selection

For comparison additional geometrical features were tested, see Table 4. For optimization results of the SVM model selection refer to Table 5.

Table 4. True positive rate of additional geometric features.

Feature histogram	LDA	cSVM	oSVM	Feature histogram	LDA	cSVM	oSVM
A.1 Centroid distances	0.417	0.465	0.153	A.6 Gauss curvature	0.214	0.241	0.021
A.2 Cube cell count	0.340	0.449	0.358	A.7 Curvatures ratio	0.286	0.302	0.035
A.3 K-Median points	0.164	0.240	0.015	A.8 Curvature change	0.260	0.277	0.014
A.4 Principal curvatures	0.271	0.301	0.034	A.9 Shape index	0.296	0.302	0.034
A.5 Mean curvature	0.237	0.268	0.030				

Table 5. SVM parameters C and γ for best classification results for each feature.

Feature histogram	C	γ	Feature histogram	C	γ
F.1 Point angles	10000	0.4	F.12 Simple surflet combinations	1000	0.1
F.2 Point distances	10000	0.6	F.13 Extended surflet combinations	50	0.1
F.3 Triangle areas	10000	0.25	A.1 Centroid distances	20000	0.01
F.4 Tetrahedron volumes	1000	2	A.2 Cube cell count	100	0.1
F.5 Normal angles	20000	0.25	A.3 K-Median points	0.01	0.01
F.6 Normal directions	50	0.5	A.4 Principal curvatures	0.1	2
F.7 Curvature angles	100	0.1	A.5 Mean curvature	0.1	5
F.8 Curvature directions	5	0.1	A.6 Gaussian curvature	0.1	20
F.9 Curvature differences	100	0.6	A.7 Curvatures ratio	0.1	5
F.10 Surflet pairs	10000	0.3	A.8 Curvature change	0.01	1.25
F.11 Triple combinations	50	0.1	A.9 Shape index	0.1	5

A.1 *Centroid distances* are computed as distance of random points to the bounding box centroid, [OFCD02].

A.2 *Cube cell count* is computed by subdividing the point cloud's bounding box into equally sized cells and counting the points in each cell. This feature is not invariant to rotation.

A.3 *K-Median points* are computed by clustering the points into k clusters, such that the sum of distances of points in the cluster to their median is minimized. The coordinates of the resulting medians are concatenated to one feature vector of size $3k$.

A.4 The moduli of *principal curvatures* κ_1, κ_2 as in Sect. 2.1 yield two concatenated histograms.

A.5, A.6, and A.7 *mean curvatures* $H = (\kappa_1 + \kappa_2)/2$, *Gauss curvatures* $K = \kappa_1 \kappa_2$, and *curvature ratios* $|\kappa_1/\kappa_2|$.

A.8 *Curvature changes* are computed as the absolute difference between a random point's principal curvatures and those of its nearest neighbor and yield two concatenated histograms.

A.9 *Shape indices* as in [KvD92].

The histograms of the geometric features A.5, A.6, A.7, and A.8 are cropped to range between the 0.05 and 0.95 percentiles. All these features are less effective in the experiments with a true-positive-rate below 0.5.

References

[AFB+12] Arbeiter, G., Fuchs, S., Bormann, R., Fischer, J., Verl, A.: Evaluation of 3D feature descriptors for classification of surface geometries in point clouds. In: International Conference on Intelligent Robots and Systems, pp. 1644–1650 (2012)

[CGA] CGAL, Computational Geometry Algorithms Library. www.cgal.org

[CP05] Cazals, F., Pouget, M.: Estimating differential quantities using polynomial fitting of osculating jets. Comput. Aided Geom. Des. **22**(2), 121–146 (2005)

[CST00] Cristianini, N., Shawe-Taylor, J.: An Introduction to Support Vector Machines and other kernel-Based Learning Methods. Cambridge University Press, Cambridge (2000)

[DHR+13] Denker, K., Hagel, D., Raible, J., Umlauf, G., Hamann, B.: On-line reconstruction of CAD geometry. In: International Conference on 3D Vision, pp. 151–158 (2013)

[DHS01] Duda, R.O., Hart, P.E., Stork, D.G.: Pattern Classification. Wiley, New York (2001)

[EYO12] Endoh, M., Yanagimachi, T., Ohbuchi, R.: Efficient manifold learning for 3D model retrieval by using clustering-based training sample reduction. In: International Conference on Acoustics, Speech and Signal Processing pp. 2345–2348 (2012)

[FB81] Fischler, M.A., Bolles, R.C.: Random sample consensus: a paradigm for model fitting with applications to image analysis and automated cartography. Commun. ACM **24**(6), 381–395 (1981)

[GKF09] Golovinskiy, A., Kim, V.G., Funkhouser, T.: Shape-based recognition of 3D point clouds in urban environments. In:International Conference on Computer Vision (2009)

[HHYR12] Hwang, H., Hyung, S., Yoon, S., Roh, K.: Robust descriptors for 3D point clouds using geometric and photometric local feature. In: International Conference on Intelligent Robots and Systems (IROS), pp. 4027–4033 (2012)

[HLLS01] Hetzel, G., Leibe, B., Levi, P., Schiele, B.: 3D object recognition from range images using local feature histograms. In: CVPR, pp. 394–399 (2001)

[IHMG08] Igel, C., Heidrich-Meisner, V., Glasmachers, T.: Shark. J. Mach. Learn. Res. **9**, 993–996 (2008)

[KvD92] Koenderink, J.J., van Doorn, A.J.: Surface shape and curvature scales. Image Vis. Comput. **10**(8), 557–565 (1992)

[LC87]   Lorensen, W.E., Cline, H.E.: Marching cubes: a high resolution 3D surface construction algorithm. In: SIGGRAPH 1987, pp. 163–169 (1987)

[MS09]   Mahmoudi, M., Sapiro, G.: Three-dimensional point cloud recognition via distributions of geometric distances. Graph. Models **71**(1), 22–31 (2009)

[OFCD02] Osada, R., Funkhouser, T., Chazelle, B., Dobkin, D.: Shape distributions. ACM Trans. Graph. **21**(4), 807–832 (2002)

[RMBB08] Rusu, R.B., Marton, Z.C., Blodow, N., Beetz, M.: Persistent point feature histograms for 3D point clouds. In: 10th International Conference on Intelligent Autonomous Systems (2008)

[SS01]   Schoelkopf, B., Smola, A.J.: Learning with Kernels. MIT Press, Cambridge (2001)

[WHH03] Wahl, E., Hillenbrand, U., Hirzinger, G.: Surflet-pair-relation histograms: a statistical 3D-shape representation for rapid classification. In: 3DIM, pp. 474–482. IEEE (2003)

[YTSO08] Yamamoto, A., Tezuka, M., Shimizu, T., Ohbuchi, R.: SHREC 2008 entry: semi-supervised learning for semantic 3D model retrieval. In: International Conference on Shape Modeling and Applications (2008)

Computing Topology Preservation of RBF Transformations for Landmark-Based Image Registration

Roberto Cavoretto[1], Alessandra De Rossi[1]([✉]), Hanli Qiao[1],
Bernhard Quatember[2], Wolfgang Recheis[2], and Martin Mayr[3]

[1] Department of Mathematics "G. Peano", University of Torino,
Via Carlo Alberto 10, 10123 Torino, Italy
{roberto.cavoretto,alessandra.derossi,hanli.qiao}@unito.it
[2] Innsbruck Medical University, Anichstrasse 35, 6020 Innsbruck, Austria
Bernhard.Quatember@uibk.ac.at, Wolfgang.Recheis@i-med.ac.at
[3] University of Applied Science,
J. Gutenberg Straße 3, 2700 Wiener Neustadt, Austria
Martin.Mayr@fhwn.ac.at

Abstract. In image registration, a proper transformation should be topology preserving. Especially for landmark-based image registration, if the displacement of one landmark is larger enough than those of neighbourhood landmarks, topology violation will be occurred. This paper aims to analyse the topology preservation of some Radial Basis Functions (RBFs) which are used to model deformations in image registration. Matérn functions are quite common in the statistic literature (see, e.g. [9,13]). In this paper, we use them to solve the landmark-based image registration problem. We present the topology preservation properties of RBFs in one landmark and four landmarks model respectively. Numerical results of three kinds of Matérn transformations are compared with results of Gaussian, Wendland's, and Wu's functions.

Keywords: Matérn functions · Elastic registration · Radial basis functions · Topology preservation

1 Introduction

Over the last years, one of the largest areas of research in medical image processing has been the development of methods in image registration. The scope is to provide a point by point correspondence between the data sets. This means to find a suitable transformation between two images, called *source* and *target* images, taken either at different times or from different sensors or viewpoints. The scope is to determine a transformation such that the transformed version of the source image is similar to the target one. There is a large number of applications demanding image registration, for an overview see e.g. [10–12,17]. Paper [17] points out that Radial Basis Functions (RBFs) are powerful tools that could

© Springer International Publishing Switzerland 2015
J.-D. Boissonnat et al. (Eds.): Curves and Surfaces 2014, LNCS 9213, pp. 96–108, 2015.
DOI: 10.1007/978-3-319-22804-4_8

be applied to registration problem of local and global deformations. They have special good property for which values of these functions only depend on the distance of points from the center. We apply them to landmark-based registration, in which the basic idea is to determine the transformation mapping the source image onto the target image using corresponding landmarks.

RBFs can be divided into two categories that are *globally supported* and *compactly supported*, respectively. In general, using globally supported RBFs, such as thin plate spline (TPS), a single landmark pair change may influence the whole registration result but, mostly, they can keep the bending energy small. Otherwise, the Compactly Supported RBFs (CSRBFs), such as Wendland's and Wu's transformations, can circumvent this disadvantage (see [1,3]), but usually they cannot guarantee that the bending energy is small. Papers [4,5] analyse different CSRBF properties for image registration which are based on global deformations.

No matter which kind of RBFs we choose to transform images, the deformed images should preserve topology. In this paper we consider the Matérn functions which are positive definite, and we compare the characters of topology preservation with Gaussian, Wendland's and Wu's functions (see [2,6,8,14,15]). The latter two kinds of functions are CSRBFs, whereas the former two, Matérn and Gaussian, are globally supported. However, we point out that they have similar behavior and their function values approach zeros with growing distance from their centers; therefore they could be truncated as CSRBFs, being able to deal with local deformations well and allowing deformation fields to be controlled and adjusted locally using a number of landmarks points. Support size is an important index to evaluate the topology preservation property of different CSRBFs, since it modifies the influence of landmarks. In general, with a small support CSRBFs can be used to deal with local image warping, whereas with a large support they can be used to deform large regions or entire images. In particular, here we focus our study on two situations, i.e. the cases of topology preservation with one and four landmarks. This choice has been done because these test examples have already been considered in the papers [7,16].

We arrange this paper as follows. Section 2 introduces the landmark-based registration problem for RBFs. In Sect. 3, three kinds of Matérn functions and their transformations are introduced. Section 4 compares results of topology preservation using Matérn transformations with results of Gaussian, Wendland's and Wu's transformations in one landmark model. Four landmarks-based registration using transformations mentioned before are analysed in Sect. 5. We conclude reviewing the main results of this paper and outlining possible future work in Sect. 6.

2 Landmark-Based Image Registration and Some RBFs

We can interpolate the displacements defined at the landmarks using RBFs to model the deformation between a pair of objects in landmark-based image registration. To do this, we define a pair of landmark sets $\mathcal{S}_N = \{\mathbf{x}_j \in \mathbb{R}^2, j =$

$1, 2, ..., N\}$ and $\mathcal{T}_N = \{\mathbf{t}_j \in \mathbb{R}^2, j = 1, 2, ..., N\}$ corresponding to the *source* and the *target* images, respectively. The displacement can be displayed by $F_k : \mathbb{R}^2 \to \mathbb{R}, k = 1, 2$, which has the following form

$$F_k(\mathbf{x}) = \sum_{j=1}^{N} \alpha_{jk} \Psi \left(\| \mathbf{x} - \mathbf{x}_j \| \right), \tag{1}$$

where Ψ stands for a radial basis function, $\| \mathbf{x} - \mathbf{x}_j \|$ is the Euclidean distance between $\mathbf{x}$ and $\mathbf{x}_j$, and the coefficient α_{jk} can be calculated by solving two linear systems. The deformation $\mathbf{f} : \mathbb{R}^2 \to \mathbb{R}^2$ can be written as

$$\mathbf{f}(\mathbf{x}) = \mathbf{x} + F_k(\mathbf{x}).$$

In this paper, we mainly refer to topology preservation property of the following RBFs:

Gaussian G: $e^{-\|r\|^2/\sigma^2}, \sigma > 0,$

Wendland $\varphi_{3,1} : (1 - \dfrac{\| r \|}{c})_+^4 (4\dfrac{\| r \|}{c} + 1), \dfrac{\| r \|}{c} \leqslant 1,$

Wu $\psi_{1,2} : (1 - \dfrac{\| r \|}{c})_+^4 (1 + 4\dfrac{\| r \|}{c} + 3(\dfrac{\| r \|}{c})^2 + \dfrac{3}{4}(\dfrac{\| r \|}{c})^3), \dfrac{\| r \|}{c} \leqslant 1.$

Here c indicates the size of the support of Wendland's and Wu's functions and σ is the locality (shape) parameter of Gaussian.

3   Matérn Transformations

Matérn functions are strictly positive definite and quite common in the statistics literature [6]. Matérn family has recently received a great deal of attention and has the following form [8]

$$M(r \mid v, c) = \frac{2^{1-v}}{\Gamma(v)} \left(\frac{\| r \|}{c} \right)^v K_v \left(\frac{\| r \|}{c} \right). \tag{2}$$

Here K_v is the *Modified Bessel Function of the second kind of order* v and c is the coefficient to determine the width or the support of such functions. Therefore $\beta = v + \frac{d}{2}$, where d is the dimension of image space, here $d = 2$. The Fourier transform of the Matérn functions is given by the *Bessel kernels*

$$\hat{M}(w) = \left(1 + \| w \|^2 \right)^{-\beta} > 0. \tag{3}$$

Therefore the Matérn functions are strictly positive definite, which is an important condition to ensure interpolation problem (1) has a unique solution, and radial on $\mathbb{R}^d$ for all $d < 2\beta$ (see [6]). We consider these three Matérn functions

$$M \left(r \mid \frac{1}{2}, c \right) = \frac{2^{\frac{1}{2}}}{\Gamma(\frac{1}{2})} \left(\frac{\| r \|}{c} \right)^{\frac{1}{2}} K_{\frac{1}{2}} \left(\frac{\| r \|}{c} \right) \doteq e^{-\|r\|/c}, \tag{4}$$

$$M\left(r \mid \frac{3}{2}, c\right) = \frac{2^{-\frac{1}{2}}}{\Gamma(\frac{3}{2})}\left(\frac{\parallel r \parallel}{c}\right)^{\frac{3}{2}} K_{\frac{3}{2}}\left(\frac{\parallel r \parallel}{c}\right) \doteq \left(1 + \frac{\parallel r \parallel}{c}\right)e^{-\parallel r \parallel/c}, \quad (5)$$

$$M\left(r \mid \frac{5}{2}, c\right) = \frac{2^{-\frac{3}{2}}}{\Gamma(\frac{5}{2})}\left(\frac{\parallel r \parallel}{c}\right)^{\frac{5}{2}} K_{\frac{5}{2}}\left(\frac{\parallel r \parallel}{c}\right) \doteq \left(1 + \frac{\parallel r \parallel}{c} + \frac{1}{3}\frac{\parallel r \parallel^2}{c^2}\right)e^{-\parallel r \parallel/c}, \quad (6)$$

where c is the locality parameter. We title the above formulas (4), (5) and (6) as $M_{1/2}$, $M_{3/2}$ and $M_{5/2}$, respectively. With regard to the landmark-based image registration context we define the Matérn transformation as follows.

Definition 1. *Given a set of source landmark points* $S_N = \{x_j \in \mathbb{R}^2, j = 1, 2, \ldots, N\}$, *and the corresponding set of target landmark points* $T_N = \{t_j \in \mathbb{R}^2, j = 1, 2, \ldots, N\}$, *the Matérn transformation* $M : \mathbb{R}^2 \to \mathbb{R}^2$ *is such that each its component*

$$M_k(\mathbf{x}) : \mathbb{R}^2 \to \mathbb{R}, \quad k = 1, 2,$$

assumes the following form

$$M_k(\mathbf{x}) = M_k(x_1, x_2) = \sum_{j=1}^{N} \alpha_{jk} M_v\left(\parallel \mathbf{x} - \mathbf{x}_j \parallel_2\right), \quad (7)$$

with $\mathbf{x} = (x_1, x_2)$ *and* $\mathbf{x}_j = (x_{j1}, x_{j2}) \in \mathbb{R}^2$.

Through Definition 1, we obtain transformation functions $M_k(\mathbf{x}) : \mathbb{R}^2 \to \mathbb{R}$, and the coefficients α_{jk} are then obtained by solving two systems of linear equations.

4   Analysis of Topology Preservation in One-Landmark Matching

Necessary conditions to have topology preservation are continuity of the function **H** and positivity of the Jacobian determinant at each point. This is achieved to get injectivity of the map [7].

Suppose that the source landmark **p** is shifted by Δ_x along the x-axis direction and by Δ_y along the y-axis direction to the target landmark **q**. The coordinates of transformation are

$$H_1(\mathbf{x}) = x + \Delta_x \Phi(||\mathbf{x} - \mathbf{p}||),$$

$$H_2(\mathbf{x}) = y + \Delta_y \Phi(||\mathbf{x} - \mathbf{p}||),$$

where Φ is any RBF.

Requiring the determinant of the Jacobian is positive, we obtain

$$\det(J(x, y)) = 1 + \Delta_x \frac{\partial \Phi}{\partial x} + \Delta_y \frac{\partial \Phi}{\partial y} > 0, \quad (8)$$

i.e.

$$\Delta_x \frac{\partial \Phi}{\partial x} + \Delta_y \frac{\partial \Phi}{\partial y} > -1,$$

or, equivalently,

$$\Delta_x \frac{\partial \Phi}{\partial r} \cos \theta + \Delta_y \frac{\partial \Phi}{\partial r} \sin \theta > -1,$$

where Φ stands for $\Phi(||\mathbf{x} - \mathbf{p}||)$ and $r = ||\mathbf{x} - \mathbf{p}||$.

If we set $\Delta = \max(\Delta_x, \Delta_y)$, the value of θ minimizing the determinant in 2D is $\frac{\pi}{4}$; thus we get

$$\Delta \frac{\partial \Phi}{\partial r} > -\frac{1}{\sqrt{2}}. \tag{9}$$

With the condition (9) one can show that all principal minors of the Jacobian are positive. It follows that the transformations defined by equation (7) preserve the topology if (9) holds. The minimum of $\frac{\partial \Phi}{\partial r}$ depends on the localization parameter and therefore on the support size of the parameter c of Matérn functions.

In the next subsections we compute the minimum support size of locality parameter, for which (9) is satisfied, of the three functions of Matérn family.

4.1 Matérn $M_{1/2}$

Now we are considering the Matérn function (4). Clearly, we cannot obtain the minimum value of c as the above mentioned method, so we search for it through numerical experiments. Here we have

$$\frac{\partial \Phi}{\partial r} = -\frac{1}{c} e^{-r/c}, \tag{10}$$

while the value of r minimizing (10) is given by $r = 0.25c$. Then computing (10) and substituting its result in (9), we obtain

$$c > \frac{\sqrt{2}}{e^{1/4}} \Delta \approx 1.1\Delta. \tag{11}$$

4.2 Matérn $M_{3/2}$

According to the Matérn function (5), we find the minimum value of c satisfying (9). Now we get

$$\frac{\partial \Phi}{\partial r} = -\frac{r}{c^2} e^{-r/c}, \tag{12}$$

while the value of r minimizing (12) is given by $r = c$. Evaluating (12) at $r = c$ and substituting its result in (9), we have

$$c > \frac{\sqrt{2}\Delta}{e} \approx 0.52\Delta. \tag{13}$$

4.3 Matérn $M_{5/2}$

Considering the Matérn function (6), similarly as the Matérn function (5), we have

$$\frac{\partial \Phi}{\partial r} = -\left(\frac{r}{3c^2} + \frac{r^2}{3c^2}\right)e^{-r/c}, \tag{14}$$

while the value of r minimizing (14) is given by $r = \frac{\sqrt{5}+1}{2}c$. Calculating (14) and substituting its result in (9), we obtain

$$c > \frac{(2\sqrt{2} + \sqrt{10})\Delta}{3e^{(\sqrt{5}+1)/2}} \approx 0.3960\Delta. \tag{15}$$

4.4 Analysis of the Results

Table 1 summarizes the minimum support sizes of locality parameter for $M_{1/2}$, $M_{3/2}$ and $M_{5/2}$, which are compared with Gaussian, Wendland's and Wu's functions (see [7, 16]). The advantage of having small supports is that the influence area of each landmark turns out to be small. This allows us to have a greater local control. From Table 1, we can see that Matérn functions have smaller supports, especially the $M_{5/2}$ function. This means that in one landmark model, the deformed field of $M_{5/2}$ is the smallest among these six transformations.

Table 1. Minimum support size for various RBFs, where $c = 2\sigma$ and $d = 2$.

G	$\varphi_{3,1}$	$\psi_{1,2}$	$M_{1/2}$	$M_{3/2}$	$M_{5/2}$
$\sigma > 1.21\Delta$	$c > 2.98\Delta$	$c > 2.80\Delta$	$c > 1.10\Delta$	$c > 0.52\Delta$	$c > 0.3960\Delta$

4.5 Numerical Results

In this section, we report the numerical experiments obtained on a grid $[0, 1] \times [0, 1]$ and compare then the distortion outcomes of the grid in the shift case of the landmark $\{(0.5, 0.5)\}$ in $\{(0.6, 0.7)\}$. In Fig. 1 we show results assuming as a support size the minimum c and σ such that (9) is satisfied, with $\Delta = 0.2$. Figure 1 shows that, for the minimum value of c and σ, all transformations can preserve topology well. In this case, the $M_{5/2}$ transformation has the smallest optimal support size around the landmark while Wendland's, Wu's and Gaussian functions have relatively larger ones, as outlined in Table 1.

Furthermore, in Table 2 we report the root mean squares errors (RMSE) obtained by using different RBFs with their own optimal locality parameters. These errors are found by computing the distances between the displacements of a grid formed by 40×40 points $\mathbf{x} \in \mathcal{X}$ and the values obtained by the transformations. They assume the following form

$$RMSE = \sqrt{\frac{\sum_{\mathbf{x} \in \mathcal{X}} \|\mathbf{x} - \mathbf{F}(\mathbf{x})\|_2^2}{\sum_{\mathbf{x} \in \mathcal{X}}}}. \tag{16}$$

However, results in Table 2 are quite similar and $M_{1/2}$ turns out to have the smallest RMSE. This means that the transformation $M_{1/2}$ produces the smallest deformation if compared with the other RBFs, even if from Fig. 1 we can see that the whole images are only slightly deformed. On the contrary, if the topology preservation condition (9) is not satisfied, the transformed image is deeply misrepresented above all around the shifted point.

Table 2. RMSE for various RBFs with one landmark.

RBFs	RMSEs
$\varphi_{3,1}\ c = 0.64$	6.1652E-002
$\psi_{1,2}\ c = 0.58$	6.5944E-002
$G\ c = 0.25$	6.8309E-002
$M_{1/2}\ c = 0.22$	5.9007E-002
$M_{3/2}\ c = 0.105$	6.0769E-002
$M_{5/2}\ c = 0.08$	6.3062E-002

5 Topology Preservation for More Extended Deformations

If we consider much larger supports which are able to cover the whole domain, the influence of each landmark extends on the entire image, thus generating global deformations. In the following we compare topology preservation properties for globally supported transformations. For this aim, we consider four inner landmarks in a grid, located so as to form a rhombus at the center of the figure, and we suppose that only the lower vertex is downward shifted of Δ [16]. The landmarks of source and target images are $P = \{(0,1),(-1,0),(0,-1),(1,0)\}$ and $Q = \{(0,1),(-1,0),(0,-1-\Delta),(1,0)\}$, respectively, with $\Delta > 0$.

Let us now consider components of a generic transformation $\mathbf{H} : \mathbb{R}^2 \to \mathbb{R}^2$ obtained by a transformation of four points P_1, P_2, P_3 and P_4, namely

$$H_1(\mathbf{x}) = x + \sum_{i=1}^{4} c_{1,i}\Phi(\|\mathbf{x} - P_i\|), \tag{17}$$

and

$$H_2(\mathbf{x}) = y + \sum_{i=1}^{4} c_{2,i}\Phi(\|\mathbf{x} - P_i\|). \tag{18}$$

The coefficients $c_{1,i}$ and $c_{2,i}$ are obtained so that the transformation maps P_i to Q_i, with $i = 1,\ldots,4$. To do that, we need to solve two systems of four equations in four unknowns, whose solutions are

$$c_{1,1} = 0, \quad c_{1,2} = 0, \quad c_{1,3} = 0, \quad c_{1,4} = 0,$$

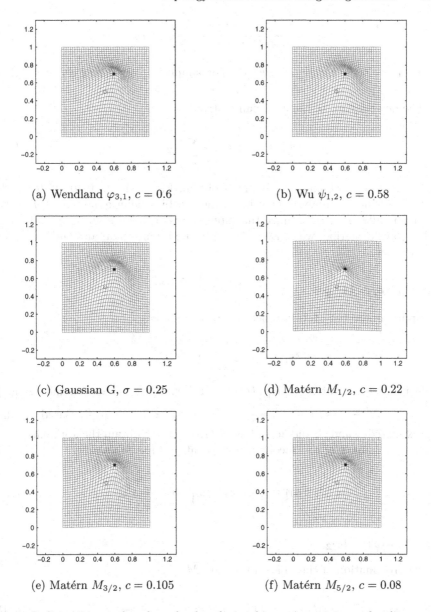

(a) Wendland $\varphi_{3,1}$, $c = 0.6$ (b) Wu $\psi_{1,2}$, $c = 0.58$

(c) Gaussian G, $\sigma = 0.25$ (d) Matérn $M_{1/2}$, $c = 0.22$

(e) Matérn $M_{3/2}$, $c = 0.105$ (f) Matérn $M_{5/2}$, $c = 0.08$

Fig. 1. Deformation results of one-landmark matching using minimum locality parameters satisfying the topology preservation condition. The source landmark is marked by a circle ($\circ$), while the target one by a star ($*$).

and

$$c_{2,1} = \frac{\beta^2 + \beta - 2\alpha^2}{(1 - \beta)[(1 + \beta)^2 - 4\alpha^2]}\Delta, \quad c_{2,2} = \frac{\alpha}{(1 + \beta)^2 - 4\alpha^2}\Delta, \qquad (19)$$

$$c_{2,3} = -\frac{1 + \beta - 2\alpha^2}{(1 - \beta)[(1 + \beta)^2 - 4\alpha^2]}\Delta, \quad c_{2,4} = c_{2,2}, \tag{20}$$

where $\alpha = \Phi\left(\frac{\sqrt{2}}{c}\right)$ and $\beta = \Phi\left(\frac{2}{c}\right)$. For simplicity, we denote $\Phi_i = \Phi(\|\mathbf{x} - P_i\|/c)$, $i = 1, \ldots, 4$.

The determinant of the Jacobian matrix is

$$\det(J(x, y)) = 1 + \sum_{i=1}^{4} c_{2,i}\frac{\partial \Phi_i}{\partial y}. \tag{21}$$

The minimum Jacobian determinant is obtained at position $(0, y)$, with $y > 1$. In the following, we analyse the value of the Jacobian determinant at $(0, y)$, with $y > 1$ for different RBFs. Since the support c is very large, in order to have a global transformation, we consider $\|\cdot\|/c$ to be infinitesimal and omit terms of higher order.

5.1 Matérn $M_{1/2}$

We approximate the Matérn function $M_{1/2}$ as follows

$$M_{1/2}(r) = e^{-r} \approx 1 - r + \frac{r^2}{2}, \tag{22}$$

while its first derivative is $M'_{1/2}(r) \approx -1 + r$. Now, we approximate α and β as $\alpha = \Phi\left(\frac{\sqrt{2}}{c}\right) \approx 1 - \frac{\sqrt{2}}{c} + \frac{1}{c^2}$, $\beta = \Phi\left(\frac{2}{c}\right) \approx 1 - \frac{2}{c} + \frac{2}{c^2}$. Based on the approximated α and β, and according to (19)–(21) we obtain the approximation of the determinant of the Jacobian of $M_{1/2}$ at $(0, y)$ as follows

$$\det(J(0, y)) \approx 1 - 2.4142\Delta\left(-1 + \frac{y}{\sqrt{y^2 + 1}}\right). \tag{23}$$

5.2 Matérn $M_{3/2}$

The approximation of the Matérn function $M_{3/2}$ is

$$M_{3/2}(r) = (1 + r)e^{-r} \approx 1 - \frac{r^2}{2} + \frac{r^3}{2}, \tag{24}$$

its first derivative is $M'_{3/2}(r) = -re^{-r} \approx -r + r^2$, α and β can be approximated by $\alpha = \Phi\left(\frac{\sqrt{2}}{c}\right) \approx 1 - \frac{1}{c^2} + \frac{\sqrt{2}}{c^3}$, $\beta = \Phi\left(\frac{2}{c}\right) \approx 1 - \frac{2}{c^2} + \frac{4}{c^3}$. Then, the $M_{3/2}$ Jacobian determinant can be formed:

$$\det(J(0, y)) \approx 1 - 1.7071\Delta\left(y^2 + 1 - y\sqrt{y^2 + 1}\right). \tag{25}$$

5.3   Matérn $M_{5/2}$

The Matérn function $M_{5/2}$ can be approximated as

$$M_{5/2}(r) = \left(1 + r + \frac{r^2}{3}\right) e^{-r} \approx 1 - \frac{r^2}{6} + \frac{2}{9}r^3, \tag{26}$$

its first derivative is $M'_{5/2}(r) = -\frac{1}{3}r + \frac{2}{3}r^2$. Approximation of α and β is $\alpha = \Phi\left(\frac{\sqrt{2}}{c}\right) \approx 1 - \frac{1}{3c^2} + \frac{4\sqrt{2}}{9c^3}$, $\beta = \Phi\left(\frac{2}{c}\right) \approx 1 - \frac{2}{3c^2} + \frac{16}{9c^3}$. Under formulas (19)–(21), we can get the Jacobian determinant function of $M_{5/2}$, i.e.

$$\det\left(J(0,y)\right) \approx 1 - 2.5607\Delta\left(y^2 + 1 - y\sqrt{y^2 + 1}\right). \tag{27}$$

5.4   Analysis of the Results

The obtained results, if compared with the ones acquired by the work [16], show same values of $\det(J(0,y))$, with $y > 1$, when one uses CSRBF transformations based on Wendland's and Wu's functions. This indicates that the functions $\varphi_{3,1}$ and $\psi_{2,1}$ have a very similar behavior. The equations obtained in [16] using Wendland's and Wu's functions guarantee the Jacobian determinant positivity for any $y > 1$. When one uses Gaussian, $M_{1/2}$, $M_{3/2}$ and $M_{5/2}$ functions, we can find $M_{1/2}$ and $M_{3/2}$ functions guarantee the positivity of the Jacobian determinant for any $y > 1$. Using the $M_{1/2}$ function, the Jacobian determinant is the closest to 1. This means that it is the best transformation in this case. While that obtained for the Gaussian, still in [16], presents a negative determinant for some values of y, as shown in Fig. 2. We also see that the $M_{5/2}$ function always presents a negative determinant for different y. Therefore, $\varphi_{3,1}$, $\psi_{2,1}$, and the first two kinds of Matérn functions ensure more easily the topology preservation, unlike the Gaussian and $M_{5/2}$ function. We can conclude that the two functions do not lead to good results in case of higher landmarks density, i.e. when distance among landmarks is very little. Moreover, each of them influences the whole image, which might produce a topology violation.

5.5   Numerical Results

Let us consider $[0,1] \times [0,1]$ and compare results obtained by its distortion, which is created by the shift of one of the four landmarks distributed in rhomboidal position. The source landmarks are $\{(0.5, 0.65), (0.35, 0.5), (0.65, 0.5), (0.5, 0.35)\}$ and are respectively transformed in the following target landmarks $\{(0.5, 0.65), (0.35, 0.5), (0.65, 0.5), (0.5, 0.25)\}$. Taking $\sigma = 50$ and $c = 100$ as support size, we obtain Fig. 3 and we show RMSEs in Table 3.

In agreement with theoretical results, Fig. 3 and Table 3 confirm that Gaussian and $M_{5/2}$ function turn out to be those which worse preserve topology, whereas all other functions present very similar deformations. In particular, the $M_{1/2}$ function provides the best transformation. In fact, although support size is large, the deformed field at landmarks is very small in the $M_{1/2}$ transformation.

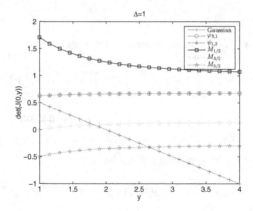

Fig. 2. Value of $\det(J(0,y))$, with $y > 1$, by varying RBFs.

Table 3. RMSE for various RBFs with four landmarks.

RBFs	RMSEs
$\varphi_{3,1}\ c = 100$	1.1311E-001
$\psi_{1,2}\ c = 100$	1.1352E-001
$G\ c = 50$	1.5965E-001
$M_{1/2}\ c = 100$	4.2735E-002
$M_{3/2}\ c = 100$	1.1369E-001
$M_{5/2}\ c = 100$	1.5942E-001

6   Conclusions and Future Work

We compared the topology preservation property of three kinds of Matérn functions with Gaussian, Wendland's and Wu's functions in one and four landmarks cases, respectively. No matter in which case, the $M_{1/2}$ transformation showed us the best advantage. We must note that, although in one landmark model $M_{1/2}$ deformation does not have the smallest locality parameter among the six transformations, it guarantees images only deformed around the landmark instead of in the whole images which is a good property in the case of local deformations. In a future work we will evaluate the topology property and other characters of these Matérn functions in real life cases, such as x-ray images of patients. Also, we are going to check the outcomes of Matérn functions in case of a large number of landmarks.

Observing the formula (2), we found that when v is large enough, Matérn function can be approximated as Gaussian function. Also in this paper, we can see that the $M_{5/2}$ function has a character more similar to the Gaussian than $M_{1/2}$ and $M_{3/2}$. Therefore another future work might be analysing and comparing behaviors of Matérn and Gaussian functions in landmark-based image registration.

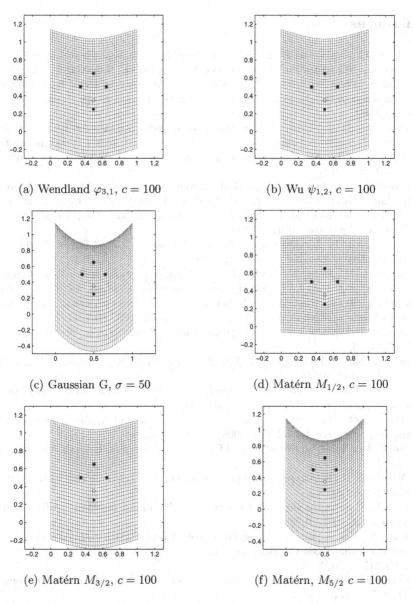

(a) Wendland $\varphi_{3,1}$, $c = 100$

(b) Wu $\psi_{1,2}$, $c = 100$

(c) Gaussian G, $\sigma = 50$

(d) Matérn $M_{1/2}$, $c = 100$

(e) Matérn $M_{3/2}$, $c = 100$

(f) Matérn, $M_{5/2}$ $c = 100$

Fig. 3. Deformation results of four landmarks; the source landmarks are marked by a circle ($\circ$), while the target ones by a star ($*$).

Acknowledgments. The work of the first author is partially supported by the University of Torino via grant "Approssimazione di dati sparsi e sue applicazioni". The second author acknowledges financial support from the GNCS–INdAM. The authors also sincerely thank the anonymous referee for helping to significantly improve this paper.

References

1. Allasia, G., Cavoretto, R., De Rossi, A.: Local interpolation schemes for landmark-based image registration: a comparison. Math. Comput. Simul. **106**, 1–25 (2014)
2. Arad, N., Dyn, N., Reisfeld, D., Yeshurun, Y.: Warping by radial basis functions: application to facial expressions. CVGIP Graph. Models Image Process. **56**, 161–172 (1994)
3. Cavoretto, R., De Rossi, A.: Landmark-based image registration using Gneiting's compactly supported functions. In: Simos, T.E., et al. (eds.) Proceedings of the ICNAAM 2012, AIP Conference Proceedings, vol. 1479, Melville, New York, pp. 1335–1338 (2012)
4. Cavoretto, R., De Rossi, A.: Analysis of compactly supported transformations for landmark-based image registration. Appl. Math. Inf. Sci. **7**, 2113–2121 (2013)
5. De Rossi, A.: Medical image registration using compactly supported functions. Commun. Appl. Ind. Math. **4**, 1–12 (2013)
6. Fasshauer, G.E.: Meshfree Approximation Methods with MATLAB. World Scientific Publishers Co., Inc., River Edge (2007)
7. Fornefett, M., Rohr, K., Stiehl, H.: Radial basis functions with compact support for elastic registration of medical images. Image Vis. Comput. **19**, 87–96 (2001)
8. Gneiting, T., Kleiber, W., Schlather, M.: Matérn cross-covariance functions for multivariate random fields. J. Amer. Statist. Assoc. **105**, 1167–1177 (2010)
9. Matérn, B.: Spatial Variation. Lecture Notes in Statistics, vol. 36. Springer, New York (1986)
10. Modersitzki, J.: FAIR: Flexible Algorithms for Image Registration, vol. 6. SIAM, Philadelphia (2009)
11. Rohr, K.: Landmark-Based Image Analysis, Using Geometric and Intensity Models. Kluwer Academic Publishers, Norwell (2001)
12. Scherzer, O.: Mathematical Models for Registration and Applications to Medical Imaging. Springer, Heidelberg (2006)
13. Stein, M.L.: Interpolation of Spatial Data: Some Theory for Kriging. Springer, New York (1999)
14. Wendland, H.: Scattered Data Approximation, vol. 17. Cambridge Univsity Press, Cambridge (2005)
15. Wu, Z.: Compactly supported positive definite radial functions. Adv. Comput. Math **4**, 283–292 (1995)
16. Yang, X., Xue, Z., Lia, X., Xiong, D.: Topology preservation evaluation of compact-support radial basis functions for image registration. Pattern Recogn. Lett. **32**, 1162–1177 (2011)
17. Zitová, B., Flusser, J.: Image registration methods: a survey. Image Vis. Comput. **21**, 977–1000 (2003)

New Bounds on the Lebesgue Constants of Leja Sequences on the Unit Disc and on ℜ-Leja Sequences

Moulay Abdellah Chkifa[1,2]([✉])

[1] UPMC Univ Paris 06, UMR 7598, Laboratoire Jacques-Louis Lions,
75005 Paris, France
[2] CNRS, UMR 7598, Laboratoire Jacques-Louis Lions, 75005 Paris, France
chkifa@ann.jussieu.fr

Abstract. In the papers [6,7] we have established linear and quadratic bounds, in k, on the growth of the Lebesgue constants associated with the k-sections of Leja sequences on the unit disc $\mathcal{U}$ and ℜ-Leja sequences obtained from the latter by projection into $[-1,1]$. In this paper, we improve these bounds and derive sub-linear and sub-quadratic bounds. The main novelty is the introduction of a "quadratic" Lebesgue function for Leja sequences on $\mathcal{U}$ which exploits perfectly the binary structure of such sequences and can be sharply bounded. This yields new bounds on the Lebesgue constants of such sequences, that are almost of order $\sqrt{k}$ when k has a sparse binary expansion. It also yields an improvement on the Lebesgue constants associated with ℜ-Leja sequences.

1 Introduction

The growth of the Lebesgue constant of Leja sequences on the unit disc and ℜ-Leja sequences was first studied in [3,4]. The main motivation was the study of the stability of Lagrange interpolation in multi-dimension based on intertwining of block unisolvent arrays. Such sequences, more particularly ℜ-Leja sequences, were also considered in many other works in the framework of structured hierarchical interpolation in high dimension. Although not always referred to as such, they are typically considered in the framework of sparse grids for interpolation and quadrature [10,11]. Indeed, the sections of length $2^n + 1$ of ℜ-Leja sequences coincide with the Clenshaw-Curtis abscissas $\cos(2^{-n}j\pi), j = 0, \ldots, 2^n$ which are de facto used, thanks to the logarithmic growth of their Lebesgue constant.

Motivated by the development of cheap and stable non-intrusive methods for the treatment of parametric PDEs in high dimension, we have also used these sequences in [5,9] with a highly sparse hierarchical polynomial interpolation procedure. The multi-variate interpolation process is based on the Smolyak formula and the sampling set is progressively enriched in a structured infinite grid $\otimes_{j=0}^d Z$ together with the polynomial space by only one element at a time. The Lebesgue constant that quantifies the stability of the interpolation process depends naturally on the sequence Z. We have shown in [7] that it has quadratic and cubic

© Springer International Publishing Switzerland 2015
J.-D. Boissonnat et al. (Eds.): Curves and Surfaces 2014, LNCS 9213, pp. 109–128, 2015.
DOI: 10.1007/978-3-319-22804-4_9

bounds in the number of points of interpolation when Z is a Leja sequence on $\mathcal{U}$ or an $\Re$-Leja sequence, thanks to the linear and quadratic bounds on the growth of the Lebesugue constant of such sequences, also established in [6,7]. We refer to the introduction and Sect. 2 in [7] for a concise description on the construction of the interpolation process and the study of its stability.

The present paper is also concerned with the growth of the Lebesgue constant of Leja and $\Re$-Leja sequences. We improve the linear and quadratic bounds obtained in [7]. In particular, we show that for $\Re$-Leja sequences, the bound is logarithmic for many values of k which may be useful for proposing cheap and stable interpolation scheme in the framework of sparse grids [11].

1.1   One Dimensional Hierarchical Interpolation

Let X be a compact domain in $\mathbb{C}$ or $\mathbb{R}$, typically the complex unit disc $\mathcal{U} := \{|z| \leq 1\}$ or the unit interval $[-1,1]$, and $Z = (z_j)_{j \geq 0}$ a sequence of mutually distinct points in X. We denote by I_{Z_k} the univariate interpolation operator onto the polynomials space $\mathbb{P}_{k-1}$ associated with the section of length k, $Z_k := (z_0, \cdots, z_{k-1})$. The interpolation operator is given by Lagrange interpolation formula: for $f \in C(X)$ and $z \in X$

$$I_{Z_k} f(z) = \sum_{j=0}^{k-1} f(z_j) l_{j,k}(z), \qquad l_{j,k}(z) := \prod_{\substack{i=0 \\ i \neq j}}^{k-1} \frac{z - z_j}{z_i - z_j}, \quad j = 0, \ldots, k-1. \quad (1.1)$$

Since the sections Z_k are nested, it is convenient to give the operator I_{Z_k} using Newton interpolation formula which amounts essentially in writing: $\Delta_0 f := I_{Z_1} f \equiv f(z_0)$ and

$$I_{Z_{k+1}} = I_{Z_k} + \Delta_k = \sum_{l=0}^{k} \Delta_l, \quad \text{where} \quad \Delta_l(Z) := I_{Z_{l+1}} - I_{Z_l}, \quad l \geq 1, \quad (1.2)$$

and computing the operators Δ_l using divided differences, see [12, Chap. 2] or equivalently the following formula which are differently normalized, see [7,9],

$$\Delta_l f = (f(z_l) - I_{Z_l} f(z_l)) \prod_{j=0}^{l-1} \frac{(z - z_j)}{(z_l - z_j)}, \quad l \geq 1, \quad (1.3)$$

The stability of the operator I_{Z_k} depends on the positions of the elements of Z_k on X, in particular through the Lebesgue constant associated with Z_k, defined by

$$\mathbb{L}_{Z_k} := \max_{f \in C(X) - \{0\}} \frac{\|I_{Z_k} f\|_{L^\infty(X)}}{\|f\|_{L^\infty(X)}} = \max_{z \in X} \lambda_{Z_k}(z), \quad (1.4)$$

where λ_{Z_k} is the so-called Lebesgue function associated with Z_k defined by

$$\lambda_{Z_k}(z) := \sum_{i=0}^{k-1} |l_{i,k}(z)|, \qquad z \in X. \quad (1.5)$$

We also introduce the notation

$$\mathbb{D}_k(Z) := \max_{f \in C(X)-\{0\}} \frac{\|\Delta_k f\|_{L^\infty(X)}}{\|f\|_{L^\infty(X)}}. \tag{1.6}$$

In the case of the unit disk or the unit interval, it is known that $\mathbb{L}_k$ the Lebesgue constant associated with any set of k mutually distinct points can not grow slower than $\frac{2\log(k)}{\pi}$ and it is well known that such growth is fulfilled by the set of k-roots of unity in the case $X = \mathcal{U}$ and the Tchybeshev or Gauss-Lobatto abscissas in the case $X = [-1, 1]$, see e.g. [2]. However such sets of points are not the sections of a fixed sequence Z.

In [3,4], the authors considered for Z Leja sequences on $\mathcal{U}$ initiated at the boundary $\partial \mathcal{U}$ and $\Re$-Leja sequences obtained by projection onto $[-1, 1]$ of the latter when initiated at 1. They showed that the growth of $\mathbb{L}_{Z_k}$ is controlled by $\mathcal{O}(k \log(k))$ and $\mathcal{O}(k^3 \log(k))$ respectively. In our previous works [6,7], we have improved these bounds to $2k$ and $8\sqrt{2}k^2$ respectively. We have also established in [7] the bound $\mathbb{D}_k \le (1 + k)^2$ for the difference operator, which could not be obtained directly from $\mathbb{D}_k \le \mathbb{L}_{Z_{k+1}} + \mathbb{L}_{Z_k}$ and which is essential to prove that the multivariate interpolation operator using $\Re$-Leja sequences has a cubic Lebesgue constant, see [7, Formula 25].

1.2   Contributions of the Paper

In this paper, we improve the bounds of the previous paper [3,4,6,7]. Our techniques of proof share several points with those developed in [6,7], yet they are shorter and relies notably on the binary pattern of Leja sequences on the unit disk. The novelty in the present paper is the introduction of the "quadractic" Lebesgue constant

$$\lambda_{Z_k,2}(z) := \left(\sum_{i=0}^{k-1} |l_{i,k}(z)|^2 \right)^{\frac{1}{2}}, \quad z \in X, \tag{1.7}$$

where $l_{i,k}$ are the Lagrange polynomials as defined in (1.5). We study this function and its maximum

$$\mathbb{L}_{Z_k,2} := \max_{z \in X} \lambda_{Z_k,2}(z). \tag{1.8}$$

We establish in Sect. 2 in the case where Z is any Leja sequence on $\mathcal{U}$ initiated on the boundary $\partial \mathcal{U}$ the "sharp" inequality

$$\lambda_{Z_k}(z_k) \le \mathbb{L}_{Z_k,2} \le 3\lambda_{Z_k}(z_k), \quad and \quad \lambda_{Z_k}(z_k) := \sqrt{2^{\sigma_1(k)} - 1}, \tag{1.9}$$

where $\sigma_1(k)$ denote the number of ones in the binary expansion of k. Cauchy-Schwatrz inequality applied to the Lebesgue function λ_{Z_k} defined in (1.5) yields $\lambda_{Z_k} \le \sqrt{k}\, \lambda_{Z_k,2}$. This shows that we also establish

$$\mathbb{L}_{Z_k} \le 3\sqrt{k}\, \sqrt{2^{\sigma_1(k)} - 1}, \tag{1.10}$$

for Leja sequences on $\mathcal{U}$, which improves considerably the linear bound $2k$ established in [6] when the binary expansion of k is very sparse. For example, for $k = 2^n + 3$ with n large, we get $\mathbb{L}_{Z_k} \leq 3\sqrt{7k} << k$. Using the bound (1.9), we establish in Sect. 3 a new bound on the growth of Lebesgue constants of $\mathfrak{R}$-Leja sequences that implies

$$\mathbb{L}_k \leq 6\sqrt{5}\, k\, 2^{\sigma_1(l)}, \quad \text{where} \quad l = k - (2^n + 1), \tag{1.11}$$

where n is the integer such that $2^n + 1 \leq k < 2^{n+1} + 1$. Again, we remark that the previous bound improves the bound $8\sqrt{2}k^2$ established in [7] when l is small compared to 2^n or very sparse in the sense of binary expansion. We actually prove a bound that is logarithmic for many values of k other than the values $2^n + 1$, see Theorem 3.2.

Finally, we provide in Sect. 4 new bounds on the growth of $\mathbb{D}_k$ the norm of the difference operators. We provide the bounds

$$\mathbb{D}_k \leq 1 + \sqrt{k(2^{\sigma_1(k)} - 1)}, \quad \mathbb{D}_k \leq 2^{\sigma_1(k)}2^n, \quad k \geq 1, \tag{1.12}$$

in the case of Leja sequences on $\mathcal{U}$ and the case of $\mathfrak{R}$-Leja sequences respectively where for the latter n is defined as above.

1.3 Notation

In the remainder of the paper, we work with the following notation: For an infinite sequence $Z := (z_j)_{j \geq 0}$ on X, we introduce the section $Z_{l,m} := (z_l, \cdots, z_{m-1})$ for any $l \leq m - 1$. Given two finite sequences $A = (a_0, \ldots, a_{k-1})$ and $B = (b_0, \ldots, b_{l-1})$, we denote by $A \wedge B$ the concatenation of A and B, i.e. $A \wedge B = (a_0, \ldots, a_{k-1}, b_0, \ldots, b_{l-1})$. For any finite set $S = (s_0, \cdots, s_{k-1})$ of complex numbers and $\rho \in \mathbb{C}$, we introduce the notation

$$\rho S := (\rho s_0, \cdots, \rho s_{k-1}), \quad \mathfrak{R}(S) := (\mathfrak{R}(s_0), \cdots, \mathfrak{R}(s_{k-1})), \quad \overline{S} := (\overline{s_0}, \cdots, \overline{s_{k-1}}). \tag{1.13}$$

Throughout this paper, to any finite set S of numbers, we associate the polynomial

$$w_S(z) := \prod_{s \in S}(z - s), \quad \text{with the convention} \quad w_\emptyset(z) := 1. \tag{1.14}$$

Any integer $k \geq 1$ can be uniquely expanded according to

$$k = \sum_{j=0}^{n} a_j 2^j, \quad a_j \in \{0, 1\}. \tag{1.15}$$

We denote by $\sigma_1(k)$, $\sigma_0(k)$ the number of ones and zeros in the binary expansion of k and by $p(k)$ the largest integer p such that 2^p divide k. For $k = 2^n, \ldots, 2^{n+1} - 1$ with binary expansion as above, one has

$$\sigma_1(k) = \sum_{j=0}^{n} a_j \quad \text{and} \quad \sigma_0(k) = \sum_{j=0}^{n}(1 - a_j) = n + 1 - \sigma_1(k). \tag{1.16}$$

We should finally note that, unless stated otherwise, we only work with complex numbers z belonging to the unit circle $\partial \mathcal{U}$. This is because in the complex setting we investigate supremums of sub-harmonic functions, λ_{Z_k} and $\lambda_{Z_k,2}$, which are always attained on the boundary.

2   Leja Sequences on the Unit Disk

Leja sequences $E = (e_j)_{j \geq 0}$ on $\mathcal{U}$ considered in [3,4,6,7] have all their initial value $e_0 \in \partial \mathcal{U}$ the unit circle. They are defined inductively by picking $e_0 \in \partial \mathcal{U}$ arbitrary and defining e_k for $k \geq 1$ by

$$e_k = \mathrm{argmax}_{z \in \mathcal{U}} |z - e_{k-1}| \ldots |z - e_0|. \tag{2.1}$$

The maximum principle implies that $e_j \in \partial \mathcal{U}$ for any $j \geq 1$. Also, the previous argmax problem might admit many solutions and e_k is one of them. We call a k-Leja section every finite sequence $(e_0, \ldots, e_{k-1})$ obtained by the same recursive procedure. In particular, when $E := (e_j)_{j \geq 1}$ is a Leja sequence then the section $E_k = (e_0, \ldots, e_{k-1})$ is k-Leja section.

In contrast to the interval $[-1, 1]$ where Leja sequences cannot be computed explicitly, Leja sequences on $\partial \mathcal{U}$ are much easier to compute. For instance, if $e_0 = 1$ then we can immediately check that $e_1 = -1$ and $e_2 = \pm i$. Assuming that $e_2 = i$ then e_3 maximizes $|z^2 - 1||z - i|$, so that $e_3 = -i$ because $-i$ maximizes jointly $|z^2 - 1|$ and $|z - i|$. Then e_4 maximizes $|z^4 - 1|$, etc. We observe a "binary patten" on the distribution of the first elements of E.

By radial invariance, an arbitrary Leja sequence $E = (e_0, e_1, \ldots)$ on $\mathcal{U}$ with $e_0 \in \partial \mathcal{U}$ is merely the product by e_0 of a Leja sequence with initial value 1. The latter are completely determined according to the following theorem, see [1,3,6].

Theorem 2.1. *Let* $n \geq 0$, $2^n < k \leq 2^{n+1}$ *and* $l = k - 2^n$. *The sequence* $E_k = (e_0, \ldots, e_{k-1})$, *with* $e_0 = 1$, *is a* k-*Leja section if and only if* $E_{2^n} = (e_0, \ldots, e_{2^n-1})$ *and* $U_l = (e_{2^n}, \ldots, e_{k-1})$ *are respectively* 2^n-*Leja and* l-*Leja sections and* e_{2^n} *is any* 2^n-*root of* -1.

In the light of the previous theorem, a natural construction of a Leja sequence $E := (e_j)_{j \geq 0}$ in $\mathcal{U}$ follows by the recursion

$$E_1 := (e_0 = 1) \quad \text{and} \quad E_{2^{n+1}} := E_{2^n} \wedge e^{\frac{i\pi}{2^n}} E_{2^n}, \quad n \geq 0. \tag{2.2}$$

This recursive construction of the sequence E yields an interesting distribution of its elements. Indeed, by an immediate induction, see [1], it can be shown that the elements e_k are given by

$$e_k = \exp\left(i\pi \sum_{j=0}^{n} a_j 2^{-j}\right) \quad \text{for} \quad k = \sum_{j=0}^{n} a_j 2^j, \quad a_j \in \{0,1\}. \tag{2.3}$$

The construction yields then a low-discrepancy sequence on $\partial \mathcal{U}$ based on the bit-reversal Van der Corput enumeration.

As already mentioned above, Theorem 2.1 characterizes completely Leja sequences on the unit circle. It has also many implications that turn out to be very useful in the analysis of the growth of Lebesgue constants.

Theorem 2.2. *Let $E := (e_j)_{j \geq 0}$ be a Leja sequence on $\mathcal{U}$ initiated at $e_0 \in \partial\mathcal{U}$. We have:*

- *For any $n \geq 0$, $E_{2^n} = e_0 \mathcal{U}_{2^n}$ in the set sense where $\mathcal{U}_{2^n}$ is the set of 2^n-root of unity.*
- *For any $k \geq 1$, $|w_{E_k}(e_k)| = \sup_{z \in \partial\mathcal{U}} |w_{E_k}(z)| = 2^{\sigma_1(k)}$.*
- *For any $n \geq 0$, $E_{2^n, 2^{n+1}} := (e_{2^n}, \cdots, e_{2^{n+1}-1})$ is a 2^n-Leja section.*
- *The sequence $E^2 := (e_{2j}^2)_{j \geq 0}$ is a Leja sequence on $\partial\mathcal{U}$.*

Such properties can be easily checked for the simple sequence defined in (2.3) and are given in [3,6] for more general Leja sequences.

2.1 Analysis of the Quadratic Lebesgue Function

It is proved in [6] that given two k-Leja sections E_k and F_k, one has $F_k = \rho E_k$ in the set sense for some $\rho \in \partial\mathcal{U}$. This means that the sequence F_k can be obtained from E_k by a permutation and the product by ρ. By inspection of the quadratic Lebesgue function (1.8), we have then that

$$\lambda_{F_k, 2}(z) = \lambda_{E_k, 2}(z/\rho), \quad z \in \mathcal{U} \implies \mathbb{L}_{F_k, 2} = \mathbb{L}_{E_k, 2}. \tag{2.4}$$

In order to compute the growth of $\mathbb{L}_{E_k, 2}$ for arbitrary Leja sequences E, it suffices then to consider E to be the simple sequence given by (2.3). Unless stated otherwise, for the rest of this section, E is exclusively used for this notation. Let us note that

$$E^2 := (e_{2j}^2)_{j \geq 0} = E. \tag{2.5}$$

In order to study the functions $\lambda_{E_k, 2}$, we adopt the methodology that we introduced in [6]. Namely, we study the implication of E being a Leja sequence in general, on the growth of $\lambda_{E_k, 2}$, then we use the implication of the particular binary distribution of E to derive such growth.

Lemma 2.3. *Let Z be a Leja sequence on a real or complex compact X. For any $k \geq 1$ and any $z \in X$, it holds*

$$\lambda_{Z_{k+1}, 2}(z) \leq \lambda_{Z_k, 2}(z) + \lambda_{Z_k, 2}(z_k) + 1. \tag{2.6}$$

Proof: We fix $k \geq 1$ and denote by $l_0, \ldots, l_{k-1}$ the Lagrange polynomials associated with the section Z_k and by $L_0, \ldots, L_k$ the Lagrange polynomials associated with the section Z_{k+1}. By Lagrange interpolation formula, for $j = 0, \ldots, k-1$

$$l_j = \sum_{i=0}^{k} l_j(z_i) L_i = L_j + l_j(z_k) L_k \quad \Rightarrow \quad L_j = l_j - l_j(z_k) L_k.$$

We have then for any $z \in X$

$$\left(\sum_{j=0}^{k-1} |L_j(z)|^2\right)^{1/2} \leq \left(\sum_{j=0}^{k-1} |l_j(z)|^2\right)^{1/2} + |L_k(z)| \left(\sum_{j=0}^{k-1} |l_j(z_k)|^2\right)^{1/2}.$$

where we have merely applied triangular inequality with the euclidean norm in $\mathbb{C}^k$. This also writes

$$\left(|\lambda_{Z_{k+1},2}(z)|^2 - |L_k(z)|^2\right)^{\frac{1}{2}} \leq \lambda_{Z_k,2}(z) + |L_k(z)|\lambda_{Z_k,2}(z_k).$$

We conclude the proof using $a \leq \sqrt{a^2 - b^2} + b$ for $a \geq b \geq 0$, and the inequality

$$|L_k(z)| = \frac{|w_{Z_k}(z)|}{|w_{Z_k}(z_k)|} \leq 1,$$

which follows from the Leja definition (2.1).   ∎

The previous result shows that given Z a Leja sequence over X, the growth of $\mathbb{L}_{Z_k,2}$ is monitored by the growth of $\lambda_{Z_k,2}(z_k)$. In particular, it is easily checked using induction on k that

$$\lambda_{Z_k,2}(z_k) = \mathcal{O}(\log(k)) \implies \mathbb{L}_{Z_k,2} = \mathcal{O}(k \log(k)), \tag{2.7}$$

and

$$\lambda_{Z_k,2}(z_k) = \mathcal{O}(k^\theta) \implies \mathbb{L}_{Z_k,2} = \mathcal{O}(k^{\theta+1}). \tag{2.8}$$

In the following, we show basically that the previous implication holds with $\theta = 1/2$ for Leja sequences on $\mathcal{U}$. However, we use the particular structure of such sequences in order to show that the exponent $\theta = 1/2$ is not deteriorated and that it is also valid for $\mathbb{L}_{E_k,2}$. We recall that we work with the simple sequence E given in (2.3) for which $E^2 = E$. The binary patten of the distribution of E on the unit disc yields the following result.

Lemma 2.4. *Let E be as in (2.3). For any $N \geq 1$, one has*

$$\lambda_{E_{2N},2}(z) = \lambda_{E_N,2}(z^2), \qquad z \in \partial\mathcal{U}. \tag{2.9}$$

Proof: Let $l_0, \ldots, l_{2N-1}$ be the Lagrange polynomials associated with E_{2N} and $L_0, \ldots, L_{N-1}$ be the Lagrange polynomials associated with E_N. Since $e_{2j+1} = -e_{2j}$ for any $j \geq 0$, then in view of (2.5)

$$w_{E_{2N}}(z) = w_{E_N^2}(z^2) = w_{E_N}(z^2).$$

Deriving with respect to z and using $(e_{2j+1})^2 = (e_{2j})^2 = e_j$ for any $j \geq 0$, we deduce that

$$|w'_{E_{2N}}(e_{2j+1})| = |w'_{E_{2N}}(e_{2j})| = 2|w'_{E_N}(e_{2j}^2)| = 2|w'_{E_N}(e_j)|, \quad j \geq 0. \tag{2.10}$$

We have for any $j = 0, \ldots, N - 1$

$$|l_{2j}(z)| = \frac{|w_{E_{2N}}(z)|}{|w'_{E_{2N}}(e_{2j})||z - e_{2j}|}, \qquad |l_{2j+1}(z)| = \frac{|w_{E_{2N}}(z)|}{|w'_{E_{2N}}(e_{2j+1})||z - e_{2j+1}|}.$$

Therefore in view of the previous equalities

$$|l_{2j}(z)|^2 + |l_{2j+1}(z)|^2 = \frac{|w_{E_N}(z^2)|^2}{4|w'_{E_N}(e_j)|^2}\Big[\frac{1}{|z - e_{2j}|^2} + \frac{1}{|z + e_{2j}|^2}\Big] = \frac{|w_{E_N}(z^2)|^2}{|w'_{E_N}(e_j)|^2|z^2 - e_j|^2} = |L_j(z^2)|^2, \tag{2.11}$$

where we have used $|a - b|^2 + |a + b|^2 = 4$ for $a, b \in \partial \mathcal{U}$ and $e_{2j}^2 = e_j$. Summing the previous identities for the indices $j = 0, \ldots, N - 1$, we get the result. ∎

We note that the previous result combined with $E_{2^n} = \mathcal{U}_{2^n}$ in the set sense implies that

$$\sum_{j=0}^{2^n - 1}\Big|\frac{z^{2^n} - 1}{2^n(z - e_j)}\Big|^2 = \lambda_{E_{2^n},2}(z) = \lambda_{E_1,2}(z^{2^n}) = 1, \tag{2.12}$$

for any $z \in \partial \mathcal{U}$. We now turn to the growth of $\lambda_{E_k,2}(e_k)$, which as mentioned earlier monitor the growth of $\mathbb{L}_{E_k,2}$.

Lemma 2.5. *For the Leja sequence E defined in (2.3), we have for any $k \geq 1$,*

$$\lambda_{E_k,2}(e_k) = \sqrt{2^{\sigma_1(k)} - 1} \tag{2.13}$$

Proof: First, by Lemma 2.4 and $e_{2N}^2 = e_N$, one has

$$|\lambda_{E_{2N},2}(e_{2N})|^2 = |\lambda_{E_N,2}(e_N)|^2, \quad N \geq 1. \tag{2.14}$$

Let now k be an odd number and we write $k = 2N + 1$ with $N \geq 1$. Let $l_0, \ldots, l_{2N}$ be the Lagrange polynomials associated with E_k and $L_0, \ldots, L_{N-1}$ be the Lagrange polynomials associated with E_N. For any $m = 0, \ldots, 2N$, one has

$$l_m(e_k) = \frac{w_{E_k}(e_k)}{(e_k - e_m)w'_{E_k}(e_m)} = \frac{w'_{E_{k+1}}(e_k)}{w'_{E_{k+1}}(e_m)} \quad \Rightarrow \quad |l_m(e_k)| = \frac{|w'_{E_{N+1}}(e_k^2)|}{|w'_{E_{N+1}}(e_m^2)|},$$

where we have used $k + 1 = 2(N + 1)$ and (2.10). Using $e_k^2 = e_N$ and $(e_{2j+1})^2 = (e_{2j})^2 = e_j$ for any j, we get for $m = 2j$ or $m = 2j + 1$ with $j = 0, \ldots, N - 1$

$$|l_m(e_k)| = \frac{|w'_{E_{N+1}}(e_N)|}{|w'_{E_{N+1}}(e_j)|} = |L_j(e_N)| \quad \text{and also} \quad |l_{2N}(e_k)| = \frac{|w'_{E_{N+1}}(e_N)|}{|w'_{E_{N+1}}(e_N)|} = 1.$$

Summing the numbers $|l_m(e_k)|^2$ over $m = 0, \ldots, 2N$, we infer

$$|\lambda_{E_{2N+1},2}(e_{2N+1})|^2 = 2|\lambda_{E_N,2}(e_N)|^2 + 1. \tag{2.15}$$

In view of the above and $\lambda_{E_1,2}(e_1)=1$, the sequence $\alpha := (\alpha_k := |\lambda_{E_k,2}(e_k)|^2)_{k \geq 1}$ satisfies:

$$\alpha_1 = 1 \quad \text{and} \quad \alpha_{2N} = \alpha_N, \quad \alpha_{2N+1} = 2\alpha_N + 1, \quad N \geq 1.$$

We have $\sigma_1(1) = 1$ and $\sigma_1(2N) = \sigma_1(N)$, $\sigma_1(2N+1) = \sigma_1(N)+1$ for any $N \geq 1$. It is then easily checked that $(2^{\sigma_1(k)} - 1)_{k \geq 1}$ satisfies the same recursion as α. This shows that $\alpha_k = 2^{\sigma_1(k)} - 1$ for any $k \geq 1$ and finishes the proof. ∎

We are now able to conclude the main result of this section, which states basically that for the sequence E or more generally any Leja sequence on $\mathcal{U}$ initiated at the boundary $\partial\mathcal{U}$, the value of $\mathbb{L}_{E_k,2} = \max_{z \in \mathcal{U}} \lambda_{E_k,2}(z)$ is almost equal to $\lambda_{E_k,2}(e_k)$.

Theorem 2.6. *For the Leja sequence E defined in (2.3), we have for any $k \geq 1$*

$$1 \leq \frac{\mathbb{L}_{E_k,2}}{\lambda_{E_k,2}(e_k)} = \frac{\mathbb{L}_{E_k,2}}{\sqrt{2^{\sigma_1(k)} - 1}} \leq 3 \qquad (2.16)$$

Proof: The first part of the inequality is immediate from the definition of $\mathbb{L}_{E_k,2}$. Also in view Lemma 2.4 and formula (2.14), we only need to show (2.16) when k is an odd number. Let $k = 2N+1$ with $N \geq 1$. Using Lemmas 2.3 and 2.4 and formula (2.14), we have

$$\lambda_{E_k,2}(z) \leq \lambda_{E_{2N},2}(z) + \lambda_{E_{2N},2}(e_{2N}) + 1 = \lambda_{E_N,2}(z^2) + \lambda_{E_N,2}(e_N) + 1.$$

If we assumes that $\lambda_{E_N,2}(z^2) \leq 3\lambda_{E_N,2}(e_N)$, we get

$$\lambda_{E_k,2}(z) \leq 4\lambda_{E_N,2}(e_N) + 1 \leq 3\sqrt{2|\lambda_{E_N,2}(e_N)|^2 + 1},$$

where we have used the elementary inequality $4t + 1 \leq 3\sqrt{2t^2 + 1}$ for any $t \geq 0$. In view of (2.15), one then gets $\lambda_{E_k,2}(z) \leq 3\lambda_{E_k,2}(e_k)$. The verification $\mathbb{L}_{E_1,2} = \lambda_{E_1,2}(e_1) = 1$ shows that the result follows using an induction on $k \geq 1$. ∎

2.2 Implications on the Lebesgue Constant

The methodology we have provided so far for bounding $\mathbb{L}_{E_k,2}$ is not new, we have developed it in [6] in order to give linear estimate for $\mathbb{L}_{E_k}$, namely $\mathbb{L}_{E_k} \leq 2k$. Theorem 2.6 has also implications on the growth of the Lebesgue constant $\mathbb{L}_{E_k}$. Indeed, Cauchy Schwartz inequality applied to the Lebesgue function λ_{E_k} implies $\lambda_{E_k} \leq \sqrt{k}\, \lambda_{E_k,2}$, so that

$$\mathbb{L}_{E_k} \leq \sqrt{k}\, \mathbb{L}_{E_k,2} \leq 3\sqrt{k(2^{\sigma_1(k)} - 1)}. \qquad (2.17)$$

The Cauchy Schwartz formula $\lambda_{E_k} \leq \sqrt{k}\, \lambda_{E_k,2}$ is possibly not very pessimistic. It has been recently proved that the Lagrange polynomials are uniformly bounded, see [14]. We shall observe in particular, see Fig. 1, that the binary pattern observed for the exact value of $\mathbb{L}_{E_k}$ is captured by the previous bound. Moreover, we are able to provide a lower bound for $\mathbb{L}_{E_k}$, that is comparable to the previous upper bound for values of k with full binary expansion.

Proposition 2.7. *For the Leja sequence E defined in (2.3), we have for any* $k \geq 1$

$$2^{\sigma_1(k)} - 1 \leq \lambda_{E_k}(e_k) \leq \mathbb{L}_{E_k}. \tag{2.18}$$

Proof: We let $N \geq 1$ and we use the notation of the proof of Lemma 2.4. As for formula (2.11) and since $|a - b| + |a + b| \geq 2$ for any $a, b \in \partial \mathcal{U}$, one has

$$|l_{2j}(z)| + |l_{2j+1}(z)| = \frac{|w_{E_N}(z^2)|}{2|w'_{E_N}(e_j)|} \frac{|z - e_{2j}| + |z + e_{2j}|}{|z - e_j|} \geq |L_j(z^2)|.$$

This implies $\lambda_{E_{2N}}(z) \geq \lambda_{E_N}(z^2)$ and more particularly $\lambda_{E_{2N}}(e_{2N}) \geq \lambda_{E_N}(e_N)$. As in the proof of Lemma 2.5, we have also $\lambda_{E_{2N+1}}(e_{2N+1}) = 2\lambda_{E_N}(e_N) + 1$. The sequence $(b_k := \lambda_{E_k}(e_k))_{k \geq 1}$ satisfies:

$$b_1 = 1 \quad \text{and} \quad b_{2N} \geq b_N, \quad b_{2N+1} = 2b_N + 1, \quad N \geq 1.$$

The sequence b then satisfies $b_k \geq 2^{\sigma_1(k)} - 1$ for any $k \geq 1$.   ∎

The previous theorem combined with Theorem 2.6 and Eq. (2.17) implies

$$\frac{\sqrt{2^{\sigma_1(k)} - 1}}{3} \mathbb{L}_{E_k,2} \leq \mathbb{L}_{E_k} \leq \sqrt{k}\, \mathbb{L}_{E_k,2}. \tag{2.19}$$

Cauchy-Schwartz inequality is then satisfactory when $k \simeq 2^{\sigma_1(k)}$, that is when k has a full binary expansion.

Remark 2.8. *For integers* $k = 2^n, \ldots, 2^{n+1} - 1$, *if* $k = 2^{n+1} - 1$ *in which case* $\sigma_1(k) = n + 1$ *is the largest possible, the bound (2.17) merely implies* $\mathbb{L}_{E_k} \leq 3k$ *which is worse than the bound* $2k$ *established [6] and the exact value* $\mathbb{L}_{E_k} = k$ *of this case, see [3]. However, since* $\sigma_1(k) = n + 1 - \sigma_0(k)$ *for any* $k = 2^n, \ldots 2^{n+1} - 1$, *then by (2.17)*

$$\mathbb{L}_{E_k} \leq \sqrt{\frac{18}{2^{\sigma_0(k)}}} \sqrt{2^n k} \leq \sqrt{\frac{18}{2^{\sigma_0(k)}}}\, k. \tag{2.20}$$

This shows in particular that $\mathbb{L}_{E_k} \leq k$ *whenever* $\sigma_0(k) \geq 5$. *This last result answers partly the conjecture raised in [3] and which states that* $\mathbb{L}_{E_k} \leq k$ *for any* $k \geq 1$.

For the purpose of the next section, we improve the bound (2.17) in the case where k is an even number. We recall that we have shown in [6, Theorem 2.8]

$$\mathbb{L}_{E_{2^p l}} \leq \mathbb{L}_{2^p} \mathbb{L}_{E_l}, \qquad p \geq 0, \ l \geq 1, \tag{2.21}$$

where $\mathbb{L}_{2^p}$ is the Lebesgue constant associated with the set of 2^p-roots of unity. The value $\mathbb{L}_{2^p}$ can be computed easily for small values of p and it grows logarithmically in 2^p, see e.g. [6, Formula 2.25],

$$\mathbb{L}_1 = 1, \quad \mathbb{L}_2 = \sqrt{2}, \quad \text{and} \quad \mathbb{L}_{2^p} \leq \frac{2}{\pi} \left(\log(2^p) + 9/4 \right), \quad p \geq 2. \tag{2.22}$$

Since $\sigma_1(k) = \sigma_1(k/2^{p(k)})$, we have then in view of (2.17) and (2.21) the following theorem

Theorem 2.9. *Let E be the Leja sequence defined in (2.3) or any Leja sequence on $\mathcal{U}$ initiated at $\partial\mathcal{U}$. We have*

$$\mathbb{L}_{E_k} \le 3\sqrt{\frac{k}{2^{p(k)}}(2^{\sigma_1(k)} - 1)} \ \mathbb{L}_{2^{p(k)}}, \quad k \ge 1. \tag{2.23}$$

We should mention that our primary interest in studying $\lambda_{E_k,2}$ was the improvement of the results of [7] concerned with the Lebesgue constants of $\Re$-Leja sequences. This will be made clear in the proof of Theorem 3.2. For the sake of the same theorem, we need also to provide a growth property of Leja sequences on the unit disc.

We let $E = (e_j)_{j\ge 0}$ be the simple Leja sequence defined by (2.3). For $m \ge 0$ and $1 \le l \le 2^{m-1}$, we introduce the notation $K = 2^m + l$ and $F_{m,l} = E_{2^m,K}$ and define the quantity

$$\gamma_{m,l} = \frac{1}{4^m}\sum_{j=0}^{K-1}\frac{4}{|w_{F_{m,l}}(\overline{e_j})|^2}. \tag{2.24}$$

The quantity $\gamma_{m,l}$ is well defined. Indeed, by the particular structure of the sequence E, we have $E_{2^m+2^{m-1}} = E_{2^m} \wedge e^{\frac{i\pi}{2^m}}E_{2^{m-1}}$, so that $E_{2^m+2^{m-1}} = \mathcal{U}_{2^m} \wedge e^{\frac{i\pi}{2^m}}\mathcal{U}_{2^{m-1}}$ in the set sense. We have then for $j = 0,\dots,2^m + l - 1$, $\overline{e_j}$ is in $\mathcal{U}_{2^m} \wedge e^{\frac{-i\pi}{2^m}}\mathcal{U}_{2^{m-1}}$ which does not intersect with $F_{m,l} \subset e^{\frac{i\pi}{2^m}}\mathcal{U}_{2^{m-1}}$. We have the following growth for $\gamma_{m,l}$.

Lemma 2.10. *For any $m \ge 1$ and any $1 \le l \le 2^{m-1}$, we have*

$$\gamma_{m,l} \le \frac{5}{2^{\sigma_1(l)+p(l)+1}} \tag{2.25}$$

Proof: Since $(e_0, e_1, e_2) = (1, -1, i)$, it can be checked that $\gamma_{1,1} = 5/4$. We then fix $m \ge 2$. We define $\rho = e_{2^m} = e^{i\pi/2^m}$, so that $F_{m,1} = \{\rho\}$. We have

$$\gamma_{m,1} = \sum_{j=0}^{2^m}\frac{4}{(2^m|e_j - \overline{\rho}|)^2} = |\lambda_{E_{2^m},2}(\overline{\rho})|^2 + \frac{4}{(2^m|\rho - \overline{\rho}|)^2} = 1 + \frac{1}{|2^m\sin(\pi/2^m)|^2}$$

where we have used (2.12) and used that $\overline{\rho}$ is a 2^m-root of -1. Since $2^m\sin(\pi/2^m) \ge 2$ then $\gamma_{m,1} \le 5/4$. For the other values of $l = 2,\dots,2^{m-1}$, we have

- If $l = 2N$, we have for any $j \ge 0$ that $w_{F_{m,l}}(\overline{e_{2j+1}}) = w_{F_{m,l}}(\overline{e_{2j}}) = w_{E_{2^{m-1},2^{m-1}+N}}(\overline{e_j})$. Pairing the indices in (2.24) as $2j$ and $2j + 1$ with $j = 0,\dots,2^{n-1} + N - 1$, we deduce

$$\gamma_{m,l} = \frac{\gamma_{m-1,N}}{2}.$$

- If $l = 2N + 1$ with $N \geq 1$, we may write

$$\gamma_{m,l} = \frac{1}{4^{m-1}} \sum_{j=0}^{K-1} \frac{|e_K - \overline{e_j}|^2}{|w_{F_{m,l+1}}(\overline{e_j})|^2} \leq \frac{1}{4^{m-1}} \sum_{j=0}^{K} \frac{|e_K - \overline{e_j}|^2}{|w_{F_{m,l+1}}(\overline{e_j})|^2} = \gamma_{m-1,N+1},$$

where we have again paired the indices by $2j$ and $2j + 1$ for $j = 0, \ldots, 2^n + (N+1) - 1$ and used $e_{2j+1} = -e_{2j}$ and the identity $|a + b|^2 + |a - b|^2 = 4$ for any $a, b \in \partial\mathcal{U}$.

Therefore

$$\gamma_{m,l} \leq \frac{5}{4} a_{m,l}, \qquad 1 \leq m, \ 1 \leq l \leq 2^{m-1},$$

where $(a_{m,l})_{\substack{1 \leq m \\ 1 \leq l \leq 2^{m-1}}}$ is the sequence that saturates the previous inequalities and hence is defined by the following recursion:

$$a_{m,1} = 1, \ m \geq 1 \quad \text{and} \quad \begin{cases} a_{m,2N} = a_{m-1,N}/2 & n \geq 1, N = 1, \ldots, 2^{m-2}, \\ a_{m,2N+1} = a_{m-1,N+1} & n \geq 1, N = 1, \ldots, 2^{m-2} - 1. \end{cases}$$

The sequence $(a_{m,l})$ has no dependance on m and it is equal, in the sense $a_{m,l} = a_l$, to the sequence $(a_l)_{l \geq 1}$ which satisfies the recursion: $a_1 = 1$, $a_{2N} = a_N/2$, $a_{2N+1} = a_{N+1}$. Since $\sigma_1(1) + p(1) = 1, \sigma_1(2N) + p(2N) = \sigma_1(N) + p(N) + 1$ and

$$\sigma_1(2N + 1) + p(2N + 1) = \sigma_1(2N + 1) = \sigma_1(N) + 1 = \sigma_1(N + 1) + p(N + 1),$$

then an immediate induction shows that $a_l = 2^{1-\sigma_1(l)-p(l)}$, which finishes the proof. ∎

3 $\Re$-Leja Sequences on $[-1, 1]$

$\Re$-Leja sequences were introduced and studied in [4]. Such sequences are simply defined as the projection, element-wise but without repetition, into [-1,1] of Leja sequences on $\mathcal{U}$ initiated at 1. More precisely, given $E = (e_j)_{j \geq 0}$ a Leja sequence on $\mathcal{U}$ initiated at 1, the $\Re$-Leja sequence $R = (r_j)_{j \geq 0}$ associated with E is obtained progressively by: $r_0 = \Re(e_0) = 1$, $J(0) = 0$ and

$$r_k = \Re(e_{J(k)}) \quad \text{where} \quad J(k) = \min\{j > J(k-1) : \Re(e_j) \notin R_k\}, \quad k \geq 1. \quad (3.1)$$

This means one projects e_j if and only if $e_j \neq \overline{e_i}$ for all $i < j$. The projection rule that prevents the repetition is provided in [4, Theorem 2.4]. One has

$$R = \Re(\Xi), \quad \text{with} \quad \Xi := (1, -1) \wedge \bigwedge_{j=1}^{\infty} E_{2^j, 2^j + 2^{j-1}}. \quad (3.2)$$

Using a simple cardinality argument, see [4, Theorem 2.4] or [7, Formula 40], this implies that the function J used in (3.1) is given by: $J(0) = 0$, $J(1) = 1$ and

$$J(k) = 2^n + k - 1, \qquad n \geq 0, \ 2^n + 1 \leq k < 2^{n+1} + 1. \quad (3.3)$$

In view of (3.2) and the properties of Leja sequences on $\mathcal{U}$, any $\Re$-Leja sequence R satisfies $r_0 = 1$, $r_1 = -1$, $r_2 = 0$ and $r_{2j-1} = -r_{2j}$ for any $j \geq 2$. An accessible example of an $\Re$-Leja sequence is the one associated with the simple Leja sequence given by the bit-reversal enumeration (2.3). We have shown in [6] that $R = (\cos(\phi_j))_{j \geq 0}$ where the sequence of angles $(\phi_k)_{k \geq 0}$ is defined recursively by $\phi_0 = 0$, $\phi_1 = \pi$, $\phi_2 = \pi/2$ and

$$\phi_{2j-1} = \frac{\phi_j}{2}, \qquad \phi_{2j} = \phi_{2j-1} + \pi, \quad j \geq 2. \tag{3.4}$$

This recursion provides a simple process to compute an $\Re$-Leja sequence. We can also construct a Leja sequence by simply using the recursion $r_0 = 1$, $r_1 = -1$, $r_2 = 0$ and

$$r_{2j-1} = \sqrt{\frac{r_j + 1}{2}}, \qquad r_{2j} = -r_{2j-1}, \quad j \geq 2. \tag{3.5}$$

One can check that the last sequence is obtained from the Leja sequence F which is constructed recursively by $F_1 = \{1\}$ and $F_{2^{n+1}} = F_{2^n} \wedge e^{\frac{i\pi}{2^n}} \overline{F_{2^n}}$. Both $\Re$-Leja sequences R satisfies $2r_0^2 - 1 = 1$, $2r_2^2 - 1 = -1$ and more generally $2r_{2j}^2 - 1 = 2r_{2j-1}^2 - 1 = r_j$ for any $j \geq 2$, thanks to the trigonometric identity $2\cos^2(\theta/2) - 1 = \cos(\theta)$. This shows that in both cases R satisfies the property

$$R^2 = R \quad \text{where} \quad R^2 := (2r_{2j}^2 - 1)_{j \geq 0}, \tag{3.6}$$

In general, given a Leja sequence E in $\mathcal{U}$ initiated at 1 and R the associated $\Re$-Leja sequence, we have that R^2 is an $\Re$-Leja sequence and it is associated with E^2 which, in view of Theorem 2.2, is also a Leja sequence initiated at 1. This result is given in [7, Lemma 3.4] and it has many useful implications that we have exploited in order to prove that $\mathbb{D}_k(R)$ grows at worse quadratically.

For all Leja sequences E on $\mathcal{U}$ initiated at 1, the section $E_{2^{n+1}}$ is equal in the set sense with $\mathcal{U}_{2^{n+1}}$ the set of 2^{n+1}-roots of unity, therefore for all $\Re$-Leja sequences R, the section R_{2^n+1} is equal to the set of Gauss-Lobatto abscissas of order 2^n, i.e.

$$R_{2^n+1} = \left\{ \cos(j\pi/2^n) : j = 0, \ldots, 2^n \right\}, \tag{3.7}$$

in the set sense. This set of abscissas is optimal as far as the Lebesgue constant is concerned, in the sense $\mathbb{L}_{R_{2^n+1}} \simeq \frac{2\log(2^n+1)}{\pi}$. More precisely, we have the bound

$$\mathbb{L}_{R_{2^n+1}} \leq 1 + \frac{2}{\pi} \log(2^n), \tag{3.8}$$

see [13, Formulas 5 and 13]. This suggests that the sequence R might have a moderate growth of the Lebesgue constant of its sections R_k.

In the paper [4], it has been proved that $\mathbb{L}_{R_k} = \mathcal{O}(k^3 \log(k))$. We have improved this bound in [6, 7] and showed that $\mathbb{L}_{R_k} \leq 8\sqrt{2}\, k^2$ for any $k \geq 2$. Here we again exploit our approach of [7], using simple calculatory arguments, relate the analysis of the Lebesgue function associated with R_k to that of the Lebesgue function associated with the smaller Leja section that yields R_k

by projection. This approach allows us to circumvent cumbersome real trigonometric functions which arise in the study λ_{R_k}, see [4,6], and to take full benefit from the machinery developed for Leja sequence on $\mathcal{U}$.

Remark 3.1. *Without loss of generality, we assume for the remainder of this section that E is the simple Leja sequence in (2.3) and R the associated $\Re$-Leja sequence. All our arguments hold in the more general case, the assumption is essentially for notational clearness. It allows us, in view of (2.5), to use E instead for E^2 and more generally instead of E^{2^p} which is defined by $E^{2^p} := ((e_{2^p j})^{2^p})_{j \geq 0}$.*

The bound (3.8) is sharp and we are only interested in bounding $\mathbb{L}_{R_k}$ when $k - 1$ is not a power of 2. For the remainder of this section, we use the notation

$$n \geq 0, \quad 2^n < k - 1 < 2^{n+1}, \quad 0 < l := k - (2^n + 1) < 2^n \atop K := 2^{n+1} + l, \quad G_k := E_K, \quad F_K := E_{2^{n+1}, K}. \tag{3.9}$$

We should note that in [7] we have used k' and F_k to denote l and F_K. In view of (3.3), we have $K = J(k)$, so that E_K is the smallest section that yields R_k by projection into $[-1, 1]$. We denote by $L_0, L_1, L_2, \cdots, L_{K-1}$ the Lagrange polynomials associated with E_K. The inspection of the the proof of [7, Lemma 6] shows that for $z \in \partial \mathcal{U}$ and $x = \Re(z)$,

$$\lambda_{R_k}(x) \leq \gamma_K(z) + \gamma_K(\bar{z}), \quad \gamma_K(z) := |w_{F_K}(\bar{z})| \sum_{j=0}^{K-1} \frac{|L_j(z)|}{|w_{F_K}(\bar{e}_j)|}. \tag{3.10}$$

In the proof of [7, Lemma 6], we have bounded the functions $|w_{F_k}|/|w_{F_k}(\bar{e}_j)|$ in the previous sum by $2^{n+\frac{1}{2}-p(l)}$. This implied the result of [7, Theorem 5], namely $\mathbb{L}_{R_k} \leq 2^{n+\frac{3}{2}-p(l)} \mathbb{L}_{E_K}$. In view of the new bound (2.23) and the facts that $p(K) = p(l)$, $\sigma_1(K) = 1 + \sigma_1(l)$ and $K = 2^n + k - 1 \leq 3 \times 2^n$, the previous bound implies

$$\mathbb{L}_{R_k} \leq 12\sqrt{3}\, 2^{\frac{3n - 3p(l) + \sigma_1(l)}{2}} \mathbb{L}_{2^{p(l)}}, \quad k \geq 1, \tag{3.11}$$

where $\mathbb{L}_{2^p}$ is bounded as in (2.22). We propose to improve slightly the previous inequality by applying rather Cauchy-Schwartz inequality when bounding the function γ_K.

Theorem 3.2. *Let R be an $\Re$-Leja sequence and n, k and l as in (3.9). We have*

$$\mathbb{L}_{R_k} \leq 6\sqrt{5}\, 2^{n+\sigma_1(l)-p(l)} \mathbb{L}_{2^{p(l)}}, \tag{3.12}$$

where $\mathbb{L}_{2^p}$ is bounded as in (2.22).

Proof: In order to lighten the notation, we use the shorthand p in order to denote $p(l)$. We introduce l' and K' and $F_{K'}$ defined by

$$l' := l/2^p, \quad K' := K/2^p = 2^{n-p+1} + l', \quad F_{K'} := E_{2^{n-p+1}, K'}.$$

The sequence E satisfies $E^2 = E$ and one can check that $w_{F_K}(z) = w_{F_{K'}}(z^{2^p})$. Also by $e_{2j}^2 = e_{2j+1}^2 = e_j$, one has $(e_{2^p j + q})^{2^p} = e_j$ for any $q = 0, \ldots, 2^p - 1$. Moreover, if $M_1, \ldots, M_{K'-1}$ are the Lagrange polynomials associated with $E_{K'}$, then

$$\sum_{q=0}^{2^p - 1} |L_{2^p j + q}(z)| \leq \mathbb{L}_{2^p} M_j(z^{2^p}), \qquad j = 0, \ldots, K' - 1,$$

see the proof of [6, Theorem 2.8]. Therefore by pairing the indices in the sum giving γ_K by $2^p j + q$ for $j = 0, \ldots, K' - 1$ and $q = 0, \ldots, 2^p - 1$, we infer

$$\gamma_K(z) \leq \left(|w_{F_{K'}}(\overline{\xi})| \sum_{j=0}^{K'-1} \frac{|M_j(\xi)|}{|w_{F_{K'}}(\overline{e_j})|} \right) \mathbb{L}_{2^p} = \mathbb{L}_{2^p} \gamma_{K'}(\xi), \qquad \text{with} \quad \xi = z^{2^p}.$$

In view of (3.10), this implies that $\mathbb{L}_{R_k} \leq 2\mathbb{L}_{2^p} \sup_{\xi \in \mathcal{U}} \gamma_{K'}(\xi)$. Applying Cauchy Schwatrz inequality to $\gamma_{K'}$ and using that $F_{K'}$ is an l'-Leja sequence, we have for any $\xi \in \mathcal{U}$

$$\gamma_{K'}(\xi) \leq 2^{\sigma_1(l')} \left(\sum_{j=0}^{K'-1} \frac{1}{|w_{F_{K'}}(\overline{e_j})|^2} \right)^{1/2} \left(\sum_{j=0}^{K'-1} |M_j(\xi)|^2 \right)^{1/2} = 2^{\sigma_1(l')+n-p} \sqrt{\gamma_{n-p+1,l'}} \, \lambda_{E_{K'},2}(\xi),$$

where $\gamma_{n-p+1,l'}$ is defined as in (2.24) with $m = n - p + 1$ and $\lambda_{E_{K'},2}$ is the quadratic Lebesgue function associated with $E_{K'}$. In view of the bounds we have for these quantities and in view of $\sigma_1(K') = 1 + \sigma_1(l')$ and $\sigma_1(l') = \sigma_1(l)$, we get

$$\gamma_{K'}(\xi) \leq 2^{\sigma_1(l)+n-p} \sqrt{\frac{5}{2^{\sigma_1(l')+1}}} \, 3\sqrt{2^{1+\sigma_1(l')} - 1} \leq 3\sqrt{5} \, 2^{\sigma_1(l)+n-p}.$$

The proof is then complete.   ∎

The bound in (3.12) improves the bound in (3.11) by $2^{\frac{\sigma_1(l)+p(l)-n}{2}}$. The bound can also yield linear estimates for $\mathbb{L}_{R_k}$, for instance when l is such that $2^{\sigma_1(l)-p(l)} \mathbb{L}_{2^{p(l)}} \leq 1$, which is the case if for example $p(l) \geq 2\sigma_1(l)$. However, if $0 < l < 2^n$ is the integer with the most number of ones in the binary expansion, i.e. $\sigma_1(l) = n$ or $l = 2^n - 1$ and $k = 2^{n+1}$, we merely get the quadratic bound

$$\mathbb{L}_{R_k} \leq 6\sqrt{5} \, 2^{2n} = \frac{3\sqrt{5}}{2} k^2. \tag{3.13}$$

In [4], Sect. 3.4, it is shown that for the values $k = 2^n$, in other words R_k is the set of Gauss-Lobatto abscissas (3.7) missing one abscissa, one has $\mathbb{L}_{R_k} \geq \lambda_{R_k}(r_k) = k-1$. As a consequence, the growth of $\mathbb{L}_{R_k}$ for $k \geq 1$ can not be slower than k. However, for this case, we can prove $\mathbb{L}_{R_k} \leq 3k$, see (4.11), showing that (3.13) is rather pessimistic.

The estimate in (3.12) is logarithmic for many values of the integer k. For instance, if $k = (2^n + 1) + 2^{n-p} k'$ for some $p = 1, \ldots, n$ and some $k' = 0, \ldots, 2^p - 1$, then we have $l = 2^{n-p}k'$, so that $n - p \leq p(l) \leq n$ and $\sigma_1(l) = \sigma_1(k') \leq p$ implying that

$$\mathbb{L}_{R_k} \leq 6\sqrt{5}\, 2^{2p}\, \mathbb{L}_{2^{p(l)}} \leq 6\sqrt{5}\, 2^{2p}\, \mathbb{L}_{2^n} \leq 6\sqrt{5}\, 2^{2p}\frac{2}{\pi}\left(\log(2^n) + 9/4\right). \quad (3.14)$$

For a small value of p, the previous estimate is as good as the optimal logarithmic estimate $\frac{2\log(k)}{\pi}$ for large values of n. Given then p fixed, one has 2^p intermediate values between $2^n + 1$ and $2^{n+1} + 1$, which are the numbers $k = (2^n + 1) + 2^{n-p}k'$ for $k' = 0, \ldots, 2^p - 1$, for which the Lebesgue constant is logarithmic. This observation can be used in order to modify the doubling rule with Clemshaw-Curtis abscissas in the framework of sparse grids, see [11].

4   Growth of the Norms of the Difference Operators

In this section, we discuss the growth of the norms of the difference operators $\Delta_0 = I_{Z_1}$ and $\Delta_k = I_{Z_{k+1}} - I_{Z_k}$ for $k \geq 1$, associated with interpolation on Leja or $\Re$-Leja sequences. We are interested in estimating their norms $\mathbb{D}_k$ defined in (1.6). Elementary arguments, see [7], show that

$$\mathbb{D}_k(Z) = \left(1 + \lambda_{Z_k}(z_k)\right) \sup_{z \in X} \frac{|w_{Z_k}(z)|}{|w_{Z_k}(z_k)|}, \quad k \geq 1. \quad (4.1)$$

In particular if Z is a Leja sequence on the compact X, then

$$\mathbb{D}_k(Z) = 1 + \lambda_{Z_k}(z_k). \quad (4.2)$$

In [6], we have established that $\lambda_{E_k}(e_k) \leq k$ if E is a Leja sequence on $\mathcal{U}$ initiated at $\partial\mathcal{U}$, which implies $\mathbb{D}_k(E) \leq 1 + k$. Here, we improve slightly this bound. As for the improvement of (2.17) into (2.23), we have

Theorem 4.1. *Let E be a Leja sequence on the unit disk initiated at $e_0 \in \partial\mathcal{U}$, One has $\mathbb{D}_0(E) = 1$ and*

$$\mathbb{D}_k(E) \leq 1 + \sqrt{\frac{k}{2^{p(k)}}\left(2^{\sigma_1(k)} - 1\right)}\, \mathbb{L}_{2^{p(k)}} \quad (4.3)$$

For $\Re$-Leja sequences R on $[-1, 1]$, we have shown in [7] using a recursion argument based on the fact that R^2 defined as in (3.6) is also an $\Re$-Leja sequence, that

$$\mathbb{D}_k(R) \leq (1 + k)^2, \quad k \geq 0. \quad (4.4)$$

In view of the new bounds obtained in this paper for Lebesgue constant of $\Re$-Leja sections, the previous bound is not sharp. Indeed, we have $\mathbb{D}_k \leq \mathbb{L}_k + \mathbb{L}_{k-1} \leq 12\sqrt{5}\, k^{3/2}$, for k such that $l = k - (2^n + 1) \leq 2^{n/2}$. We give here a sharper bound for $\mathbb{D}_k(R)$. We recall that up to a rearrangement in the formula (4.1), see [7] for justification, we may write the quantities $\mathbb{D}_k(R)$ in a more convenient form for $\Re$-Leja sequences. We introduce the polynomial $W_{R_k} := 2^k w_{R_k}$, we have

$$\mathbb{D}_k(R) = 2\beta_k(R) \sup_{x\in[-1,1]} |W_{R_k}(x)|, \quad \beta_k(R) := \frac{1+\lambda_{R_k}(r_k)}{2|W_{R_k}(r_k)|}, \qquad (4.5)$$

We have already proved in [7, Lemma 7] that

$$\beta_{2^n}(R) = 1/4 \quad \text{and} \quad \beta_k(R) \leq 2^{\sigma_0(k)-p(k)-1}, \quad \text{for} \quad k \neq 2^n. \qquad (4.6)$$

Here we provide a sharper bound for $\mathbb{D}_k(R)$ by slightly improving the estimate $4^{\sigma_1(k)+p(k)-1}$ that we have established in [7] for $\sup_{x\in[-1,1]} |W_{R_k}(x)|$.

Lemma 4.2. Let R be an $\Re$-Leja sequence in $[-1,1]$, $n \geq 1$, $2^n + 1 \leq k < 2^{n+1}+1$ and $l = k - (2^n+1)$. One has $\sup_{x\in[-1,1]} |W_{R_k}(x)| \leq 2^{n+3}$ if $k = 2^{n+1}$, else

$$\sup_{x\in[-1,1]} |W_{R_k}(x)| \leq 2^{2\sigma_1(k)+p(k)-1}. \qquad (4.7)$$

Proof: We use the notation K, G_k and F_K as in (3.9) and introduce $G_{k+1} := E_{K+1}$ and $F_{K+1} := E_{2^{n+1},K+1}$. In view of [7, Lemma 5], one has for $z \in \partial \mathcal{U}$ and $x = \Re(z)$

$$|W_{R_k}(x)| = |z^2 - 1||w_{G_K}(z)||w_{F_K}(\overline{z})| = |z-\overline{z}||w_{G_K}(z)||w_{F_K}(\overline{z})|.$$

Also since $|z - \overline{z}| \leq |z - e_K| + |\overline{z} - e_K|$, then

$$|W_{R_k}(x)| \leq |w_{G_{k+1}}(z)||w_{F_K}(\overline{z})| + |w_{G_k}(z)||w_{F_{K+1}}(\overline{z})|.$$

In the two previous inequalities, one has $F_K = \emptyset$ and $w_{F_K} \equiv 1$ in the case $k = 2^n + 1$. We have that G_k, G_{k+1}, F_K and F_{K+1} are all Leja sections with length K, $K+1$, l and $l+1$ respectively. Therefore, by the second property in Theorem 2.2

$$|W_{R_k}(x)| \leq \min\left(2^{1+\sigma_1(K)+\sigma_1(l)}, 2^{\sigma_1(K+1)+\sigma_1(l)} + 2^{\sigma_1(K)+\sigma_1(l+1)}\right) = 2^{2+\sigma_1(l)}\min(2^{\sigma_1(l)}, 2^{\sigma_1(l+1)}),$$

where we have used $\sigma_1(K) = 1 + \sigma_1(l)$ and $\sigma_1(K+1) = 1 + \sigma_1(l+1)$ since $K = 2^{n+1} + l$ and $l < 2^n$. If $k = 2^{n+1}$, i.e. $l = 2^n - 1$, then $|W_{R_k}(x)| \leq 2^{3+n}$. Else by $k = 2^n + (l+1)$ and $0 \leq l < 2^n - 1$,

$$\sigma_1(k) - 1 = \sigma_1(l+1) \quad \text{and} \quad \sigma_1(k) - 2 + p(k) = \sigma_1(k-1) - 1 = \sigma_1(l).$$

Therefore

$$|W_{R_k}(x)| \leq 2^{2\sigma_1(k)+p(k)-1}\min(2^{-1+p(k)}, 1),$$

which completes the proof. ∎

By injecting the estimate of the previous lemma and the estimate of (4.6) in formula (4.5) and by using the identity $\sigma_0(k) + \sigma_1(k) = n+1$ for $2^n \leq k < 2^{n+1}$, we are able to conclude the following result.

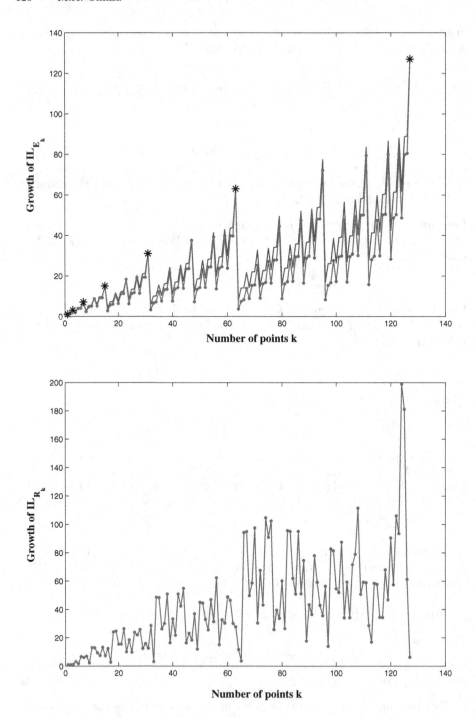

Fig. 1. Exact Lebesgue constants associated to the k-sections of the Leja sequence E and the assciated $\Re$-Leja sequence R for $k = 1, 3, \ldots, 129$ (Color figure online).

Corollary 4.3. *Let R be an $\mathfrak{R}$-Leja sequence in $[-1,1]$. The norms of the difference operators satisfy, $\mathbb{D}_0(R) = 1$ and for $2^n \leq k < 2^{n+1}$*

$$\mathbb{D}_k(R) \leq 2^{\sigma_1(k)} 2^n \tag{4.8}$$

The previous estimates can be used in order to provide estimates for $\mathbb{L}_{R_k}$ that can be sharper than (3.12). We have $\Delta_k = I_k - I_{k-1}$, therefore

$$|\mathbb{L}_{R_{k+1}} - \mathbb{L}_{R_k}| \leq \mathbb{D}_k(R), \quad k \geq 1. \tag{4.9}$$

In particular, the estimate in the previous corollary combined with the sharp bound (3.8) implies that for the value $k = 2^n$, we get

$$\mathbb{L}_{R_k} \leq 1 + \frac{2}{\pi} \log(2^n) + 2^{n+1} \leq 3k \tag{4.10}$$

This shows that in the case $k = 2^n$ which corresponds to R_k being the set of Gauss-Lobatto abscissas (3.7) missing one abscissas and for which $\mathbb{L}_{R_k} \geq k$, the previous bound is satisfactory. This also confirm that the estimates (3.12) is indeed pessimistic in this case, see the inequality (3.13). This added to the observed growth of $\mathbb{L}_{R_k}$ for values $k \leq 128$, Fig. 1, suggests that the bound

$$\mathbb{L}_{R_k} \leq 3k, \quad k \geq 1, \tag{4.11}$$

might be valid for any $\mathfrak{R}$-Leja sequence R. We conjecture its validity.

In Fig. 1, we also represent for the values $k \leq 128$, the growth of the Lebesgue constant $\mathbb{L}_{E_k}$ (in blue) and the estimate $\sqrt{k(2^{\sigma_1(k)} - 1)}$ (in red) which multiplied by 3 bounds $\mathbb{L}_{E_k}$, see (2.17). We observe that the regular patterns in the graph of $k \mapsto \mathbb{L}_{E_k}$, which reveals the particular role of divisibility by powers of 2 in k, is caught by the estimate. The worst values of $\mathbb{L}_{E_k}$ appear for the values $k = 2^n - 1$ for which it was proved in [3] that $\mathbb{L}_{E_k} = k$ and which is also equal to $\sqrt{k(2^{\sigma_1(k)} - 1)}$ since $\sigma_1(k) = n$.

References

1. Bialas-Ciez, L., Calvi, J.P.: Pseudo Leja sequences. Ann. Mat. Pura Appl. **191**, 53–75 (2012)
2. Bernstein, S.: Sûr la limitation des valeurs d'un polynôme $P_n(x)$ de degré n sûr tout un segment par ses valeurs en $n+1$ points du segment. Isv. Akad. Nauk SSSR **7**, 1025–1050 (1931)
3. Calvi, J.P., Phung, V.M.: On the Lebesgue constant of Leja sequences for the unit disk and its applications to multivariate interpolation. J. Approximation Theor. **163**(5), 608–622 (2011)
4. Calvi, J.P., Phung, V.M.: Lagrange interpolation at real projections of Leja sequences for the unit disk. Proc. Am. Math. Soc. **140**(12), 4271–4284 (2012)
5. Chkifa, A., Cohen, A., Passaggia, P.Y., Peter, J.: A comparative study between kriging and adaptive sparse tensor-product methods for multi-dimensional approximation problems in aerodynamics design. ESAIM Proc. Surv. **48**, 248–261 (2015)

6. Chkifa, M.A.: On the Lebesgue constant of Leja sequences for the complex unit disk and of their real projection. J. Approximation Theor. **166**, 176–200 (2013)
7. Cohen, A., Chkifa, M.A.: On the stability of polynomial interpolation using hierarchical sampling. In: Sampling Theory, a Renaissance, Birkhaeuser (2015, to appear)
8. Chkifa, M.A.: Méthodes polynomiales parcimonieuses en grande dimension: application aux EDP paramétriques. Ph.D. thesis, Laboratoire Jacques Louis Lions (2014)
9. Chkifa, A., Cohen, A., Schwab, C.: High-dimensional adaptive sparse polynomial interpolation and applications to parametric PDEs. Found. Comput. Math. **14**, 601–633 (2013)
10. Bungartz, H.J., Griebel, M.: Sparse grids. Acta Numerica **13**, 147–269 (2004)
11. Gunzburger, M.D., Webster, C.G., Guannan, Z.: Stochastic finite element methods for partial differential equations with random input data. Acta Numerica **23**, 521–650 (2014)
12. Davis, P.J.: Interpolation and Approximation. Blaisdell Publishing Company, New York (1963)
13. Dzjadyk, V.K., Ivanov, V.V.: On asymptotics and estimates for the uniform norms of the Lagrange interpolation polynomials corresponding to the Chebyshev nodal points. Anal. Math. **9–11**, 85–97 (1983)
14. Irigoyen, A.: A uniform bound for the Lagrange polynomials of Leja points for the unit disk. http://arxiv.org/pdf/1411.5527.pdf
15. Devore, R.A., Lorentz, G.G.: Constructive Approximation. Springer, New York (1993)

A Curvature Smooth Lofting Scheme for Singular Point Treatments

Elaine Cohen[1]([✉]), Robert Haimes[2], and Richard Riesenfeld[1]

[1] School of Computing, University of Utah, Salt Lake City, UT 84112, USA
cohen@cs.utah.edu
[2] Department of Aeronautics and Astronautics,
Massachusetts Institute of Technology, Cambridge, Massachusetts 02139, USA

Abstract. This paper presents a new *end condition scheme* for lofting through singular points, one that is easily specified and can generate shapes that are appropriate for lofting to the noses of subsonic/transonic aerodynamic and some hydrodynamic crafts. A degeneracy in the tensor product parameterization is typical in representations of spheres and ellipsoids, or topological equivalents. In that case a grid of data typically sets a whole row to a single value, so it becomes difficult and cumbersome to specify shape characteristics in different directions emanating from that point. Some standardly used *end conditions* result in shapes that are C^0 at the singular point, an undesirable outcome in nose regions of subsonic aircraft, so the straightforward nose-to-tail lofts are intractable for these vehicles. The proposed method overcomes this problem (See Figs. 1 and 2 for examples).

Keywords: Tensor product lofting · Singular point

1 Introduction

Aircraft design methodologies have evolved with the advent of improved design tools. This growth commenced from an "Edisonian" era, where experimentation was required due to little known theory, and subsequently moved to an era of analysis-and-test, followed by physics-based modeling, simulation-based design and multidisciplinary design optimization (MDO). A broadly accepted design process spans various design methodologies in use today. This general process is principally divided into three *phases*, denoted *Conceptual Design*, *Preliminary Design*, and *Detailed Design*, respectively, where the amount of quantitative geometry information defining an aircraft concept increases from beginning to end. (Other labels are timetimes used, but they are simply semantic variations for the same phenomenon.) As the level of fidelity in the geometry improves, the selection of design and analysis methods must comply with corresponding fidelity. Typically the underlying geometric representation changes from one design phase to the next, therefore a gap in analysis methodology arises across design phase transitions due to inconsistent geometry definitions.

© Springer International Publishing Switzerland 2015
J.-D. Boissonnat et al. (Eds.): Curves and Surfaces 2014, LNCS 9213, pp. 129–150, 2015.
DOI: 10.1007/978-3-319-22804-4_10

Fig. 1. Accurate renderings of final representation of helicopter. The upper left is a view from the diagonal front of the model, while the lower right is a closeup of the nose region. The elliptically specified radii of normal curvature at the data points are attained within $\epsilon = 10^{-6}$ of relative error.

Thus, the intended seamless design process encompassing the entire lifecycle development of an aircraft actually reduces to a set of disjoint modular design phases, wherein independent and disparate design methodologies are defined within the scope of a given level of analysis and geometry fidelity.

By expanding upon the technology found within the ESP [3] environment, a methodology is possible that unifies low- and high-fidelity representations of aircraft geometry into a single model definition. The approach combines elements from Constructive Solid Modeling (CSG) techniques and direct (*bottom-up*) construction approaches, resulting in a model definition suitably parameterized for both low- and high-fidelity design spaces. This yields a multi-fidelity geometry representation where low-fidelity design variables are mapped to the dimensioning of sectional primitive geometry in the 3D model and subsequently high-fidelity surface or body meshes can be extracted from that same model. Moreover, the model provides a multidisciplinary geometry representation by adding the features or parts pertaining to subsystems formed from discipline-specific geometric requirements. This multi-fidelity, multidisciplinary geometry approach can result in models that exhibit properties of malleability, robustness, and flexibility for use in a design optimization setting. Once such models are developed, they enable the implementation of multi-fidelity design frameworks, where a consistent geometric

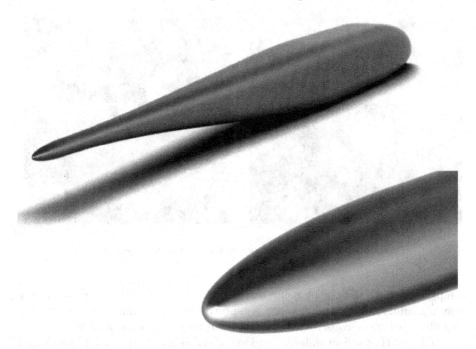

Fig. 2. The upper left is a view from the diagonal rear of the model, while the lower right is a closeup of the tail region. The tail point was specified to have spherical radius of normal curvature at all the data points. It was attained within $\epsilon = 10^{-6}$ of relative error.

representation is critical for invoking both low- and high-fidelity analysis models in a seamless manner throughout the design process. Having a single unified model with the ability to generate geometry commensurate with the analysis at-hand allows introducing high-fidelity analysis earlier in the design process. This can help identify and remediate problems that would require engineering workarounds if discovered later in the process.

In order to realize this lofty goal it is imperative to have construction primitives that are easy and simple to define and can generate the desired shapes with a minimal suite of operations. *Lofting* is a commonly used operation to generate wing, tail and fuselage component representations of aircraft. Both the beginning (at the nose) and the aft portion (the tail) come to a point unless traditional lofting is modified to avoid it. When using a typical lofting scheme that terminates at a singular point, a conical shape is generated with a singular point at the apex. But this is not what is usually seen in subsonic and transonic transport vehicles (as is depicted in Fig. 3) where the shape at the noses appear more elliptical.

If the construction of the fuselage uses a traditional loft, it is usually terminated at a section before the nose, leaving the fuselage flat at the end. Then the nose region is generated by a separate set of construction primitives. Finally, the

Fig. 3. Examples of noses of current transsonic transports.

complete fuselage is built by using the union operator. It should be noted that from a manufacturing/maintenance perspective this procedure, in fact, reflects how the fuselage is constructed (separate surfaces can be seen in both examples of Fig. 3 that reflect different regions of the skin). This is where radar and other avionics equipment requiring easy access are housed. But from a design perspective this is an unnecessary and added complication. This existing overall procedure has the following specific issues:

- Complex construction. It is not clear what the best suite of operations that would result in the appropriate nose portion.
- Continuity. Unless carefully constructed, only C^0 smoothness occurs at the join of the fuselage and nose regions. Aerodynamically C^2 is indicated for this region.

Therefore, without a lofting procedure that can produce these desired shapes at the nose and aft terminal points, a special construction scheme would be required to fit a surface to the nose region that has the desired characteristics of producing the appropriate nose shape, and, additionally, can *feather* into a loft with good continuity. This paper presents a new loft *end condition scheme* so that the loft can be applied nose-to-tail, as shown in Figs. 1 and 2, without any additional complex construction operations.

2   Statement of Problem

The problem we consider is specified by a transfinite interpolation. That is, given a sequence of non-intersecting parametric cubic B-spline curves in 3-D,

$$\{\gamma_i(u) = (x_i(u), y_i(u), z_i(u))\}_{i=0}^n;$$

all sharing the same knot vector,

$$\boldsymbol{\mu} = \{\mu_k\}_{k=0}^{M+4},$$

we construct the supporting surface. While the solution approach is valid for floating and periodic cases, we develop the algorithms for the open knot vector case. In this case the knot vector has unique ascending values $\{u_\ell\}_{\ell=0}^{M-2}$. Although the problem of creating a tensor product B-spline surface $\boldsymbol{\sigma}(u, v)$, such that $\boldsymbol{\sigma}(u, v_i) = \boldsymbol{\gamma}_i(u)$, $j = 0, \ldots, n$ for a given ascending parameter $\boldsymbol{v} = \{v_i\}_{i=0}^n$ has many solutions, our methodology deals with the case for which:

- $\boldsymbol{\gamma}_0(u) = \boldsymbol{p}$, for all u, i.e., the curve degenerates to the point $\boldsymbol{p}$.
- The surface $\boldsymbol{\sigma}$ should be G^1 at $\boldsymbol{p}$, and the normal curvature along each of the curves $\boldsymbol{\sigma}(u_j, v)$ is a prescribed (possibly different) value and is normal curvature continuous around $\boldsymbol{p}$.

Since we will need to specify tangent vectors along each curve to meet the interpolation requirements, we use complete cubic spline interpolation methodology, and generalize it appropriately.

The generated approach is appropriate when the curves are either closed (look periodic) or not, or even when the initial data is just a grid of data points, as we will derive.

3   Related Work

The problem that this paper deals with arises when the goal is to create a lofted surface either from a logically rectangular grid of points (whose first and/or last rows of data degenerate to a single point), or from a sequence of curves that are used to build a lofted surface where, the beginning or end coalesce to a single point. The desire is for the interpolating surface, unlike traditional (conical-like degenerate treatments), to embody *rounded* shapes at the degenerate point.

Cohen proposed a method in [1] to ensure that the final lofted surface is tangent plane continuous at the degenerate point. Higher order smoothness was not dealt with.

It has been noted that subdivision surfaces can behave in undesirable ways around extraordinary points, particularly those with many adjacent facets. In [7] it is shown how to combine Catmull-Clark subdivision and polar subdivision surfaces for a better result around the extraordinary point. In a related approach to creating a smooth surface through an extraordinary point [5], it is assumed that position, first, and second derivatives are known along the *tensor border regions* of the cap, and a guide surface of at least degree 2 for the cap is given, and develops four approaches to determining a curvature continuous surface at the singular point. The one approach that does not require further subdivision of triangle fan regions around a singular point results in cap of surface patches that are degree 6×5, and meet the given piecewise bicubic surrounding surfaces with C^2 continuity. The other approaches also use guide surfaces and tensor borders. One creates triangular polynomial patches of degree 9 with 3 times the number of patche, while another requires further patch subdivision to attain a mixture of degree 6 triangles and degree 6×5 quadrilateral patches. Myles *et al.* [8] deals with the problem of creating a smooth B-spline representation in the triangle regions around a singular point that has C^2 smoothness with the uniform floating bicubic tensor product B-spline surface derived on the regular part of the mesh. The surface created at the fan is a uniform tensor product that is degree 6 in one direction and degree 3 in the other direction,

While intriguing, these approaches assume several conditions that make it not applicable to the problem treated herein. In particular, the new approach articulated in this paper assumes no existing mesh. It does require a that the section curves be interpolated, but there are no known surfaces through such curves. Also a single representation that can be designed and modified simultaneously around the degenerate point and across the design is necessary. Moreover, our approach is aimed at allowing the singular point to exhibit elliptical or even more general normal curvature behavior around the degenerate point.

Other approaches tend to deal with trying to create smooth surfaces at extraordinary points where multiple tensor product surfaces meet. In that case, the surfaces are not degenerate, but attaining smoothness between patches, in turn, becomes the problem. A recent example of that technique is found in [6].

We have found little else that has been aimed at solving this or related problems, either commercially or in academic research.

4   Mathematical Background

To solve the problem posed in Sect. 2, we first discuss curve complete spline interpolation and then the general surface complete spline interpolation problem. Then we recall the definition of the *normal curvature of a curve in a surface*. Finally, this section presents the steps in the algorithm that lead to a solution.

4.1   Curve Complete Spline Interpolation

Given a sequence of data points $\{d_i\}_{i=0}^n$, and a corresponding sequence of increasing parametric values $\{v_i\}_{i=0}^n$, the problem is to create a knot vector ν and a corresponding cubic B-spline curve $c(v)$ such that the curve interpolates the data when the curve parameter takes on the knot values, respectively.

The knot vector $\boldsymbol{\nu}$ is created by setting

$$\nu_i = \begin{cases} v_0 & i = 0, 1, 2, 3 \\ v_{i-3} & 4 \le i \le n+2 \\ v_n & i = n+3, n+4, n+5, n+6. \end{cases} \tag{1}$$

This gives rise to $n + 3$ degrees of freedom with only $n + 1$ data locations. The last two degrees of freedom are specified by setting the tangents at v_0 and v_n.

This results in $(n + 3)$ linearly independent equations in $(n + 3)$ unknowns. Further, the matrix form is tridiagonal and readily solvable [2].

Now we generalize to the case in which the data is a sequence of non-intersecting parametric cubic B-spline curves

$$\{\gamma_i(u) = (x_i(u), y_i(u), z_i(u))\}_{i=0}^n$$

all with the same knot vector

$$\mu = \{\mu_k\}_{k=0}^{M+4}$$

so that

$$\gamma_i(u) = \sum_{j=0}^M g_{i,j} B_{j,\mu}(u).$$

This time the problem is to create a bicubic tensor product B-spline surface σ so that $\sigma(u, v_i) = \gamma_i(u)$. Once again it is required to use the knot vector ν, so the surface has the representation $\sigma(u, v) = \sum_i \sum_j c_{i,j} B_{j,\mu}(u) N_{i,\nu}(v)$. It is deliberately chosen to use B for the B-splines over the knot vector μ and N for the B-splines over the knot vector ν. Thus, without loss of generality, we henceforth we leave implicit the subscript designating the knot vector over which the bases are computed.

Since

$$\sigma(u, v_\ell) = \sum_i \sum_j c_{i,j} B_j(u) N_i(v_\ell)$$

$$= \sum_j \left(\sum_i c_{i,j} N_i(v_\ell) \right) B_j(u)$$

$$= \gamma_\ell(u),$$

$$g_{\ell,j} = \begin{cases} c_{0,j}, & \ell = 0, \\ \sum_{i=\ell}^{\ell+2} c_{i,j} N_i(v_\ell), & 0 < \ell < n, \\ c_{n+2,j}, & \ell = n. \end{cases}$$

Now it is required to specify the tangent curves at v_0 and v_n. Note that in terms of the final surface, formally we have the tangent curve at v_0 is

$$\frac{\partial \sigma}{\partial v}(u, v_0) = \frac{3}{\nu_4 - \nu_1} \sum_{m=0}^M (c_{1,m} - c_{0,m}) B_m(u). \tag{2}$$

A similar equation is valid at v_n. If $\gamma_n(u)$ is also degenerate to a point, the solution approach can be applied at that end as well. In this discussion, we derive the tangent curve at only v_0, the nose.

At this point, we still need to determine a tangent curve at v_0 such that, when the spline interpolant is generated, the normal curvatures at the knot values of μ will have the input specified values. We see there are $n + 1$ data curves but

$n + 3$ curves of coefficients to be specified. The extra degrees of freedom at the respective ends provide the opportunity to set them in a way that yields the goal radii of normal curvature. Our approach derives the tangent directions and magnitudes with an iterative approach, as we shall show.

Note that instead of having $n + 1$ data B-spline curves for input, one could have a grid of data, one data point for each distinct value of μ as well as one for each distinct value of ν. If the data is periodic as a function of u, then a corresponding periodic knot vector is appropriate. Otherwise, either an open knot vector is used so additional tangent values for the beginning and ending values of μ must be generated, if they are not specified as part of the data, or a different interpolating technique for each cross section curve of data must be used.

4.2   Normal Curvature

Suppose $\sigma(u, v)$ is a C^2 parametric surface with image $S \subset \Re^3$. Let $(\bar{u}, \bar{v})$ be a point in the domain of σ and for $t \in (-a, a)$. Suppose $\phi(t) = (u(t), v(t))$ is a C^2 curve in the domain of σ such that $\phi(0) = (\bar{u}, \bar{v})$. Then $\alpha(t) = \sigma(\phi(t)) = \sigma(u(t), v(t))$ is a C^2 curve in S.

Recall the definition of the curvature $\kappa(t)$ of a curve α at a parameter t.

$$\kappa(t) = \frac{\|\alpha'(t) \times \alpha''(t)\|}{\|\alpha'(t)\|^3}$$

The normal to the curve α is defined as

$$N = \frac{\|\alpha'\|}{\|\alpha' \times \alpha''\|}\alpha'' - \frac{\alpha' \cdot \alpha''}{\|\alpha'\|\|\alpha' \times \alpha''\|}\alpha'.$$

Then the *curvature vector* for α is defined as

$$\kappa(t)N(t) = \frac{1}{\|\alpha'(t)\|^2}\alpha''(t) - \frac{\alpha'(t) \cdot \alpha''(t)}{\|\alpha'(t)\|^4}\alpha'(t)$$

Suppose the surface $\sigma(u, v)$ has surface normal $n(u, v)$. The *normal curvature of α in σ* is defined as the inner product of the curvature vector of the curve with the surface normal,

$$\kappa(t)N(t) \cdot n(\phi(t)) = \frac{1}{\|\alpha'(t)\|^2}\alpha''(t) \cdot n(\phi(t)), \tag{3}$$

since the curve tangent $(\alpha'(t))$ is orthogonal to the surface normal $n(\phi(t))$.

4.3   Normal Curvature at Singular Point

Given a bicubic parametric B-spline surface $\sigma = \sum_j \sum_i c_{i,j} B_j(u) N_i(v)$ with knot vectors μ and ν in u and v, respectively, as above, if $\sigma(u, v_0) = p$ for all u over which σ is defined, then the surface has a singularity at that boundary.

Since the parameterization is not regular at that point, one cannot compute the surface normal curvature and the surface principal curvatures. Instead we can compute one sided normal curvatures of curves at p if we know the proposed surface normal at p. Let

$$a = \{a_i\} = \{\mu_3, (\mu_3 + \mu_4)/2, \mu_4, \mu_5, \ldots, u_{M-4}, u_M, (\mu_M + \mu_{M-+1})/2, \mu_{M+1}\}$$
$$= \{u_0, (u_0 + u_1)/2, u_1, u_2, \ldots, u_{M-4}, u_{M-3}, (u_{M-3} + u_{M-2})/2, u_{M-2}\}. \tag{4}$$

This vector contains $M+1$ different values of u and includes all the distinct knot values. It will be used to generate the tangent vector directions and magnitudes for specifying interpolant.

We consider isoparametric curves $\beta_\ell(v) = \sigma(a_\ell, v), \ell = 0, 1, \ldots, M$. If the surface is smooth at the singular point with normal vector n and κ_ℓ^n denotes the goal normal curvature of $\beta_\ell(v_0)$, then we first solve for both $\beta_\ell''(v_0) \cdot n$ and $\|\beta_\ell'(v_0)\|^2$ using one-sided derivatives.

$$\frac{\partial \sigma}{\partial v}(u, v_0) = \frac{3}{\nu_4 - \nu_1} \sum_{m=0}^{M} (c_{1,m} - c_{0,m}) B_m(u) \tag{5}$$

Since the surface is smooth at (u, v_0) for all u, and n is the normal to its tangent plane, $0 = \frac{\partial \sigma}{\partial v}(u, v_0) \cdot n$,

$$\frac{\partial^2 \sigma}{\partial v^2}(u, v_0) \cdot n = \left(6 \sum_{m=0}^{M} \frac{\frac{c_{2,m} - c_{1,m}}{\nu_5 - \nu_2} - \frac{c_{1,m} - c_{0,m}}{\nu_4 - \nu_1}}{\nu_4 - \nu_2} B_m(u) \right) \cdot n$$

$$= \frac{6}{\nu_4 - \nu_2} \left(\sum_{m=0}^{M} \frac{c_{2,m} - c_{1,m}}{\nu_5 - \nu_2} B_m(u) - \sum_{m=0}^{M} \frac{c_{1,m} - c_{0,m}}{\nu_4 - \nu_1} B_m(u) \right) \cdot n$$

$$= \frac{6}{(\nu_4 - \nu_2)(\nu_5 - \nu_2)} \sum_{m=0}^{M} (c_{2,m} - c_{1,m}) \cdot n \, B_m(u). \tag{6}$$

Substitution into Eqs. 3, 5, and 6, results in

$$\beta_\ell'(v_0) = \begin{cases} \frac{3}{\nu_4 - \nu_1}(c_{1,0} - p) & \ell = 0 \\ \frac{3}{\nu_4 - \nu_1} \sum_{m=0}^{3}(c_{1,m} - p)B_m(a_1) & \ell = 1, \\ \frac{3}{\nu_4 - \nu_1} \sum_{m=\ell-1}^{\ell+1}(c_{1,m} - p)B_m(a_\ell) & \ell = 2, \ldots, M-2 \\ \frac{3}{\nu_4 - \nu_1} \sum_{m=M-3}^{M}(c_{1,m} - p)B_m(a_{M-1}) & \ell = M-1 \\ \frac{3}{\nu_4 - \nu_1}(c_{1,M} - p) & \ell = M \end{cases} \tag{7}$$

and

$$\beta_\ell''(v_0) \cdot n = \begin{cases} \frac{6}{(\nu_4 - \nu_2)(\nu_5 - \nu_2)}(c_{2,0} - c_{1,0}) \cdot n & \ell = 0 \\ \frac{6}{(\nu_4 - \nu_2)(\nu_5 - \nu_2)} \sum_{m=0}^{3}(c_{2,m} - c_{1,m}) \cdot n B_m(a_1) & \ell = 1, \\ \frac{6}{(\nu_4 - \nu_2)(\nu_5 - \nu_2)} \sum_{m=\ell-1}^{\ell+1}(c_{2,m} - c_{1,m}) \cdot n B_m(a_\ell) & \ell = 2, \ldots M-2 \\ \frac{6}{(\nu_4 - \nu_2)(\nu_5 - \nu_2)} \sum_{m=M-3}^{M}(c_{m,2} - c_{m,1}) \cdot n B_m(a_{M-1}) & \ell = M-1 \\ \frac{6}{(\nu_4 - \nu_2)(\nu_5 - \nu_2)}(c_{2,M} - c_{1,M}) \cdot n & \ell = M. \end{cases} \tag{8}$$

Thus, the normal curvature at v_0 at each parameter value in $\boldsymbol{a}$ is

$$
\kappa_\ell^n = \begin{cases}
\dfrac{6}{9}\dfrac{(\nu_4-\nu_1)^2}{(\nu_4-\nu_2)(\nu_5-\nu_2)}\dfrac{(\boldsymbol{c}_{2,0}-\boldsymbol{c}_{1,0})\cdot\boldsymbol{n}}{\|\boldsymbol{c}_{1,0}-\boldsymbol{p}\|^2} & \ell = 0 \\[2ex]
\dfrac{6}{9}\dfrac{(\nu_4-\nu_1)^2}{(\nu_4-\nu_2)(\nu_5-\nu_2)}\dfrac{\sum_{m=0}^{3}(\boldsymbol{c}_{2,m}-\boldsymbol{c}_{1,m})\cdot\boldsymbol{n}B_m(a_1)}{\|\sum_{m=0}^{3}(\boldsymbol{c}_{1,m}-\boldsymbol{p})B_m(a_1)\|^2}, & \ell = 1 \\[2ex]
\dfrac{6}{9}\dfrac{(\nu_4-\nu_1)^2}{(\nu_4-\nu_2)(\nu_5-\nu_2)}\dfrac{\sum_{m=\ell-1}^{\ell+1}(\boldsymbol{c}_{2,m}-\boldsymbol{c}_{1,m})\cdot\boldsymbol{n}B_m(a_\ell)}{\|\sum_{m=\ell-1}^{\ell+1}(\boldsymbol{c}_{1,m}-\boldsymbol{p})B_m(a_\ell)\|^2}, & \ell = 2,\ldots,M-2 \\[2ex]
\dfrac{6}{9}\dfrac{(\nu_4-\nu_1)^2}{(\nu_4-\nu_2)(\nu_5-\nu_2)}\dfrac{\sum_{m=M-3}^{M}(\boldsymbol{c}_{2,m}-\boldsymbol{c}_{1,m})\cdot\boldsymbol{n}B_m(a_{M-1})}{\|\sum_{m=M-3}^{M}(\boldsymbol{c}_{1,m}-\boldsymbol{p})B_m(a_{M-1})\|^2}, & \ell = M-1 \\[2ex]
\dfrac{6}{9}\dfrac{(\nu_4-\nu_1)^2}{(\nu_4-\nu_2)(\nu_5-\nu_2)}\dfrac{(\boldsymbol{c}_{2,M}-\boldsymbol{c}_{1,M})\cdot\boldsymbol{n}}{\|(\boldsymbol{c}_{1,M}-\boldsymbol{p})\|^2} & \ell = M.
\end{cases} \tag{9}
$$

If the normal curvatures have been specified, the coefficients $\boldsymbol{c}_{1,m}{}_{m=0}^{M}$, are the solutions to a nonlinear (nonrational) system of equations. The chosen approach is nonlinear and iterative.

5   Solution

Since the problem is nonlinear, our approach is to determine a set of tangent vector directions for each parameter value in $\boldsymbol{a}$ (see Eq. 4), and an initial set of corresponding magnitudes so that the interpolating surface's normal curvature in those directions approximates the specified goal normal curvatures. This is done by estimating the magnitude of the tangent vector that approximates the related radius of normal curvature. Then the next steps are to iterate the tangent magnitudes, keeping the directions and all the other surface coefficients fixed, until the resulting surface has the specified normal curvature to within a prespecified ϵ in each parameter in $\boldsymbol{a}$. At that time, since the surface no longer interpolates all the data curves, the current iteration of the computed tangent vector is used to interpolate the data once again with complete spline interpolation. The process is repeated until both the normal curvature along the prespecified curves in the interpolated lofted surface is within ϵ.

We hypothesize that the data and vector $\boldsymbol{a}$ is as in Sect. 4.3.

In addition to being valid for the standard cases, this approach is valid for more general situations that include,

- $\boldsymbol{p} = \boldsymbol{\gamma}_0(u)$, for all u. $\boldsymbol{p}$ is called the *nose point*.
- each curve in the dataset may be planar but need not be, although when $\boldsymbol{\gamma}_1$ and $\boldsymbol{\gamma}_2$ are *close to* planar the solution has better aesthetics.
- an axis called the *central axis* is almost orthogonal to $\boldsymbol{\gamma}_1$, and is one of the coordinate axes x, y, or z (in vector notation, $\boldsymbol{e}_1$, $\boldsymbol{e}_2$, $\boldsymbol{e}_3$).
- $\boldsymbol{\gamma}_1$ is star shaped with respect to the central axis.
- the nose point may be on the central axis, but may also be offset.
- the tangent plane to the nose point may be orthogonal to the central axis, but need not be. However, its normal, $\boldsymbol{n}$, must make a smaller angle with the central axis than with the other two coordinate axes.

Throughout this paper it has been assumed that there is a nose point, indicated as $\boldsymbol{\gamma}_0$ and n other parametric cubic B-spline curves.

The algorithm requires as input:

1. $(n+1)$ cubic parametric B-spline curves $\{\gamma_i(u)\}_{i=0}^n$, all defined with respect to the same knot vector $\boldsymbol{\mu} = \{\mu_k\}_{k=0}^{M+4}$ having distinct values $\boldsymbol{u} = \{u_k\}_{k=0}^{M-2}$, such that $\gamma_0(u) = \boldsymbol{p}$, for all u.
2. An increasing sequence $\boldsymbol{v} = \{v_i\}_{i=0}^n$, that represents distinct parametric values for each γ_i.
3. An approximation error bound ϵ.
4. Coordinate axes of the tangent plane of the resulting surface at the nose point, as well as the desired normal radii of normal curvature in those directions of the resulting surface at $\boldsymbol{p}$.

Step 1. Find the tangent directions at singular point along prespecified isoparametric curves.

Step 2. Determine goal radii of normal curvature in each of the tangent directions.

Step 3. Estimate initial tangent vector magnitudes at $\boldsymbol{p}$ along directions found in Step 1.

Step 4. Estimate tangent vectors at the tail (if required).

Step 5. Iterate on nose tangent vector curve until the computation of radius of normal curvature gives the correct value to within ϵ.

Step 6. If the radius of normal curvature curve is prescribed for the tail, iterate on that also.

Step 7. Using given data curves and iterated tangent vector curve(s), solve for the complete spline interpolated surface.

Step 8. Check the values for the radius of normal curvature at the singular point along each of the prespecified isoparametric curves.
If all the values are not within ϵ, then
    Using the current interpolated surface, return to Step 3,
else
    Return

5.1 Find the Tangent Directions

In order that the vectors $\boldsymbol{c}_{1,j}$ be reasonably oriented with respect to the other coefficients of the curve β_j, we narrow its specification to be along a line in the tangent plane at the singularity defined by other coefficients of β_j. There are many possible approaches to specifying the tangent direction for each $\boldsymbol{\beta}_j(v)$ at v_0. Each has its relative strengths and weaknesses.

For example, since $\boldsymbol{\beta}_j(v_i) = \boldsymbol{\sigma}(a_j, v_i) = \gamma_i(u_{j-1})$, for $j = 2, \ldots, M - 2$, one could try interpolating with a quadratic $i = 0, 1, 2$ and projecting its tangent at v_0 onto the tangent plane. This approach sometimes gives good results, but we have had cases in test datasets where the results with this method produce poorly formed surfaces (including flipped tangents and cusps). Using higher degree interpolation does not improve the results. This occurs in all of the generalized cases. Another possibility would be to project the directions of the lines joining $\boldsymbol{p}$ to $\gamma_1(a_j)$ onto the nose tangent plane, but that also fails in most of the generalized cases.

Therefore, we have adopted a straightforward approach that has worked for the pathological cases tested. This approach requires most of the constraints,

- γ_1 is almost planar and almost orthogonal with respect to the central axis.
- n makes a smaller angle with the central axis than with the other two coordinate axes.
- γ_1 is star-shaped with respect to the central axis.

In case another approach can provide better results with fewer constraints, it can be adopted without changing the rest of the methodology.

We ensure that the nose plane normal n points in towards the data. Next, the central axis is found. Since the coordinates of n correspond to the cosines of n with each coordinate axis, the coordinate with the largest value indicates the axis that makes the smallest angle. Suppose that coordinate is s. So e_s will be the central axis.

Next, find the unique quaternion that represents the rotation that takes e_s to n, and form the corresponding 3×3 rotation matrix. Call that matrix R, so $Re_s = n$. This will be used to rotate the tangent vector directions into the tangent plane.

Let $\hat{b}_j = \gamma_1(a_j)$, for $j = 0, 1, \ldots, M$. Now, $b_j = \hat{b}_j - \hat{b}_j \cdot e_s \, \hat{b}_j$, does an orthographic projection of $\hat{b}_j$ onto the coordinate plane perpendicular to the central axis, and $\delta_j = \frac{b_j}{\|b_j\|}$ provides a unit vector in that direction. Finally, $d_j = R\delta_j$, for all j, rotates the vectors into the tangent plane, and the vectors $\{d_j\}_{j=0}^M$ will be used as the nose plane tangent vector directions.

5.2   Determine Radii of Normal Curvature in Tangent Directions

This approach enables the radii of the normal curvatures at p along the β_j curves to range from being all the same to having more generalized curve shapes, including elliptical shape behaviors. For elliptical shapes, we decide how to compute the appropriate radius of curvature for each d_j by considering an ellipsoid of the form $\psi(u, v) = (x(u, v), y(u, v), z(u, v))$ where for $a > 0$ and $b > 0$,

$$x(u, v) = -a \cos u \sin v$$
$$y(u, v) = -b \sin u \sin v$$
$$z(u, v) = \cos v.$$

The non-unit normal to ψ is $\eta(u, v) = \frac{\partial \psi}{\partial u} \times \frac{\partial \psi}{\partial v}$, which is not defined in this parameterization at $v = \pi$, since $\frac{\partial \psi}{\partial u}(u, \pi) = (0, 0, 0)$. That is, the parameterization is not *regular* at the pole. Define the normal to be $(0, 0, 1)$ at the pole (pointing inward) and consider curves defined for arbitrary but fixed u, let $\phi_u(v) = \sigma(u, v)$. Therefore $\phi'_u(v) = (-a \cos u \cos v, -b \sin u \cos v, -\sin v)$. Then $\phi'_\pi(0) = (a, 0, 0)$ and $\phi'_{\pi/2}(\pi) = (0, b, 0)$ form an orthogonal basis for the tangent plane, once they have become unit vectors they are $\bar{e}_1 = (1,0,0)$ and $\bar{e}_2 = (0,1,0)$. Its normal curvature at $v = \pi$ is

$$
\begin{aligned}
k_u^n &= \frac{\phi_u''(\pi) \cdot (0,0,1)}{\|\phi_u'(\pi)\|^2} \\
&= \frac{-\cos \pi}{\|(-a \cos u \cos \pi, -b \sin u \cos \pi, -\sin \pi)\|^2} \\
&= \frac{1}{a^2 \cos^2 u + b^2 \sin^2 u}
\end{aligned}
$$

Thus, the radius of normal curvature in the $(\cos u, \sin u)$ direction around the degenerate point $v = \pi$ is

$$
a^2 \cos^2 u + b^2 \sin^2 u
$$

with the value a^2 in the $\bar{e}_1$ axis and value b^2 in the $\bar{e}_2$ direction. Hence, given pre-specified directions for $\bar{e}_1$ and $\bar{e}_2$ in the nose tangent plane, such that $\bar{e}_1 \times \bar{e}_2 = n$, and pre-specified values for the normal radii of curvature a and b, in those directions, it is possible to determine desired radii of curvature in directions d_j, for all j. For example, if $d_j = \cos \theta_j \bar{e}_1 + \sin \theta_j \bar{e}_2$, then $r_j = a^2 \cos^2 \theta_j + b^2 \sin^2 \theta_j$.

5.3 Estimating Initial Tangent Magnitudes at v_0

Suppose

$$
q_j(t) = p + t d_j \qquad j = 0, 1, \ldots, M.
$$

Now suppose

$$
\Delta_j(v) = p N_0(v) + q_j(t) N_1(v) + \gamma_1(a_j) N_2(v),
$$

where $\{N_i\}_{i=0}^2$ are the first 3 B-splines in the v direction. These are the only terms of curves in v that have any effect on the first and second derivatives at v_0.

We find the first estimate of the normal curvature on each β_j curve at v_0, by estimating the curve curvature of each Δ_j at v_0, and calling the value κ_j. Experimentally we have found that even though this simple approach does converge quickly, it does not result in good values for the difficult cases (See Sect. 6).

$$
\kappa_j = \frac{\left\| 3 \frac{p + t_j d_j - p}{\nu_4 - \nu_1} \times 6 \frac{\frac{\gamma_1(a_j) - (p + t_j d_j)}{\nu_5 - \nu_2} - \frac{p + t_j d_j - p}{\nu_4 - \nu_1}}{\nu_4 - \nu_2} \right\|}{\left\| 3 \frac{p + t_j d_j - p}{\nu_4 - \nu_1} \right\|^3} \tag{10}
$$

$$
= \frac{2}{3} \frac{\left\| \frac{t_j d_j}{\nu_4 - \nu_1} \times \frac{\frac{\gamma_1(a_j) - (p + t_j d_j)}{\nu_5 - \nu_2} - \frac{t_j d_j}{\nu_4 - \nu_1}}{\nu_4 - \nu_2} \right\|}{\left\| \frac{t_j d_j}{\nu_4 - \nu_1} \right\|^3} \tag{11}
$$

$$
= \frac{2(\nu_4 - \nu_1)^2}{3(\nu_5 - \nu_2)(\nu_4 - \nu_2)} \frac{\| t_j d_j \times (\gamma_1(a_j) - p) \|}{\| t_j d_j \|^3} \tag{12}
$$

$$
= \frac{1}{t_j^2} \frac{2(\nu_4 - \nu_1)^2}{3(\nu_5 - \nu_2)(\nu_4 - \nu_2)} \| d_j \times (\gamma_1(a_j) - p) \| \tag{13}
$$

Since the radius of normal curvature is specified (see Sect. 5.2), we approximate the curve radius of curvature with the specified radius of normal curvature, r_j. Set $\kappa_j = k_j^n = 1/r_j$ and solve for t_j to get

$$t_j = \sqrt{r_j \frac{2\,(\nu_4 - \nu_1)^2}{3\,(\nu_4 - \nu_2)(\nu_5 - \nu_2)} \|\mathbf{d}_j \times (\gamma_1(a_j) - \mathbf{p})\|} \tag{14}$$

So the initial tangent vector value for each β_j is

$$\frac{3}{\nu_4 - \nu_1} t_j \mathbf{d}_j$$

where t_j is given by Eq. 14.

5.4   Iterate to Improve Radii of Normal Curvatures

At this stage the method has all the position and tangent curve information needed to solve a lofting problem. We solve in order to create an interpolatory B-spline surface that is G^1. It is solved as a sequence of $M+1$ curve interpolation problems. Call the resulting control mesh $\mathbf{c}$.

Next, the method develops an iteration technique based on Eq. 9 for which the tangent vectors estimated in Sect. 5.3 are used as initial values. In this section the accurate equations for the normal curvatures at $\beta_\ell(v_0)$ are used. Notice that although the isoparametric curves that correspond to knots have only 3 items in the summation for i, at a_1 and a_{M-1} these two isoparametric curves have 4.

In this step of the iteration, current values for t_j are used, except for the one that will be improved. To distinguish the next magnitude of the tangent vector from the current one, t_ℓ, it will be named τ_ℓ, so the new control point for the next stage will be $\mathbf{p} + \tau_\ell \mathbf{d}_\ell$.

First, consider the normal curvatures for β_ℓ, $\ell = 0, \ldots, M$. By Eq. 8,

$$G_\ell = \beta_\ell''(v_0) \cdot \mathbf{n} = \begin{cases} \frac{6}{(\nu_4 - \nu_2)(\nu_5 - \nu_2)}(\mathbf{c}_{2,0} - \mathbf{c}_{1,0}) \cdot \mathbf{n} B_0(a_0) & \ell = 0, \\ \frac{6}{(\nu_4 - \nu_2)(\nu_5 - \nu_2)} \sum_{m=0}^{3}(\mathbf{c}_{2,m} - \mathbf{c}_{1,m}) \cdot \mathbf{n} B_m(a_1) & \ell = 1, \\ \frac{6}{(\nu_4 - \nu_2)(\nu_5 - \nu_2)} \sum_{m=\ell-1}^{\ell+1}(\mathbf{c}_{2,m} - \mathbf{c}_{1,m}) \cdot \mathbf{n} B_m(a_\ell) & \ell = 2, \ldots M - 2, \\ \frac{6}{(\nu_4 - \nu_2)(\nu_5 - \nu_2)} \sum_{m=M-3}^{M}(\mathbf{c}_{2,m} - \mathbf{c}_{1,m}) \cdot \mathbf{n} B_m(a_{M-1}) & \ell = M - 1, \\ \frac{6}{(\nu_4 - \nu_2)(\nu_5 - \nu_2)}(\mathbf{c}_{2,M} - \mathbf{c}_{1,M}) \cdot \mathbf{n} B_M(a_M) & \ell = M. \end{cases}$$

Notice that G_ℓ has no dependence on the magnitude of the tangent vector at the nose. Suppose

$$\hat{\ell} = \begin{cases} \{0, 2, 3\} & \ell = 1 \\ \{\ell - 1, \ell + 1\}, & \ell = 2, \ldots, M - 2 \\ \{M - 3, M - 2, M\}, & \ell = M - 1 \end{cases}$$

From Eq. 5 for $\ell \in \{1, 2, \ldots, M-1\}$, we see,

$$\|\beta_\ell'(v_0)\|^2 = \frac{9}{(v_4 - v_1)^2} \begin{cases} \|\sum_{m=0}^{3}(c_{1,m} - p)B_m(a_1)\|^2 & \ell = 1, \\ \|\sum_{m=\ell-1}^{\ell+1}(c_{1,m} - p)B_m(a_\ell)\|^2 & \ell = 2, \ldots, M-2 \\ \|\sum_{m=M-3}^{M}(c_{1,m} - p)B_m(a_{M-1})\|^2 & \ell = M-1 \end{cases}$$

$$= \frac{9}{(v_4 - v_1)^2}\left(\|\tau_\ell d_\ell B_\ell(a_\ell) + \sum_{m \in \hat{\ell}} t_m d_m B_m(a_\ell)\|^2\right)$$

$$= \frac{9}{(v_4 - v_1)^2}B_\ell(a_\ell)^2\tau_\ell^2 + \left(2\frac{9}{(v_4 - v_1)^2}\sum_{m \in \hat{\ell}} t_m d_\ell \cdot d_m B_\ell(a_\ell)B_m(a_\ell)\right)\tau_\ell$$

$$\quad + \frac{9}{(v_4 - v_1)^2}\|\sum_{m \in \hat{\ell}} t_m d_m B_m(a_\ell)\|^2$$

$$= D_\ell \tau_\ell^2 + E_\ell \tau_\ell + F_\ell$$

So,

$$\frac{1}{\kappa_\ell^n} = r_\ell = \frac{D_\ell \tau_\ell^2 + E_\ell \tau_\ell + F_\ell}{G_\ell}$$

Hence, it is only necessary to solve, for its positive root,

$$D_\ell \tau_\ell^2 + E_\ell \tau_\ell + F_\ell - r_\ell G_\ell = 0$$

and

$$\tau_\ell = \frac{-E_\ell + \sqrt{E_\ell^2 - 4D_\ell(F_\ell - r_\ell G_\ell)}}{2D_\ell} \tag{15}$$

After the values τ_j, $j = 0, \ldots, M$ have been solved, set $t_j \leftarrow \tau_j$, $j = 0, \ldots, M$. Then, iterate again on the estimated normal curvature values. When those values are within tolerance, re-solve the interpolation problem. Then, compute the normal curvatures using the revised mesh for β_j, for all j. If all the normal curvatures are not within tolerance, repeat the update process until the solution is found.

5.5 Solve the Interpolation

Once the mesh has been modified to correct the values for the radii of normal curvature, it is necessary to resolve the interpolation problem. Remember, from Eq. 2, set $T_i = \frac{3}{v_4 - v_1}(t_i d_i)$ to represent the tangent curve needed for interpolation at the nose point.

$$T(u) = \sum_{i=0}^{M} T_i B_i(u)$$

5.6   Determine if Further Computation Is Necessary

Once the surface is interpolated, the values of $c_{2,j}$ will have changed, so it is necessary see how much the values of the normal curvature have changed. At this stage the normal curvatures at $v = v_0$ for each β_j are computed. If they are within acceptable tolerance, the computation is finished. If all are not within tolerance it is necessary to continue to iterate.

Currently there is no formal proof of convergence of the method. Empirically however, all experimental cases that have been tried, even the most unrealistically poor nose plane conditions and *wiggling* of the nearby section curves, have converged in relatively few iterations. We demonstrate several in Sect. 6.

6   Results

The proposed methodology has been tested on a multiple datasets. The idea has been to simulate both realistic data, and data that is so unrealistic that it is to the point of being a pathological stress test of the algorithm. The goal has been to examine under what constraints the technique converges and how quickly. For each dataset, we discuss its characteristics, show results for the Radius of Normal Curvature (RNC) at each of the parametric values of the data points in the cross-sectional direction, and then how many iterations of curvature improvement were performed in order to attain the RNC specified. We demonstrate that even for cases in which the initial estimated RNC differs from the goal by a substantial amount, within just a few iterations, the method converges.

The curves shown in Fig. 4 are typical of the data input. The blue curves are the input section data. Either the B-spline curves themselves are input, or data points are input with parameter values from which interpolated data curves are created. The gray dashed lines connect the knot values of the B-spline curves. They will be the curve data input to the nose treatment. In the data shown, the sections are planar and orthogonal to the major axis, but the nose has been offset from its central position. The surfaces below show the result when the normal to the nose tangent plane is tilted down. This dataset is discussed below in the context of two different normal to the nose tangent plane.

In the comparison figures (Figs. 5, 7(b), 8) there are 3 curves shown. The light green dashed curve is a graph of the initial estimated RNC curve (ORNC). The dark green heavier dashed curve is a graph of the estimated RNC curve after iteration and re-interpolation (IRNC). The black curve is a piecewise linear curve that graphs the goal RNC values (GRNC) at the data points that our method is fitting. Except where the RNC values at the data points were too few to get a good idea of the curve near those parameter values, the IRNC curve tracks, and is hard to distinguish from the GRNC. As is demonstrated below, as few as 15–20 iterations and then re-interpolation gives errors less than 10^{-7} at all data points in all examples. The error is measured as relative error. That is, at each data point, the final IRNC curve is evaluated and differenced with the GRNC value, and then that difference is divided by the GRNC value. For example, if the goal value was 0.01, but the attained value was 0.009 (90 % of desired),

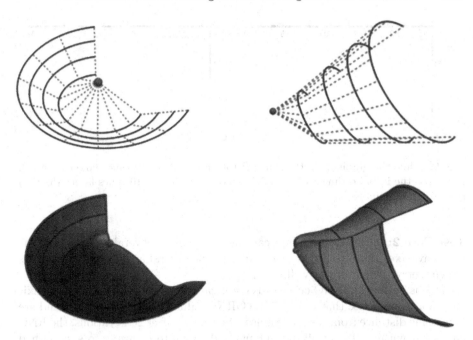

Fig. 4. The top figures show typical data from views near nose point (the red ball). The bottom figures show a cutaway of the resulting final surfaces near the nose with interpolated curves in blue (Color figure online).

then the relative error is $0.001/0.01 = 0.1$. The impetus is removing factors of scale from the problem.

Test Case 1: In this first example, conical data was studied ending with a spherical nose point with 2 different radii of curvature, one of which was 10 times larger than the other. In each case the data is the same. Note that in this problem, the data was a grid of points from which each the end tangents to each section curve were estimated and then the curves were interpolated. Each section curve is planar and orthogonal to the major axis of the model. The nose point is at the center of the complete cone and the normal to the nose plane is in the direction of the major axis.

In Fig. 5, the RNC curves are shown. In each case, the initial estimate for the tangent was made and the interpolation performed. That led to the dashed green values for the initial RNC curves. Notice that in both cases, the ORNC values are about 14 % off from the required values. Then there were 20 iterations on the RNC for the nose and then interpolation was performed again. The relative error was less than 10^{-6} at all data points. The IRNC curve values are so close to the goal RNC that they do not show any difference, except at each end of the curve where there are slight deviations from the constant value between data points. We considered making a *logarithm* scale on the vertical axis, but the curves looked the same.

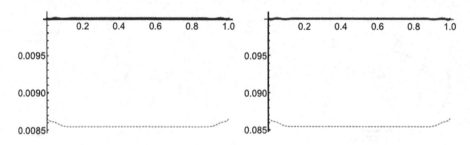

Fig. 5. While the graphs look the same, (b) is on a scale of 10 times larger than (a). Note that the initial estimate of the RNC curve in (b) is also 10 times larger than (a) (Color figure online).

Test Case 2: In this case an elliptical cone was chosen. Several sections near the nose have axes that are not perpendicular to the central axis and whose centers of symmetry are off axis as well.

In this figure the IRNC curve of is elliptical with a ration of 4/3, major axis to minor axis. Notice that in Fig. 7 the ORNC values are not symmetric and are have some distance from the goal values. Once again after 20 iterations, the RNC curve was within 10^{-7} at all data points and seems to converge to a smoothed version of the piecewise linear RNC curve from the data point values. Also, there are no undulations near the ends of the IRNC curve.

Test Case 3: In this dataset the nose has been offset from the central axis, and is a complete periodic version of the data shown in Fig. 4. We consider two cases. One that has the normal pointed upward making an angle of 20° with the central axis, whose images and data are on the left of the figures. The second has the normal pointed downward making an angle of −20° with the central axis, whose images and data are on the right of the figures. Both have RNC curves with the ratio of major radius to minor radius of 1.6. In Fig. 8 we see that the ORNC values in (b) have significant errors and do not seem likely to converge. The relative error of ORNC values of the right RNC curve at the data points was up to 50 % for Fig. 9(b), but it converged within 20 iterations to be less than 10^{-6}. Figure 8(a) had better initial estimates and, in the same number of iterations (20) the IRNC converged to have relative errors less than 10^{-7} at the data points. Also, the IRNC curve seems to converge to a smoothed version of the piecewise linear RNC curve from the data point values.

When the normal is pointed upward in the same offset direction as the nose point, the nose point remains the leading position on the surface. However, when the nose point is offset upward, but the normal is pointed downward, we see that in order to accommodate the G^1 requirement on the surface, that the nose point is not the most forward position on the surface. Because of this we see that Fig. 6 has a much more ellipsoidal and rounder looking region than that in Fig. 9.

Test Cases 4: This is the helicopter fuselage case shown in Figs. 1 and 2. An elliptical nose was specified with a major RNC to minor RNC ratio of 1.5,

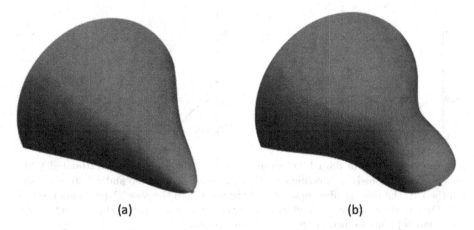

Fig. 6. Noses from circular cone data. Both have spherical nose points. The nose point is shown as a small red sphere. The right image (b) has RNC 10 times larger than that in the left image (a) (Color figure online).

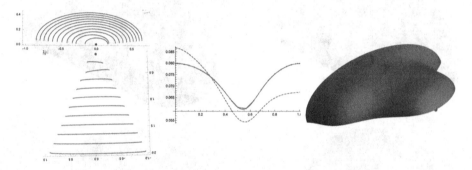

Fig. 7. (a) Data, highly eccentric elliptical cone with sections off angle and off axis; (b) RNC results: dashed line is ORNC, solid blue line is GRNC, solid green line is IRNC curve, (c) Rendered image of final nose. Nose point shown as small red sphere to emphasize the asymmetric behavior (Color figure online).

and the tail point is specified as spherical. In this case, the initial estimate for the data RNC was iterated 15 times and then reinterpolated to attain a relative error of less than 10^{-6}.

Determining good tangent directions is important to the success of the method. While our approach may seem simple, more mathematical approaches that have been tried can fail when any of these several events occur when the:

- nose is offset from the central axis,
- normal to the nose tangent plane is not orthogonal to the central axis of the loft,
- data sections are not orthogonal to the central axis,
- centers of the data sections do not align with the central axis.

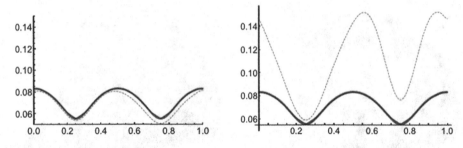

Fig. 8. Comparison of data RNC values (black) with the originally estimated RNC curves (green dashed), and estimated RNC curves after iteration and re-interpolating. On the left, the nose is offset upward and the normal to the nose plane points upward 20°. On the right, the nose is offset upwad the same amount, but the tangent to the nose plan is pointed downard 20° (Color figure online).

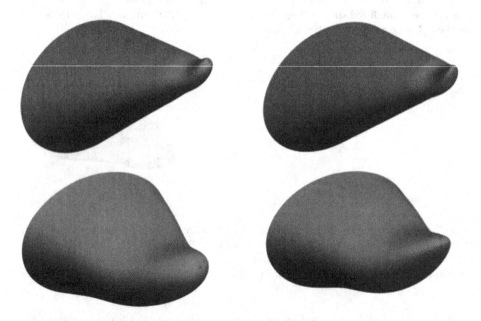

Fig. 9. Rendered images of final representation. On the bottom are tight views of the nose regions of the models above.Nose point shown in red (Color figure online).

In those cases, using interpolation along the length data parameter direction can fail to provide reasonable tangent directions.

Although our approach allows specifying ellipsoidal RNC nose conditions as well as spherical, it has limitations. Ratios larger than about 1.7 can lead to correct radii of normal curvatures, but the surface may have resulting concavities near the degenerate treatment in order to meet those specifications.

It is clear from the methodology that the RNC curve is continuous, but not C^1, although it appears smooth in the graphs. This is because it is a ratio of a scalar C^1 quartic B-spline curve and a C^0 linear scalar B-spline curve. However, the higher order discontinuities occur at the data parameter points, exactly at the locations where the fit is best.

7 Conclusion

Certain types of tensor product surface parametrization singularities occur frequently in design situations. When a lofting approach is used to create surfaces that have characteristics similar to surfaces of revolution (even though they may not be circular) and have singular points at the beginning, end, or both, a difficulty is encountered in specifying desirably smooth conditions at that point (when the same lofting methodology that is used for the rest of the representation). We have presented an algorithmic approach to creating a G^1 smooth surface at this type of singularity that enables specification of an ellipsoidal radius of normal curvature in all data parameter interpolating directions. The presented method is straightforward and empirically, with just a few iterations, has given an approximation that is within $\epsilon = 10^{-6}$ tolerance at data point directions and is close and continuously varying in between. While a formal proof of convergence has remained elusive, it is felt that inasmuch as the method produces results that can be confirmed to be satisfactory, it can be employed in actual practice without unanticipated problems occurring.

Limitations of the approach include the necessity to attain good tangent directions at the nose in each of the data parameters, a limitation on the eccentricity of the nose plane ellipse specified, a limitation on the angle of the nose plane normal direction, and a limitation on the offset of the nose point from the central axis, with respect to the other data curves.

This approach has been tested on periodic data sections parameterized as B-splines with open end conditions and non-uniform knot vectors, and data sections that are not periodic. In the latter case, when the data for a periodic surface appears as two or more surfaces to be interpolated to the nose point, our approach creates separate surfaces that give matching radii of normal curvature. However, if the individual surfaces are generated from data and the tangents must be estimated, the surfaces are only C^0 where they meet.

In summary, we present a lofting method with new end conditions of demonstrable utility. This method has been implemented in EGADS [4] which is part of ESP [3] and is freely distributed (Open Source).

Acknowledgments. This work is supported in part by NSF IIS-1117997 and in part through NASA Cooperative Agreement NNX11AI66A – Christopher Heath (NASA Glenn Research Center) is the Technical Monitor.

References

1. Cohen, E.: Some mathematical tools for a Modeler's workbench. Comput. Graph. Appl. **3**(7), 63–66 (1983)
2. Cohen, E., Riesenfeld, R.F., Elber, G.: Geometric Modeling with Splines: An Introduction. A. K. Peters Ltd, Natick (2001)
3. Haimes, R., Dannenhoffer, J.: The engineering sketch pad: a solid-modeling, feature-based, web-enabled system for building parametric geometry. In: AIAA Paper, p. 3073, June 2013
4. Haimes, R., Drela, M.: On the construction of aircraft conceptual geometry for high fidelity analysis and design. In: AIAA Paper, p. 0683, January 2012
5. Karčiauskas, K., Peters, J.: Finite curvature continuous polar patchworks. In: Hancock, E.R., Martin, R.R., Sabin, M.A. (eds.) Mathematics of Surfaces XIII. LNCS, vol. 5654, pp. 222–234. Springer, Heidelberg (2009)
6. Karciauskas, K., Peters, J.: Biquintic G^2 surfaces. In: 14th IMA Conference on the Mathematics of Surfaces: Conference Proceedings on CD-ROM (2013)
7. Myles, A., Karciauskas, K., Peters, J.: Pairs of bi-cubic surface constructions supporting polar connectivity. Comput. Aided Geom. Des. **25**(8), 621–630 (2008)
8. Myles, A., Peters, J.: C^2 splines covering polar configurations. Comput. Aided Des. **43**(11), 1322–1329 (2011)

A Consistent Statistical Framework for Current-Based Representations of Surfaces

Benjamin Coulaud and Frédéric J.P. Richard[(⊠)]

Aix-Marseille Université, CNRS, Centrale Marseille, I2M, UMR 7373,
13453 Marseille, France
benjamin.coulaud@etu.univ-amu.fr, frederic.richard@univ-amu.fr

Abstract. In this paper, we deal with the statistical analysis of surfaces. To address this issue, we extend to a stochastic setting a surface representation by currents (Glaunès and Vaillant, 2005). This extension is inspired from the theory of generalized stochastic processes due to Itô (1954) and Gelfand and Vilenkin (1964). It consists of random linear forms defined on a space of vector fields. Upon this representation, we build a probabilistic model that allows us to describe the random variability of current-based representations of surfaces. For this model, we then propose estimators of a mean representant of a surface population, and the population variance. Finally, we establish the consistency of these estimators.

1 Introduction

Surfaces are not only an important source of research in several branches of mathematics, but also objects of primary interest in many application domains. In the biomedical domain, the diffusion of imaging techniques has led to an important production of images from which surfaces of organs (heart, brain, etc.) are commonly extracted. This growing access to large populations of surfaces raises statistical issues. It calls for methods that would help analyze the surface variability within and across populations of healthy and diseased patients.

The statistical analysis of surfaces is a difficult task. Due to their Riemannian nature, surfaces can not be analyzed using classical methods that only apply to Euclidian objects. There are methods which have been specifically developed for Riemannian manifolds such as surfaces (see [3,16] for instance). Alternately, we propose a statistical framework based on an Euclidian surface representation known as "current" in the literature of Shape and Image Analysis. Introduced by Glaunès and Vaillant [18] and further developed in [4,9,12], this representation has gained popularity in its applications to various medical surfaces [5–9,13,17].

Following [18], a current s representing a surface $\mathcal{S}$ is a continuous linear form defined on a Hilbert space V composed of vector fields of $\mathbb{R}^3$. For instance, the representant s of a continuously differentiable surface $\mathcal{S}$ is defined as

$$s(w) = \int_{\mathcal{S}} \langle e_1(u) \wedge e_2(u); w(u) \rangle_{\mathbb{R}^3} d\sigma(u), \forall\, w \in V, \tag{1}$$

J.-D. Boissonnat et al. (Eds.): Curves and Surfaces 2014, LNCS 9213, pp. 151–159, 2015.
DOI: 10.1007/978-3-319-22804-4_11

where $\wedge$ is the cross-product of $\mathbb{R}^3$, $\langle e_1(u)\wedge\cdot;\cdot\rangle_{\mathbb{R}^3}$ its inner product, σ the surfacial measure on $\mathcal{S}$, and vectors $e_1(u)$, $e_2(u)$ a direct basis of the plane tangent to $\mathcal{S}$ at u. Under mild assumptions on V, the linear form s is continuous, and belongs to the topological dual space V^* of V (see Sect. 2). Hence, elements we deal with belong to a space of functions defined on an infinite-dimensional function space.

The first contribution of this paper is the definition of a stochastic setting where these elements can be viewed as random realizations. Using concepts of the theory established by Itô [14,15] and Gelfand and Vilenkin [11], we define a random linear form as a generalized stochastic process indexed on the space V. We further give appropriate definitions of expectation, autocovariance and noise associated to this random form. This allows us to set a model describing random variations of surface representants around a mean.

Using this model, we then construct estimators of the mean representant and the variance. This construction is based on approximations of vector fields in finite-dimensional subspaces of V. Finally, we show the consistency of these estimators as both the dimension of the approximation subspaces and the population size increase.

2   Statistical Framework

2.1   Surface Representation

Let $\mathcal{D}$ be a bounded cube from $\mathbb{R}^3$ and Θ be a set of $\mathcal{C}^1$ compact surfaces included in $\mathcal{D}$. Let V_0 be a vector subspace of $\mathcal{L}^2(\mathcal{D}, \mathbb{R})$. We assume that V_0 is a RKHS with a bounded reproducing kernel K_0 [2]. We consider the Hilbert space $V := (V_0)^3$ equipped with the inner product

$$\langle v; w\rangle_V := \sum_{i=1}^{3} \langle v_i; w_i\rangle_{V_0}, \forall v := (v_1, v_2, v_3), w := (w_1, w_2, w_3) \in V$$

where $\langle \cdot; \cdot\rangle_{V_0}$ is the inner product of V_0. A typical example of RKHS with bounded kernel is the Sobolev space $\mathcal{H}^2(\mathcal{D})$.

Proposition 1. *Let $\mathcal{S} \in \Theta$. The representant s defined by Eq. (1) is an element of the topological dual V^* of V.*

Proof. As the linearity of s is obvious, we focus on its continuity. Since V is an RKHS, $\|w(u)\|_{\mathbb{R}^3} \leq \|K_0(u,.)\|_{V_0}\|w\|_V$, for all $w \in V$ and $u \in \mathcal{D}$. Hence, using Cauchy-Schwartz inequality, we get

$$\forall w \in V, |s(w)| \leq \|w\|_V \int_{\mathcal{S}} \|K_0(u,.)\|_{V_0} d\sigma(u).$$

As K_0 is bounded and $\mathcal{S}$ is compact, s is continuous.

2.2   Random Linear Forms

In this section, we give some definitions concerning random linear forms. These definitions are inspired from works on generalized stochastic processes pioneered by Itô [14,15] and Gelfand and Vilenkin [11], and also developed in [10].

Let $\mathcal{E} := (\Omega, \mathcal{F}, \mathbb{P})$ be a probability space. We denote $\mathcal{L}^2(\mathcal{E})$ the space of square-integrable random variables on $\mathcal{E}$.

Definition 1. *We call random linear form a continuous linear form mapping the vector field space V into $\mathcal{L}^2(\mathcal{E})$.*

We denote $V_{\mathcal{E}}^*$ the space of these random linear forms. Generalized stochastic processes introduced in [11,14] correspond to linear forms defined on the space V of infinitely differentiable functions with compact support, whereas, here, it is a RKHS with a bounded kernel.

Definition 2. *Let S be a random linear form on V. The expectation $\mathbb{E}^*[S]$ of S is defined as the mapping*

$$\mathbb{E}^*[S] : w \in V \longrightarrow \mathbb{E}[S(w)] \in \mathbb{R},$$

and its autocovariance γ_S as the mapping

$$\gamma_S : (v, w) \in V \times V \longrightarrow \mathrm{Cov}(S(v), S(w)) \in \mathbb{R},$$

where $\mathbb{E}$ and Cov stand for the usual expectation and covariance of random variables, respectively.

Let us notice that $\mathbb{E}^*[S]$ is a continuous linear form on V and γ_S is a continuous bilinear form on $V \times V$. Hence, as a consequence of the Riesz representation theorem, there exist a unique element v_S in V and a unique continuous endomorphism Λ_S on V such that $\mathbb{E}^*[S](w) = \langle v_S; w \rangle_V$ and $\gamma_S(v, w) = \langle \Lambda_S(v), w \rangle_V$ for all $v, w \in V$.

Definition 3. *We say that $S \in V_{\mathcal{E}}^*$ is a Gaussian linear form if, for all $N \in \mathbb{N}^*$, $w_1, \cdots, w_N \in V$, the random vector $(S(w_1), \cdots, S(w_N))$ is Gaussian.*

Let us quote that a Gaussian random linear form is entirely characterized by its expectation and autocovariance.

Definition 4. *We say that two random linear forms S_1 and S_2 are independent if, for all $N \in \mathbb{N}^*$, $w_1, \cdots, w_N \in V$, random vectors $(S_1(w_1), \cdots, S_1(w_N))$ and $(S_2(w_1), \cdots, S_2(w_N))$ are independent.*

Definition 5. *A white noise B in $V_{\mathcal{E}}^*$ is a random linear form which satisfies $\mathbb{E}^*[B](v) = 0$, and $\gamma_B(v, w) = \sigma^2 \langle v; w \rangle_V$ for all $v, w \in V$.*

The parameter σ^2 is called the variance of the white noise.

2.3 Surface Variability

Let $\{\mathcal{S}_k, k \in [\![1, K]\!]\}$ be a population of surfaces, and $\{s_k, k \in [\![1, K]\!]\}$ their respective representants in V^*. We describe the variability of these surfaces through their representants. We consider each representant s_k as a realization of a random linear form S_k in $V_{\mathcal{E}}^*$. We assume that random linear forms $S_1, \cdots, S_K$ are independent and identically distributed according to the model

$$S_k = \mu + B_k, \forall\, k \in [\![1, K]\!], \tag{2}$$

in which $\mu \in V^*$, and B_k are independent Gaussian white noises of $V_{\mathcal{E}}^*$. The mean surface representant μ and the noise variance σ^2 are both unknown, and to be estimated from $S_1, \cdots, S_K$ (see the next section).

3 Estimation

This section is devoted to the estimation of parameters μ and σ^2 in Model (2).

3.1 Approximation

The parameter μ is a linear form of V^*, which is an infinite-dimensional subspace of $\mathcal{L}^2(\mathcal{D}, \mathbb{R})$. We deal with its estimation using approximations in finite-dimensional subspace of V. Let $(\phi_i)_{i \in I}$ be a countable total linearly independent family of V. Let $(I_N)_{N \in \mathbb{N}^*}$ be a sequence of embedded finite subsets of I such that $\underset{N \in \mathbb{N}^*}{\cup} I_N = I$. For $N \in \mathbb{N}^*$, we consider the approximation subspaces $V_N := \mathrm{Vect}(\phi_i, i \in I_N)$ equipped with the inner product induced by V. These subspaces are embedded and their union is dense in V. For $N \in \mathbb{N}^*$, we also define the vector subspace $\tilde{V}_N^*$ of V^*

$$\tilde{V}_N^* := \{\mu \in V^* | v_\mu \in V_N\},$$

where v_μ denotes the Riesz representant of the linear form μ. We will denote $\mu^N = \underset{\nu \in \tilde{V}_N^*}{\mathrm{argmin}} \|\mu - \nu\|_{V^*}$ (resp. $v_\mu^N = \underset{w \in V_N}{\mathrm{argmin}} \|v_\mu - w\|_V$) the orthogonal projection of μ (resp. v_μ) on $\tilde{V}_N^*$ (resp. V_N). Each element μ of V^* can be approximated by an element of $\tilde{V}_N^*$ using an orthogonal projection v_μ^N of its Riesz representant v_μ into V_N. One can show that v_μ^N is the element of V_N which satisfies $\langle v_\mu^N; w \rangle_V = \mu(w), \forall w \in V_N$. Equivalently, $v_\mu^N = \sum_{i \in I_N} \alpha_i \phi_i$, where $(\alpha_i)_i$ is the solution of the linear system $\sum_{j \in I_N} g_{i,j} \alpha_j = \sum_{i \in I_N} \mu(\phi_i)$, for all $i \in I_N$ with $g_{i,j} = \langle \phi_i, \phi_j \rangle_V$. For $i, j \in I_N$, we will denote $h_{i,j}^N$ the terms which satisfy $\sum_{k \in I_N} g_{(i,k)} h_{(k,j)}^N = \delta_{(i,j)}, \forall i, k \in I_N$. We have

$$v_\mu^N = \sum_{i \in I_N} \alpha_i \phi_i, \quad \text{with } \alpha_i = \sum_{j \in I_N} h_{(i,j)}^N \mu(\phi_j). \tag{3}$$

3.2   Maximum Likelihood Estimators

The parameter μ of Model (2) is estimated via its Riesz representant v_μ. Let v_μ^N be the orthogonal projection of μ into V_N given by Eq. (3). For each $k \in [\![1, K]\!]$, we define the random vectors $Z_k^N := (S_k(\phi_i))_{i \in I_N}$, $\eta_k^N := (B_k(\phi_i))_{i \in I_N}$, and $\zeta^N := (\mu(\phi_i))_{i \in I_N}$. According to Model (2), we can write

$$Z_k^N = \zeta^N + \eta_k^N, \forall k \in [\![1, K]\!]. \tag{4}$$

For $k \in [\![1, K]\!]$, random vectors $Z_k^N = (Z_{k,i}^N)_{i \in I_N}$ are *i.i.d.* according to a multi-variate Gaussian law with an expectation ζ^N and a covariance matrix Γ^N whose terms are given by

$$\Gamma_{i,j}^N = \mathrm{Cov}(Z_{k,i}^N, Z_{k,j}^N) = \gamma_B(\phi_i, \phi_j) = \sigma^2 \langle \phi_i, \phi_j \rangle_V, \forall i, j \in I_N.$$

The maximum likelihood estimator of ζ^N is then

$$\overline{Z_K^N} = \frac{1}{K} \sum_{k=1}^{K} Z_k^N. \tag{5}$$

Accordingly, we define the estimator of the Riesz representant v_μ of μ as

$$\widehat{v_{\mu,K}^N} = \sum_{m \in I_N} \sum_{l \in I_N} h_{m,l}^N \overline{Z_{K,l}^N} \phi_m. \tag{6}$$

In other words, the linear form μ is estimated by the random linear form

$$\widehat{\mu_K^N}(w) = \langle \widehat{v_{\mu,K}^N}; w \rangle_V, \forall\, w \in V. \tag{7}$$

Besides, the maximum likelihood estimator of σ^2 is

$$\widehat{\nu_K^N} = \frac{1}{K|I_N|} \sum_{i,j \in I_N} \sum_{k=1}^{K} (Z_{k,i}^N - \overline{Z_{K,i}^N}) h_{i,j}^N (Z_{k,j}^N - \overline{Z_{K,j}^N}),$$

where $|I_N|$ is the cardinal of I_N.

3.3   Consistency

We now state the consistency of our estimators as the dimension N of the approximation space and the sample size K tend to $+\infty$.

Theorem 1 (Consistency of the mean representant estimator). *The linear form* $\mathbb{E}^*[\widehat{\mu_K^N}]$ *converges pointwise to* μ *as* N *and* K *both tend to* $+\infty$, *i.e.*

$$\lim_{N,K \to +\infty} \mathbb{E}^*[\widehat{\mu_K^N}](w) = \mu(w), \forall w \in V.$$

The quadratic error converges pointwise to 0 *as* N *and* K *tend to* $+\infty$, *i.e.*

$$\lim_{N,K \to +\infty} \mathbb{E}[(\widehat{\mu_K^N}(w) - \mu(w))^2] = 0, \forall w \in V.$$

Theorem 2 (Convergence of the variance estimator). *The estimator $\widehat{\nu_K^N}$ of the variance σ^2 is asymptotically unbiased, i.e.*

$$\lim_{N,K\to+\infty} \mathbb{E}[\widehat{\nu_K^N}] = \sigma^2.$$

Proof (Theorem 1). First, we prove the pointwise convergence. It suffices to show that, for any $w \in V$, $\lim_{N,K\to\infty} \mathbb{E}[\widehat{\mu_K^N}(w) - \mu(w)] = 0$. Using notations of Sect. 3, we notice that

$$\mathbb{E}[\widehat{\mu_K^N}(w) - \mu(w)] = \mathbb{E}[\widehat{\mu_K^N}(w) - \mu^N(w)] + \mu^N(w) - \mu(w),$$
$$= \mathbb{E}[\langle \widehat{v_{\mu,K}^N} - v_\mu^N ; w\rangle_V] + \langle v_\mu^N - v_\mu ; w\rangle_V,$$
$$= \mathbb{E}[\sum_{m,l\in I_N} h_{m,l}^N \langle (\overline{Z_{K,l}^N} - \zeta_l^N); w\rangle_V] + \langle v_\mu^N - v_\mu ; w\rangle_V.$$

But, $\overline{Z_{K,l}^N} - \zeta_l^N = \eta_l^N$, and $\mathbb{E}(\eta_l^N) = 0$. Thus, $\mathbb{E}[\widehat{\mu_K^N}(w) - \mu(w)] = \langle v_\mu^N - v_\mu ; w\rangle_V$. Since $\lim_{N\to+\infty} \|v_\mu - v_\mu^N\|_V = 0$, it follows that $\mathbb{E}[\widehat{\mu_K^N}(w) - \mu(w)] \overset{N\to\infty}{\longrightarrow} 0$.

We now focus on the mean square convergence. We will use the following lemma whose proof is postponed at the end of the section.

Lemma 1. *For all w in V, we have*

$$\mathbb{E}[(\widehat{\mu_1^N}(w) - \mu(w))^2] \overset{N\to\infty}{\longrightarrow} \sigma^2 \|w\|_V^2.$$

For $K \in \mathbb{N}^*$,

$$\mathbb{E}[(\widehat{\mu_K^N}(w) - \mu(w))^2] = \mathbb{E}\left[(\frac{1}{K}\sum_{k=1}^K (\widehat{\mu_k^N}(w) - \mu^N(w)) + (\mu^N(w) - \mu(w)))^2\right],$$
$$= \text{Var}\left[\frac{1}{K}\sum_{k=1}^K \widehat{\mu_k^N}(w)\right] + (\mu^N(w) - \mu(w))^2.$$

Since $(\widehat{\mu_k^N}(w))_{k\in[\![1,K]\!]}$ are i.i.d., we obtain

$$\mathbb{E}[(\widehat{\mu_K^N}(w) - \mu(w))^2] = \frac{1}{K}\text{Var}[\widehat{\mu_1^N}(w)] + (\mu^N(w) - \mu(w))^2,$$
$$= \frac{1}{K}\mathbb{E}[(\widehat{\mu_1^N}(w) - \mu^N(w))^2] + (\mu^N(w) - \mu(w))^2.$$

But, $\mathbb{E}[(\widehat{\mu_1^N}(w) - \mu^N(w))^2] \le \mathbb{E}[(\widehat{\mu_1^N}(w) - \mu(w))^2]$. Hence, applying Lemma 1, we get

$$\lim_{N,K\to\infty} \mathbb{E}[(\widehat{\mu_K^N}(w) - \mu(w))^2] = 0.$$

Proof (Theorem 2). Let us notice that $\mathbb{E}[\widehat{\nu_K^N}]$ is equal to

$$\frac{1}{K|I_N|}\sum_{m,l\in I_N} h_{m,l}^N \sum_{k=1}^{K}\mathbb{E}\left[(B_k(\phi_m)-\frac{1}{K}\sum_{k=1}^{K}B_k(\phi_m))(B_k(\phi_l)-\frac{1}{K}\sum_{k=1}^{K}B_k(\phi_l))\right].$$

As the sample is *i.i.d.*, we thus get

$$\mathbb{E}[\widehat{\nu_K^N}]=\frac{1}{|I_N|}\sum_{m,l\in I_N} h_{m,l}^N \mathbb{E}\left[(B_1(\phi_m)-\frac{1}{K}\sum_{k=1}^{K}B_k(\phi_m))(B_1(\phi_l)-\frac{1}{K}\sum_{k=1}^{K}B_k(\phi_l))\right].$$

Due to the independence of $(B_k)_{k\in\{1,\ldots,K\}}$, we have, for all $m,l\in I_N$,

$$\mathbb{E}\left[(B_1(\phi_m)-\frac{1}{K}\sum_{k=1}^{K}B_k(\phi_m))(B_1(\phi_l)-\frac{1}{K}\sum_{k=1}^{K}B_k(\phi_l))\right]=\sigma^2\langle\phi_m,\phi_l\rangle_V(1-\frac{1}{K}).$$

Therefore, $\mathbb{E}[\widehat{\nu_K^N}]=\sigma^2(1-\frac{1}{K})$, which concludes the proof.

Proof (Lemma 1). Let $m,n\in I$, choose N' be such that ϕ_n and ϕ_m are in $V_{N'}$, and set $N\geq N'$. $N\in\mathbb{N}^*$. We first develop $\mathbb{E}[(\widehat{\mu_1^N}(\phi_j)-\mu(\phi_m))(\widehat{\mu_1^N}(\phi_n)-\mu(\phi_n))$. Writing $\widehat{\mu_1^N}(\phi_j)-\mu(\phi_m)=\widehat{\mu_1^N}(\phi_j)-\mu^N(\phi_j)+\mu^N(\phi_j)-\mu(\phi_j)$, and noticing that $\mu^N-\mu$ is null on V_N, we have

$$\mathbb{E}[(\widehat{\mu_1^N}(\phi_m)-\mu(\phi_m))(\widehat{\mu_1^N}(\phi_n)-\mu(\phi_n))]$$
$$=\mathbb{E}[(\widehat{\mu_1^N}(\phi_m)-\mu^N(\phi_m))(\widehat{\mu_1^N}(\phi_n)-\mu^N(\phi_n))],$$
$$=\mathbb{E}[\langle\widehat{v_{\mu,1}^N}-v_\mu^N;\phi_m\rangle_V\,\langle\widehat{v_{\mu,1}^N}-v_\mu^N;\phi_n\rangle_V],$$
$$=\mathbb{E}\left[\langle\sum_{i\in I_N}\phi_i\sum_{l\in I_N}h_{l,i}^N\eta_{1,l}^N;\phi_m\rangle_V\,\langle\sum_{j\in I_N}\phi_j\sum_{p\in I_N}h_{p,j}^N\eta_{1,p}^N;\phi_n\rangle_V\right],$$
$$=\sigma^2\sum_{l\in I_N}\sum_{p\in I_N}(\sum_{i\in I_N}g_{m,i}h_{l,i}^N)(\sum_{j\in I_N}g_{n,j}\,h_{p,j}^N)\langle\phi_l;\phi_p\rangle_V.$$

But, $\sum_{j\in I_N}g_{n,j}\,h_{p,j}^N=\delta_{n,p}$, Therefore,

$$\lim_{N\to\infty}\mathbb{E}[(\widehat{\mu_1^N}(\phi_m)-\mu(\phi_m))(\widehat{\mu_1^N}(\phi_n)-\mu(\phi_n))]=\sigma^2\langle\phi_m;\phi_n\rangle_V.$$

Now, when $w:=\sum_{i\in I_M}\alpha_i\phi_i\in V_M$ for some $M\in\mathbb{N}^*$, we have

$$\mathbb{E}[(\widehat{\mu_1^N}(w)-\mu(w))^2]=\sum_{i\in I_M}\sum_{j\in I_M}\alpha_i\alpha_j\mathbb{E}[(\widehat{\mu_1^N}(\phi_i)-\mu(\phi_i))(\widehat{\mu_1^N}(\phi_j)-\mu(\phi_j))].$$

It follows that

$$\lim_{N\to+\infty}\mathbb{E}[(\widehat{\mu_1^N}(w)-\mu(w))^2]=\sum_{i\in I_M}\sum_{j\in I_M}\sigma^2\alpha_i\alpha_j\langle\phi_i;\phi_j\rangle_V=\sigma^2\|w\|_V^2.$$

Now, let us consider an arbitrary w in V, and its orthogonal projection w^M into V_M. By continuity, $(\widehat{\mu_1^N}(w^M) - \mu(w^M))^2 \overset{M\to\infty}{\longrightarrow} (\widehat{\mu_1^N}(w) - \mu(w))^2$ a.s.. Moreover,

$$(\widehat{\mu_1^N}(w^M) - \mu(w^M))^2 \le \|\widehat{v_{\mu,1}^N} - v_\mu\|_V^2 \|w^M\|_V^2 \le \|\widehat{v_{\mu,1}^N} - v_\mu\|_V^2 \|w\|_V^2.$$

Therefore, using the dominated convergence theorem,

$$\mathbb{E}[(\widehat{\mu_1^N}(w^M) - \mu(w^M))^2] \overset{M\to\infty}{\longrightarrow} \mathbb{E}[(\widehat{\mu_1^N}(w) - \mu(w))^2].$$

As a consequence,

$$\lim_{N\to\infty} \mathbb{E}[(\widehat{\mu_1^N}(w) - \mu(w))^2] = \lim_{N,M\to\infty} \mathbb{E}[(\widehat{\mu_1^N}(w^M) - \mu(w^M))^2] = \sigma^2 \|w\|_V^2.$$

4 Discussion

We constructed a statistical framework which is suitable for the analysis of surfaces represented by currents (*i.e.* linear continuous forms on a vector field space V) [18]. This framework was built upon a stochastic extension of the current-based representation of surfaces. This representation consisted of random linear forms defined as generalized stochastic processes indexed on the functional space V. Viewing surface representants as realizations of these random linear forms, we could set up a model that described the random variability of surfaces. For this model, we proposed estimators of a mean representant of a surface population, and the population variance. These estimators were derived from approximations of vector fields in finite-dimensional subspaces of V. Eventually, we established the consistency of these estimators.

In our variation model (see Eq. (2)), it is possible to include the action of deformations so as to explicitly account for geometric variations. For instance, we can define a probabilistic model

$$S_k = \psi_k \cdot \mu + B_k, \forall\, k \in [\![1, K]\!], \tag{8}$$

where $\psi_k \cdot \mu$ represents the action of a deformation ψ_k on a mean surface representant μ [18]. The mean representant μ of such a model is known as a deformable template in the literature of Shape and Image Analysis (see [1] for a review). Probabilistic models of the form (8) were already proposed for images and other Euclidian objects, but are not well-established for objects such as surfaces or their current-based representants. Hence, the model proposed here could fill a gap of probabilistic frameworks for surface deformable templates. In the framework of this model, we plan to extend our estimation method to the computation of a template.

References

1. Allassonnière, S., Bigot, J., Glaunès, J., Maire, F., Richard, F.: Statistical models for deformable templates in image and shape analysis. Annales Mathématiques de l'Institut Blaise Pascal **20**(1), 1–35 (2013)

2. Aronszajn, N.: Theory of reproducing kernels. Trans. Am. Math. Soc. **68**(3), 337–404 (1950)
3. Bhattacharya, R., Patrangenaru, V.: Statistics on manifolds and landmarks based image analysis: a non-parametric theory with applications. J. Stat. Plann. Infer. **145**(1), 1–22 (2014)
4. Charon, N., Trouvé, A.: The varifold representation of non-oriented shapes for diffeomorphic registration. SIAM J. Imaging Sci. **6**(4), 2547–2580 (2013)
5. Cury, C., Glaunès, J.A., Colliot, O.: Template estimation for large database: a diffeomorphic iterative centroid method using currents. In: Nielsen, F., Barbaresco, F. (eds.) GSI 2013. LNCS, vol. 8085, pp. 103–111. Springer, Heidelberg (2013)
6. Durrleman, S., Ayache, N., Trouvé, A., Fillard, P., Pennec, X.: Registration, atlas estimation and variability analysis of white matter fiber bundles modeled as currents. Neuroimage **55**(3), 1073–1090 (2011)
7. Durrleman, S., Pennec, X., Trouvé, A., Ayache, N., et al.: A forward model to build unbiased atlases from curves and surfaces. In: 2nd MICCAI Workshop on Mathematical Foundations of Computational Anatomy, pp. 68–79 (2008)
8. Durrleman, S., Pennec, X., Trouvé, A., Thompson, P., Ayache, N.: Inferring brain variability from diffeomorphic deformations of currents: an integrative approach. Med. Image Anal. **12**(5), 626–637 (2008)
9. Durrleman, S., Trouvé, A., Pennec, X., Ayache, N.: Statistical models of sets of curves and surfaces based on currents. Med. Image Anal. **13**(5), 793–808 (2009)
10. Fernique, X.: Processus linéaires généralisés. Annales de l'Institut Fourier **17**(1), 1–92 (1967)
11. Gelfand, I., Vilenkin, N.Y.: Generalized Functions Applications to Harmonical Analysis, vol. 4. Academic Press, New York (1964)
12. Glaunès, J., Joshi, S.: Template estimation from unlabeled point set data and surfaces for computational anatomy. In: Pennec, X., Joshi, S. (eds) Proceedings of the International Workshop on the Mathematical Foundations of Computational Anatomy (MFCA-2006), pp. 29–39 (2006)
13. Gorbunova, V., Durrleman, S., Lo, P., Pennec, X., De Bruijne, M., et al.: Curve and surface-based registration of lung CT images via currents. In: Proceedings of Second International Workshop on Pulmonary Image, Analysis, pp. 15–25 (2009)
14. Itô, K.: Stationary random distributions. Mem. Col. Sci. Kyoto Imp. Univ., Ser. A **28**, 223–229 (1954)
15. Itô, K.: Isotropic random current. In: Proceedings of the Third Berkeley Symposium on Mathematical Statistics and Probability, vol. 2, pp. 125–132 (1956)
16. Pennec, X.: Intrinsic statistics on Riemannian manifolds: basic tools for geometric measurements. J. Math. Imaging Vision **25**, 127–154 (2006)
17. Tilotta, F., Glaunès, J., Richard, F., Rozenholc, Y.: A local technique based on vectorized surfaces for craniofacial reconstruction. Forensic Sci. Int. **200**(1), 50–59 (2010)
18. Vaillant, M., Glaunès, J.: Surface matching via currents. In: Christensen, G.E., Sonka, M. (eds.) IPMI 2005. LNCS, vol. 3565, pp. 381–392. Springer, Heidelberg (2005)

Isotropic Möbius Geometry and i-M Circles on Singular Isotropic Cyclides

Heidi E.I. Dahl[✉]

Department of Applied Mathematics, SINTEF ICT,
Forskningsveien 1, 0373 Oslo, Norway
heidi.dahl@sintef.no

Abstract. The Möbius geometry of $\mathbb{R}^3$ has an isotropic counterpart in $\mathbb{R}^3_{++0}$. We describe the isotropic Möbius model of surfaces in $\mathbb{R}^3_{++0}$ and show how the degree of a surface changes under i-M inversions while the number of families of i-M circles remain constant. This gives us a generalization of the classification of families of lines and i-M circles on quadratic surfaces in $\mathbb{R}^3_{++0}$ to isotropic cyclides with real singularities, containing up to 4 such families.

1 Introduction

The geometry of un-oriented spheres in $\mathbb{R}^3$ is known as Möbius geometry [1]. Möbius transformations map spheres to spheres, considering planes as spheres with infinite radius. Any Möbius transformation can be expressed as compositions of translations, similarities, and inversions [5, Theorem 68.2 and 68.3].

Möbius geometry is used in the study of surfaces in $\mathbb{R}^3$ containing families of quadratic curves, and in particular families of circles. Recent papers include the study of Darboux cyclides in [11], and it is used in [6] to describe *celestial surfaces*: surfaces containing at least two families of circles.

In [4], Krasauskas et al. describe the isotropic counterpart to Möbius geometry, i.e., Möbius geometry in $\mathbb{R}^3_{++0}$ (see also [7–9]). In this paper we will extend this description, defining the isotropic Möbius (i-M) model of a surface in $\mathbb{R}^3_{++0}$ and using it to show how i-M transformations affect the degree of a surface and preserve the number of families of i-M circles it contains (Sect. 2). This is an adaption of [6, Sect. 2.5][1] from Euclidean $\mathbb{R}^3$ to isotropic space $\mathbb{R}^3_{++0}$. We use these results to show how our classification of families of i-M circles and lines on quadratic surfaces in $\mathbb{R}^3_{++0}$ [2] can be extended to isotropic cyclides with real singularities (Sect. 3).

2 Möbius Geometry in $\mathbb{R}^3_{++0}$

The three-dimensional *isotropic space*, denoted $\mathbb{R}^3_{++0}$, is $\mathbb{R}^3$ equipped with the scalar product with signature $(++0)$ [4,9,10]. The signature defines the scalar

---

[1] Based on Sect. 5 of v1 of the paper, see http://arxiv.org/abs/1302.6710.

© Springer International Publishing Switzerland 2015
J.-D. Boissonnat et al. (Eds.): Curves and Surfaces 2014, LNCS 9213, pp. 160–168, 2015.
DOI: 10.1007/978-3-319-22804-4_12

product $\langle \boldsymbol{v}, \boldsymbol{v}' \rangle = v_1 v_1' + v_2 v_2'$ of the vectors $\boldsymbol{v} = (v_1, v_2, v_3)$, $\boldsymbol{v}' = (v_1', v_2', v_3') \in \mathbb{R}^3_{++0}$ and the metric

$$|\boldsymbol{p} - \boldsymbol{p}'|^2 = \langle \boldsymbol{p} - \boldsymbol{p}', \boldsymbol{p} - \boldsymbol{p}' \rangle = (p_1 - p_1')^2 + (p_2 - p_2')^2, \quad \boldsymbol{p}, \boldsymbol{p}' \in \mathbb{R}^3_{++0}. \quad (1)$$

Thus the distance between the points $\boldsymbol{p}$ and $\boldsymbol{p}'$ is measured horizontally, parallel to the plane $x_3 = 0$.

The isotropic counterpart to Möbius geometry, where isotropic Möbius (i-M) spheres are vertical rotational elliptic paraboloids and non-isotropic (non-vertical) planes (cf. spheres and planes in Möbius geometry), was described in [4,7–9]. The intersection of two i-M spheres is an i-M circle, and is either a non-isotropic line, a vertical parabola, or an ellipse whose *top view* (the orthogonal projection onto the plane $x_3 = 0$) is a circle. In [4, Lemma 1] they show that all i-M transformations can be expressed in terms of the following four generating transformations:

- *uniform scalings* $S_r(\boldsymbol{x}) = (x_0, r x_1, r x_2, r x_3)$, $r \in \mathbb{R}$,
- *translations* $T_{\boldsymbol{p}}(\boldsymbol{x}) = (x_0, x_1 + p_1 x_0, x_2 + p_2 x_0, x_3 + p_3 x_0)$, $\boldsymbol{p} = (p_1, p_2, p_3) \in \mathbb{R}^3_{++0}$,
- *inversions* $\mathrm{inv}(\boldsymbol{x}) = (x_1^2 + x_2^2, x_0 x_1, x_0 x_2, x_0 x_3)$,
- *vertical reflections* $\mathrm{inv}^0(\boldsymbol{x}) = (x_0, x_1, x_2, -x_3)$,

where $\boldsymbol{x} = (x_0, x_1, x_2, x_3) \in \mathbb{P}^3$ is a point in the projective space over $\mathbb{R}^3_{++0}$. For example, the inversion $\mathrm{inv}_{\boldsymbol{p}}$ in an arbitrary point $\boldsymbol{p}$ is a composition of a translation of $\boldsymbol{p}$ to the origin, an inversion, and the translation back:

$$\mathrm{inv}_{\boldsymbol{p}} = T_{\boldsymbol{p}} \circ \mathrm{inv} \circ T_{\boldsymbol{p}}^{-1}. \quad (2)$$

The isotropic counterpart to the sphere $\mathcal{S}^3$ in Möbius geometry is the Blaschke cylinder in $\mathbb{P}^4$ [4]:

$$\mathcal{B} : x_1^2 + x_2^2 - 2 x_0 x_\infty = 0, \quad \boldsymbol{x} = (x_0, x_1, x_2, x_3, x_\infty) \in \mathbb{P}^4. \quad (3)$$

As in the Euclidean setting, we define a stereographic projection π [4] from the Blaschke cylinder to the projective space over $\mathbb{R}^3_{++0}$

$$\pi : \mathbb{P}^4 \to \mathbb{P}^3, \quad (x_0, x_1, x_2, x_3, x_\infty) \mapsto (x_0, x_1, x_2, x_3), \quad (4)$$

and its inverse $\sigma = \pi^{-1}$

$$\sigma : \mathbb{P}^3 \to \mathcal{B} \subset \mathbb{P}^4, \quad (x_0, x_1, x_2, x_3) \mapsto \left(x_0^2, x_0 x_1, x_0 x_2, x_0 x_3, \frac{1}{2} (x_1^2 + x_2^2) \right). \quad (5)$$

In the projective space $\mathbb{P}^3$ over $\mathbb{R}^3_{++0}$, the *isotropic absolute conic* is defined by

$$x_1^2 + x_2^2 = 0, \quad x_0 = 0, \quad \boldsymbol{x} \in \mathbb{P}^3. \quad (6)$$

The *i-Möbius type* (d, c) of a surface in $\mathbb{R}^3_{++0}$ is defined by its degree d and the multiplicity c of the isotropic absolute conic on the projective model of the surface in $\mathbb{P}^3$. For example, a vertical circular cylinder

$$x_1^2 + x_2^2 = r^2 x_0^2, \quad \boldsymbol{x} \in \mathbb{P}^3, \quad r \in \mathbb{R} \quad (7)$$

is of degree 2 and contains the isotropic absolute conic with multiplicity 1, as $x_1^2 + x_2^2$ occurs with multiplicity 1 if we set $x_0 = 0$ in (7). Thus the i-M type of a vertical circular cylinder is $(2, 1)$. The *i-Möbius model* of a surface $W \subset \mathbb{P}^3$ is the closure of its inverse projection in the Blaschke cylinder $\overline{\sigma(W)} \subset \mathcal{B}$, and its *i-Möbius degree* is the degree of the i-M model.

In the following three propositions we prove the isotropic counterparts to [6, Proposition 27–29], to a large extent using the same approach as in [6].

Proposition 1 (Factorization of i-M Transformations). *Let $\mu : \mathbb{P}^3 \to \mathbb{P}^3$ be an i-M transformation. Then there exists a unique linear isomorphism $\beta : \mathcal{B} \to \mathcal{B}$ such that the* i-M transformation diagram *of μ commutes:*

$$
\begin{array}{ccc}
\mathcal{B} & \xrightarrow{\ \beta\ } & \mathcal{B} \\[4pt]
{\scriptstyle\sigma}\big\uparrow & & \big\downarrow{\scriptstyle\pi} \\[4pt]
\mathbb{P}^3 & \xrightarrow{\ \mu\ } & \mathbb{P}^3
\end{array}
\tag{8}
$$

Proof. We have $\pi \circ \sigma (x_0, x_1, x_2, x_3) = (x_0, x_1, x_2, x_3)$, so we can write $\mu = \pi \circ (\sigma \circ \mu \circ \pi) \circ \sigma$. Thus the i-M transformation diagram commutes if β can be written as $\beta = \sigma \circ \mu \circ \pi$. We can explicitly calculate β for each of the four generators of i-M transformations.

Let $\boldsymbol{x} = (x_0, x_1, x_2, x_3, x_\infty) \in \mathcal{B}$, $r \in \mathbb{R}$, and $\boldsymbol{p} = (p_1, p_2, p_3) \in \mathbb{R}_{++0}^3$. If $\mu = S_r$ is a uniform scaling, then $\beta_r = \sigma \circ S_r \circ \pi$ and

$$
\begin{aligned}
\beta_r(\boldsymbol{x}) &= (x_0^2, rx_0x_1, rx_0x_2, rx_0x_3, \tfrac{1}{2}r^2(x_1^2 + x_2^2)) \\
&= (x_0, rx_1, rx_2, rx_3, r^2 x_\infty).
\end{aligned}
\tag{9}
$$

If $\mu = T_{\boldsymbol{p}}$ is a translation, then $\beta_{\boldsymbol{p}} = \sigma \circ T_{\boldsymbol{p}} \circ \pi$ and

$$
\begin{aligned}
\beta_{\boldsymbol{p}}(\boldsymbol{x}) = &\ (x_0^2, x_0(x_1 + p_1 x_0), x_0(x_2 + p_2 x_0), x_0(x_3 + p_3 x_0), \\
&\ \tfrac{1}{2}((x_1 + p_1 x_0)^2 + (x_2 + p_2 x_0)^2)) \\
= &\ (x_0, x_1 + p_1 x_0, x_2 + p_2 x_0, x_3 + p_3 x_0, \\
&\ (p_1^2 + p_2^2)x_0 + p_1 x_1 + p_2 x_2 + x_\infty).
\end{aligned}
\tag{10}
$$

If $\mu = \mathrm{inv}$ is an inversion, then $\beta_{\mathrm{inv}} = \sigma \circ \mathrm{inv} \circ \pi$ and

$$
\begin{aligned}
\beta_{\mathrm{inv}}(\boldsymbol{x}) = &\ ((x_1^2 + x_2^2)^2, (x_1^2 + x_2^2)x_0x_1, (x_1^2 + x_2^2)x_0x_2, \\
&\ (x_1^2 + x_2^2)x_0x_3, \tfrac{1}{2}(x_1^2 + x_2^2)x_0^2) \\
= &\ (2x_\infty, x_1, x_2, x_3, \tfrac{1}{2}x_0).
\end{aligned}
\tag{11}
$$

If $\mu = $ is a vertical reflection, then $\beta_{\mathrm{inv}^0} = \sigma \circ \mathrm{inv}^0 \circ \pi$ and

$$
\begin{aligned}
\beta_{\mathrm{inv}^0}(\boldsymbol{x}) &= (x_0^2, x_0x_1, x_0x_2, -x_0x_3, \tfrac{1}{2}(x_1^2 + x_2^2)) \\
&= (x_0, x_1, x_2, -x_3, x_\infty).
\end{aligned}
\tag{12}
$$

In each case, we simplify the last term using (3), as the point x lies on the Blaschke cylinder. For each of these transformations, β is linear, and as the associated matrices are invertible, β is bijective and thus a linear isomorphism. This proves the existence and uniqueness of β, thus proving the proposition.

Proposition 2 (i-M Degree). *Let* $W \subset \mathbb{P}^3$ *be a variety.*

(a) The i-M degree of W *is equal to the number of intersection points with* $\dim(W)$ *i-M spheres, outside the isotropic absolute conic.*

(b) If W *is a surface of i-M type* (d, c), *then the i-M degree of* W *is* $2(d - c)$.

Proof. $\sigma(W) \subset \mathcal{B} \subset \mathbb{P}^4$ is the i-Möbius model of W. Its degree is equal to the number of intersection points with $\dim(\sigma(W)) = \dim(W)$ hyperplanes in $\mathbb{P}^4$. Hyperplanes in $\mathbb{P}^4$ pulls back along σ onto i-M spheres in $\mathbb{P}^3$ [4, Theorem 3]. As σ is defined outside the isotropic absolute conic, this proves (a).

When W is a surface, (a) gives its degree as the number of intersection points with two i-M spheres. The intersection of two i-M spheres is an i-M circle, which in general has $2d$ intersection points with W. $2c$ of these lie on the isotropic absolute conic, as an i-M circle intersects the isotropic absolute conic in two points (counting multiplicities) and the isotropic absolute conic has multiplicity c in W. Thus the number of intersection points of two i-M spheres with W outside the isotropic absolute conic is $2(d - c)$, proving (b).

Proposition 3 (i-M Type). *Let* $W \in \mathbb{P}^3$ *be a surface of i-M type* (d, c).

(a) The i-M degree is invariant under i-M transformations.

(b) For any point $p \in \mathbb{P}^3$ *outside the isotropic absolute conic, of multiplicity* m *with respect to* W, *there exists an i-M transformation* μ *such that the i-M type of* $\mu(W)$ *is*

$$(2(d - c) - m, (d - c) - m). \tag{13}$$

(c) The number of families of i-M circles on W *is invariant under i-M transformation.*

Proof. As degree is invariant under linear isomorphisms, (a) follows from Proposition 1.

We prove (b) by tracing the changes in degrees in the i-M transformation diagram. W is of i-M type (d, c), so the degree of $\sigma(W)$ is $2(d - c)$, after Proposition 2 (b). Applying (a), this is also the degree of $\beta \circ \sigma(W)$.

Consider the i-M transformation inv_p inverting the surface through the point $p \in \mathbb{R}^3_{++0}$, with the i-M transformation diagram

$$\begin{array}{ccc} \mathcal{B} & \xrightarrow{\ \beta\ } & \mathcal{B} \\ {\scriptstyle\sigma}\uparrow & & \downarrow{\scriptstyle\pi} \\ \mathbb{P}^3 & \xrightarrow{\ \mathrm{inv}_p\ } & \mathbb{P}^3 \end{array} \tag{14}$$

If $\boldsymbol{x} = (x_0, x_1, x_2, x_3) \in \mathbb{P}^3$, then

$$
\begin{aligned}
\mathrm{inv}_{\boldsymbol{p}}(\boldsymbol{x}) &= T_{\boldsymbol{p}} \circ \mathrm{inv} \circ T_{\boldsymbol{p}}^{-1}(x_0, x_1, x_2, x_3) \\
&= T_{\boldsymbol{p}}\left((x_1 - p_1 x_0)^2 + (x_2 - p_2 x_0)^2, x_0(x_1 - p_1 x_0), x_0(x_2 - p_2 x_0), x_0(x_3 - p_3 x_0)\right) \\
&= \begin{bmatrix}
(x_1 - p_1 x_0)^2 + (x_2 - p_2 x_0)^2 \\
x_0(x_1 - p_1 x_0) + \left((x_1 - p_1 x_0)^2 + (x_2 - p_2 x_0)^2\right) p_1 \\
x_0(x_2 - p_2 x_0) + \left((x_1 - p_1 x_0)^2 + (x_2 - p_2 x_0)^2\right) p_2 \\
x_0(x_3 - p_3 x_0) + \left((x_1 - p_1 x_0)^2 + (x_2 - p_2 x_0)^2\right) p_3
\end{bmatrix}^T = \boldsymbol{y}
\end{aligned}
\tag{15}
$$

To find the intersection with the isotropic absolute conic, we eliminate the first coordinate $\boldsymbol{y}_0$, and find that $\frac{x_1}{x_0} = p_1$ and $\frac{x_2}{x_0} = p_2$. This also eliminates $y_1^2 + y_2^2$, so the intersection with the isotropic absolute conic is exactly the image of the isotropic (vertical) line through the point $\boldsymbol{p}$. Given two i-M spheres passing through $\boldsymbol{p}$, they have no other intersection points with the isotropic line through $\boldsymbol{p}$. As i-M transformations map i-M spheres to i-M spheres [4, Lemma 2], the images of these two i-M spheres under $\mathrm{inv}_{\boldsymbol{p}}$ are two i-M spheres passing through the point $\mathrm{inv}_{\boldsymbol{p}}(\boldsymbol{p})$ on the isotropic absolute conic. By [4, Theorem 3], their images under σ are two planes in $\mathbb{P}^4$. The planes intersect $\beta \circ \sigma(W)$ in $2(d - c)$ points. To obtain the degree of $\mathrm{inv}_{\boldsymbol{p}}(W) = \pi \circ \beta \circ \sigma$ we need to subtract the intersections with the absolute quadric, counting multiplicities, thus if m is the multiplicity of $\boldsymbol{p}$ with respect to W (see, e.g., [3] for the definition of the multiplicity of a point with respect to a surface) then the degree of $\mathrm{inv}_{\boldsymbol{p}}(W)$ is $2(d - c) - m$.

Proposition 2 gives the relationship between the i-M type of a surface and the degree of its Möbius model. Knowing that the degree of $\mathrm{inv}_{\boldsymbol{p}}(W)$ is $2(d - c) - m$ and the degree of its Möbius model is $2(d - c)$, we find that the multiplicity c of the isotropic absolute conic in $\mathrm{inv}_{\boldsymbol{p}}(W)$ is $(d - c) - m$. Thus the i-M type of the surface is $(2(d - c) - m, (d - c) - m)$, proving (b).

(c) is immediate for uniform scalings, translations, and vertical reflections. From [4, Lemma 2], i-M spheres and circles are preserved under inversions. As no two i-M circles are inverted onto the same i-M circle, the number of families of i-M circles on W is invariant for inversions. This concludes the proof of the proposition.

3   Families of i-M Circles on Singular Isotropic Cyclides

In [2] we classify quadratic surfaces in $\mathbb{R}^3_{++0}$ up to isometric equivalence, and give canonical forms for each of the 23 classes. For each of the canonical forms, we describe the families of lines and i-M circles they contain. In particular, we find that one-sheeted hyperboloids contain four such families and cones contain three. By Proposition 3(c), the number of such families are preserved under i-M transformations, and Proposition 3(b) express the change in the i-M type of the surface under inversion. In this section we will study the change in i-M type of quadratic surfaces in $\mathbb{R}^3_{++0}$ under the inversion $\mathrm{inv}_{\boldsymbol{p}}$, and show that these inverse surfaces include singular isotropic cyclides. Our results for quadratic surface can therefore be extended to a classification of i-M circles on isotropic cyclides with real singularities.

Consider the i-M sphere

$$x_1^2 + x_2^2 = x_0 x_3, \tag{16}$$

a vertical rotational elliptic paraboloid. Its i-M type is $(2,1)$, so $d - c = 1$. Thus the inversion inv_p in a point p changes the i-M type to $(2 - m, 1 - m)$, where m is the multiplicity of p with respect to the i-M sphere. If p is not on the surface, then $m = 0$ and the i-M type is unchanged by inversion. As i-M spheres are inverted onto i-M spheres [4, Lemma 2], this is expected. If p lies on the surface, $m = 1$, and the i-M type of the inverse is $(1,0)$, i.e., a non-isotropic plane.

The circular cylinder in (7) is also of i-M type $(2,1)$, resulting in the same changes in i-M type as for the i-M sphere. In this case the inverse of i-M type $(2,1)$ remains a vertical circular cylinder, as inversions preserve isotropic (vertical) lines and circular top views. This also means that an inverse of i-M type $(1,0)$ is an isotropic plane.

Conversely, if we start from a plane, then $d - c = 1$ and the i-M type of its inverse is $(2 - m, 1 - m)$: it is inverted onto either a plane, or an i-M sphere or a vertical circular cylinder if p is respectively on or outside the plane. If the plane is isotropic, its inverse is either an isotropic plane or a vertical circular cylinder. If it is non-isotropic, its inverse is either a non-isotropic plane or an i-M sphere.

For the other quadratic surfaces in $\mathbb{R}^3_{++0}$, the multiplicity of the isotropic absolute conic is zero, so their i-M type is $(2,0)$. This gives us $d - c = 2$, and inversion in a point p changes the i-M type to $(4 - m, 2 - m)$. If p is outside the quadric, $m = 0$ and the inverse is a quartic surface containing the absolute conic with multiplicity 2. If p is a non-singular point on the quadric, $m = 1$ and the inverse is a cubic surface containing the absolute conic with multiplicity 1. If p is a singular point, then $m = 2$ and the inverse remains a quadratic surface of i-M type $(2,0)$.

An *isotropic cyclide* is defined by the equation

$$a \left(x_1^2 + x_2^2\right)^2 + L(\boldsymbol{x}) \left(x_1^2 + x_2^2\right) + Q(\boldsymbol{x}) = 0, \quad \boldsymbol{x} = (x_1, x_2, x_3) \in \mathbb{R}^3_{++0}, \tag{17}$$

where L and Q are respectively linear and quadratic polynomials. If $a \neq 0$, the cyclide is of degree 4, and if $a = 0$ and $L \neq 0$, it is of degree 3. This class of surfaces include all quadratic surfaces, for $a = L = 0$. If we homogenize (17) and set $x_0 = 0$, we see that the isotropic cyclide contains the isotropic absolute conic with multiplicity 2 when it is quartic, and 1 if it is cubic. Thus the i-M type of a quartic isotropic cyclide is $(4,2)$, and the i-M type of a cubic isotropic cyclide is $(3,1)$. In both cases $d - c = 2$ and the i-M type of the inverse surface is $(4 - m, 2 - m)$.

If p is outside the cyclide the i-M type of the inverse is $(4,2)$, in fact the inverse is a quartic isotropic cyclide. If p is a non-singular point, then the i-M type of the inverse is $(3,1)$, i.e., a cubic isotropic cyclide. If we invert through a singular point of multiplicity 2, the resulting i-M type is $(2,0)$, i.e., a quadratic surface that is neither an i-M sphere nor a vertical circular cylinder.

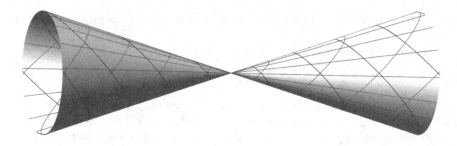

Fig. 1. A cone with its three families of i-M circles.

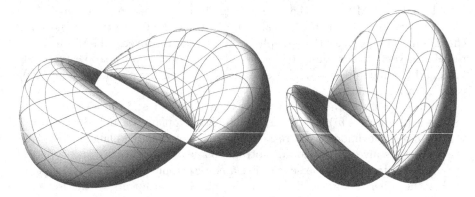

Fig. 2. Two quartic isotropic cyclides, inverse images of a horizontal cone, and their three families of i-M circles.

The inverse relation between quadratic surfaces and singular isotropic cyclides implies that our classification of families of lines and i-M circles on quadrics in [2] can be extended to a classification of families of lines and i-M circles on singular isotropic cyclides. In particular, a consequence of Proposition 3(c) is that the total number of such families is at most 4, in the case of inversions of one-sheeted hyperboloids. As a one-sheeted hyperboloid has no real singularities, its inverse isotropic cyclide has a single finite singularity corresponding to the ideal points of the hyperboloid. An isotropic cyclide with two finite real singularities is inverse to a cone, and thus contains 3 families of lines and/or i-M circles.

Example: The Inverse Surfaces of a Cone

We illustrate the change in i-M type under the inversion $\text{inv}_{\boldsymbol{p}}$ by an example. Consider the cone defined by

$$x_2^2 + x_3^2 = a^2 x_1^2, \tag{18}$$

i.e., the canonical form $\mathscr{C}_h$ of horizontal cones in $\mathbb{R}^3_{++0}$ [2]. This cone contains one family of lines and two families of elliptic i-M circles, shown in Fig. 1. In general the inverse of the cone is the isotropic cyclide defined by the implicit equation

$$\alpha_1 \left(x_1^2 + x_2^2\right)^2 + L \left(x_1^2 + x_2^2\right) + Q + L\alpha_2 + \alpha_1 \alpha_2^2 = 0 \qquad (19)$$

where

$$\begin{aligned}
&\alpha_1 = a^2 p_1^2 - p_2^2 - p_3^2, \qquad \alpha_2 \;=\; p_1^2 + p_2^2 - 1, \\
&L = -2 \left(p_1 \left(2\alpha_1 - a^2\right) x_1 + p_2 \left(2\alpha_1 + 1\right) x_2 + p_3 x_3\right), \\
&Q = - \left(2\alpha_1 \left(2p_2^2 - 3\alpha_2\right) + 4 \left(p_2^2 + p_3^2\right) - a^2\right) x_1^2 + \left(2\alpha_1 \left(2p_2^2 + \alpha_2\right) + 4p_2^2 - 1\right) x_2^2 \\
&\quad - x_3^2 + 4 p_1 p_2 \left(2\alpha_1 + 1 - a^2\right) x_1 x_2 + 4 \left(p_1 x_1 + p_2 x_2\right) p_3 x_3. \qquad (20)
\end{aligned}$$

If the centre of inversion $\boldsymbol{p} = (p_1, p_2, p_3)$ lies outside the cone, we have $\alpha_1 \neq 0$ and the isotropic cyclide is quartic of i-M type $(4, 2)$. Two examples of quartic isotropic cyclides are shown in Fig. 2.

If $\boldsymbol{p}$ lies on the cone, then $\alpha_1 = 0$. We find that $L = 0$ if and only if $\boldsymbol{p} = (0, 0, 0)$, thus the cyclide is a quadratic surface of i-M type $(2, 0)$ if and only if the centre of inversion is the apex of the cone, i.e., a singular point of multiplicity 2. The implicit equation of this inverse surface is identical to (18).

If $\boldsymbol{p}$ is a non-singular point on the cone, $\alpha_1 = 0$, $L \neq 0$, and the cyclide is cubic. Thus the i-M type of the isotropic cyclide is $(3, 1)$. An example of a cubic isotropic cyclide is shown in Fig. 3.

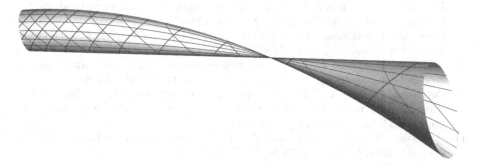

Fig. 3. A segment of a cubic isotropic cyclide, inverse image of a horizontal cone, and its three families of i-M circles.

4   Conclusions

We have outlined the isotropic counterpart of the Möbius model described in [6], and shown how the degree of a surface changes under i-M inversions while the number of families of i-M circles remain constant. We find a generalization of the classification of families of i-M circles and lines on quadratic surfaces in

$\mathbb{R}^3_{++0}$ [2] to families on isotropic cyclides with real singularities. In particular, we find that singular isotropic cyclides contain up to 4 families of lines and/or i-M circles.

This paper is the starting point for the study of families of i-M circles on non-singular isotropic cyclides, and for a generalization of Lubbes' celestial surfaces [6] to isotropic space.

Acknowledgements. We would like to thank Rimvydas Krasauskas, Niels Lubbes, Torgunn Karoline Moe, and Severinas Zubė for our discussions about and their feedback on this paper. We would also like to thank the reviewers for their constructive comments.

References

1. Cecil, T.E.: Lie Sphere Geometry. Springer, New York (2008)
2. Dahl, H.E.I., Krasauskas, R.: Quadrics in isotropic space and applications (2014)
3. Hartshorne, R.: Algebraic Geometry. Graduate texts in Mathematics, vol. 52. Springer, New York (1977)
4. Krasauskas, R., Zubė, S., Cacciola, S.: Bilinear Clifford-Bézier patches on isotropic cyclides. In: Floater, M., Lyche, T., Mazure, M.-L., Mørken, K., Schumaker, L.L. (eds.) MMCS 2012. LNCS, vol. 8177, pp. 283–303. Springer, Heidelberg (2014)
5. Kreyszig, E.: Differential Geometry. Dover Publications, New York (1991)
6. Lubbes, N.: Families of circles on surfaces. Accepted for publication in Contributions to Algebra and Geometry (2013)
7. Peternell, M., Pottmann, H.: A Laguerre geometric approach to rational offsets. Comput. Aided Geom. Des. **15**(3), 223–249 (1998)
8. Peternell, M., Pottmann, H.: Applications of Laguerre geometry in CAGD. Comput. Aided Geom. Des. **15**(2), 165–186 (1998)
9. Pottmann, H., Grohs, P., Mitra, N.J.: Laguerre minimal surfaces, isotropic geometry and linear elasticity. Adv. Comput. Math. **31**(4), 391–419 (2008)
10. Pottmann, H., Liu, Y.: Discrete surfaces in isotropic geometry. In: Martin, R., Sabin, M.A., Winkler, J.R. (eds.) Mathematics of Surfaces 2007. LNCS, vol. 4647, pp. 341–363. Springer, Heidelberg (2007)
11. Pottmann, H., Shi, L., Skopenkov, M.: Darboux cyclides and webs from circles. Comput. Aided Geom. Des. **29**(1), 77–97 (2012)

Symbolic Computation of Equi-affine Evolute for Plane B-Spline Curves

Éric Demers[(✉)], François Guibault, and Christophe Tribes

École Polytechnique de Montréal, 2900 Boulevard Edouard-Montpetit,
Montréal QC H3T 1J4, Canada
eric.demers@polymtl.ca

Abstract. In Euclidean plane geometry, the evolute of both B-spline and NURBS curves are NURBS curves. Moreover, this evolute can be computed symbolically. This article extends those results to the equi-affine plane geometry. Analogously to Euclidean geometry, the equi-affine evolute of both B-spline and NURBS curves are NURBS curves and an algorithm for the symbolic computation is given for B-spline curves. This results in a new method to analyze the global affine differential properties of B-spline curves and assess B-spline curve quality in an affine invariant context.

Keywords: B-spline · NURBS · Polynomial · Rational · Equi-affine · Evolute

1 Introduction

1.1 Motivation

In Euclidean geometry, curvature and arc length are invariant under the group of isometric transformations of space, which consists of translations and rigid rotations. In equi-affine, affine and projective geometries, it is also possible to identify properties of curves that are invariant under the special affine group, affine group and projective linear group respectively. This hierarchy of geometries is important to sight, computer visualization and image recognition. These geometries among others were organized in a coherent whole by Felix Klein [17].

Our working hypothesis is that notions of this hierarchy of geometries, and not exclusively notions of Euclidean geometry, can be useful to improve the appearance, simplify and better control computer generated curves. Among the above hierarchy of geometries, the equi-affine geometry is the closest to Euclidean geometry, which makes it a natural starting point for investigation. Wherefore the decision to explore the possibilities offered by equi-affine geometry.

An equi-affine transformation of the plane preserves areas, straight lines, ellipses, parabolas and hyperbolas. However it does not necessarily preserve distances and angles [12]. Consequently, the differential invariants of this geometry are of higher order.

© Springer International Publishing Switzerland 2015
J.-D. Boissonnat et al. (Eds.): Curves and Surfaces 2014, LNCS 9213, pp. 169–180, 2015.
DOI: 10.1007/978-3-319-22804-4_13

Among relevant notions of equi-affine geometry, the evolute was selected since it provides a mechanism to analyze global differential properties of a curve. The focus of this article is the equi-affine evolute of a parametric plane curve and its symbolic computation.

1.2 Literature Review

The Euclidean version of the evolute and its inverse operation, the involute, are frequently mentioned in the literature [5,8,21]. They played an important role in the early days of differential geometry [18,22]. The evolute has a close geometric link with medial axis, the symmetry set and parallel curves. The medial axis is a subset of the symmetry set. An important link between the evolute and the symmetry set is that an endpoint of the symmetry set lies at a cusp of the evolute [10]. While the link between evolute and parallel curve is that parallel curves share the same evolute [7]. The evolute is also a subject of study in singularity theory [2].

The equi-affine evolute is less frequently mentioned in the literature. One can study the equi-affine evolute with singularity theory where the correspondence between the equi-affine evolute cusps and sextactic points of the curve is proven [16]. Furthermore, the endpoints of the affine distance symmetry set lie in the cusps of the equi-affine evolute [15].

Authors of classic books that address equi-affine differential geometry include W.S. Blaschke [1], E. Cartan [3,4] and H. W. Guggenheimer [13].

1.3 Outline

Section 2 derives the equi-affine differential invariants using the method of the moving frame. Section 3 proves that the equi-affine evolute of both polynomial and rational curves are rational curves. Consequently, the equi-affine evolute of both B-spline and NURBS curves are NURBS curves. Section 4 discusses properties of the equi-affine evolute. Section 5 applies the results of Sect. 3 to compute symbolically the equi-affine evolute of a plane B-spline curve.

2 Equi-affine Differential Geometry and the Extended Affine Plane

2.1 The Frenet-Serret Equations of Equi-affine Geometry

We denote $v \wedge w = v_1 w_2 - v_2 w_1$ the area of the parallelogram spaned by two vectors of the plane. In equi-affine geometry, the moving frame R is no longer composed of orthogonal vectors of unit length as in Euclidean geometry, but rather spans a parallelogram of unit area. For this purpose $i = \frac{d\gamma}{d\sigma}$ and $j = \frac{d^2\gamma}{d\sigma^2}$ are selected. Then we look for a parameter σ such that $det(R) = 1$, that is, $\frac{d\gamma}{d\sigma} \wedge \frac{d^2\gamma}{d\sigma^2} = 1$ all along the curve.

Equi-affine geometry has thus an arc length σ different than the one for Euclidean geometry. The expression of the equi-affine arc length gives to a curve its natural equi-affine parametrization.

To express this natural parametrization σ, we use the chain rule to obtain the two equations:

$$\frac{d\gamma}{du} = \frac{d\gamma}{d\sigma}\frac{d\sigma}{du}$$

$$\frac{d^2\gamma}{du^2} = \frac{d^2\gamma}{d\sigma^2}\left(\frac{d\sigma}{du}\right)^2 + \frac{d\gamma}{d\sigma}\frac{d^2\sigma}{du^2}$$

Which are used to find the expression:

$$\left(\frac{d\gamma}{du} \wedge \frac{d^2\gamma}{du^2}\right) = \left(\frac{d\gamma}{d\sigma} \wedge \frac{d^2\gamma}{d\sigma^2}\right)\left(\frac{d\sigma}{du}\right)^3$$

For the sake of conciseness, henceforward the notation $\gamma_u = \frac{d\gamma}{du}$ and $\gamma_\sigma = \frac{d\gamma}{d\sigma}$ will be used. By applying the constraint $\gamma_\sigma \wedge \gamma_{\sigma\sigma} = 1$ we get the result:

$$d\sigma = (\gamma_u \wedge \gamma_{uu})^{1/3} du \tag{1}$$

Which is the expression of the equi-affine arc length for a curve with a general parameter u.

If this general parameter happen to be the Euclidean arc length s, knowing that the Euclidean curvature is $k = \gamma_s \wedge \gamma_{ss}$, the last equation becomes:

$$d\sigma = k^{1/3} ds \tag{2}$$

The Frenet-Serret equations are as follows:

$$\frac{d\gamma}{d\sigma} = i \tag{3}$$

$$\frac{di}{d\sigma} = j \tag{4}$$

$$\frac{dj}{d\sigma} = -\kappa i \tag{5}$$

where:

$$\kappa = \gamma_{\sigma\sigma} \wedge \gamma_{\sigma\sigma\sigma} \tag{6}$$

And since $i = \gamma_\sigma$ and $j = \gamma_{\sigma\sigma}$, Eq. (5) can be written:

$$\gamma_{\sigma\sigma\sigma} = -\kappa\gamma_\sigma \tag{7}$$

Drawing an analogy with Euclidean geometry, the vector $i = \gamma_\sigma$ is called the equi-affine tangent vector, the vector $j = \gamma_{\sigma\sigma}$ is called the equi-affine normal and $\kappa = \gamma_{\sigma\sigma} \wedge \gamma_{\sigma\sigma\sigma}$ is called the equi-affine curvature.

2.2   The Extended Affine Plane

To define the extended affine plane let us first add one dimension to the plane. If the coordinates of the space are (x, y, z), then a point on the affine plane is obtained by the projection $(x/z, y/z)$. This projection operation maps lines on the extended (x, y, z) space to points in the affine plane located at $z = 1$. A point at infinity is expressed by the coordinates $(x, y, 0)$. By definition the extended affine plane is the affine plane to which are added the points at infinity.

3   Equi-affine Evolute

The equation for the equi-affine evolute with respect to the natural parameter σ can be found in [1,13].

$$\epsilon = \gamma + \frac{\gamma_{\sigma\sigma}}{\kappa} \tag{8}$$

When $\kappa = 0$ the evolute has a point located at infinity which is part of the extended affine plane.

To the best of our knowledge, the following result that, in the plane, the equi-affine evolute of both polynomial and rational curves are rational curves is new.

Proposition 1. *The equi-affine evolute ϵ of a parametric curve $\gamma(u)$ where u is a general parameter is obtained from the following equation:*

$$\epsilon = \gamma + \frac{(\gamma_u \wedge \gamma_{uu})^2 \gamma_{uu} - \frac{1}{3}(\gamma_u \wedge \gamma_{uu})(\gamma_u \wedge \gamma_{uuu})\gamma_u}{\frac{1}{3}(\gamma_u \wedge \gamma_{uu})(4\gamma_{uu} \wedge \gamma_{uuu} + \gamma_u \wedge \gamma_{uuuu}) - \frac{5}{9}(\gamma_u \wedge \gamma_{uuu})^2} \tag{9}$$

Proof. The chain rule and Eq. (1) yield:

$$\gamma_\sigma = \gamma_u \frac{du}{d\sigma} = \gamma_u(\gamma_u \wedge \gamma_{uu})^{-1/3} \tag{10}$$

Differentiating a second time:

$$\gamma_{\sigma\sigma} = \gamma_{uu}(\gamma_u \wedge \gamma_{uu})^{-2/3} - \frac{1}{3}\gamma_u(\gamma_u \wedge \gamma_{uuu})(\gamma_u \wedge \gamma_{uu})^{-5/3} \tag{11}$$

Differentiating a third time:

$$\begin{aligned}
\gamma_{\sigma\sigma\sigma} = {} & \gamma_{uuu}(\gamma_u \wedge \gamma_{uu})^{-1} - \gamma_{uu}(\gamma_u \wedge \gamma_{uu})^{-2}(\gamma_u \wedge \gamma_{uuu}) \\
& + \frac{5}{9}\gamma_u(\gamma_u \wedge \gamma_{uu})^{-3}(\gamma_u \wedge \gamma_{uuu})^2 \\
& - \frac{1}{3}\gamma_u(\gamma_u \wedge \gamma_{uu})^{-2}(\gamma_{uu} \wedge \gamma_{uuu}) \\
& - \frac{1}{3}\gamma_u(\gamma_u \wedge \gamma_{uu})^{-2}(\gamma_u \wedge \gamma_{uuuu})
\end{aligned}$$

This allows to obtain the equi-affine curvature:

$$\kappa = \gamma_{\sigma\sigma} \wedge \gamma_{\sigma\sigma\sigma} = \frac{4\gamma_{uu} \wedge \gamma_{uuu} + \gamma_u \wedge \gamma_{uuuu}}{3(\gamma_u \wedge \gamma_{uu})^{5/3}} - \frac{5}{9} \frac{(\gamma_u \wedge \gamma_{uuu})^2}{(\gamma_u \wedge \gamma_{uu})^{8/3}} \tag{12}$$

Hence we find that:

$$\frac{\gamma_{\sigma\sigma}}{\kappa} = \frac{\gamma_{uu}(\gamma_u \wedge \gamma_{uu})^{-2/3} - \frac{1}{3}\gamma_u(\gamma_u \wedge \gamma_{uuu})(\gamma_u \wedge \gamma_{uu})^{-5/3}}{\frac{4\gamma_{uu}\wedge\gamma_{uuu}+\gamma_u\wedge\gamma_{uuuu}}{3(\gamma_u\wedge\gamma_{uu})^{5/3}} - \frac{5}{9}\frac{(\gamma_u\wedge\gamma_{uuu})^2}{(\gamma_u\wedge\gamma_{uu})^{8/3}}}$$

The expected result is obtained by multiplying the numerator and the denominator by $(\gamma_u \wedge \gamma_{uu})^{8/3}$.   □

Corollary 1. *In the plane, the equi-affine evolute of both polynomial and rational curves are rational curves.*

Proof. This corollary is deduced from the facts that the derivative of a rational curve is a rational curve, that the multiplication of two rational functions is a rational function and that there is no fractional exponent in Eq. 9 of the last proposition. The same conclusion applies to polynomial curves which are a subset of rational curves.   □

Corollary 2. *In the plane, the equi-affine evolute of both B-spline and NURBS curves are NURBS curves.*

Proof. This corollary is a consequence of corollary 1 since a B-spline curve is a piecewise polynomial curve and since a NURBS curve is a piecewise rational polynomial curve.   □

Corollary 3. *The equi-affine evolute of a polynomial curve γ of degree $n \geq 2$ is a rational curve of degree $(5n - 8)$.*

Proof. Suppose a polynomial curve $\gamma(u)$ of degree n and its derivatives of degree $(n-1), (n-2), \ldots$.
The term $\gamma_u \wedge \gamma_{uu}$ gives a function of degree $(n-1) + (n-2) = 2n - 3$.
The term $(\gamma_u \wedge \gamma_{uu})^2 \gamma_{uu}$ gives a curve of degree $2(2n-3)+(n-2) = (5n-8)$. Other terms in Eq. (9) have degree less or equal to this one.

□

Corollary 4. *The equi-affine evolute of a rational curve γ of degree $n \geq 2$ is a rational curve of degree $(25n - 8)$.*

Proof. Suppose a rational curve $\gamma(u)$ of degree n.

$$\gamma(u) = \frac{p(u)}{w(u)}$$

where $p(u)$ is a polynomial curve and $w(u)$ a polynomial function both of degree n.

The numerator of γ_u is of degree $(2n - 1)$.

$$\gamma_u(u) = \frac{\boldsymbol{p}_u w - \boldsymbol{p} w_u}{w^2}$$

The numerators of $\gamma_{uu}, \gamma_{uuu}, \gamma_{uuuu}$ are respectively of degree $(4n-2), (8n-3), (16n-4)$.

The term $\gamma(\gamma_u \wedge \gamma_{uu})(\gamma_u \wedge \gamma_{uuuu})$ gives a curve of degree $(n) + (2n-1) + (4n-2) + (2n-1) + (16n-4) = 25n-8$. Other terms in the Eq. (9) have degree less or equal to this one.   □

Corollary 5. *The equi-affine evolute of a rational Bézier curve of degree 2 is a rational Bézier curve of degree 42 with all coincident control points.*

This result follows from the last corollary and from the fact that a conic has for equi-affine evolute a single point. This is a good test for a NURBS symbolic library!

In comparison, the equi-affine evolute of a polynomial (nonrational) curve of degree 2 is a rational curve of degree 2 (see corollary 3). To avoid high degrees we will limit ourselves to equi-affine evolute of (nonrational) B-spline curves.

4   Some Properties of the Equi-affine Evolute

4.1   Six Sextactic Points Theorem

The analog of the four vertex theorem [19] of Euclidean geometry is the six sextactic points theorem [14] in equi-affine geometry. A sextactic point on a curve corresponds to a local maximum or minimum of equi-affine curvature. A convex closed curve that is not a conic has a minimum of six sextactic points. To a sextactic point of a curve γ will correspond a cusp point on the equi-affine evolute ϵ [16]. Under the same circumstances, the equi-affine evolute will therefore have a minimum of six cusp points.

4.2   Equi-affine Evolute and Affine W-Curves

The equi-affine evolute of an affine w-curve is an affine w-curve [1]. Affine w-curves are a set that contains conic, powers (like $y = x^3$), exponential and logarithmic spiral curves. An affine w-curve equi-affine curvature κ is either constant or $\kappa = \frac{c}{\sigma^2}$ where c is a constant and σ is the equi-affine arc length.

4.3   Curvature Interpretation

The role played by the circle in Euclidean geometry is played by the conic in equi-affine geometry. The variation of the Euclidean curvature expresses how a curve differs from a circle at a given point. Likewise the variation of the equi-affine curvature expresses how a curve differs from a conic at a given point. The Euclidean evolute of a circle is a single point that can be called the center of Euclidean curvature. Likewise, the equi-affine evolute of a conic is a single point that can be called the center of equi-affine curvature.

4.4 Euclidean, Affine and Projective Invariance

Euclidean curvature extrema called vertex have corresponding cusps on the evolute. The Euclidean evolute shape is invariant under the isometry and similarity group of transformations. Moreover, the curvature extrema, hence the number of cusp points on this evolute, are invariant under the more general Möbius transformations. Hence, the number of cusps on the evolute provides a mechanism to analyze global differential properties of a plane curve invariant under isometry, similarity and Möbius transformations.

Analogously equi-affine curvature extrema called sextactic points have corresponding cusps on the equi-affine evolute [16]. The equi-affine evolute shape is invariant under the group of isometry, similarity, equi-affine and affine transformations. Moreover, the sextactic points, hence the number of cusp points on this evolute, are invariant under projective transformations. Hence, the number of cusps on the equi-affine evolute provides a mechanism to analyze global differential properties of a plane curve invariant under isometry, similarity, equi-affine, affine and projective transformations.

5 Symbolic Computation of a B-Spline Equi-affine Evolute

Equation 9 of Sect. 3 has allowed to establish that the equi-affine evolute of a polynomial curve is a rational curve. This section uses this result to compute symbolically the equi-affine evolute of a B-spline curve.

A NURBS can be considered as a variable on which algebraic operations can be performed to obtain a new NURBS [6,20]. Symbolic computation for the equi-affine evolute is a direct application of this principle.

The first step of Algorithm 1, to compute a B-spline equi-affine evolute, is to decompose a B-spline curve in Bézier components using a knot insertion algorithm. Algebraic operations of Eq. (9) are then applied to Bézier curves as explained below. See also [23] for more details on how to build a library of symbolic computation of polynomials in Bernstein form.

5.1 Bézier Curve

The notation used here for a Bézier curve γ of degree n is:

$$\gamma(u) = \sum_{i=0}^{n} B_{i,n}(u)\boldsymbol{P}_i \quad 0 \leq u \leq 1$$

where the functions $B_{i,n}(u)$ are the Bernstein basis polynomials [9]:

$$B_{i,n}(u) = \binom{n}{i} u^i (1-u)^{n-i}$$

and where the points $\boldsymbol{P}_i$ define the control polygon.

Algorithm 1. B-spline equi-affine evolute computation
Input: B-spline curve
Output : Rational Bézier curves

1: Decompose a B-spline curve in Bézier components
2: Compute the derivatives γ_u, γ_{uu}, γ_{uuu} and γ_{uuuu}
3: Compute the expressions $\gamma_u \wedge \gamma_{uu}$, $\gamma_u \wedge \gamma_{uuu}$, $\gamma_{uu} \wedge \gamma_{uuu}$ and $\gamma_u \wedge \gamma_{uuuu}$
4: Compute the numerator N of the fraction from Eq. (9)
5: Compute the denominator D of the same fraction
6: Add the term γ to the fraction N/D:

$$\epsilon = \frac{\gamma D + N}{D}$$

7: Increase the degree of the denominator by n so that the numerator and the denominator are both of degree $(5n - 8)$

5.2 The Derivative of a Bézier Curve

The derivative of a Bézier curve γ with respect to the parameter u is another Bézier curve γ_u that is obtained with the following equation:

$$\gamma_u(u) = \sum_{i=0}^{n-1} n B_{i,n-1}(u) \Delta P_i$$

where $\Delta P_i = P_{i+1} - P_i$
Bézier curve derivatives are needed at step 2 of Algorithm 1.

5.3 The Multiplication of Two Bézier Functions

The operation $\gamma_u \wedge \gamma_{uu} = (\gamma_1)_u(\gamma_2)_{uu} - (\gamma_1)_{uu}(\gamma_2)_u$ involves two function multiplications and a substraction.

Let $f(u)$ and $g(u)$ be two Bézier functions of degree m and n.

$$f(u) = \sum_{i=0}^{m} B_{i,m}(u) a_i$$

$$g(u) = \sum_{i=0}^{n} B_{i,n}(u) b_i$$

The multiplication of these two functions gives the Bézier function of degree $m + n$

$$f(u)g(u) = \sum_{i=0}^{m+n} B_{i,m+n}(u) c_i$$

where

$$c_i = \sum_{j=max(0,i-n)}^{min(m,i)} \frac{\binom{m}{j}\binom{n}{i-j}}{\binom{m+n}{i}} a_j b_{i-j}, \quad i = 0, \cdots, m+n$$

In the same way, we can multiply a Bézier function f and a Bézier curve $\gamma = (\gamma_1, \gamma_2)$ to get the result $f\gamma = (f\gamma_1, f\gamma_2)$.
Bézier function multiplications are needed at step 3 of Algorithm 1.

5.4 Rational Bézier Curves

The result of the computation is the equi-affine evolute ϵ that consists of rational Bézier curves :

$$\epsilon = \frac{\gamma D + N}{D}$$

The denominator takes the form:

$$D = \sum_{i=0}^{n} B_{i,n}(u) w_i$$

where the w_i are scalar values. A control point i with a zero w_i component is located at infinity. See for example Goldman [11, p.257] for more details on how to handle those points.

5.5 Degree Elevation

The numerator and denominator of a rational Bézier curve must have the same degree. In fact it must have the same basis functions $B_{i,n}(u)$. In our case, to get the same basis functions, the degree of the denominator must be increased.

We can express the same Bézier function f with basis functions of higher degree. Every time the degree is increased, more control points w_i will be added and they will be closer to the function f. New control points w_i^{n+r} corresponding to a degree $n + r$ are:

$$w_i^{n+r} = \sum_{j=max(0,i-r)}^{min(n,i)} \frac{\binom{r}{i-j}\binom{n}{j}}{\binom{n+r}{i}} w_j^n, \quad i = 0 \ldots, n+r$$

Bézier degree elevation is needed at step 7 of Algorithm 1.

5.6 Example

This example is a closed curve of degree $n = 6$. The control points of the B-spline have coordinates:

$$P_i = \{(2,0), (\sqrt{2}, \sqrt{2}/2), (0,1), (-\sqrt{2}, \sqrt{2}/2),$$

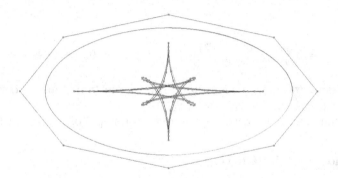

Fig. 1. An oval curve and its equi-affine evolute.

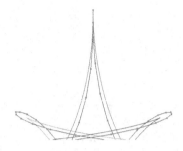

Fig. 2. Zoom on the equi-affine evolute control polygon.

$$(-2,0), (-\sqrt{2}, -\sqrt{2}/2), (0,0), (\sqrt{2}, -\sqrt{2}/2)\}$$

And the cyclic nodal vector is:

$$U = \{0.0, 0.25, 0.25, 0.5, 0.5, 0.75, 0.75, 1.0\}$$

In Fig. 1, the control polygon of a B-spline of degree $n = 6$ is found on the outside and defines an oval curve. Inside the oval curve, the equi-affine evolute of degree $5n - 8 = 22$ is found with its control polygon which was obtained symbolically. Figure 2 is a zoom on the top of the equi-affine evolute to better show its control polygon.

6   Conclusion

The exploration of the possibilities offered by the combination of equi-affine differential geometry and B-spline curves yielded some positive results. The expression of the equi-affine evolute was derived for a general parameter. It was unexpected to find that this expression contains no root. Consequently, the equi-affine evolute of both polynomial and rational curves are rational curves. The same conclusion applies to B-spline and NURBS curves and it was shown in details how to compute the equi-affine evolute of a B-spline symbolically.

Some properties of the equi-affine evolute were also discussed. The equi-affine evolute of an affine w-curve is an affine w-curve. The role that the circle holds to analyze the properties of curves in Euclidean geometry is played by conics in equi-affine geometry. Vertex points in Euclidean geometry become sextactic points in equi-affine geometry. In particular, the number of sextactic points, hence the number of cusps on the equi-affine evolute, is not only an affine invariant but also a projective invariant.

References

1. Blaschke, W.S.: Vorlesungen über Differentialgeometrie und geometrische Grundlagen von Einsteins Relativitätstheorie, ii: Affine Differentialgeometrie, Springer (1923)
2. Bruce, J.W., Giblin, P.J.: Curves and Singularities: A Geometrical Introduction to Singularity Theory. Cambridge University Press, Cambridge (1992)
3. Cartan, E.: La théorie des groupes finis et continus et la géométrie différentielle traitées par la méthode du repère mobile, Jacques Gabay, 1992. Original edition, Gauthiers-Villars (1937)
4. Cartan, E: Leçons sur la théorie des espaces à connexion projective, Jacques Gabay, 1992. Original edition, Gauthiers-Villars (1937)
5. Chen, X.: An application of singularity theory to robust geometric calculation of interactions among dynamically deforming geometric objects, Ph.D. thesis, University of Utah, May 2008
6. Chen, X., Riesenfeld, R.F., Cohen, E.: Complexity reduction for symbolic computation with rational B-splines. Int. J. Shape Model. **13**(1), 25–49 (2007)
7. Elber, G., Lee, I.K., Kim, M.S.: Comparing offset curve approximation methods. IEEE Comput. Graphics Appl. **17**(3), 62–71 (1997). (English)
8. Farouki, R.T.: Pythagorean-hodograph curves, vol. 1. Springer, Heidelberg (2008)
9. Farouki, R.T.: The bernstein polynomial basis: a centennial retrospective. Comput. Aided Geom. Des. **29**(6), 379–419 (2012). (English)
10. Giblin, P.J., Kimia, B.B.: On the local form and transitions of symmetry sets medial axes and shocks. Int. J. Comput. Vis. **54**(1–2), 143–156 (2003). (English)
11. Goldman, R.: Pyramid Algorithms: A Dynamic Programming Approach to Curves and Surfaces for Geometric Modeling. Morgan Kaufmann, Massachusetts (2002)
12. Gowers, T., June, B.-G., Imre, L.: The princeton companion to mathematics, pp. 38–43, Princeton University Press, NJ (2008)
13. Guggenheimer, H.W.: Differential Geometry. Dover Publication, New York (1977)
14. Guieu, L., Mourre, E., Ovsienko, V.Y.: Theorem on six vertices of a plane curve via sturm theory. In: Arnold, V.I., Gelfand, I.M., Retakh, V.S., Smirnov, M (eds.) The Arnold-Gelfand Mathematical Seminars, pp. 257–266. Springer (1997)
15. Holtom, P.A.: Affine-invariant symmetry sets. Ph.D. thesis, University of Liverpool (2000)
16. Izumiya, S., Sano, T.: Generic affine differential geometry of plane curves. Proc. Edinb. Math. Soc. **41**(2), 315–324 (1998). (English)
17. Klein, F.: Elementary Mathematics from an Advanced Standpoint, Geometry (1908). Reprinted Dover, New York (1939)
18. Levien, R.L.: From spiral to spline: Optimal techniques in interactive curve design. Ph.D. thesis, University of California, Berkeley (2009)

19. Osserman, R.: The four-or-more vertex theorem. Am. Math. Monthly **92**(5), 332–337 (1985)
20. Piegl, L., Tiller, W.: Symbolic operators for NURBS. Comput.-Aided Des. **29**(5), 361–368 (1997). (English)
21. Porteous, I.R.: Geometric Differentiation: for the Intelligence of Curves and Surfaces. Cambridge University Press, Cambridge (2001)
22. Stillwell, J.: Mathematics and its history, Second ed., Springer, New York (2002)
23. Tsai, Y.F., Farouki, R.T.: Algorithm 812: BPOLY: An object-oriented library of numerical algorithms for polynomials in Bernstein form. ACM Trans. Math. Softw. (TOMS) **27**(2), 267–296 (2001). (English)

On-line CAD Reconstruction with Accumulated Means of Local Geometric Properties

Klaus Denker[1][✉], Bernd Hamann[2], and Georg Umlauf[1]

[1] Institute for Optical Systems, University of Applied Science Constance,
Konstanz, Germany
{kdenker,umlauf}@htwg-konstanz.de
[2] Institute for Data Analysis and Visualization (IDAV), Department
of Computer Science, University of California, Davis, CA 95616, USA
hamann@cs.ucdavis.edu

Abstract. Reconstruction of hand-held laser scanner data is used in industry primarily for reverse engineering. Traditionally, scanning and reconstruction are separate steps. The operator of the laser scanner has no feedback from the reconstruction results. On-line reconstruction of the CAD geometry allows for such an immediate feedback.

We propose a method for on-line segmentation and reconstruction of CAD geometry from a stream of point data based on means that are updated on-line. These means are combined to define complex local geometric properties, e.g., to radii and center points of spherical regions. Using means of local scores, planar, cylindrical, and spherical segments are detected and extended robustly with region growing. For the on-line computation of the means we use so-called accumulated means. They allow for on-line insertion and removal of values and merging of means. Our results show that this approach can be performed on-line and is robust to noise. We demonstrate that our method reconstructs spherical, cylindrical, and planar segments on real scan data containing typical errors caused by hand-held laser scanners.

1 Introduction

Hand-held laser scanners in industry are primarily used for reverse engineering. A human operator moves the laser scanner along the surface of a physical object to sample its geometry. Usually the point set generated by the scanning process is shown to the operator. The reconstruction of geometric objects is computed afterward in a separate step, where a manual extraction of the geometric parameters is done. When data is missing in the scan the complete process fails. Thus, the whole scanning and reconstruction process must be repeated.

When using an on-line reconstruction algorithm, the reconstruction is done simultaneously with the acquisition of the scan data. Missing data can immediately be detected and corrected by the operator: one simply scans the critical region again. Thus, all data structures of the reconstruction are updated with the new data to improve the quality of the reconstruction immediately.

© Springer International Publishing Switzerland 2015
J.-D. Boissonnat et al. (Eds.): Curves and Surfaces 2014, LNCS 9213, pp. 181–201, 2015.
DOI: 10.1007/978-3-319-22804-4_14

In this context *on-line computation* means that the data is already processed while the operator is scanning the object. Therefore, the operator can interact with the results of this reconstruction process. In contrast to real-time computing, no strict time constraints have to be guaranteed.

We present an on-line algorithm for segmentation and reconstruction of CAD geometry from a stream of point data. It is based on accumulated means of local geometric quantities and classification scores for different surface types. Accumulated means allow for on-line insertion, removal, and merging of data.

Data structures for local geometry, edge detection, and segmentation are described in Sect. 3. Accumulated means of local geometric properties and classification scores are presented in Sect. 4. A segmentation strategy using the advantages of accumulated means is described in Sect. 5. The performance of the method with simulated and real scan data is analyzed in Sect. 7.

2   Related Work

One of the first approaches of reconstructing data from triangulation laser scanners is presented in [1], where cylinders are fitted to point data from a laser scanner. To reconstruct complex objects a segmentation of the surface is necessary. An extensive overview of surface segmentation methods is provided in [24]. The underlying data structures are either meshes, e.g. [29], or a k-nearest neighbor graph, e.g. [30], for an efficient local surface analysis. For CAD applications usually a segmentation by surface type is computed. The Gaussian sphere is used in [3] to segment surfaces by their dimensionality, whereas variational surfaces are used in [29].

Early methods for the reconstruction of geometry are discussed in [27]. They are based on the reconstruction of polygonal boundary models and the fitting of simple surfaces and free-form geometry. The method presented in [2] focuses on the reconstruction of rotational and translational surfaces and blends. In [29] variational surfaces are used to reconstruct implicit representations of geometric primitives. These methods do not work for on-line computations of point streams.

Iterative methods for geometry reconstruction often apply stochastic algorithms like Random Sample Consensus (RANSAC) fitting. They are used for the reconstruction of geometric primitives [23] or super-quadrics [4]. These RANSAC based methods are applied to random sub-sets of the data, whereas our segmentation approach is based on all available data. Iteration between segmentation and reconstruction is used in [26]. Here, based on quadric fitting the segmentation of an unorganized point cloud is iteratively improved. Complete data sets are required for each iteration.

Stream processing of point data is discussed in [6,18]. A sweep line approach sequentially processes large point sets with a set of geometric operators. Thus, the point sets have to be pre-sorted along one spatial direction.

Hand-held laser scanners generate unorganized streams of point data. Such streams can be triangulated on-line [5]. The data is reduced to a set of vertices that are almost uniformly distributed. A surface mesh is generated on-line connecting these vertices. This approach was extended by [10,11] with a multi-level data structure. Thus, the mesh can be adapted to non-uniform point densities.

An incremental computation of means is introduced in [22,28]. These methods are compared with other non-incremental methods to compute means and variances with respect to numerical performance in [8,16]. The application of these methods for variance computations is also discussed in [15]. These incremental means are constructed to add single values to the mean. The removal of values or merging of multiple means is not considered.

Clustering methods like k-means [19] or *unweighted pair group method with arithmetic mean* (UPGMA) [25] use mean positions of multi-dimensional data. On-line k-means methods like the ones covered in [31] allow the incremental addition of data to the mean centroids of the clusters. New values are added with a fixed or decreasing learning rate. Data changing the cluster is not removed from the old cluster centroids. We note that on-line k-means is not an on-line method in the sense we define in the Introduction. Efficient implementations of UPGMA, as in [12] based on arithmetic distance means, use a reduction formula to merge means. Though, no higher order geometric properties are used.

A method that combines local geometric properties to more stable estimations is presented in [17]. Discrete estimates for normals and curvatures are combined over an area using a voting algorithm. However, this is not an on-line method and no geometric parameters derived from the local geometry are used.

The method presented in this paper is partially based on [9]. It differs in the used methods for segmentation and reconstruction. In [9] the segmentation is based only on local geometric properties and the reconstruction of geometric primitives is based on quadric fitting.

In this paper we use a segmentation based on accumulated means. The segmentation also provides the information for the reconstruction of geometric primitives. Thus, there is no separate reconstruction step or quadrics fitting.

3 Data Structures

The on-line reconstruction is based on a ball tree data structure as proposed in [11]. It is used to store and process the data stream from the laser scanner and to prepare it for the determination of surface segments.

3.1 Ball Tree

The data stream generated by a laser scanner consists of noisy raw point data in three-dimensional (3d) space and orientation data of the laser probe. The raw points are assigned to $n(eighborhood)$-balls β that are defined as

$$\beta(\boldsymbol{c}, r) = \{\boldsymbol{x} \in \mathbb{R}^3 : \|\boldsymbol{x} - \boldsymbol{c}\| < r\},$$

where $\boldsymbol{c}$ denotes the center and r the radius of the n-ball. The n-balls might overlap. Every new raw point $\boldsymbol{q}$ is added to the n-ball $\beta(\boldsymbol{c}, r)$ that contains

q with minimal distance $\|q - c\|$. If no such n-ball exists, a new n-ball $\beta(q, r)$ is constructed. The radius r is the maximum r such that $\beta(q, r)$ does not contain any other n-ball centers. For detail we refer to [11].

For fast access, the n-balls are organized in a *ball tree*, which is an octree data structure. An n-ball is added to the octree region that contains its center. The edge length of this region is the radius of its n-balls. Thus, n-ball radii are dyadic fractions of the initial scan area's edge length e_O, i.e., $r = e_O/2^n, n \in \mathbb{N}$.

Each n-ball β stores estimates for the local normal and principal curvatures. The *principal component analysis* (PCA) [14] is used to compute the normal n of the raw points of a neighborhood of n-balls in the ball tree. The Weingarten map [7] of a quadratic approximation P_β to these raw points based on the local tangent plane defined by n is used to compute estimates on the principal curvatures κ_1, κ_2 and principal directions d_1, d_2. These estimates are only used, when the ratio of the smaller eigenvalues of the PCA is less than $1/2$.

For each n-ball β the arithmetic mean of its raw points is computed and projected onto P_β. The resulting point p represents the local geometry of β and is used for the reconstruction. Additionally, each n-ball holds a list of closest neighbor points, similar to [30].

3.2 Sharp Feature Detection

The quadratic polynomial P_β approximates the geometry of the raw points of β accurately only when the geometry is sufficiently smooth. Sharp features cannot be approximated accurately by a single quadratic polynomial. Thus, we use two polynomial approximations to reconstruct sharp features. More than two approximations are possible but their reconstruction is not sufficiently stable.

To compute P_β we use a *least absolute deviation approximation* instead of a least squares approximation. It is based on the L^1 norm instead of the L^2 norm and is better suited for partial reconstructions. For the computation *iteratively re-weighted least squares* (IRWLS) [13] are used. For IRWLS compute the weighted least squares approximation with weights $w_i = 1/\sqrt{d_i}$, where d_i is the distance of the approximation to the raw points. These weights are adjusted in every iteration. The weight adjustment compensates for the difference between the L^1 and L^2 norm. Usually, after approximately five iterations P_β is sufficiently close to a least absolute deviation approximation.

During the IRWLS computation the number of outliers with distance $d_i > \varepsilon_{sd}r$ is recorded. If there are more than 20% outliers, we assume a sharp feature in the scanned geometry and approximate a second polynomial P_β^2. This second polynomial is initially IRWLS approximated to the outliers. Both polynomials $P_\beta^1 = P_\beta$ and P_β^2 are IRWLS approximated to all raw points with weights

$$w_i^1 = \max\left(d_i^2 - \varepsilon_{sd}, 0\right)/\sqrt{d_i^1}, \quad \text{and} \quad w_i^2 = \max\left(d_i^1 - \varepsilon_{sd}, 0\right)/\sqrt{d_i^2}.$$

Thus, raw points close to one of the approximations obtain a small weight for the other one. The superscript k refers to quantities related to $P_\beta^k, k = 1, 2$.

The ratio of outliers not close to any surface to the number of approximated raw points defines a quality score q_{ls} for the approximation.

For sharp features, the point p is projected onto the intersection curve of P_β^1 and P_β^2 by iterative projection to the intersection line of local tangent planes.

3.3 Segment Data Structure

A *segment* is a set of n-balls of approximately the same geometric type. The implementation is based on a hash table [20]. Thus, look-up and insertion operations have amortized computational complexity $O(1)$. For the on-line CAD reconstruction segments are classified by their geometry. Currently the supported *segment types* are planar, cylindrical, spherical, and unknown geometry.

Each segment maintains a set of its neighbor segments. This set contains all segments containing n-balls that are neighbors to n-balls in the segment. Neighbor segments are added to this set when n-balls are added or modified. At the same time a back link is added to the neighbor segment's set. Neighbor entries are only removed upon segment deletion.

For the association of n-balls to segments also a hash table is used. Furthermore, a hash table for each n-balls is used to maintain all geometric quantities that are associated to this n-ball in order to guarantee consistently accumulated means, as described in the next section.

4 Accumulated Means

The advantage of using means of geometric properties instead of purely local geometric properties for the detection of surface segments is the increased stability in the computations. Thus, we use means throughout our on-line approach.

Weighted and unweighted arithmetic means of a set of n data x_i with constant weights w_i are defined as

$$\bar{x}_n = \frac{\sum_{i=1}^n x_i w_i}{\sum_{i=1}^n w_i} \quad \text{and} \quad \tilde{x}_n = \frac{1}{n} \sum_{i=1}^n x_i.$$

For on-line computations means are computed by incremental addition of data. Besides, it is necessary to remove data from a mean or to merge two means. We call a quantity *accumulated* when it supports these three *accumulation operations*, e.g., $\bar{x}_n$ is an accumulated weighted arithmetic mean.

4.1 Accumulated Arithmetic Means

The incremental computation of an unweighted arithmetic mean $\tilde{x}_n$ when adding a value $x_n, n > 1$, is defined as

$$\tilde{x}_n = \tilde{x}_{n-1} + \frac{1}{n}\left(x_n - \tilde{x}_{n-1}\right), \qquad \tilde{x}_1 = x_1.$$

For numerical reasons we avoid adding $(n-1)\tilde{x}_{n-1}$, see [16]. To compute a weighted arithmetic mean $\bar{x}_n$, the average weight $\tilde{w}_n$ is used

$$\bar{x}_n = \bar{x}_{n-1} + \frac{w_n}{(n-1)\tilde{w}_{n-1} + w_n}(x_n - \bar{x}_{n-1}), \qquad \bar{x}_1 = x_1.$$

A value $x_{n+1}, n \geq 1$, can be removed from an unweighted arithmetic mean by

$$\tilde{x}_n = \tilde{x}_{n+1} - \frac{1}{n}(x_{n+1} - \tilde{x}_{n+1})$$

and from a weighted arithmetic mean by

$$\bar{x}_n = \bar{x}_{n+1} - \frac{w_{n+1}}{(n+1)\tilde{w}_{n+1} - w_{n+1}}(x_{n+1} - \bar{x}_{n+1}).$$

Two arithmetic means $\tilde{a}_n$ and $\tilde{b}_m$ with $m \leq n$ are merged to one combined arithmetic mean $\tilde{x}_{n+m}$ by

$$\tilde{x}_{n+m} = \tilde{a}_n + \frac{m}{n+m}\left(\tilde{b}_m - \tilde{a}_n\right). \tag{1}$$

The combined weighted arithmetic mean is computed using the arithmetic means of the weights $\tilde{u}_n, \tilde{v}_m$ of the partial weighted arithmetic means $\bar{a}_n$ and $\bar{b}_m$

$$\bar{x}_{n+m} = \bar{a}_n + \frac{m\tilde{v}_m}{n\tilde{u}_n + m\tilde{v}_m}\left(\bar{b}_m - \bar{a}_n\right).$$

The weight mean $\tilde{w}_{n+m}$ for the combined mean $\bar{x}_{n+m}$ is computed from $\tilde{u}_n, \tilde{v}_m$ by Equation (1). We note that the computational complexity of all accumulation operations is $O(1)$, which is optimal for an on-line algorithm.

4.2   Accumulated Arithmetic Means in the Ball Tree

For each n-ball in a ball tree we obtain sets of values that are combined in accumulated arithmetic means. Unweighted arithmetic means are used for position p, normal n, and ball radius r. Their means $\tilde{p}, \tilde{n}, \tilde{r}$ are used to detect planar segments. Accumulated weighted arithmetic means $\bar{\kappa}_1, \bar{\kappa}_2, \bar{H}$ are computed for principal curvatures κ_1, κ_2, and mean curvature H. The respective weights are given by the quality score q_{ls} of the least squares approximation.

4.3   Indirect Accumulated Means

For each segment accumulated weighted arithmetic means for both principal curvatures and the mean curvature yield *indirect accumulated weighted means* for the hypothetical cylinder and sphere radii $\bar{r}_1, \bar{r}_2$, and $\bar{r}_s$

$$\bar{r}_i = -\bar{\kappa}_i^{-1}, i = 1, 2, \quad \text{and} \quad \bar{r}_s = -\bar{H}^{-1}.$$

Other geometric quantities are defined by normalized, non-oriented, vector-valued data, e.g., principal curvature directions d_1, d_2. Thus, arithmetic means cannot be used to combine principal directions. Due to normalization, this data lies on the unit-sphere. Their distribution on the unit-sphere is estimated by a PCA step. For both principal curvature directions $d_i, i = 1, 2$, the 3×3 co-variance matrix C_i is computed. The eigenvalues $\lambda_{i,j}, j = 1, 2, 3$, of C_i character-ize these distributions. The eigenvectors of the largest eigenvalues $\lambda_{i,3}$ are used as unweighted means $\tilde{d}_i$ for the non-oriented directions. Co-variance matrices C_i are accumulated, because data d can be added or removed by

$$C_i^{\text{new}} = C_i^{\text{old}} \pm d_i \cdot d_i^t,$$

and two co-variance matrices C_i^1, C_i^2 are merged by

$$C_i = C_i^1 + C_i^2,$$

yielding the co-variance of the union of both sets of non-oriented directions. Thus, $\tilde{d}_i, i = 1, 2$, are *indirect accumulated means*. They provide information for cylinder detection: the cylinder axis has the same direction as the principal direction of the smaller absolute principal curvature.

The complexity for the accumulation operations for a 3×3 co-variance matrix C is $O(1)$. Due to the constant size of C, the computation of its eigenvalues and eigenvectors has also complexity $O(1)$. Thus, the overall complexity is $O(1)$.

4.4   Accumulated Means Depending on Means

For each segment hypothetical cylinder and sphere center points are estimated by locally approximated center points

$$p_i = p - \bar{r}_i \cdot n, i = 1, 2, \quad \text{and} \quad p_s = p - \bar{r}_s \cdot n.$$

These local estimates accumulate to the weighted arithmetic means $\bar{p}_1, \bar{p}_2$, and $\bar{p}_s$ with weights $|\bar{\kappa}_1|, |\bar{\kappa}_2|$, and $|\bar{H}|$, respectively. We use weighted arithmetic means to account for the non-linear distribution of center points and the fact that there is no multi-dimensional harmonic mean. Since the local estimates and the weights vary with time, they are stored for the accumulation operations.

Using $\bar{p}_1, \bar{p}_2$, and $\bar{p}_s$ a more stable local estimate for the radii can be com-puted, see Fig. 1,

$$R_i = \|(p - \bar{p}_i) - ((p - \bar{p}_i)\tilde{d}_{3-i})\tilde{d}_{3-i}\|, \qquad i = 1, 2,$$
$$R_s = \|p - \bar{p}_s\|.$$

The accumulated weighted means of corrected radii $\bar{R}_1, \bar{R}_2$, and $\bar{R}_s$ are computed again with weights $|\bar{\kappa}_1|, |\bar{\kappa}_2|$, and $|\bar{H}|$.

4.5   Scores for Geometric Primitives

For segmentation we define scores to measure the fitting quality of an n-ball to planar, spherical, cylindrical, or unknown geometry. We use multiplicative scores

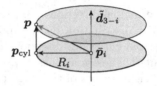

Fig. 1. Points and directions of a cylindrical segment.

that are centered at 1 using a normalization weight ω, see Table 1. Values smaller that 1 denote high fitting quality. Partial scores s_i are combined by weighted multiplication using weights w_i to yield a combined score

$$s = \prod_{i=1}^{n} \left((s_i - 1)\, w_i + 1 \right) \text{ with } w_i \in (0,1).$$

For each segment all scores are maintained at all times. Currently we compute scores for planar, cylindrical, spherical and unknown geometries. Each of these scores consists of a partial score for the segment size, a distance score, an angle score, and a curvature score. The latter is used only for cylindrical and spherical geometries. In the computations all means belong to the same segment and all local data belongs to the corresponding n-ball. The weights in the multiplicative combination are 3/4 for the distance and angle sores and 1/2 otherwise.

The score for planar segments s_{pln} is computed based on these quantities:

Distance score: Normal distance $\tilde{n}(\tilde{p} - p)/(\bar{r}\omega_{\text{nd}})$.
Angle score: $\sphericalangle(n, \tilde{n})$.
Size score: Reciprocal of number of n-balls in the segment.

The score for cylindrical segments s_{cyl} is based of the direction of the cylinder curvature $\bar{\kappa}_\delta$ that is orthogonal to the cylinder axis $\tilde{d}_\Delta$ with

$$\delta = \operatorname*{argmin}_{i=1,2} |\lambda_{i,3}| \quad \text{and} \quad \Delta = \operatorname*{argmax}_{i=1,2} |\lambda_{i,3}|.$$

Thus, the index δ is an indirect accumulated quantity characterizing the type of the cylinder. To simplify radius computations we compute a point on the cylinder surface in the plane normal to $\tilde{d}_\Delta$ containing $\bar{p}_\delta$, see Fig. 1,

$$p_{\text{cyl}} = p - \left((p - \bar{p}_\delta)\, \tilde{d}_\Delta \right) \tilde{d}_\Delta.$$

The partial scores for cylindrical segments are:

Distance score: Radius difference $\left| \|p_{\text{cyl}} - \bar{p}_\delta\| - |\bar{R}_\delta| \right| / (\omega_{\text{cyl}} |\bar{R}_\delta|)$.
Angle score: Average of $\sphericalangle(d_\Delta, \tilde{d}_\Delta)$ and $\sphericalangle(n, p_{\text{cyl}} - \bar{p}_\delta)$.
Curvature score: $\max(|\kappa_\delta|, |\bar{\kappa}_\delta|)/\min(|\kappa_\delta|, |\bar{\kappa}_\delta|)$.
Size score: Reciprocal of number of n-balls in the segment.

The score for spherical segments s_{sph} is computed from:

Distance score: Radius difference $\big|\|p - \bar{p}_s\| - |\bar{R}_s|\big|/(\omega_{\mathrm{sph}}|\bar{R}_s|)$.
Angle score: $\sphericalangle(n, p - \bar{p}_s)$.
Curvature score: $\max(|H|, |\bar{H}|)/\min(|H|, |\bar{H}|)$.
Size score: Reciprocal number of n-balls in the segment.

For unknown geometry a score s_{unk} is computed based mostly on local values of two neighbored n-balls β and β_{nb}:

Distance score: Normal distance $n_{\mathrm{nb}}(p_{\mathrm{nb}} - p)/(\bar{r}\omega_{\mathrm{nd}})$.
Angle score: $\sphericalangle(n, n_{\mathrm{nb}})$.
Size score: Reciprocal of number of n-balls in the segment.

Since the scores are combined from different geometric quantities, the ranges for the overall scores s_{pln}, s_{cyl}, s_{sph}, and s_{unk} differ significantly, e.g., good values of s_{pln} tend to be much smaller than good values of s_{cyl}. To compensate for this effect and to avoid a time-consuming re-balancing of the partial scores, the overall scores are re-balanced by correction factors. These factors are 2.5 for s_{pln}, 0.7 for s_{cyl}, 0.9 for s_{sph}, and 6.0 for s_{unk}.

For s_{pln}, s_{cyl}, and s_{sph} accumulated means $\bar{s}_{\mathrm{pln}}$, $\bar{s}_{\mathrm{cyl}}$, and $\bar{s}_{\mathrm{sph}}$ are computed for each segment. These means are weighted by the quality q_{ls} and are used to determine the overall segment type. The smallest mean score determines the segment type. If it is larger than 1 the segment is classified as unknown geometry.

5   Segmentation Strategy

Our segmentation algorithm is implemented as a separate layer on top of the ball tree described in Sect. 3. The n-balls can be added, removed or modified at any time. For each of these updates we need to update the segmentation.

5.1   Changing Single n-Balls

When a new n-ball is generated, it is assigned to the best-fitting segment. To find the best-fitting segment the score s_{pln}, s_{cyl}, s_{sph}, or s_{unk} of this n-ball for each hypothetical segment that contains an n-ball in the local neighborhood is computed. The segment type of the hypothetical segment determines which score for the n-ball is computed. Then it is added to the segment with the minimal score by adding its data and scores to the means of the segment. The n-ball is also inserted into the segment data structures.

To remove an n-ball its old values are removed from the means of its segment. The n-ball is removed from the segment data structure. If a segments contains no n-balls after a removal, the segment is deleted.

For modifications of n-balls the same minimal scores as for adding n-balls are computed. When a score indicates a change of segments, the n-ball is removed from the old segment and added to the new one. Otherwise, the new values of the n-ball replace the old values in the means of its segment.

Accumulating the means of a segment is of complexity $O(1)$. Assuming that the size of the local neighborhood of an n-ball is constant, the search for the best-fitting segment has also complexity $O(1)$. Updating the set of n-balls of a segment and updating the global hash table of n-balls have complexity $O(1)$.

5.2 Merging Segments

After several updates for one segment, a hypothetical merged segment with each of its neighbor segments is computed. This hypothetical segment contains the data of both segments using the accumulation formula for the means and covariance matrices. We note that in the sequel quantities with superscript $k = 1, 2$ refer to data associated with one of the merged segments and quantities without superscript to data of the resulting hypothetical segment.

The segment type of the hypothetical segment is determined by the accumulated mean scores $\bar{s}_{\mathrm{pln}}$, $\bar{s}_{\mathrm{cyl}}$, $\bar{s}_{\mathrm{sph}}$. Depending on the resulting segment type further merge conditions based on thresholds ε are evaluated, see Table 1. For planar segments the normal distance between the center points $\tilde{p}^1$, $\tilde{p}^2$ of both original segments and the angle between their normals $\tilde{n}^1$, $\tilde{n}^2$ are evaluated, i.e.

$$|\tilde{n}^1(\tilde{p}^1 - \tilde{p}^2)| + |\tilde{n}^2(\tilde{p}^2 - \tilde{p}^1)| < \varepsilon_{\mathrm{nds}}(\tilde{r}^1 + \tilde{r}^2) \quad \text{and} \quad \sphericalangle(\tilde{n}^1\tilde{n}^2) < \varepsilon_{\sphericalangle}.$$

For cylindrical segments the angle between the cylinder axes, the deviation of the cylinder axes are evaluated, and the difference of radii, i.e.

$$\sphericalangle(\tilde{d}_\Delta^1 \tilde{d}_\Delta^2) < \varepsilon_{\mathrm{ca}}, \qquad \|(\bar{p}_\delta^2 - \bar{p}_\delta^1) - ((\bar{p}_\delta^2 - \bar{p}_\delta^1)\tilde{d}_\Delta)\tilde{d}_\Delta\| < \varepsilon_{\mathrm{cc}}|\bar{R}_\delta|, \quad \text{and}$$
$$|\bar{R}_\delta^1 - \bar{R}_\delta^2| < \varepsilon_{\mathrm{cr}}|\bar{R}_\delta|.$$

For spherical segments the difference of radii and the distance of their center points are evaluated, i.e.

$$|\bar{R}_s^1 - \bar{R}_s^2| < \varepsilon_{\mathrm{sr}}|\bar{R}_s| \quad \text{and} \quad \|\bar{p}_s^1 - \bar{p}_s^2\| < \varepsilon_{\mathrm{sc}}|\bar{R}_s|.$$

Segments classified as unknown geometry cannot be merged.

When the hypothetical segment satisfies the combination of the merge conditions, the merge operation is realized: The means and data of the smaller segment are merged to the means and the data of the larger segment, the neighbor information is updated, and then the smaller segment is deleted. After a merge operation, further checks for merges of the involved segments are dropped.

The complexity to compute a hypothetical segment is $O(1)$. Hypothetical segments are computed for each of the k neighbors, so the complexity of checking for merges is $O(k)$. The complexity of a merge operation is $O(1)$ for merging the means. Merging the sets of n-balls of the segments has an amortized complexity of $O(m)$, where m is the number of n-balls in the smaller segment. The global hash table associating n-balls to segments is updated with complexity $O(m)$. Thus, the overall complexity to merge two segments is $O(k + m)$.

5.3 Consistency Checks

Every time a segment changes its type, all its n-balls are checked for consistency. This is also done when a cylindrical segment changes the direction of its axis. For each n-ball in the segment the segment with the minimal score s_{pln}, s_{cyl}, s_{sph}, or s_{unk} is determined. If it differs from the current segment, the n-ball is removed from the current segment and added to the new segment.

The computational complexity of this consistency check is $O(m)$ where m is the number of n-balls in the segment. Each of these n-balls determines its best-fitting neighboring segment with $O(1)$ time complexity. It is removed from the current segment with $O(1)$ and added to another one with $O(1)$, if necessary.

6 Segment Boundaries

In CAD applications one must handle trimmed surfaces. A boundary curve is constructed that represents the outer contour of the surface. To render the results of the on-line CAD reconstruction a similar approach is used. Polygonal boundaries are constructed surrounding the areas of the primitives. A polygonal boundary B is an ordered sequence of n-balls β_i with $1 \leq i \leq n_B$. By definition, it is ordered counterclockwise when seen from the outside.

The segmentation in Sect. 5 uses a region growing approach. A segment starts with a single n-ball and is extended by adding n-balls. When it consists of three n-balls, a first initial boundary B with $n_B = 3$ can be constructed. A local planar projection is used to choose a counterclockwise orientation of B.

6.1 Boundary Expansion

All the following operations use a local planar projection based on the segment data, e.g., in the direction towards the center point of a spherical segment. The orientation of a segment is defined by the sign of its average normal $\bar{n}$ or radius $\bar{r}_1$, $\bar{r}_2$, or $\bar{r}_s$.

For each n-ball β_{new} that is added, modified or removed from a segment, B is updated. When a newly added n-ball is outside of B, it is inserted into the closest edge $\langle \beta_i, \beta_{i+1} \rangle$ of B. The neighbors β_i of β_{new} on the boundary B are determined. Both edges $\langle \beta_{i-1}, \beta_i \rangle$ and $\langle \beta_i, \beta_{i+1} \rangle$ are candidates for inserting β_{new}. In the projection, β_{new} has to lie within the sector bounded by the half-angle lines to its two neighboring edges, see Fig. 2. The edge that fulfills this condition and whose center point is closest to β_{new} is chosen for insertion. After the insertion, consistency checks improve the shape of the boundary.

Changes of the values of an n-ball on the boundary trigger only the consistency checks. If it is not on the boundary, it is treated like a newly added n-ball. When removing an n-ball from the segment, it is also removed from the boundary. If it was on the boundary, we attempt to re-connect the boundary by inserting neighboring n-balls to the boundary. In the projection space, the neighboring n-ball β_{nb} with the smallest sum of inner angles to the half-angle lines next to the deleted n-ball is searched, see Fig. 2. If such a β_{nb} is found and it lies outside of the chosen boundary edge, it is inserted into the boundary.

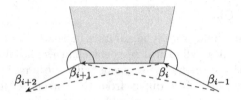

Fig. 2. Sector (gray) between half-angle lines (red) of a boundary edge $\langle \beta_i, \beta_{i+1} \rangle$.

6.2   Multiple Boundaries

A segment can have multiple boundaries B_i, e.g., a complete cylinder mantle has two boundaries around the caps. When scanning such a cylindrical segment, a single boundary B_1 is created that is split into two boundaries B_1, B_2, when a full scan around the cylinder was completed. Whenever two outsides of the boundary are close to each other, such a split operation is performed.

Merging of segments, see Sect. 5.2, requires a merging of their boundaries. After the segments are merged, all boundaries of the segment are checked for neighboring n-balls on different boundaries of the same segment. If a part of both boundaries is close together and runs in opposite direction, the boundaries are merged. The boundary parts that are close to each other are removed and their end points are connected to create a single boundary.

6.3   Local Consistency Checks

After connections or n-balls of B are modified, a local consistency check is performed for a single n-ball β_i. If the outer angle between both its edges $\langle \beta_{i-1}, \beta_i \rangle$ and $\langle \beta_i, \beta_{i+1} \rangle$ is smaller than 150°, β_i is removed from the boundary. For all neighbors β_{nb} of β_i on the same boundary, intersections are computed between all connected boundary edges. If two intersections are found, the boundary is split by connecting the two segments before and after the intersections.

If in the local neighborhood of β_i a point on another boundary on the same segment is found, a merge operation of these boundaries is initiated.

Finally, for segments of two edges on B that are close to the edges next to β_i with have opposite directions a split operation is done, see Sect. 6.2.

7   Results

For our method we demonstrate the effectiveness, robustness to noise, and performance for real scan data. It runs at interactive speed, where the segmentation takes only little processing power. All results are computed using the thresholds in Table 1, which are scale invariant because they are angles or relative distances.

Table 1. Used scale invariant thresholds.

Symbol	Threshold	Value	Symbol	Threshold	Value
ε_{sd}	Distance to surface	0.14	ω_{nd}	Normal distance	0.8
ω_{cyl}	Cylinder radius difference	0.1	ω_{sph}	Sphere radius difference	0.1
ε_{nds}	Normal distance between $\tilde{p}$	0.4	$\varepsilon_{\sphericalangle}$	Angle between $\tilde{n}$	$20°$
ε_{ca}	Angle between $\tilde{d}_\Delta$	$20°$	ε_{cr}	Difference between $\bar{R}_\delta$	0.2
ε_{cc}	Deviation of cylinder axes	0.4	ε_{sr}	Difference between $\bar{R}_s$	0.2
ε_{sc}	Distance of $\bar{p}_s$	0.4			

7.1 Synthetic Scan Data

To demonstrate the robustness to noise, we use synthetic scan data, where the true segment types are known. These data sets are shown in Fig. 3. The simulated plane has an edge length of 20 cm. The cylinder height is 20 cm and the radius is 10 cm; the sphere radius is 10 cm. The added noise is generated from normal-distributed random numbers. It is scaled according to different levels of standard deviation σ. We use two different types of noise:

Laser noise: Noise of the laser scanner is directed along the scanning direction. It is applied to each raw point separately.

Tracking noise: Noise of scanner tracking is added to the scanner position and the raw points in all three spatial directions. It is applied to each scan line.

The plots in Fig. 4a to c show the results of our experiments with synthetic planar, cylindrical, and spherical data, respectively. For each level of noise with standard deviation $\sigma = i/10[mm], i = 0, \ldots, 10$, we have computed 25 experimental segmentations and reconstructions. For each experiment new noise is generated. The plots show the average distance of the parameters of plane, cylinder, and sphere to the known ground truth parameters. For planes the average normal distance is plotted. For cylinders and spheres the average radius difference and the average distances of the center lines or points are plotted.

The evaluation shows that tracking noise has the largest effect on the reconstruction. The plane reconstruction even improves when some laser noise is added to the tracking noise. Cylindrical and spherical reconstructions lead to better results for the radius than for the center line or point. Planar reconstruction is least affected by noise followed by spherical reconstruction. The added noise has the most influence on the cylindrical reconstruction.

Planar reconstruction is based on positions p and normals n that are quite robust to noise. Spheres also use the mean curvatures H for the radius. The cylinder radius depends on principal curvatures κ_1, κ_2 and principal directions d_1, d_2. Especially the principal directions are effected substantially by local noise. The reconstruction errors of cylinder radii are around 10 times larger than the errors of sphere radii for the same levels of tracking noise. Both segment types use radii $\bar{r}_1$, $\bar{r}_2$, $\bar{r}_s$ and normals n for center reconstruction. Here, the error for cylinders is six times larger than for spheres in the presence of tracking noise.

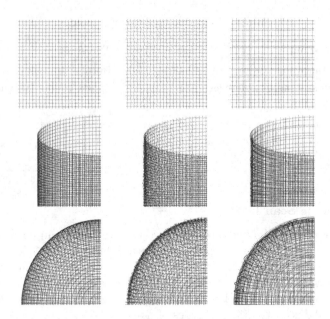

Fig. 3. Sections of simulated scans without noise (left), with laser noise (middle), and with tracking noise (right).

While the precision of the reconstruction is affected by the noise none of the reconstructions has failed. No outliers with extremely high error rates appear in the data of these plots. Noise with standard deviation $\sigma > 1$ mm would cause the segments to split into smaller segments, reconstructing parts of the data. Most of these would be planar as a curvature based reconstruction would fail.

The sharp feature detection is shown in Fig. 5. The n-balls near the edge have two different normals and the points p are projected onto the feature line.

7.2   Hand-Tracked Synthetic Scan Data

The synthetic scan data simulates linear movements of the scanner that cause a rasterized structure of the scan lines. To avoid this effect it is necessary to track hand movements. We use a magnetic *six degrees of freedom* (6DoF) input controller called *Razer Hydra* [21] to track interactive scans of the simulated objects, see Fig. 6. The sphere and the cylinder were scanned only from one side to demonstrate the reconstruction of incomplete objects.

For all synthetic data-sets with noise level $\sigma < 0.5$ mm the segmentation produces perfect results: a single segment containing all n-balls is reconstructed.

For the evaluation of these scans we plot in Fig. 4d the same parameters as in Fig. 4a to c for both types of noise. The non-uniform distribution of the scan lines does not influence the noise robustness for planar reconstruction much. The partial scans of the cylinder and sphere are much more affected by noise. Their reconstructions have errors roughly twice as large as those for the complete

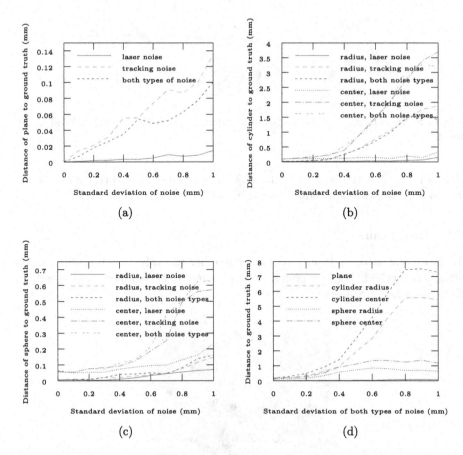

Fig. 4. Plots of the average normal distance of 25 experiments for noise levels $\sigma = i/10$[mm], $i = 0, \ldots, 10$, for plane, cylinder, and sphere reconstruction. Plot (d) shows the result for all segment types using a 6DoF input controller.

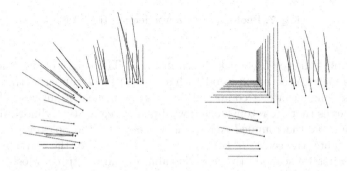

Fig. 5. N-ball points p and normals n near a sharp edge computed without and with sharp feature detection.

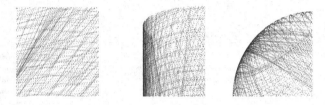

Fig. 6. Sections of simulated scans scanned interactively with tracking of hand movements.

data-set. Since only one half of the objects is scanned, the corrected radii R_i, R_s do not have the same stabilizing effect as for complete scans.

7.3   Real Scans

We use point streams from a real laser scanner to evaluate the on-line reconstruction method. For the presented experiments, we scanned objects with a *Faro Edge Laser ScanArm*.

Fig. 7. Photograph of a wooden fire truck toy.

We scanned the toy fire truck shown in Fig. 7. The length of the truck is 0.6 meters. Its material is coated wood with stripes of black rubber as tires around the wheels. The surface is composed of basic geometric primitives. There are many flat areas that are reconstructed with planar segments. Cylindrical regions can be found at the rounded edges, the wheels, and the holes at the ladder. The front lights, the center of the wheels, and the head of the driver figure are spherical. First we scanned the driver's cabin of the toy truck. Most segments are detected correctly, see Fig. 8. Because of the dark material, the tires are not detected as single cylindrical segments. They split up into multiple smaller flat and cylindrical segments. At the rims, flat and spherical segments are correctly reconstructed. Cylinders with small radii are reconstructed at the rounded edges of the driver's cabin. Even a small spherical segment can be seen at the front left corner of the hood. At the radiator grill, a flat area is reconstructed that contains two holes for the lights. The reconstruction shows some remaining problems with the boundary reconstruction that result in incomplete and jagged shapes. The floor below the toy truck in Fig. 8d shows that missing regions are covered by n-ball vertices. They are correctly segmented but not enclosed by the reconstructed boundary.

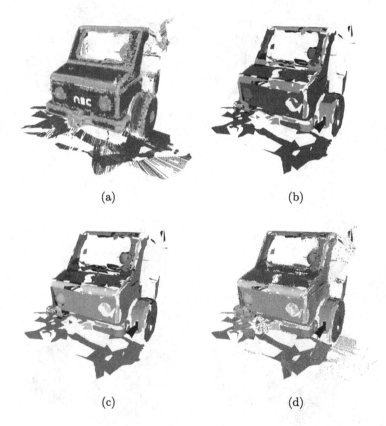

(a)   (b)

(c)   (d)

Fig. 8. Raw data and reconstruction from a scan of the driver's cabin of the wooden fire truck toy: 1 271 342 points, 12 325 scan lines, 15 013 n-balls, 1 011 segments. Raw data colored by curvature magnitude (a), segments colored by segment type: planar, cylindrical, and spherical (b), or each segment individually (c), additionally the n-ball vertices are shown in the same colors in (d) (Color figure online).

For our second example, we scanned the rear end and the ladder of the toy fire truck, see Fig. 9. The bottom of the ladder platform consists of large cylinders. They demonstrate the capabilities of the accumulated means reconstruction to correctly classify segments with large radii. Four cylindrical holes at the upper

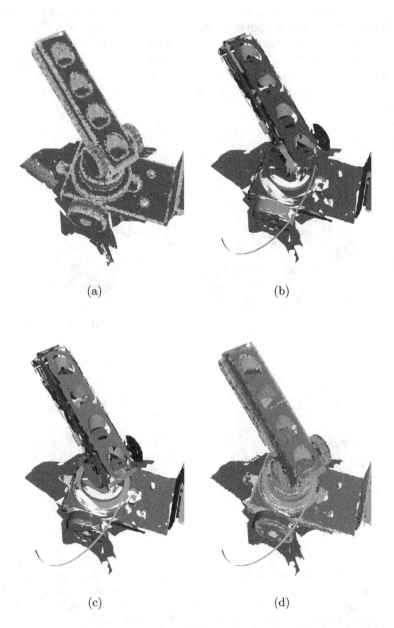

(a)   (b)

(c)   (d)

Fig. 9. Raw data and reconstruction from a scan of the ladder of the wooden fire truck toy: 3 128 055 points, 26 749 scan lines, 32 002 n-balls, 1 733 segments. Raw data colored by curvature magnitude (a), segments colored by segment type: planar, cylindrical, and spherical (b), or each segment individually (c), additionally the n-ball vertices are shown in the same colors in (d) (Color figure online).

part of the ladder are scanned from the inside. They are correctly reconstructed and rendered as cylinders with surface normals pointing to the inside. The signs of principal curvatures are preserved during the accumulated means reconstruction. The black rubber tire visible in this scan was captured better than the one in the last scan and is reconstructed by a single cylinder mantle. Another cylinder mantle with slightly smaller radius reconstructs the visible part of the wooden wheel. Figure 9c and d show the tire in purple and the outside of the wheel in blue. The narrow green cylinder in Fig. 9b to d shows a typical error of the boundary rendering. Temporarily, the outer parts of the cylinder contained boundary points. Therefore, it is assumed that these parts lie within the boundary. Instead of just a small area that really is inside the boundary, almost the complete cylinder mantle is rendered.

7.4   Performance

A large number of segments was reconstructed for these scans. Though, the implementation always performs at interactive speed, processing 30 scan lines with up to 5 640 points per second. We used an Intel Core 2 Quad Q6600 2.4 GHz computer with 8 GB of RAM. The most processing power is used to update the ball tree data structure. The thread responsible for the computation of the accumulated means and the segmentation only uses 18 % on average and 28 % at maximum of a single processor core.

8   Conclusions

We presented an on-line segmentation and reconstruction method for CAD geometry from a stream of point data. This method is based on accumulated means that are used for on-line computation of means of geometric properties. An advantage of this approach is that the results can be evaluated immediately by the operator of the hand-held laser scanner.

Currently the approach is limited to planar, cylindrical, and spherical shapes. We plan to extend the approach to other geometric primitives like cones, ellipsoids, tori, and rolling ball blends. Furthermore, an on-line boundary reconstruction of segments could be used to improve the results of the geometry reconstruction. Such boundaries are necessary to reconstruct trimming curves and intersections between geometric primitives.

Acknowledgments. This work was supported by DFG grant UM 26/5-1.

References

1. Agin, G., Binford, T.: Computer description of curved objects. IEEE Trans. Comput. **C-25**(4), 439–449 (1976)
2. Benkő, P., Martin, R.R., Várady, T.: Algorithms for reverse engineering boundary representation models. Comput. Aided Des. **33**(11), 839–851 (2001)

3. Benkő, P., Várady, T.: Direct segmentation of smooth, multiple point regions. In: Geometric Modeling and Processing, pp. 169–178. IEEE (2002)
4. Biegelbauer, G., Vincze, M.: Efficient 3D object detection by fitting superquadrics to range image data for robot's object manipulation. In: International Conference on Robotics and Automation, pp. 1086–1091. IEEE (2007)
5. Bodenmüller, T., Hirzinger, G.: Online surface reconstruction from unorganized 3d-points for the DLR hand-guided scanner system. In: 2nd Symposium on 3D Data Processing, Visualization and Transmission, pp. 285–292 (2004)
6. Boesch, J., Pajarola, R.: Flexible configurable stream processing of point data. In: WSCG 2009, pp. 49–56 (2009)
7. do Carmo, M.P.: Differential Geometry of Curves and Surfaces. Pearson, New York (1976)
8. Chan, T., Golub, G., Leveque, R.: Algorithms for computing the sample variance: analysis and recommendations. Am. Stat. **37**(3), 242–247 (1983)
9. Denker, K., Hagel, D., Raible, J., Umlauf, G., Hamann, B.: On-line reconstruction of cad geometry. In: International Conference on 3D Vision, pp. 151–158 (2013)
10. Denker, K., Lehner, B., Umlauf, G.: Online triangulation of laser-scan data. In: Garimella, R., (ed.) 17th International Meshing Roundtable, pp. 415–432 (2008)
11. Denker, K., Lehner, B., Umlauf, G.: Real-time triangulation of point streams. Eng. Comput. **27**(1), 67–80 (2011)
12. Gronau, I., Moran, S.: Optimal implementations of upgma and other common clustering algorithms. Inf. Process. Lett. **104**(6), 205–210 (2007)
13. Holland, P.W., Welsch, R.E.: Robust regression using iteratively reweighted least-squares. Com. Stat.: Theory Methods **A6**, 813–827 (1977)
14. Jolliffe, I.: Principal Component Analysis. Springer-Verlag, Heidelberg (2002)
15. Knuth, D.E.: Seminumerical Algorithms, The Art of Computer Programming, 3rd (edn.), chap. 4.2.2, vol. 2, p. 232. Addison-Wesley, Boston (1998)
16. Ling, R.F.: Comparison of several algorithms for computing sample means and variances. J. Am. Stat. Assoc. **69**(348), 859–866 (1974)
17. Page, D., Sun, Y., Koschan, A., Paik, J., Abidi, M.: Normal vector voting: crease detection and curvature estimation on large, noisy meshes. Graph. Models **64**, 199–229 (2002)
18. Pajarola, R.: Stream-processing points. In: IEEE Visualization, pp. 239–246 (2005)
19. Press, W., Teukolsky, S., Vetterling, W., Flannery, B.: Numerical Recipes: The Art of Scientific Computing. Cambridge University Press, Cambridge (2007)
20. Qt Project Hosting: Qt Documentation (2013). http://qt-project.org/doc/qt-5/containers.html#algorithmic-complexity
21. Razer: Hydra Motion Sensing Controller (2013). http://www.razerzone.com/gaming-controllers/razer-hydra
22. van Reeken, A.J.: Letters to the editor: dealing with Neely's algorithms. Commun. ACM **11**(3), 149–150 (1968)
23. Schnabel, R., Wahl, R., Klein, R.: Efficient RANSAC for point-cloud shape detection. Comput. Graph. Forum **26**(2), 214–226 (2007)
24. Shamir, A.: A survey on mesh segmentation techniques. Comput. Graph. Forum **27**(6), 1539–1556 (2008)
25. Sokal, R.R., Michener, C.D.: A statistical method for evaluating systematic relationships. Univ. Kans. Sci. Bull. **28**, 1409–1438 (1958)
26. Vančo, M., Hamann, B., Brunnett, G.: Surface reconstruction from unorganized point data with quadrics. Comput. Graph. Forum **27**(6), 1593–1606 (2008)
27. Várady, T.: Reverse engineering of geometric models - an introduction. Comp. Aided Des. **29**(4), 255–268 (1997)

28. Welford, B.P.: Note on a method for calculating corrected sums of squares and products. Technometrics **4**(3), 419–420 (1962)
29. Wu, J., Kobbelt, L.: Structure recovery via hybrid variational surface approximation. Comput. Graph. Forum **24**(3), 277–284 (2005)
30. Yamazaki, I., Natarajan, V., Bai, Z., Hamann, B.: Segmenting point-sampled surfaces. Vis. Comput. **26**(12), 1421–1433 (2010)
31. Zhong, S.: Efficient online spherical k-means clustering. In: Proceedings of 2005 IEEE International Joint Conference on Neural Networks, vol. 5, pp. 3180–3185 (2005)

Analysis of Intrinsic Mode Functions Based on Curvature Motion-Like PDEs

El Hadji S. Diop[1]($\boxtimes$) and Radjesvarane Alexandre[2]

[1] Department of Mathematics, University of Thiès,
Cité Malick SY, 967 Thiès, Sénégal
ehsdiop@hotmail.com

[2] Arts et Métiers ParisTech, 12 rue Edouard Manet, 75013 Paris, France
Radjesvarane.ALEXANDRE@paristech.fr

Abstract. We provide here some contributions for both the comprehension and theoretical modeling of the well known Empirical Mode
Decomposition (EMD) method in $2D$. This is achieved by reconsidering
the so-called local mean which is formulated now with the morphological
median operator. Doing this helps us derive curvature motion-like partial differential equations (PDEs) that mimic the $2D$ sifting process. In
addition to the mathematical framework brought out herein, preliminary
results show also that our proposed approach behaves like the $2D$ EMD
and in a better way.

Keywords: Partial differential equations · Image analysis · Empirical
mode decomposition · Multiscale analysis

1 Introduction

Let us recall that the Empirical Mode Decomposition (EMD) was introduced
by Huang et al. [9] for analyzing linear and non-stationary time series. The
EMD thus decomposes any data into several basic components called intrinsic
mode functions (IMFs). It was originally dedicated for signal processing, and
decomposes any signal into a sum of IMFs, each of them are then extracted at
every scale going from fine to coarse in an iterative procedure called the sifting
process (SP). It is important to keep in mind that EMD was originally formulated
as an algorithm, without any mathematical framework as we recall in Sect. 2.
The EMD have then been extended in image processing in a straightforward way,
and some algorithms were proposed with various applications [1,2,6,12,13]. Let
us point out the fact that all existing $2D$ EMD algorithms nearly differ from
one to another just by interpolates used to build envelopes [3,13]; moreover,
using a surface interpolation approach is very time consuming. Bhuiyan et al.
[1] used order statistics filters to reduce the computation time, though. The
main criticism of the EMD has been for a while its lack of a mathematical
framework. The principal difficulty in studying the EMD is the mean envelope
estimation which is very important due to the fact that it governs the crucial

© Springer International Publishing Switzerland 2015
J.-D. Boissonnat et al. (Eds.): Curves and Surfaces 2014, LNCS 9213, pp. 202–209, 2015.
DOI: 10.1007/978-3-319-22804-4_15

SP step where IMFs are extracted. Almost all existing $2D$ EMD algorithms are just a straightforward extension of the $1D$ case, in the sense that the way the mean envelope is estimated is the same. It is then normal to encounter the same or worst side effects of the $1D$ case such as boundary effects, overshoots, undershoots, etc. .... We have recently provided some interesting theoretical contributions for both in $1D$ [4,5] and $2D$ [7,8]. We pursue the same goal in this article by providing a mathematical framework for both the comprehension and modeling of EMD in $2D$. The paper is organized as follows: basics on EMD are recalled in Sect. 2. We present our approach in Sect. 3, and numerical results are displayed in Sect. 4.

2   Basics on EMD

The EMD could be summarized for any signal S, as followings [9]:

1. Find all the extrema of $S(x)$
2. Interpolate the maxima of $S(x)$ (*resp.* the minima of S(x)), denoted by $E_{max}(x)$ (*resp.* $E_{min}(x)$)
3. Compute the *local mean*:

$$m(x) = \frac{1}{2}(E_{max}(x) + E_{min}(x)) \tag{1}$$

4. Extract the detail $d(x) = S(x) - m(x)$
5. Iterate on $m(x)$

The $2D$ EMD, as well as $2D$ IMF, are defined as extensions of the $1D$ case without any theoretical justification. Let us recall the following definitions of $2D$ IMFs [1,10]:

Definition 1. *A function f is called a $2D$ IMF if it satisfies the two following conditions:*
(i) the local mean of f regarding (1) is equal to zero,
(ii) f has exactly one zero between any two consecutive local extrema.

Another definition of $2D$ IMF was proposed by Damerval et al. [3], based on numerical simulations:

Definition 2. *A function f is a $2D$ IMF if and only if the following conditions are satisfied:*

(i) the local mean of f computed with (1) is null,
(ii) maxima of f are positive, and the minima of f are negative,
(iii) the number of maxima of f and the number of minima of f are equal.

In practice, the $2D$ EMD procedure is refined using an iterative loop going from steps (1)-(4), and is called the $2D$ sifting process (SP). One iterates on the detail $d(x)$ until the local mean is equal to zero to get a $2D$ IMF (*cf.* Definitions 1 and 2). In [3], the procedure is stopped after a number of iterations. The image $I(x)$ is first decomposed in the inner loop as $I(x) = d_1(x) + m_1(x)$. Then, the first residual $m_1(x)$ is decomposed as $m_1(x) = d_2(x) + m_2(x)$, and so on. Finally, the EMD decomposes $I(x)$ as:

$$I(x) = \sum_{k=1}^{K} d_k(x) + r(x), \tag{2}$$

where d_k is the kth 2D IMF and $r(x)$ is the residual.

3   Mathematical Modeling of Our Approach

Let $I : \Omega \longrightarrow \mathbb{R}, I = I(x), x \in \Omega$ be a continuous image, where Ω is the domain of I which is assumed to be an open bounded set of $\mathbb{R}^2$. Let us denote by $\partial\Omega$ the boundary of Ω; $\partial\Omega$ is assumed to be Lipschitz.

We introduced in [7,8] the sequence $(h_n)_{n\in\mathbb{N}}$, and showed that the whole SP is fully determined by it. The sequence $(h_n)_{n\in\mathbb{N}}$ was defined by:

$$\begin{cases} h_{n+1} = h_n - \frac{1}{2}(\hat{h}_n + \check{h}_n) \\ h_0 = \quad I. \end{cases} \tag{3}$$

where $\hat{h}_n$ (resp. $\check{h}_n$) denotes the continuous interpolate of h_n maxima (resp. minima). The main idea of our preceding works was to define the operator Φ s.t. $\forall n \in \mathbb{N}$, $\Phi(h_n) = \frac{1}{2}(\hat{h}_n + \check{h}_n)$. Then, $\forall n \in \mathbb{N}$, one has: $h_n = (I_d - \Phi)^n h_0$, where f^n is an application s.t. $f^0 = I_d$, with I_d denoting the identity operator, and $f^n = \underbrace{f \circ f \circ \cdots \circ f}_{n \text{ times}}$.

Let us recall the median operator. So, let the structuring element B be a bounded set in Ω, and for $x \in \Omega$ and $\lambda \in \mathbb{R}$, let us consider the following sets $\check{E}_\lambda(x) = \{y \in B; I(x+y) < \lambda\}$ and $\hat{E}_\lambda(x) = \{y \in B; I(x+y) > \lambda\}$.

Definition 3. *The median of the image I w.r.t. the structuring element B, denoted by med_B, is the function*

$$med_B \, I : x \mapsto \left\{ \lambda \in \mathbb{R}; |\check{E}_\lambda(x)| \le \frac{|B|}{2} \text{ and } |\hat{E}_\lambda(x)| \le \frac{|B|}{2} \right\}, \tag{4}$$

where $| \cdot |$ the Lebesgue measure or counting measure associated to Ω.

The median operator is a morphological filter and has interesting properties, such as being a rank filter and a contrast -invariant operator. Also, it commutes with thresholding, and in that sense, it can be seen as a stack filter. In addition, it preserves quite well image shape, and has found interesting applications in various image processing problems; *e.g.*, image reconstruction, denoising, inpainting, ... Due to that, we replace it in (3), and formulate now the 2D SP in this manner:

$$\begin{cases} h_{n+1} = (I_d - med_B)h_n \\ h_0 = \quad I, \, \forall \, x \in \Omega. \end{cases} \tag{5}$$

In preceding works [7,8], we took the $2D$ local mean $\varPhi$ as the half some of the inf and sup operators; *i.e.*, $\varPhi = \frac{1}{2}(\sup + \inf)$, which is truly not morphological, despite each of sup and inf filters is morphological.

We now aim at providing a continuous formulation of the $2D$ SP, instead of in a discrete manner (5). This will be done by using an asymptotic expansion of med_B (the proof of that result is not provided here due to paper length limitation). Let B be a disc of radius $r > 0$ and centered at the origin, *i.e.*, $B = B_r = B(O, r)$:

Theorem 1. *Let $h : \mathbb{R}^2 \to \mathbb{R}$ be a C^2 function s.t. $\nabla h(x) \neq 0, \forall x \in \mathbb{R}^2$. Then, for $r > 0$ small, one has the following asymptotic development of med_{B_r}:*

$$med_{B_r} h(x) = h(x) + \frac{r^2}{6}|\nabla h(x)| \; div\left(\frac{\nabla h(x)}{|\nabla h(x)|}\right) + o(r^2). \qquad (6)$$

Let $\tau > 0$ small, and let us define the function $\tilde{h}$ in the following:

$$\begin{cases} \tilde{h} : \Omega \times \{\tau\mathbb{N}\} \longrightarrow \mathbb{R} \\ (x, n\tau) \longmapsto h_n(x). \end{cases} \qquad (7)$$

Let us consider a smoothed interpolate of $\tilde{h}$, and let us denote it by $\tilde{h}$ for simplicity. Taylor expansion yields:

$$h_{n+1}(x) = \tilde{h}(x, n\tau + \tau) = \tilde{h}(x, n\tau) + \tau\frac{\partial\tilde{h}}{\partial t}(x, n\tau) + o(\tau^2). \qquad (8)$$

By considering (5) and Theorem 1, one has:

$$h_{n+1}(x) = -\frac{r^2}{6}|\nabla h(x)|div\left(\frac{\nabla h(x)}{|\nabla h(x)|}\right) + o(r^2). \qquad (9)$$

Thus, taking $\tau = r^2$ yields:

$$\frac{\partial\tilde{h}}{\partial t}(x, n\tau) = -\frac{\tilde{h}(x, n\tau)}{r^2} - \frac{|\nabla h(x)|}{6}div\left(\frac{\nabla h(x)}{|\nabla h(x)|}\right) + o(1). \qquad (10)$$

The preceding Eqs. (7)-(10) show that, if the structuring element is a disk of radius $r > 0$ small, then, the discrete $2D$ SP achieved through (5) is consistent with the PDE:

$$\begin{cases} \dfrac{\partial h}{\partial t} + \dfrac{h}{r^2} + \dfrac{|\nabla h|}{6} \; div\left(\dfrac{\nabla h}{|\nabla h|}\right) = & 0 \\ h(x,0) & =I(x), \forall \, x \in \Omega \qquad (11) \\ \dfrac{\partial h}{\partial n} & = \; 0 \text{ on } \partial\Omega. \end{cases}$$

Running (11) with a well defined *stopping criterion* is definitely equivalent to perform the $2D$ SP. In fact, at an initial step $h_0 = I$, one computes (11) until the *stopping criterion* is fulfilled, we get then the first mode h_1. To get h_2, one computes again (11) with the initial condition as being equal to $I - h_1$, and so on. The following proposition (not proven here again due to pages length limitation) helps set up the stopping criterion:

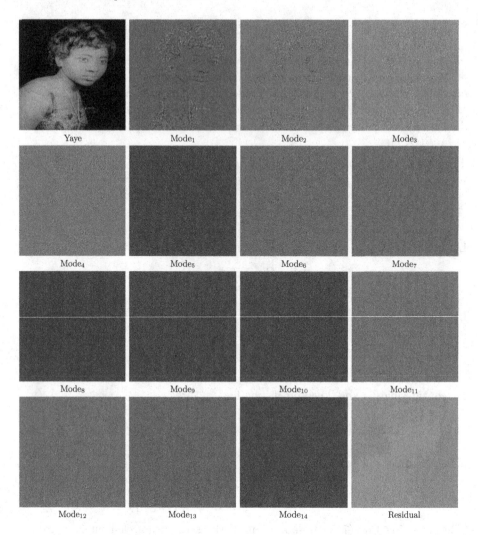

Fig. 1. Curvature motion-like image decomposition using our approach

Proposition 1. *Let h be solution of PDE (11). Then, $\forall\ \varepsilon > 0$, there exists $T \in \mathbb{R}_+$ s.t. $\forall\ t > T$, one has:*

$$\int_{\Omega} h(x,t)\ dx < \varepsilon. \tag{12}$$

Finally, we define our decomposition modes as follow:

Definition 4. *A function h is a mode if and only if: h is a solution of the PDE (11), and there exists some $T > 0$ s.t. $h(.,T)$ is a null mean function.*

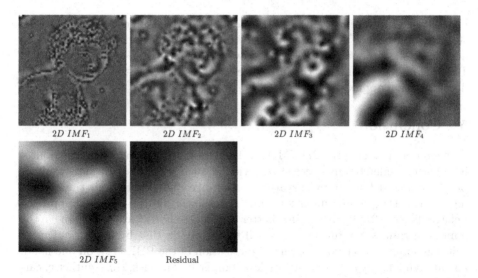

2D IMF₁   2D IMF₂   2D IMF₃   2D IMF₄

2D IMF₅   Residual

Fig. 2. Decomposition using BEMD with thine-plate smoothing spline [11]

4   Numerical Results

PDE (11) is implemented using an explicit scheme; *i.e.*, $h^{n+1} = h^n + \triangle t \cdot F^n(h^n)$. Derivatives are obtained with the following schemes:

$$\left(\frac{\partial h}{\partial x}\right)^+_{i,j} = \frac{h_{i+1,j} - h_{i,j}}{\triangle x}, \quad \left(\frac{\partial h}{\partial x}\right)^-_{i,j} = \frac{h_{i,j} - h_{i-1,j}}{\triangle x}$$

$$\left(\frac{\partial h}{\partial y}\right)^+_{i,j} = \frac{h_{i,j+1} - h_{i,j}}{\triangle x}, \quad \left(\frac{\partial h}{\partial y}\right)^-_{i,j} = \frac{h_{i,j} - h_{i,j-1}}{\triangle y}$$

Obtained decomposition results are showed in Fig. 1; comparisons with former $2D$ EMD which is implemented using thine-plate smoothing splines as described in [11], are provided in Fig. 2. Obtained results first show that our proposed approach behaves in the same manner as the classical EMD. Let us recall that in $1D$, EMD decomposes any signal as a sum of IMFs, each of them being extracted during the SP from fine to coarse. For an image, the highest image frequency holds information related to edges, texture, or noise. Thus, $2D$ EMD should be a multiscale image analysis tool where image details will less and less appear as one goes further in scales. In that sense, our approach truly behaves as $2D$ EMD. On another hand, comparison results with former $2D$ EMD show what we gain from our approach. In fact, transitions Modes gained from the presented PDE-based technique hold more intrinsic image information than $2D$ IMFs. This is mainly due, firstly, to formulating the $2D$ mean envelope by means of the morphological median operator (*cf.* Eq. (5)), secondly, to taking more advantage of the continuous (PDE) formulation of the $2D$ SP (*cf.* Eq. (11)), and thirdly, to totally avoiding the interpolation stage necessary in existing $2D$

EMD algorithms. For instance, $Mode_1$ and IMF_1 both hold most of image details; starting from the 3rd stage, resulting IMF_k, $k \geq 3$, does not have anymore to deal with the original image (Fig. 1-Yaye), in comparison to $Mode_k$, $k \geq 3$. The same remark remains true for the residuals obtained with the two different decomposition approaches.

5 Conclusion

A theoretical model of the $2D$ EMD has been provided herein, which contributes in a better understanding of the behavior of such a method. In fact, we have proposed a model driven by curvature motion-like PDEs that mimic the $2D$ sifting process. Despite preliminary results have shown that the proposed approach behaves like the $2D$ EMD and in a better way, more efforts have to be done in the numerical results. So, future works will be mainly on that by finding out more suitable numerical schemes for the discretization of the PDE (11), on one hand, and applying firstly such schemes on less complicated images; *e.g.*, synthetic, cartoon images, $\cdots$, which are smooth or at least piecewise smooth, before working on real images, on the other hand.

References

1. Bhuiyan, S.M.A., Adhami, R.R., Khan, J.F.: A novel approach of fast and adaptative bidimensional empirical mode decomposition. In: IEEE ICASSP, pp. 1313–1316 (2008)
2. Boudraa, A.O., Cexus, J.C., Salzenstein, F., Beghdadi, A.: Emd-based multibeam echosounder images segmentation. In: IEEE ISCCSP (Marocco) (2006)
3. Damerval, C., Meignen, S., Perrier, V.: A fast algorithm for bidimensional emd. IEEE Sign. Process. Lett. **12**(10), 701–704 (2005)
4. Diop, E.H.S., Alexandre, R., Boudraa, A.-O.: Analysis of intrinsic mode functions: a PDE approach. IEEE Sign. Process. Lett. **17**(4), 398–401 (2010)
5. Diop, E.H.S., Alexandre, R., Perrier, V.: A PDE based and interpolation-free framework for modeling the sifting process in a continuous domain. Adv. Comput. Math. **38**, 801–835 (2013)
6. Diop, E.H.S., Boudraa, A.O., Khenchaf, A., Thibaud, R., Garlan, T.: Multiscale characterization of bathymetric images by empirical mode decomposition. In: MARID III, Leeds, UK, 1–3 April 2008
7. Diop, E.H.S., Alexandre, R., Boudraa, A.-O.: A PDE model for 2D intrinsic mode functions. In: IEEE ICIP, Cairo, Egypt, pp. 3961–3964 (2009)
8. Diop, E.H.S., Alexandre, R., Moisan, L.: Intrinsic nonlinear multiscale image decomposition: a $2D$ empirical mode decomposition-like tool. Comput. Vis. Image Underst. **116**(1), 102–119 (2012)
9. Huang, N.E., Shen, Z., Long, S.R., Wu, M.C., Shih, H.H., Zheng, Q., Yen, N.-C., Tung, C.C., Liu, H.H.: The empirical mode decomposition and the hilbert spectrum for nonlinear and non-stationary time series analysis. Roy. Soc. **454**, 903–995 (1998)
10. Linderhed, A.: 2-d empirical mode decompositions- in the spirit of image compression. In: Proceedings of SPIE, WICAA IXI, Florida, USA, vol. 4738, 1–5 April 2002

11. Linderhed, A.: Compression by image empirical mode decomposition. In: IEEE ICIP, Genoa, Italy, vol. I, pp. 553–556, 11–14 September 2005
12. Liu, Z., Peng, S.: Estimation of image fractal dimension based on empirical mode decomposition. In: ACIVS, Brussels, Belgium (2004)
13. Xu, Y., Liu, B., Liu, J., Riemenschneider, S.: Two-dimensional empirical mode decomposition by finite elements. The Royal Society **A**, 1–17 (2006)

Optimality of a Gradient Bound for Polyhedral Wachspress Coordinates

Michael S. Floater[✉]

Department of Mathematics, University of Oslo,
PO Box 1053, Blindern, 0316 Oslo, Norway
michaelf@math.uio.no

Abstract. In a recent paper with Gillette and Sukumar an upper bound was derived for the gradients of Wachspress barycentric coordinates in simple convex polyhedra. This bound provides a shape-regularity condition that guarantees the convergence of the associated polyhedral finite element method for second order elliptic problems. In this paper we prove the optimality of the bound using a family of hexahedra that deform a cube into a tetrahedron.

Keywords: Barycentric coordinates · Wachspress coordinates · polyhedral finite element method

1 Introduction

There is growing interest in using generalized barycentric coordinates for finite element methods on polygonal and polyhedral meshes [1,3,7–9,13]. In order to establish the convergence of such methods one needs to derive an upper bound on the gradients of the coordinate functions in each polygonal or polyhedral cell. Such an upper bound was derived recently in [1] for Wachspress' rational coordinates in simple convex polytopes in $\mathbb{R}^d$. By 'simple' here we mean a polytope in which every vertex has exactly d incident faces. Thus in the important cases $\mathbb{R}^2$ and $\mathbb{R}^3$, this means any convex polygon in $\mathbb{R}^2$, but in $\mathbb{R}^3$, it means a convex polyhedron in which every vertex has three incident faces. For example, among the Platonic solids, the tetrahedron, cube, and dodecahedron are simple polyhedra, while the octahedron and icosahedron are not.

In the 2-D polygonal case, the upper bound of [1] was shown there to be optimal. The purpose of this note is to prove that the bound is also optimal in 3-D. The proof uses a one-parameter family of hexahedra (which are simple) in which the three faces incident on one vertex approach coplanarity. The family continuously deforms a cube, until, in the limiting case, it becomes a tetrahedron.

It is hoped that this work might motivate further research on this topic. For example, finding a similar, simple gradient bound for Wachspress coordinates in non-simple polyhedra is an open question, as is a corresponding bound for other types of coordinates such as mean value coordinates.

© Springer International Publishing Switzerland 2015
J.-D. Boissonnat et al. (Eds.): Curves and Surfaces 2014, LNCS 9213, pp. 210–215, 2015.
DOI: 10.1007/978-3-319-22804-4_16

2   Barycentric Coordinates on Polyhedra

Let $P \subset \mathbb{R}^3$ be a convex polyhedron, viewed as an open set, with V and F its vertices and faces; see Fig. 1. We call a set of functions $\phi_{\mathbf{v}} : P \to \mathbb{R}$, $\mathbf{v} \in V$, *barycentric coordinates* if they are non-negative, and if, for any $\mathbf{x} \in P$,

$$\sum_{\mathbf{v} \in V} \phi_{\mathbf{v}}(\mathbf{x}) \mathbf{v} = \mathbf{x}, \qquad \sum_{\mathbf{v} \in V} \phi_{\mathbf{v}}(\mathbf{x}) = 1. \tag{1}$$

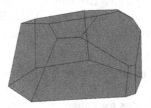

Fig. 1. A simple convex polyhedron

It has been shown [2] that such functions extend continuously to the boundary ∂P of P. They have the Lagrange property at the vertices, i.e., $\phi_{\mathbf{v}}(\mathbf{u}) = \delta_{\mathbf{vu}}$, $\mathbf{v}, \mathbf{u} \in V$, and they are linear along each edge of P. The function $\phi_{\mathbf{v}}$ is zero on any face f that does not have $\mathbf{v}$ as a vertex. If $f \in F$ is a face and $\mathbf{x}$ is any point in f,

$$\sum_{\mathbf{v} \in V_f} \phi_{\mathbf{v}}(\mathbf{x}) \mathbf{v} = \mathbf{x}, \qquad \sum_{\mathbf{v} \in V_f} \phi_{\mathbf{v}}(\mathbf{x}) = 1, \tag{2}$$

where $V_f \subset V$ is the set of vertices of f.

Suppose next that $\phi_{\mathbf{v}} \in C^1(P)$ for all $\mathbf{v} \in V$, which is, for example, the case for Wachspress coordinates [5,6,10–12] and mean value coordinates [2,4], both of which are in $C^\infty(P)$. Then we could consider using the $\phi_{\mathbf{v}}$ as shape functions over a domain partitioned into convex polyhedra, as in [1,3,7]. In order to establish a convergence theory for such a method we need to consider the vertex-based interpolation operator I of a function $u : \overline{P} \to \mathbb{R}$ given by

$$I(u) := \sum_{\mathbf{v} \in V} u(\mathbf{v}) \phi_{\mathbf{v}}. \tag{3}$$

With $|\cdot|$ the Euclidean norm in $\mathbb{R}^3$, we need to derive an upper bound on

$$|\nabla I(u)(\mathbf{x})|, \qquad \mathbf{x} \in P,$$

in terms of u and the geometry of P. Observe that for $\mathbf{x} \in P$,

$$|\nabla I(u)(\mathbf{x})| \le \sum_{\mathbf{v} \in V} |u(\mathbf{v}) \nabla \phi_{\mathbf{v}}(\mathbf{x})| \le \lambda(\mathbf{x}) \max_{\mathbf{v} \in V} |u(\mathbf{v})|, \tag{4}$$

where

$$\lambda(\mathbf{x}) := \sum_{\mathbf{v} \in V} |\nabla \phi_{\mathbf{v}}(\mathbf{x})|, \tag{5}$$

and so

$$\sup_{\mathbf{x} \in P} |\nabla I(u)(\mathbf{x})| \leq \Lambda \max_{\mathbf{v} \in V} |u(\mathbf{v})|,$$

where

$$\Lambda := \sup_{\mathbf{x} \in P} \lambda(\mathbf{x}).$$

Thus the function $\lambda : P \to \mathbb{R}$ plays a role similar to that of the Lebesgue function in the theory of polynomial interpolation, and Λ acts like the Lebesgue constant. The goal is to derive an upper bound on Λ in terms of the geometry of P.

3   Wachspress Coordinates for Polyhedra

Wachspress' rational coordinates for convex polygons and polyhedra were developed and studied by Wachspress, Warren and others [5,6,10–12]. In [11], Warren generalized the 2-D coordinates of Wachspress to simple convex polyhedra. The derivation was based on the so-called 'adjoint' of the polyhedron. In [12], Warren et al. derived the same coordinates in a different way, avoiding the adjoint, as follows. For each face $f \in F$, let $\mathbf{n}_f \in \mathbb{R}^3$ denote its unit outward normal, and for any $\mathbf{x} \in P$, let $h_f(\mathbf{x})$ denote the perpendicular distance of $\mathbf{x}$ to f, which can be expressed as the scalar product

$$h_f(\mathbf{x}) = (\mathbf{v} - \mathbf{x}) \cdot \mathbf{n}_f,$$

for any vertex $\mathbf{v} \in V$ belonging to f. For each vertex $\mathbf{v} \in V$, let f_1, f_2, f_3 be the three faces incident to $\mathbf{v}$, and for $\mathbf{x} \in P$, let

$$w_{\mathbf{v}}(\mathbf{x}) = \frac{\det(\mathbf{n}_{f_1}, \mathbf{n}_{f_2}, \mathbf{n}_{f_3})}{h_{f_1}(\mathbf{x}) h_{f_2}(\mathbf{x}) h_{f_3}(\mathbf{x})}, \tag{6}$$

where it is understood that f_1, f_2, f_3 are ordered such that the determinant in the numerator is positive. Here, for vectors $\mathbf{a}, \mathbf{b}, \mathbf{c} \in \mathbb{R}^3$,

$$\det(\mathbf{a}, \mathbf{b}, \mathbf{c}) := \begin{vmatrix} a_1 & b_1 & c_1 \\ a_2 & b_2 & c_2 \\ a_3 & b_3 & c_3 \end{vmatrix}.$$

Thus the ordering of f_1, f_2, f_3 must be anticlockwise around $\mathbf{v}$, seen from outside P. In this way, $w_{\mathbf{v}}(\mathbf{x}) > 0$, and it was shown in [12] that the functions

$$\phi_{\mathbf{v}}(\mathbf{x}) := \frac{w_{\mathbf{v}}(\mathbf{x})}{\sum_{\mathbf{u} \in V} w_{\mathbf{u}}(\mathbf{x})} \tag{7}$$

are barycentric coordinates, i.e., satisfy (1). Some matlab code for evaluating these coordinates and their gradients can be found in [1].

4   Gradient Bound

The following bound on Λ was derived in [1] for Wachspress coordinates $\phi_{\mathbf{v}}$ for a simple polyhedron P: if

$$h_* := \min_{f \in F} \min_{\mathbf{v} \in V \setminus V_f} h_f(\mathbf{v}), \tag{8}$$

the minimal perpendicular distance between vertices and faces of P, then

$$\Lambda \le \frac{6}{h_*}. \tag{9}$$

The significance of this is that it implies convergence for the polyhedral finite element method applied to second order elliptic problems with shape-regularity parameter $\kappa := \operatorname{diam}(P)/h_*$.

We will establish the optimality of (9) by constructing a family of hexahedra $\{P_\epsilon\}$ containing a vertex $\mathbf{v}$ such that

$$\lim_{\epsilon \to 0} h_* \lambda(\mathbf{v}) = 6. \tag{10}$$

In fact we do this for λ defined with respect to *any* barycentric coordinates that are C^1 at the vertices of P. Though the particular constant 6 in (10) is not essential, (10) clearly shows the dependency of Λ on h_*.

We recall first a preliminary result of [1]. Let $\mathbf{v}$ be some vertex of P, and let $\mathbf{w}_1, \mathbf{w}_2, \mathbf{w}_3$ be its three neighbours, and define the three vectors $\mathbf{e}_k = \mathbf{w}_k - \mathbf{v}$, k=1,2,3. It was shown in [1] that for any barycentric coordinates $\phi_{\mathbf{u}}$, $\mathbf{u} \in V$, that are C^1 at $\mathbf{v}$,

$$\lambda(\mathbf{v}) = |\nabla \phi_{\mathbf{v}}(\mathbf{v})| + \sum_{i=1}^{3} |\nabla \phi_{\mathbf{w}_i}(\mathbf{v})|, \tag{11}$$

with

$$\nabla \phi_{\mathbf{w}_1}(\mathbf{v}) = \frac{1}{D} \mathbf{e}_2 \times \mathbf{e}_3, \quad \nabla \phi_{\mathbf{w}_2}(\mathbf{v}) = \frac{1}{D} \mathbf{e}_3 \times \mathbf{e}_1, \quad \nabla \phi_{\mathbf{w}_3}(\mathbf{v}) = \frac{1}{D} \mathbf{e}_1 \times \mathbf{e}_2,$$

$$\nabla \phi_{\mathbf{v}}(\mathbf{v}) = -\frac{1}{D}(\mathbf{e}_2 \times \mathbf{e}_3 + \mathbf{e}_3 \times \mathbf{e}_1 + \mathbf{e}_1 \times \mathbf{e}_2), \quad \text{and} \quad D = \det(\mathbf{e}_1, \mathbf{e}_2, \mathbf{e}_3).$$

This is a consequence of the linearity of the coordinates on the edges of P, their Lagrange property at the vertices, and the assumption of C^1 regularity at $\mathbf{v}$.

From these formulas we obtain

Lemma 1. *Suppose the vertex* $\mathbf{v}$ *is such that*

$$|\mathbf{e}_1| = |\mathbf{e}_2| = |\mathbf{e}_3|, \qquad \mathbf{e}_2 \cdot \mathbf{e}_3 = \mathbf{e}_3 \cdot \mathbf{e}_1 = \mathbf{e}_1 \cdot \mathbf{e}_2,$$

and $h_* = h_f(\mathbf{w}_3)$, *where* f *is the face containing* $\mathbf{v}$, $\mathbf{w}_1$, *and* $\mathbf{w}_2$. *Then,*

$$h_* \lambda(\mathbf{v}) = 3 + R,$$

where $R = (|\mathbf{e}_2 \times \mathbf{e}_3 + \mathbf{e}_3 \times \mathbf{e}_1 + \mathbf{e}_1 \times \mathbf{e}_2|)/|\mathbf{e}_1 \times \mathbf{e}_2|$.

This follows from the formula (11) and the fact that $h_f(\mathbf{w}_3) = -\mathbf{e}_3 \cdot \mathbf{n}_f$ and $\mathbf{n}_f = -(\mathbf{e}_1 \times \mathbf{e}_2)/|\mathbf{e}_1 \times \mathbf{e}_2|$. Thus, to prove the optimality of the bound (9) it is sufficient to find a sequence of simple polyhedra with a vertex $\mathbf{v}$ satisfying the conditions of the lemma and such that $R \to 3$.

5 A Family of Hexahedra

We achieve the convergence $R \to 3$ with the following family of hexahedra. In analogy with the unit cube, we will use the typical notation $\mathbf{v}_{ijk}$, $0 \le i, j, k \le 1$, for the eight vertices of the hexahedron, with the connectivity it implies. For any ϵ with $0 < \epsilon < 3\sqrt{2}$, let P_ϵ be the hexahedron with vertices, in increasing z coordinate,

$$\mathbf{v}_{000} = (0, 0, 0),$$
$$\mathbf{v}_{100} = (2, 0, \epsilon), \quad \mathbf{v}_{010} = (-1, \sqrt{3}, \epsilon), \quad \mathbf{v}_{001} = (-1, -\sqrt{3}, \epsilon),$$
$$\mathbf{v}_{110} = r(1, \sqrt{3}, 2\epsilon), \quad \mathbf{v}_{011} = r(-2, 0, 2\epsilon), \quad \mathbf{v}_{101} = r(1, -\sqrt{3}, 2\epsilon),$$
$$\mathbf{v}_{111} = (0, 0, 3\sqrt{2}),$$

where $r = 2\sqrt{2}/(\sqrt{2} + \epsilon)$. One can check that the six 'faces' really are faces, i.e., planar, by verifying that the two determinants

$$\det(\mathbf{v}_{100} - \mathbf{v}_{000}, \mathbf{v}_{110} - \mathbf{v}_{000}, \mathbf{v}_{010} - \mathbf{v}_{000}),$$
$$\det(\mathbf{v}_{101} - \mathbf{v}_{001}, \mathbf{v}_{100} - \mathbf{v}_{001}, \mathbf{v}_{110} - \mathbf{v}_{001}),$$

are zero. One can also verify that if $\epsilon = \sqrt{2}$ then $r = 1$ and $P_{\sqrt{2}}$ is a cube with side length $\sqrt{6}$. Figure 2 shows P_ϵ for four values of ϵ, decreasing from $\sqrt{2}$.

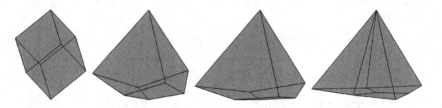

Fig. 2. Hexahedron P_ϵ, $\epsilon = \sqrt{2}(1, 0.25, 0.1, 0.01)$

Considering again the ratio R of the lemma, let

$$\mathbf{v} = \mathbf{v}_{000}, \quad \mathbf{w}_1 = \mathbf{v}_{100}, \quad \mathbf{w}_2 = \mathbf{v}_{010}, \quad \mathbf{w}_3 = \mathbf{v}_{001},$$

in which case the conditions of the lemma hold for small ϵ. We compute

$$\mathbf{e}_1 \times \mathbf{e}_2 = (2, 0, \epsilon) \times (-1, \sqrt{3}, \epsilon) = (-\sqrt{3}\epsilon, -3\epsilon, 2\sqrt{3}),$$
$$\mathbf{e}_2 \times \mathbf{e}_3 = (2\sqrt{3}\epsilon, 0, 2\sqrt{3}), \quad \mathbf{e}_3 \times \mathbf{e}_1 = (-\sqrt{3}\epsilon, 3\epsilon, 2\sqrt{3}),$$

and so

$$\mathbf{e}_1 \times \mathbf{e}_2 + \mathbf{e}_2 \times \mathbf{e}_3 + \mathbf{e}_3 \times \mathbf{e}_1 = (0, 0, 6\sqrt{3}).$$

Therefore,

$$R = \frac{6\sqrt{3}}{2\sqrt{3(1 + \epsilon^2)}} = \frac{3}{\sqrt{1 + \epsilon^2}} \to 3 \qquad \text{as } \epsilon \to 0.$$

References

1. Floater, M.S., Gillette, A., Sukumar, N.: Gradient bounds for wachspress coordinates on polytopes. SIAM J. Numer. Anal. **52**, 515–532 (2014)
2. Floater, M.S., Kos, G., Reimers, M.: Mean value coordinates in 3D. Comput. Aided Geom. Des. **22**, 623–631 (2005)
3. Gillette, A., Rand, A., Bajaj, C.: Error estimates for generalized barycentric interpolation. Adv. Comput. Math. **37**, 417–439 (2012)
4. Ju, T., Schaefer, S., Warren, J.: Mean value coordinates for closed triangular meshes. ACM Trans. on Graph. **24**, 561–566 (2005)
5. Ju, T., Schaefer, S., Warren, J., Desbrun, M.: A geometric construction of coordinates for convex polyhedra using polar duals, symposium on geometry processing. Eurographics Assoc. **2005**, 181–186 (2005)
6. Meyer, M., Barr, A., Lee, H., Desbrun, M.: Generalized barycentric coordinates for irregular polygons. J. Graph. Tools **7**, 13–22 (2002)
7. Rand, A., Gillette, A., Bajaj, C.: Interpolation error estimates for mean value coordinates over convex polygons. Adv. Comput. Math. **39**, 327–347 (2013)
8. Sukumar, N., Tabarraei, A.: Conforming polygonal finite elements. Int. J. Numer. Methods Eng. **61**, 2045–2066 (2004)
9. Talischi, C., Paulino, G.H., Le, C.H.: Honeycomb wachspress finite elements for structural topology optimization. Struct. Multidisc. Optim. **37**, 569–583 (2009)
10. Wachspress, E.L.: A Rational Finite Element Basis. Academic Press, New York (1975)
11. Warren, J.: Barycentric coordinates for convex polytopes. Adv. Comput. Math. **6**, 97–108 (1996)
12. Warren, J., Schaefer, S., Hirani, A.N., Desbrun, M.: Barycentric coordinates for convex sets. Adv. Comput. Math. **27**, 319–338 (2007)
13. Wicke, M., Botsch, M., Gross, M.: A finite element method on convex polyhedra. In: Proceedings of Eurographics **2007**, 255–364 (2007)

Differential Geometry Revisited by Biquaternion Clifford Algebra

Patrick R. Girard[1]([✉]), Patrick Clarysse[1], Romaric Pujol[2],
Liang Wang[1], and Philippe Delachartre[1]

[1] Université de Lyon, CREATIS, CNRS UMR 5220, Inserm U1044, INSA-Lyon,
Université LYON 1, Bât. Blaise Pascal, 7 Avenue Jean Capelle,
69621 Villeurbanne, France
patrick.girard@creatis.insa-lyon.fr
[2] Université de Lyon, Pôle de Mathématiques, INSA-Lyon,
Bât. Léonard de Vinci, 21 Avenue Jean Capelle,
69621 Villeurbanne, France

Abstract. In the last century, differential geometry has been expressed within various calculi: vectors, tensors, spinors, exterior differential forms and recently Clifford algebras. Clifford algebras yield an excellent representation of the rotation group and of the Lorentz group which are the cornerstones of the theory of moving frames. Though Clifford algebras are all related to quaternions via the Clifford theorem, a biquaternion formulation of differential geometry does not seem to have been formulated so far. The paper develops, in $3D$ Euclidean space, a biquaternion calculus, having an associative exterior product, and applies it to differential geometry. The formalism being new, the approach is intended to be pedagogical. Since the methods of Clifford algebras are similar in other dimensions, it is hoped that the paper might open new perspectives for a 4D hyperbolic differential geometry. All the calculi presented here can easily be implemented algebraically on Mathematica and numerically on Matlab. Examples, matrix representations, and a Mathematica work-sheet are provided.

Keywords: Clifford algebras · Quaternions · Biquaternions · Differential geometry · Rotation group $SO(3)$ · Hyperquaternion algebra

1 Introduction

Much of differential geometry is still formulated today within the $3D$ vector calculus which was developed at the end of the nineteenth century. In recent years, new mathematical tools have appeared, based on Clifford algebras [1–10] which give an excellent representation of groups, such as the rotation group $SO(3)$ or the Lorentz group, which are the cornerstones of the theory of moving frames. Since the methods of Clifford algebras can easily be transposed to other dimensions, the question naturally arises of whether it is possible to rewrite differential geometry within a Clifford algebra in order to open new perspectives for

© Springer International Publishing Switzerland 2015
J.-D. Boissonnat et al. (Eds.): Curves and Surfaces 2014, LNCS 9213, pp. 216–242, 2015.
DOI: 10.1007/978-3-319-22804-4_17

$4D$ modeling. Such an extension might proceed as follows. A $4D$ tetraquaternion calculus has already been presented in [7,8]. A moving surface $OM = f(t,u,v)$ can be viewed as a hypersurface (with normal n) in a $4D$ pseudo-euclidean space. The invariants are then obtained by diagonalizing the second fundamental form via a rotation around n combined with a Lorentz boost along n, generalizing the methods presented here. Though Clifford algebras can be presented in various ways, the originality of the paper lies in the use biquaternions. We shall first introduce quaternions and Clifford algebras together with a demonstration of Clifford's theorem relating Clifford algebras to quaternions. Then, we shall develop the biquaternion calculus (with its associative exterior product) and show how classical differential geometry can be reformulated within this new algebraic framework.

2   Clifford Algebras: Historical Perspective

2.1   Hamilton's Quaternions and Biquaternions

In 1843, W.R. Hamilton (1805–1865) discovered quaternions [11–17] which are a set of four real numbers:

$$a = a_0 + a_1 i + a_2 j + a_3 k \tag{1}$$
$$= (a_0, a_1, a_2, a_3) \tag{2}$$
$$= (a_0, \overrightarrow{a}) \tag{3}$$

where i, j, k multiply according to the rules

$$i^2 = j^2 = k^2 = ijk = -1 \tag{4}$$
$$ij = -ji = k \tag{5}$$
$$jk = -kj = i \tag{6}$$
$$ki = -ik = j. \tag{7}$$

The conjugate of a quaternion is given by

$$a_c = a_0 - a_1 i - a_2 j - a_3 k. \tag{8}$$

Hamilton was to give a $3D$ interpretation of quaternions; he named a_0 the scalar part and $\overrightarrow{a}$ the vector part. The product of two quaternions a and b is defined by

$$
\begin{aligned}
ab = &(a_0 b_0 - a_1 b_1 - a_2 b_2 - a_3 b_3) \\
&+ (a_0 b_1 + a_1 b_0 + a_2 b_3 - a_3 b_2)i \\
&+ (a_0 b_2 + a_2 b_0 + a_3 b_1 - a_1 b_3)j \\
&+ (a_0 b_3 + a_3 b_0 + a_1 b_2 - a_2 b_1)k
\end{aligned} \tag{9}
$$

and in a more condensed form

$$ab = (a_0 b_0 - \overrightarrow{a} \cdot \overrightarrow{b}, \, a_0 \overrightarrow{b} + b_0 \overrightarrow{a} + \overrightarrow{a} \times \overrightarrow{b}) \tag{10}$$

where $\vec{a} \cdot \vec{b}$ and $\vec{a} \times \vec{b}$ are respectively the usual scalar and vector products. Quaternions (denoted by $\mathbb{H}$) constitute a non commutative field without zero divisors (i.e. $ab = 0$ implies a or $b = 0$). At the end of the nineteenth century, the classical vector calculus was obtained by taking $a_0 = b_0 = 0$ and by separating the dot and vector products. Hamilton also introduced complex quaternions he called biquaternions which we shall use in the next parts.

2.2   Clifford Algebras and Theorem

About the same time Hamilton discovered the quaternions, H.G. Grassmann (1809–1877) had the fundamental idea of a calculus composed of n generators $e_1, e_2, \ldots e_n$ multiplying according to the rule $e_i e_j = -e_j e_i (i \neq j)$ [18–21]. In 1878, W.K. Clifford (1845–1878) was to give a precise algebraic formulation thereof and proved the Clifford theorem relating Clifford algebras to quaternions. Though Clifford did not claim any particular originality, his name was to become attached to these algebras [22,23].

Definition 1. *Clifford's algebra C_n is defined as an algebra (over $\mathbb{R}$) composed of n generators $e_1, e_2, \ldots, e_n$ multiplying according to the rule $e_i e_j = -e_j e_i$ $(i \neq j)$ and such that $e_i^2 = \pm 1$. The algebra C_n contains 2^n elements constituted by the n generators, the various products $e_i e_j, e_i e_j e_k, \ldots$ and the unit element 1.*

Examples of Clifford algebras (over $\mathbb{R}$) are

1. complex numbers $\mathbb{C}$ $(e_1 = i, e_1^2 = -1)$.
2. quaternions $\mathbb{H}$ $(e_1 = i, e_2 = j, e_i^2 = -1)$.
3. biquaternions $\mathbb{H} \otimes \mathbb{C}$ $(e_1 = Ii, e_2 = Ij, e_3 = Ik, I^2 = -1, e_i^2 = 1$, I commuting with $i, j, k)$. Matrix representations of biquaternions are given in the appendix.
4. tetraquaternions $\mathbb{H} \otimes \mathbb{H}$ $(e_0 = j, e_1 = kI, e_2 = kJ, e_3 = kK, e_0^2 = -1, e_1^2 = e_2^2 = e_3^2 = 1$, where the small i, j, k commute with the capital $I, J, K)$ [7,8].

All Clifford algebras are related to quaternions via the following theorem.

Theorem 1. *If $n = 2m$ (m : integer), the Clifford algebra C_{2m} is the tensor product of m quaternion algebras. If $n = 2m - 1$, the Clifford algebra C_{2m-1} is the tensor product of $m - 1$ quaternion algebras and the algebra $(1, \omega)$ where ω is the product of the $2m$ generators $(\omega = e_1 e_2 \ldots e_{2m})$ of the algebra C_{2m}.*

Proof. The above examples of Clifford algebras prove the Clifford theorem up to $n = 4$. For any n, Clifford's theorem can be proved by recurrence as follows [24, p. 378]. The theorem being true for $n = (2, 4)$, suppose that the theorem is true for $C_{2(n-1)}$, to $C_{2(n-1)}$ one adds the quantities

$$f = e_1 e_2 \ldots e_{2(n-1)} e_{2n-1}, g = e_1 e_2 \ldots e_{2(n-1)} e_{2n} \tag{11}$$

which anticommute among themselves and commute with the elements of $C_{2(n-1)}$; hence, they constitute a quaternionic system which commutes with $C_{2(n-1)}$. From the various products between f, g and the elements of $C_{2(n-1)}$ one obtains a basis of C_{2n} which proves the theorem.   □

Hence, Clifford algebras can be formulated as hyperquaternion algebras the latter being defined as either a tensor product of quaternion algebras or a subalgebra thereof.

3   Biquaternion Clifford Algebra

3.1   Definition

The algebra (over $\mathbb{R}$) has three anticommuting generators $e_1 = Ii, e_2 = Ij, e_3 = Ik$ with $e_1^2 = e_2^2 = e_3^2 = 1$ ($I^2 = -1$, I commuting with i, j, k). A complete basis of the algebra is given in the following table

$$
\begin{array}{|l|l|l|l|}
\hline
1 & i = e_3 e_2 & j = e_1 e_3 & k = e_2 e_1 \\
\hline
I = e_1 e_2 e_3 & Ii = e_1 & Ij = e_2 & Ik = e_3 \\
\hline
\end{array}
\tag{12}
$$

A general element of the algebra can be written

$$A = p + Iq \tag{13}$$

where $p = p_0 + p_1 i + p_2 j + p_3 k$ and $q = q_0 + q_1 i + q_2 j + q_3 k$ are quaternions. The Clifford algebra contains scalars p_0, vectors $I(0, q_1, q_2, q_3)$, bivectors $(0, p_1, p_2, p_3)$ and trivectors (pseudo-scalars) Iq_0 where all coefficients (p_i, q_i) are real numbers; we shall call these multivector spaces respectively V_0, V_1, V_2 and V_3. The product of two biquaternions $A = p + Iq$ and $B = p' + Iq'$ is defined by

$$AB = (pp' - qq') + I(pq' + qp') \tag{14}$$

where the products in parentheses are quaternion products. The conjugate of A is defined as

$$A_c = (p_c + Iq_c) \tag{15}$$

with p_c and q_c being the quaternion conjugates with $(AB)_c = B_c A_c$. The dual of A noted A^* is defined by

$$A^* = IA \tag{16}$$

and the commutator of two Clifford numbers by

$$[A, B] = \frac{1}{2}(AB - BA). \tag{17}$$

3.2   Interior and Exterior Products

Products Between Vectors and Multivectors. In this section we shall adopt the general approach used in [4] though our algebra differs as well as several formulas. The product of two general elements of the algebra being given, one can define interior and exterior products of two vectors $a (= a_1 iI + a_2 jI + a_3 kI)$ and b via the obvious identity

$$ab = \frac{1}{2}(ab + ba) - [-\frac{1}{2}(ab - ba)] \tag{18}$$

$$= a.b - a \wedge b \tag{19}$$

with $a.b$ being the interior product

$$a.b = \frac{1}{2}(ab + ba) \tag{20}$$

$$= a_1b_1 + a_2b_2 + a_3b_3 \in V_0 \tag{21}$$

and $a \wedge b$ the exterior product

$$a \wedge b = -\frac{1}{2}(ab - ba) \tag{22}$$

$$= (a_2b_3 - a_3b_2)i + (a_3b_1 - a_1b_3)j + (a_1b_2 - a_2b_1)k \in V_2 \tag{23}$$

which has the same components as the pseudo-vector $\vec{a} \times \vec{b}$. Next we define the interior products $a.A_p$ and $A_p.a$ (with $2 \le p \le 3$ and $A_p = v_1 \wedge v_2 \wedge \ldots \wedge v_p$, $v_i \in V_1$)

$$a.A_p = \Sigma_{k=1}^{p}(-1)^k(a.v_k)v_1 \wedge \ldots \wedge v_{k-1} \wedge v_{k+1} \wedge \ldots \wedge v_p \tag{24}$$

together with

$$A_p.a \equiv (-1)^{p-1}a.A_p. \tag{25}$$

Explicitly, we have

$$a.(v_1 \wedge v_2) = -(a.v_1)v_2 + (a.v_2)v_1 \tag{26}$$

$$a.(v_1 \wedge v_2 \wedge v_3) = -(a.v_1)(v_2 \wedge v_3) + (a.v_2)(v_1 \wedge v_3) - (a.v_3)(v_1 \wedge v_2). \tag{27}$$

The interior product $a.A_p$ allows the definition of the multivector $a \wedge A_p$ and $A_p \wedge a$ via the relations

$$aA_p = a.A_p - a \wedge A_p \tag{28}$$

$$A_pa = A_p.a - A_p \wedge a \tag{29}$$

with

$$A_p \wedge a = (-1)^p a \wedge A_p. \tag{30}$$

Multiplying both sides of Eq. (29) with $(-1)^p$ and applying Eqs. (25 and 30), we obtain

$$(-1)^p A_pa = -a.A_p - a \wedge A_p \tag{31}$$

Combining Eqs. (28 and 31), we obtain the formulas valid in all cases ($1 \le p \le 3$)

$$a.A_p = \frac{1}{2}[aA_p - (-1)^p A_pa] \in V_{p-1} \tag{32}$$

$$a \wedge A_p = -\frac{1}{2}[aA_p + (-1)^p A_pa] \in V_{p+1} \tag{33}$$

A_3 being a pseudo-scalar, commuting with any Clifford number, we have in particular $a \wedge A_3 = 0$.

Products Between Multivectors. Other interior and exterior products between two multivectors A_p and B_q are defined for $p \leq q$ [4]

$$A_p \cdot B_q \equiv (v_1 \wedge v_2 \wedge \cdots \wedge v_{p-1}) \cdot (v_p \cdot B_q) \tag{34}$$

$$A_p \wedge B_q \equiv v_1 \wedge (v_2 \wedge \cdots \wedge v_p) \wedge B_q \tag{35}$$

with

$$A_p \cdot B_q = (-1)^{p(q+1)} B_q \cdot A_p \tag{36}$$

which defines $B_q \cdot A_p$ for $q \geq p$. The various products are given in Table 1.

Table 1. Interior and exterior products with their corresponding expressions in the classical vector calculus (with $B = b \wedge c$, $B_1 = a \wedge b$, $B_2 = c \wedge d$, $T_1 = a \wedge b \wedge c$, $T_2 = f \wedge g \wedge h$ and $a, b, c, d, e, f, g, h \in V_1, T \in V_3$)

Multivector calculus	Classical vector calculus
$a.b = \frac{1}{2}(ab + ba) \in V_0$	$\vec{a} \cdot \vec{b}$
$a \wedge b = -\frac{1}{2}(ab - ba) \in V_2$	$\vec{a} \times \vec{b}$
$a.B = \frac{1}{2}(aB - Ba) \in V_1$	$\vec{a} \times \left(\vec{b} \times \vec{c} \right)$
$a \wedge B = -\frac{1}{2}(aB + Ba) \in V_3$	$\vec{a} \cdot \left(\vec{b} \times \vec{c} \right)$
$B_1.B_2 = -\frac{1}{2}(B_1 B_2 + B_2 B_1) \in V_0$	$\left(\vec{a} \times \vec{b} \right) \cdot \left(\vec{c} \times \vec{d} \right)$
$[B_1, B_2] = \frac{1}{2}(B_1 B_2 - B_2 B_1) \in V_2$	$\left(\vec{a} \times \vec{b} \right) \times \left(\vec{c} \times \vec{d} \right)$
$T_1.T_2 = -\frac{1}{2}(T_1 T_2 + T_2 T_1) \in V_0$	$\left[\vec{a} \cdot \left(\vec{b} \times \vec{c} \right) \right] \left[\vec{f} \cdot \left(\vec{g} \times \vec{h} \right) \right]$
$B.T = -\frac{1}{2}(BT + TB) \in V_1$	$\left(\vec{a} \times \vec{b} \right) \left[\vec{f} \cdot \left(\vec{g} \times \vec{h} \right) \right]$
$a.T = \frac{1}{2}(aT + Ta) \in V_2$	$\vec{a} \left[\vec{f} \cdot \left(\vec{g} \times \vec{h} \right) \right]$

Associativity. A major property of the exterior product is its associativity which is expressed as (with $v_i \in V_1$) [4].

$$(v_1 \wedge v_2) \wedge v_3 = v_1 \wedge (v_2 \wedge v_3) \tag{37}$$

Proof.

$$(v_1 \wedge v_2) \wedge v_3 = v_3 \wedge (v_1 \wedge v_2) \tag{38}$$

$$= \frac{1}{2} \left[-v_3 (v_1 \wedge v_2) - (v_1 \wedge v_2) v_3 \right] \tag{39}$$

$$= \frac{1}{4} \left[v_3 (v_1 v_2 - v_2 v_1) + (v_1 v_2 - v_2 v_1) v_3 \right] \tag{40}$$

$$v_1 \wedge (v_2 \wedge v_3) = \frac{1}{2} \left[-v_1 (v_2 \wedge v_3) - (v_2 \wedge v_3) v_1 \right] \tag{41}$$

$$= \frac{1}{4} \left[v_1 (v_2 v_3 - v_3 v_2) + (v_2 v_3 - v_3 v_2) v_1 \right]. \tag{42}$$

Since $v_3v_1v_2 - v_2v_1v_3 = -v_1v_3v_2 + v_2v_3v_1$ because of

$$(v_3v_1 + v_1v_3)\, v_2 = v_2\, (v_3v_1 + v_1v_3), \tag{43}$$

Eq. (37) is established.   □

3.3   General Formulas

Among general formulas, one has with $(a, b, c, d) \in V_1$, $(B, B_i) \in V_2$ and F, G, H being any elements

$$(a \wedge b)\,.B = a.\,(b.B) = -b.\,(a.B) \tag{44}$$

$$(a \wedge b)\,.\,(c \wedge d) = (a.c)\,(b.d) - (a.d)\,(b.c) \tag{45}$$

$$[F, [G, H]] - [G, [F, H]] = [[F, G]\,, H] \tag{46}$$

$$B_2.\,(B_1.a) - B_1.\,(B_2.a) = [B_2, B_1]\,.a \tag{47}$$

$$a.A_p = \left(a \wedge A_p^*\right)^* \tag{48}$$

$$a \wedge A_p = \left(a.A_p^*\right)^* \tag{49}$$

$$B_1 \wedge B_2^* = (B_1.B_2)^* \tag{50}$$

$$B_1.B_2 = B_1^*.B_2^* \tag{51}$$

Proof. Equation (44) results from the definition (34). Equation (45) follows from

$$(a \wedge b)\,.\,(c \wedge d) = a.\,[b.\,(c \wedge d)] \tag{52}$$

with

$$b.\,(c \wedge d) = (b.d)\,c - (b.c)\,d \tag{53}$$

hence,

$$a.\,[b.\,(c \wedge d)] = (a.c)\,(b.d) - (b.c)\,(a.d). \tag{54}$$

Equation (46) is simply the Jacobi identity which entails Eq. (47). Equation (48) is established as follows (with $n = 3$)

$$a \wedge A_p^* = a \wedge A_{n-p} \tag{55}$$

$$= -\frac{1}{2}\left[aA_{n-p} + (-1)^{n-p} A_{n-p}a\right] \tag{56}$$

$$= -\frac{1}{2}\left[aA_{n-p} - (-1)^p A_{n-p}a\right] \tag{57}$$

$$= -\frac{I}{2}\left[aA_p - (-1)^p A_p a\right] = -\left(a.A_p\right)^* \tag{58}$$

hence, we obtain since $(A^*)^* = -A$ the relation

$$\left(a \wedge A_p^*\right)^* = (a.A_p). \tag{59}$$

Equation (49) follows from

$$a.A_p^* = a.A_{n-p} \tag{60}$$

$$= \frac{1}{2}\left[aA_{n-p} - (-1)^{n-p}A_{n-p}a\right] \tag{61}$$

$$= \frac{1}{2}\left[aA_{n-p} + (-1)^p A_{n-p}a\right] \tag{62}$$

$$= \frac{I}{2}\left[aA_p + (-1)^p A_p a\right] \tag{63}$$

thus we get

$$\left(a.A_p^*\right)^* = \frac{-1}{2}\left[aA_p + (-1)^p A_p a\right] = a \wedge A_p. \tag{64}$$

Equation (50) results from

$$B_1 \wedge B_2^* = B_2^* \wedge B_1 \tag{65}$$

$$= \frac{-I}{2}\left(B_2 B_1 + B_1 B_2\right) = (B_1.B_2)^* \tag{66}$$

and Eq. (51) from

$$(B_1.B_2) = \frac{-1}{2}\left(B_1 B_2 + B_2 B_1\right) \tag{67}$$

$$= \frac{1}{2}\left(IB_1 IB_2 + IB_2 IB_1\right) = (B_1^*.B_2^*). \tag{68}$$

$\square$

4   Multivector Geometry

4.1   Analytic Geometry

The equation of a straight line parallel to the vector u and going through the point a is expressed by

$$(x - a) \wedge u = 0 \tag{69}$$

yielding the solution

$$x - a = \lambda u \tag{70}$$

$$x = \lambda u + a \qquad (\lambda \in \mathbb{R}). \tag{71}$$

Similarly, the equation of a plane going through the point a parallel to the plane $B = u \wedge v$ is expressed by

$$(x - a) \wedge (u \wedge v) = 0 \tag{72}$$

with the solution

$$x - a = \lambda u + \mu v \tag{73}$$

$$x = \lambda u + \mu v + a \qquad (\lambda, \mu \in \mathbb{R}) \tag{74}$$

4.2   Orthogonal Projections

Orthogonal Projection of a Vector on a Vector. The orthogonal projection of a vector $u = u_{\parallel} + u_{\perp}$ on a vector a with $u_{\perp} \cdot a = 0$, $u_{\parallel} \wedge a = 0$ is obtained as follows. Since

$$ua = u \cdot a - u \wedge a \tag{75}$$

one has

$$u_{\parallel}a = u_{\parallel} \cdot a = u \cdot a \tag{76}$$

$$u_{\perp}a = -u_{\perp} \wedge a = -u \wedge a \tag{77}$$

therefore

$$u_{\parallel} = (u \cdot a)a^{-1} \tag{78}$$

$$u_{\perp} = -(u \wedge a)a^{-1}. \tag{79}$$

Orthogonal Projection of a Vector on a Plane. Similarly, to obtain the orthogonal projection of a vector $u = u_{\parallel} + u_{\perp}$ on a plane $B = a \wedge b$ (with $u_{\perp} \cdot B = 0$, $u_{\parallel} \wedge B = 0$) one writes

$$uB = u \cdot B - u \wedge B \tag{80}$$

hence, the solution is (with $B^{-1} = B_c/BB_c$)

$$u_{\parallel} = (u \cdot B)B^{-1} \tag{81}$$

$$u_{\perp} = -(u \wedge B)B^{-1}. \tag{82}$$

Orthogonal Projection of a Plane on a Plane. As another example, let us give the orthogonal projection of a plane $B_1 = B_{1\parallel} + B_{1\perp}$ on the plane $B_2 = a \wedge b$ with $B_{1\perp} \cdot B_2 = 0$, and $[B_{1\parallel}, B_2] = 0$. Using the relation

$$B_1 B_2 = -B_1 \cdot B_2 + [B_1, B_2], \tag{83}$$

we obtain

$$B_{1\parallel} = -(B_1 \cdot B_2)B_2^{-1} \tag{84}$$

$$B_{1\perp} = \{[B_1, B_2]\} B_2^{-1}. \tag{85}$$

5   Differential Operators and Integrals

5.1   Differential Operators

In Cartesian coordinates, the nabla operator $\nabla = Ii\frac{\partial}{\partial x_1} + Ij\frac{\partial}{\partial x_2} + Ik\frac{\partial}{\partial x_3}$ acting on a scalar f, a vector a $(= a_1 Ii + a_2 Ij + a_3 Ik)$, a bivector B $(= B_1 i + B_2 j + B_3 k)$ and a trivector T $(= \tau I)$ yields respectively

$$\nabla f = Ii\frac{\partial}{\partial x_1} + Ij\frac{\partial}{\partial x_2} + Ik\frac{\partial}{\partial x_3} = \operatorname{grad} f \in V_1 \tag{86}$$

$$\nabla . a = \frac{\partial a_1}{\partial x_1} + \frac{\partial a_2}{\partial x_2} + \frac{\partial a_3}{\partial x_3} = \operatorname{div} a \in V_0 \tag{87}$$

$$\nabla \wedge a = \left(\frac{\partial a_3}{\partial x_2} - \frac{\partial a_2}{\partial x_3} \right) i + \left(\frac{\partial a_1}{\partial x_3} - \frac{\partial a_3}{\partial x_2} \right) j + \left(\frac{\partial a_2}{\partial x_1} - \frac{\partial a_1}{\partial x_2} \right) k \tag{88}$$

$$= \operatorname{rot} a \in V_2 \tag{89}$$

$$\nabla . B = (\nabla \wedge B^*)^* = (\operatorname{rot} B^*)^* \in V_1 \tag{90}$$

$$\nabla \wedge B = (\nabla . B^*)^* = (\operatorname{div} B^*)^* \in V_3 \tag{91}$$

$$\nabla . T = \nabla T = -(\nabla T^*)^* = -(\operatorname{grad} T^*)^* \in V_2. \tag{92}$$

Hence, the various operators can be expressed with the usual ones (grad, div, rot) and the duality. Among a few properties of the nabla operator, one has

$$\nabla^2 = \triangle, \nabla \wedge (\nabla \wedge f) = 0, \nabla \wedge (\nabla \wedge a) = 0 \tag{93}$$

$$\triangle a = \nabla (\nabla a) = \nabla (\nabla . a - \nabla \wedge a) = \nabla (\nabla . a) - \nabla . (\nabla \wedge a), \tag{94}$$

where the last equation results from

$$\nabla (\nabla \wedge a) = \nabla . (\nabla \wedge a) - \nabla \wedge (\nabla \wedge a) = \nabla . (\nabla \wedge a). \tag{95}$$

5.2   Integrals and Theorems

The length, surface and volume integrals are respectively for a curve $x(u)$, surface $x(u, v)$ and a volume $x(u, v, w)$

$$L = \int ds = \int \sqrt{(dx)^2} = \int \sqrt{\left(\frac{dx}{du} \right)^2} \, du \tag{96}$$

$$S = \int \int \sqrt{-\left(\frac{\partial x}{\partial u} \wedge \frac{\partial x}{\partial v} \right)^2} \, dudv \tag{97}$$

$$V = \int \int \int \sqrt{-\left(\frac{\partial x}{\partial u} \wedge \frac{\partial x}{\partial v} \wedge \frac{\partial x}{\partial w} \right)^2} \, dudvdw. \tag{98}$$

The formulas exhibit immediately the transformation properties under a change of coordinates.

Stokes' theorem is expressed for a vector a (with $dl = dx, dS = dl_1 \wedge dl_2$)

$$\oint a.dl = \int (\nabla \wedge a).dS; \tag{99}$$

the same formula can be used for a bivector B by taking $a = B^*$

$$\oint B^*.dl = \int (\nabla \wedge B^*).dS = -\int (\nabla . B)^*.dS \tag{100}$$

$$= \int \operatorname{rot} B^*.dS. \tag{101}$$

Ostrogradsky's theorem for a bivector B yields (with $d\tau = dl_1 \wedge dl_2 \wedge dl_3$)

$$\oint B.dS = \int (\nabla \wedge B).d\tau = \int (\operatorname{div} B^*)^*.d\tau \tag{102}$$

$$= \int (\operatorname{div} B^*) \, dV. \tag{103}$$

For a vector a, one obtains with $B = -a^*$

$$-\oint a^*.dS = \int (\operatorname{div} a) \, dV \tag{104}$$

which transforms, since $a^*.dS = -a.dS^*$, into

$$\oint a.dS^* = \int (\operatorname{div} a) \, dV. \tag{105}$$

6   Orthogonal Groups $O(3)$ and $SO(3)$

Definition 2. *The symmetric of x with respect to a plane is obtained by drawing the perpendicular to the plane and by extending this perpendicular by an equal length.*

Let x be a vector, x' its symmetric to a plane and a a unit vector perpendicular to the plane. From the geometry, $x' - x$ is perpendicular to the plane and thus parallel to a; similarly, $x' + x$ is parallel to the plane and thus perpendicular to a. Consequently, one has

$$x' = x + \lambda a, a \cdot \left(\frac{x' + x}{2} \right) = 0; \tag{106}$$

hence, one obtains (with $a \cdot a = a^2 = 1$)

$$a \cdot \left(x + \frac{\lambda a}{2} \right) = 0 \tag{107}$$

yielding $\lambda = -\frac{2(a \cdot x)}{a \cdot a}$ and

$$x' = x - \frac{2(a \cdot x)a}{a \cdot a} = x - \frac{(ax + xa)a}{a \cdot a} \tag{108}$$

$$= -axa. \tag{109}$$

Definition 3. *The orthogonal group $O(3)$ is the group of linear operators which leave invariant the quadratic form $x \cdot y = x_1 y_1 + x_2 y_2 + x_3 y_3$.*

Theorem 2. *Every rotation of $O(3)$ is the product of an even number ≤ 3 of symmetries, any reflection is the product of an odd number ≤ 3 of symmetries.*

The special orthogonal group $SO(3)$ is constituted by rotations i.e. of proper transformations $f(x)$ of determinant equal to 1 (i.e. $\alpha = f(e_1) \wedge f(e_2) \wedge f(e_3) = I$). A reflection is an improper transformation of determinant equal to -1 (i.e. $\alpha = -I$). Combining two orthogonal symmetries, we obtain

$$x' = (ba)\, x\, (ab) = rxr_c \tag{110}$$

with $r = ba$, $r_c = a_c b_c = ab$, $rr_c = 1$. One can express r as

$$r = \left(\cos\frac{\theta}{2} + u\sin\frac{\theta}{2}\right) = e^{\frac{1}{2}u\theta} \tag{111}$$

with $u = u_1 i + u_2 j + u_3 k$ $(u^2 = -1)$. Equation (110) represents a conical rotation of the vector x by an angle θ around the unit vector $u^* = Iu$. One verifies that the rotation conserves the norm $x'^2 = x^2$. The same equation holds for any element A of the algebra $A' = rAr_c$ since the product of two vectors x, y transforms as

$$x'y' = (rxr_c)(ryr_c) = r(xy)r_c \tag{112}$$

and similarly for the product of three vectors as well as a linear combination of such products. The above formulas allow to easily express the classical moving frames such as the Frenet and Darboux frames within the Grassmannian scheme.

7 Curves

7.1 Generalities

Consider a 3D curve $x(t)$ $(= x_1(t)e_1 + x_2(t)e_2 + x_3(t)e_3)$ where e_i is the canonical orthonormal basis $(e_1 = Ii, e_2 = Ij, e_3 = Ik)$. Taking the length of the curve s as parameter we have $x = f(s)$ with $ds = \sqrt{(dx)^2}$. The tangent unit vector at a point $M(x)$ is

$$T = \frac{dx}{ds}, T^2 = \left(\frac{dx}{ds}\right)^2 = 1. \tag{113}$$

The equation of the tangent at a point $M(x)$ is given by

$$(X - x) \wedge \frac{dx}{ds} = 0 \tag{114}$$

where X is a generic point of the tangent. The equations of the plane perpendicular to the curve and of the osculating plane at the point M read respectively

$$(X - x) \cdot \frac{dx}{ds} = 0 \tag{115}$$

$$(X - x) \wedge \frac{dx}{ds} \wedge \frac{d^2x}{ds^2} = 0 \tag{116}$$

where X is a generic point of the plane [25].

7.2   Frenet Frame

The Frenet frame (v_i) attached to the curve $x(t)$ is given by

$$v_i = r e_i r_c \tag{117}$$

where $r = e^{\frac{1}{2}u\theta}$ expresses the rotation of angle θ around the unit vector u^* $(rr_c = 1, u^2 = -1)$ and e_i is the canonical orthonormal basis. After differentiation, one obtains (using the relation $dr_c r = -r_c dr$ resulting from the differentiation of $rr_c = 1$)

$$dv_i = r\left(r_c dr e_i + e_i dr_c r\right) r_c \tag{118}$$
$$= r\left(d\iota_F . e_i\right) r_c \tag{119}$$
$$= r\left(De_i\right) r_c \tag{120}$$

with

$$d\iota_F = 2r_c dr = 2e^{-\frac{1}{2}u\theta} e^{\frac{1}{2}u\theta}\left(\frac{d\theta}{2}u + \frac{\theta}{2}du\right) \tag{121}$$

$$= (d\theta u + \theta du) \tag{122}$$
$$= (da)\, i + (db)\, j + (dc)\, k \in V_2 \tag{123}$$

and

$$De_i = d\iota_F . e_i = \frac{1}{2}\left(d\iota_F e_i - e_i d\iota_F\right). \tag{124}$$

We shall call $d\iota_F = 2r_c dr$ the affine connection bivector. Explicitly, one has

$$De_1 = (dc)\, e_2 - (db)\, e_3 \tag{125}$$
$$De_2 = -(dc)\, e_1 + (da)\, e_3 \tag{126}$$
$$De_3 = (db)\, e_1 - (da)\, e_2. \tag{127}$$

The Frenet frame is defined by the affine connection bivector

$$d\iota_F = 2r_c dr = (\tau ds)\, i + (\rho ds)\, k \tag{128}$$

where $\rho = 1/R$ is the curvature and $\tau = 1/T$ the torsion. This gives the Frenet equations

$$De_1 = (\rho ds)\, e_2 \tag{129}$$
$$De_2 = -(\rho ds)\, e_1 + (\tau ds)\, e_3 \tag{130}$$
$$De_3 = (-\tau ds)\, e_2. \tag{131}$$

7.3   Curvature and Torsion

To obtain the curvature and torsion we define α and β

$$\alpha = \frac{dx}{ds} \wedge \frac{d^2 x}{ds^2} = \rho v_1 \wedge v_2 \tag{132}$$

$$\beta = \frac{dx}{ds} \wedge \frac{d^2x}{ds^2} \wedge \frac{d^3x}{ds^3} = I\rho^2\tau \tag{133}$$

and using the Lagrange equation

$$(v_1 \wedge v_2)^2 = (v_1.v_2)^2 - (v_1)^2 (v_2)^2 = -1 \tag{134}$$

we obtain the invariants

$$\rho = \sqrt{-\alpha^2} = \sqrt{-\left(\frac{dx}{ds} \wedge \frac{d^2x}{ds^2}\right)^2} \tag{135}$$

$$\tau = \frac{I\beta}{\alpha^2} = \frac{I\left(\frac{dx}{ds} \wedge \frac{d^2x}{ds^2} \wedge \frac{d^3x}{ds^3}\right)}{\left(\frac{dx}{ds} \wedge \frac{d^2x}{ds^2}\right)^2}. \tag{136}$$

Under a change of parameter t, one has using $\frac{dx}{dt} = \frac{dx}{ds}\frac{ds}{dt}$

$$\frac{dx}{dt} \wedge \frac{d^2x}{dt^2} = \left(\frac{ds}{dt}\right)^3 \alpha \tag{137}$$

and thus one obtains the curvature

$$\rho = \sqrt{-\left(\frac{dx}{dt}\right)^{-6}\left(\frac{dx}{dt} \wedge \frac{d^2x}{dt^2}\right)^2}. \tag{138}$$

For the torsion, proceeding similarly, we get under a change of parameter

$$\frac{dx}{dt} \wedge \frac{d^2x}{dt^2} \wedge \frac{d^3x}{dt^3} = \left(\frac{ds}{dt}\right)^6 \beta \tag{139}$$

and thus

$$\tau = \frac{I\left(\frac{dx}{dt} \wedge \frac{d^2x}{dt^2} \wedge \frac{d^3x}{dt^3}\right)}{\left(\frac{dx}{dt} \wedge \frac{d^2x}{dt^2}\right)^2} \in V_0. \tag{140}$$

7.4   Example

As example, consider the curve $x(t) = (2\cos t)Ii + (2\sin t)Ij + (t)Ik$. The line element is $ds = \sqrt{dx^2} = \sqrt{5}dt$; writing $x' = \frac{dx}{dt}$, etc., we have

$$x' \wedge x'' = (2\sin t)i - (2\cos t)j + 4k, \tag{141}$$

$$(x' \wedge x'')^2 = -20 \tag{142}$$

$$x' \wedge x'' \wedge x''' = 4I. \tag{143}$$

The curvature and torsion are respectively

$$\rho = \sqrt{-(x')^{-6}(x' \wedge x'')^2} = \frac{2}{5} \tag{144}$$

$$\tau = \frac{I(x' \wedge x'' \wedge x''')}{(x' \wedge x'')^2} = \frac{1}{5}. \tag{145}$$

The equation of the osculating plane is (with $X = (X_1)\,Ii + (X_2)\,Ij + (X_3)\,Ik$ being a generic point of the plane)

$$(X - x) \wedge \frac{dx}{ds} \wedge \frac{d^2x}{ds^2} = \frac{I}{\sqrt{5}}\left(-2t + 2X_3 - X_2\cos t + X_1\sin t\right) = 0. \tag{146}$$

The Frenet basis v_i is

$$v_1 = \frac{1}{\sqrt{5}}\left[(-2\sin t)\,Ii + (2\cos t)\,Ij + Ik\right] \tag{147}$$

$$v_2 = (-\cos t)\,Ii - (\sin t)\,Ij \tag{148}$$

$$v_3 = (v_1 \wedge v_2)^* = \frac{1}{\sqrt{5}}\left[(\sin t)\,Ii - (\cos t)\,Ij + 2Ik\right]. \tag{149}$$

The basis v_i is obtained via the following rotations. First, the frame is brought into its initial position (at $t = 0$) via the rotation $f_0 = f_1 f_2$ with $\tan\theta = \frac{1}{2}$ and

$$f_1 = e^{k\frac{\pi}{2}} = \frac{1}{\sqrt{2}}(1 + k) \tag{150}$$

$$f_2 = e^{-j\frac{\theta}{2}} = \left(\sqrt{\frac{1}{\sqrt{2}} + \frac{1}{\sqrt{5}}} - j\sqrt{\frac{1}{\sqrt{2}} - \frac{1}{\sqrt{5}}}\right) \tag{151}$$

yielding

$$f_0 = \sqrt{\left(\frac{1}{4} + \frac{1}{2\sqrt{5}}\right)} + i\sqrt{\frac{1}{4} - \frac{1}{2\sqrt{5}}} - j\sqrt{\frac{1}{4} - \frac{1}{2\sqrt{5}}} + k\sqrt{\frac{1}{4} + \frac{1}{2\sqrt{5}}}. \tag{152}$$

Next, follows the rotation due to the affine connection bivector

$$f_3 = \cos\frac{t}{2} + \left(\frac{i}{\sqrt{5}} + \frac{2k}{\sqrt{5}}\right)\sin\frac{t}{2}. \tag{153}$$

The end result is $r = f_1 f_2 f_3$ and explicitly

$$r = \left(A\cos\frac{t}{2} - C\sin\frac{t}{2}\right) + i\left(B\cos\frac{t}{2} + D\sin\frac{t}{2}\right) \tag{154}$$

$$+ j\left(-B\cos\frac{t}{2} + D\sin\frac{t}{2}\right) + k\left(A\cos\frac{t}{2} + C\sin\frac{t}{2}\right) \tag{155}$$

with

$$A = \frac{1}{10}\sqrt{5\left(5 + 2\sqrt{5}\right)},\ B = \frac{1}{10}\sqrt{5\left(5 - 2\sqrt{5}\right)} \tag{156}$$

$$C = \frac{1}{10}\left(\sqrt{5 - 2\sqrt{5}} + 2\sqrt{5 + 2\sqrt{5}}\right), \tag{157}$$

$$D = \frac{1}{10}\left(-2\sqrt{5 - 2\sqrt{5}} + \sqrt{5 + 2\sqrt{5}}\right). \tag{158}$$

Finally, one verifies that $v_i = re_i r_c$.

8   Surfaces

8.1   Generalities

Consider in a $3D$ Euclidean space a surface $x = f(u, v)$. The tangent plane is given by $f_u \wedge f_v$ ($f_u = \frac{\partial f}{\partial u}, f_v = \frac{\partial f}{\partial v}$) and the unit normal by

$$h = \frac{(f_u \wedge f_v)^*}{\sqrt{-(f_u \wedge f_v)^2}} \tag{159}$$

with (f_u, f_v, h) being a direct trieder ($f_u \wedge f_v \wedge h = \lambda I, \lambda > 0$). Take an ortho-normal moving frame of basis vectors v_i and a vector a (of components A with respect to the moving frame) attached to the surface. This frame is obtained by rotating the canonical frame e_i; hence, one has

$$v_i = r e_i r_c, a = r A r_c. \tag{160}$$

Differentiating these relations, one obtains $dv_i = r(De_i) r_c, da = r(DA) r_c$ with

$$De_i = d\iota.e_i, DA = dA + d\iota.A \tag{161}$$

where DA is the covariant differential (dA being a differentiation with respect to the components only) and $d\iota = 2r_c dr$.

8.2   Darboux Frame

The Darboux frame (v_{iD}) is obtained for a curve on the surface from the Frenet frame by a rotation of an algebraic angle $\alpha = \angle(v_{3F}, h)$ around v_{1F}. Hence

$$v_{iD} = (r_D) e_i (r_D)_c \tag{162}$$

with $r_D = r_1 r$ and

$$r_1 = \cos\frac{\alpha}{2} - (v_{1F})^* \sin\frac{\alpha}{2} = r p r_c \tag{163}$$

where $p = \cos\frac{\alpha}{2} + i \sin\frac{\alpha}{2}$; hence $r_D = r_1 r = rp$. Consequently, the rotation bivector $d\iota_D = 2(r_D)_c dr_D$ can be expressed as follows

$$d\iota_D = 2(rp)_c d(rp) = 2p_c r_c (drp + rdp) \tag{164}$$
$$= p_c (d\iota_F) p + 2p_c dp \tag{165}$$
$$= (\tau_r ds) i - (\rho_n ds) j + (\rho_g ds) k \tag{166}$$

where $\rho_g = \rho \cos\alpha$ is the geodesic curvature, $\rho_g = -\rho \sin\alpha$ the normal curvature and $\tau_r = \frac{d\alpha}{ds} + \tau$ the relative torsion. The Darboux equations thus become

$$dv_{iD} = r_D (d\iota_D.e_i) (r_D)_c \tag{167}$$

and explicitly

$$dv_{1D} = (\rho_g v_{2D} + \rho_n v_{3D}) \, ds \qquad (168)$$
$$dv_{2D} = (-\rho_g v_{1D} + \tau_r v_{3D}) \, ds \qquad (169)$$
$$dv_{3D} = (-\rho_n v_{1D} - \tau_r v_{2D}) \, ds. \qquad (170)$$

If the Darboux frame is rotated in the tangent plane by an angle $\theta(s)$ around the unit normal h one obtains, applying the same reasoning as above, for the rotation $r = r_D f$ with $f = \cos\frac{\theta}{2} + k\sin\frac{\theta}{2}$. The new vectors v_i of the frame become $v_i = r e_i r_c$ and the affine connection bivector transforms into

$$d\iota = 2r_c dr \qquad (171)$$
$$= 2f_c (r_D)_c [(dr_D) f + r_D df] \qquad (172)$$
$$= f_c (d\iota_D) f + 2f_c df. \qquad (173)$$

Applying this formula, one obtains

$$d\iota = ds (\tau_r \cos\theta - \rho_n \sin\theta) \, i + ds (-\rho_n \cos\theta - \tau_r \sin\theta) \, j \qquad (174)$$
$$+ (\rho_g ds + d\theta) \, k. \qquad (175)$$

8.3 Integrability Conditions

The $3D$ Euclidean space being without torsion and curvature, this entails that dM and dv_i are integrable.

Integrability of dM. Consider a point on the surface, one has

$$dM = r (DM) r_c = f_u du + f_v dv \qquad (176)$$

where DM are the components expressed in the moving frame (with the moving vectors $v_i = r (e_i) r_c$). One has

$$DM = \omega_1 e_1 + \omega_2 e_2 \qquad (177)$$
$$= (A_1 du + B_1 dv) \, e_1 + (A_2 du + B_2 dv) \, e_2 \qquad (178)$$

with

$$A_1 = (r_c f_u r) . e_1 = f_u . v_1, \; B_1 = (r_c f_v r) . e_1 = f_v . v_1 \qquad (179)$$
$$A_2 = (r_c f_u r) . e_2 = f_u . v_2, \; B_2 = (r_c f_v r) . e_2 = f_v . v_2. \qquad (180)$$

The affine connection bivector $d\iota = 2r_c dr$ can be written [15, II, p. 410] as $d\iota = \iota_1 du + \iota_2 dv$ (with $\iota_1 = a_1 i + b_1 j + c_1 k$, $\iota_2 = a_2 i + b_2 j + c_2 k$). The integrability condition of dM is expressed by the condition

$$\Delta (DM) - D (\Delta M) = 0 \qquad (181)$$

with

$$\Delta (DM) = \delta (DM) + \delta \iota.DM \tag{182}$$
$$D (\Delta M) = d (\Delta M) + d\iota.\Delta M. \tag{183}$$

This leads to the relation

$$\delta (DM) - d (\Delta M) = d\iota.\Delta M - \delta\iota.DM. \tag{184}$$

Explicitly, it reads

$$\frac{\partial A_1}{\partial v} - \frac{\partial B_1}{\partial u} = -B_2 c_1 + A_2 c_2 \tag{185}$$

$$\frac{\partial A_2}{\partial v} - \frac{\partial B_2}{\partial u} = B_1 c_1 - A_1 c_2 \tag{186}$$

$$A_1 b_2 + B_2 a_1 = A_2 a_2 + B_1 b_1. \tag{187}$$

The linear Eqs. (185 and 186) determine c_1, c_2; the result is

$$c_1 = \frac{-1}{A_1 B_2 - A_2 B_1} \left[A_1 \left(\frac{\partial A_1}{\partial v} - \frac{\partial B_1}{\partial u} \right) + A_2 \left(\frac{\partial A_2}{\partial v} - \frac{\partial B_2}{\partial u} \right) \right] \tag{188}$$

$$c_2 = \frac{-1}{A_1 B_2 - A_2 B_1} \left[B_2 \left(\frac{\partial A_2}{\partial v} - \frac{\partial B_2}{\partial u} \right) + B_1 \left(\frac{\partial A_1}{\partial v} - \frac{\partial B_1}{\partial u} \right) \right] \tag{189}$$

Integrability of dv_i. The integrability conditions of dv_i are obtained similarly, with $dv_i = rDe_i r_c$. The condition $\delta dv_i - d\delta v_i = 0$ leads to the relation

$$\Delta (De_i) - D (\Delta e_i) = 0 \tag{190}$$

with $De_i = d\iota.e_i$, $\Delta e_i = \delta\iota.e_i$ and

$$\Delta (De_i) = \delta (d\iota.e_i) + \delta\iota. (d\iota.e_i) \tag{191}$$
$$D (\Delta e_i) = d (\delta\iota.e_i) + d\iota. (\delta\iota.e_i). \tag{192}$$

Applying Eq. (47), one has

$$\delta\iota. (d\iota.e_i) - d\iota. (\delta\iota.e_i) = - [d\iota, \delta\iota] .e_i \tag{193}$$

hence, the integrability condition of dv_i can be expressed as

$$(\delta d\iota - d\delta\iota) .e_i = [d\iota, \delta\iota] .e_i \tag{194}$$

and thus $(\delta d\iota - d\delta\iota) = [d\iota, \delta\iota]$ or

$$\frac{\partial \iota_1}{\partial v} - \frac{\partial \iota_2}{\partial u} = [\iota_1, \iota_2]. \tag{195}$$

Explicitly, these equations read [15, II, p. 412]

$$\frac{\partial a_1}{\partial v} - \frac{\partial a_2}{\partial u} = b_1 c_2 - b_2 c_1 \tag{196}$$

$$\frac{\partial b_1}{\partial v} - \frac{\partial b_2}{\partial u} = c_1 a_2 - c_2 a_1 \tag{197}$$

$$\frac{\partial c_1}{\partial v} - \frac{\partial c_2}{\partial u} = a_1 b_2 - a_2 b_1. \tag{198}$$

8.4   Curvature Lines and Curvature: First Method

Consider a moving frame $v_i = re_ir_c$ with $De_i = d\iota.e_i$ $(d\iota = 2r_cdr)$ and $DM = \omega_1e_1 + \omega_2e_2$ expressed in the moving frame. The fundamental form II can be expressed as

$$II = -De_3.DM = -(d\iota.e_3).DM \qquad (199)$$

$$= DM.(e_3.d\iota) = (DM \wedge e_3).d\iota. \qquad (200)$$

The affine connection bivector $d\iota = dai + dbj + dck$ can be developed on $\omega_1(= A_1du + B_1dv), \omega_2 (= A_2du + B_2dv)$ as follows [26, p. 209]

$$da = L_{21}\omega_1 + L_{22}\omega_2 = a_1du + a_2dv \qquad (201)$$

$$db = -L_{11}\omega_1 - L_{12}\omega_2 = b_1du + b_2dv. \qquad (202)$$

Identifying the coefficients of du, dv, and solving the linear system, one obtains

$$L_{11} = \frac{b_2A_2 - b_1B_2}{A_1B_2 - A_2B_1}, L_{22} = \frac{a_2A_1 - a_1B_1}{A_1B_2 - A_2B_1} \qquad (203)$$

$$L_{12} = \frac{b_1B_1 - b_2A_1}{A_1B_2 - A_2B_1}, L_{21} = \frac{a_1B_2 - a_2A_2}{A_1B_2 - A_2B_1}. \qquad (204)$$

Due to the integrability condition Eq. (187), we have $L_{12} = L_{21}$ and thus the fundamental form II becomes

$$II = L_{11}\omega_1^2 + L_{22}\omega_2^2 + 2L_{12}\omega_1\omega_2. \qquad (205)$$

If we rotate the frame by an angle Φ around v_3 we have $v_i' = fv_if_c$ with

$$f = \cos(\Phi/2) - v_3^* \sin(\Phi/2) \qquad (206)$$

$$= r[\cos(\Phi/2) - e_3^* \sin(\Phi/2)]r_c \qquad (207)$$

$$= r[\cos(\Phi/2) + k\sin(\Phi/2)]r_c. \qquad (208)$$

The total rotation is $R = fr = rp$ with $p = \cos(\Phi/2) + k\sin(\Phi/2)$ and the affine connection bivector transforms into

$$d\iota' = 2R_cdR = 2p_cr_c(drp) = p_c(d\iota)p \qquad (209)$$

$$= da'i + db'j + dc'k \qquad (210)$$

with

$$da' = \cos(\Phi)da + \sin(\Phi)db = L_{12}'\omega_1' + L_{22}'\omega_2' \qquad (211)$$

$$db' = -\sin(\Phi)da + \cos(\Phi)db = -L_{11}'\omega_1' - L_{12}'\omega_2' \qquad (212)$$

$$dc' = dc. \qquad (213)$$

The vector $DM' = \omega_1'e_1 + \omega_2'e_2$ transforms in the same way i.e.,

$$\omega_1' = \cos(\Phi)\omega_1 + \sin(\Phi)\omega_2 \qquad (214)$$

$$\omega_2' = -\sin(\Phi)\omega_1 + \cos(\Phi)\omega_2. \tag{215}$$

The coefficients L_{ij}' are obtained by expressing (da, db) of Eqs. (211 and 212) in terms of (L_{ij}, ω_i) via the Eqs. (201 and 202) and then by writing the ω_i in terms of ω_i'; one finds

$$L_{11}' = L_{11}\cos^2(\Phi) + L_{22}\sin^2(\Phi) + L_{12}\sin(2\Phi) \tag{216}$$

$$L_{22}' = L_{11}\sin^2(\Phi) + L_{22}\cos^2(\Phi) - L_{12}\sin(2\Phi) \tag{217}$$

$$L_{12}' = L_{12}\cos(2\Phi) + \frac{1}{2}\sin(2\Phi)(L_{22} - L_{11}). \tag{218}$$

The curvature lines are the lines for which the fundamental form II becomes diagonal, i.e., when $L_{12}' = 0$, or

$$\tan 2\Phi = \frac{2L_{12}}{(L_{11} - L_{22})}. \tag{219}$$

Along these lines, the curvature is defined by $De_3 = -K\,(DM)$ with

$$De_3 = db'e_1 - da'e_2 \tag{220}$$

$$da' = L_{22}'\omega_2', db = -L_{11}'\omega_1' \tag{221}$$

and $DM = \omega_1'e_1$ or $DM = \omega_2'e_2$; hence, the curvatures are given by $K_1 = L_{11}', K_2 = L_{22}'$ i.e., as a function of the angle Φ. To obtain the standard formulas, we write

$$\cos 2\Phi = \frac{L_{11} - L_{22}}{\left[(L_{11} - L_{22})^2 + 4L_{12}^2\right]^{1/2}}, \sin 2\Phi = \frac{2L_{12}}{\left[(L_{11} - L_{22})^2 + 4L_{12}^2\right]^{1/2}} \tag{222}$$

and use $\cos^2\Phi = \frac{1}{2}(1 + \cos 2\Phi), \sin^2\Phi = \frac{1}{2}(1 - \cos 2\Phi)$; after rearrangement, we get

$$K_1 = L_{11}' = \frac{1}{2}\left(L_{11} + L_{22} - \sqrt{(L_{11} - L_{22})^2 + 4L_{12}^2}\right) \tag{223}$$

$$K_2 = L_{12}' = \frac{1}{2}\left(L_{11} + L_{22} + \sqrt{(L_{11} - L_{22})^2 + 4L_{12}^2}\right). \tag{224}$$

Hence, the Gaussian curvature K is given by

$$K = K_1 K_2 = L_{11}L_{22} - L_{12}^2 \tag{225}$$

$$= \frac{a_1 b_2 - a_2 b_1}{A_1 B_2 - A_2 B_1} = \frac{\left(\frac{\partial c_1}{\partial v} - \frac{\partial c_2}{\partial u}\right)}{A_1 B_2 - A_2 B_1} \tag{226}$$

where we have made use of the integrability condition Eq. (198). Replacing c_1, c_2 by their expressions of Eqs. (188 and 189), we finally get for the Gaussian curvature

$$K = \frac{1}{(A_1 B_2 - A_2 B_1)}\left\{ \frac{\partial}{\partial u}\left[\frac{B_2\left(\frac{\partial A_2}{\partial v} - \frac{\partial B_2}{\partial u}\right) + B_1\left(\frac{\partial A_1}{\partial v} - \frac{\partial B_1}{\partial u}\right)}{(A_1 B_2 - A_2 B_1)}\right] - \frac{\partial}{\partial v}\left[\frac{A_1\left(\frac{\partial A_1}{\partial v} - \frac{\partial B_1}{\partial u}\right) + A_2\left(\frac{\partial A_2}{\partial v} - \frac{\partial B_2}{\partial u}\right)}{(A_1 B_2 - A_2 B_1)}\right] \right\}. \tag{227}$$

The Gaussian curvature thus depends only on the metric $ds = \sqrt{(dM)^2}$, as stated by Gauss' theorem. The mean curvature H is given by

$$H = \frac{1}{2}(K_1 + K_2) = \frac{1}{2}(L_{11} + L_{22}) \tag{228}$$

$$= \frac{-b_1 B_2 + b_2 A_2 + a_2 A_1 - a_1 B_1}{2(A_1 B_2 - A_2 B_1)}. \tag{229}$$

8.5   Gaussian and Mean Curvature: Second Method

The Gaussian and mean curvatures can also be derived as follows [25]. The curvature K is defined by the relation

$$dv_3 = -K dM. \tag{230}$$

Developing that equation we have

$$\frac{\partial v_3}{\partial u} du + \frac{\partial v_3}{\partial v} dv = -K(x_u du + x_v dv) \tag{231}$$

or

$$\left(K x_u + \frac{\partial v_3}{\partial u}\right) du + \left(K x_v + \frac{\partial v_3}{\partial v}\right) dv = 0. \tag{232}$$

The two vectors in parentheses are parallel and thus

$$\left(K x_u + \frac{\partial v_3}{\partial u}\right) \wedge \left(K x_v + \frac{\partial v_3}{\partial v}\right) = 0 \tag{233}$$

which gives the equation

$$K^2 (x_u \wedge x_v) + K\left(\frac{\partial v_3}{\partial u} \wedge x_v + x_u \wedge \frac{\partial v_3}{\partial v}\right) + \frac{\partial v_3}{\partial u} \wedge \frac{\partial v_3}{\partial v} = 0. \tag{234}$$

Multiplying with the exterior product on the left by $n = (x_u \wedge x_v)^*$ and using Eq. (50)

$$(x_u \wedge x_v)^* \wedge (x_u \wedge x_v) = -\left[(x_u \wedge x_v)^2\right]^* = I\left(n^2\right) \tag{235}$$

we obtain the formula [25]

$$K^2 I\left(n^2\right) + K\left[n \wedge \frac{\partial v_3}{\partial u} \wedge x_v + n \wedge x_u \wedge \frac{\partial v_3}{\partial v}\right] + n \wedge \frac{\partial v_3}{\partial u} \wedge \frac{\partial v_3}{\partial v} = 0. \tag{236}$$

Calling K_1, K_2 the two roots of the equation, one has for the Gaussian curvature

$$K = K_1 K_2 = \frac{-I}{n^2}\left(n \wedge \frac{\partial v_3}{\partial u} \wedge \frac{\partial v_3}{\partial v}\right). \tag{237}$$

The mean curvature is given by

$$H = \frac{1}{2}(K_1 + K_2) = \frac{I}{2n^2}\left(n \wedge x_u \wedge \frac{\partial v_3}{\partial v} + n \wedge \frac{\partial v_3}{\partial u} \wedge x_v\right). \tag{238}$$

8.6   Curves on Surfaces: Asymptotic, Curvature and Geodesic Lines

Consider on a surface, at a point OM, an orthonormal frame v_1, v_2, v_3 (with v_3 being the unit normal). One has (with $\omega_1 = \cos\Phi ds$, $\omega_2 = \sin\Phi ds$) [15, II, p. 413]

$$d(OM) = (v_1 \cos\Phi + v_2 \sin\Phi)\, ds \tag{239}$$
$$= \omega_1 v_1 + \omega_2 v_2 \tag{240}$$

$$d^2(OM) = \frac{d^2(OM)}{ds^2} ds = (v_2 dc - v_3 db)\cos\Phi \tag{241}$$
$$+ (-v_1 dc + v_3 da)\sin\Phi + (v_2\cos\Phi - v_1\sin\Phi)\, d\Phi \tag{242}$$
$$= -v_1 \sin\Phi(dc + d\Phi) + v_2\cos\Phi\,(dc + d\Phi) \tag{243}$$
$$+ v_3\,(\sin\Phi da - \cos\Phi db). \tag{244}$$

Asymptotic Lines. The normal curvature is defined by

$$\rho_n(s) = \frac{d^2(OM)}{ds^2}.v_3 \tag{245}$$
$$= \left(\sin\Phi\frac{da}{ds} - \cos\Phi\frac{db}{ds}\right). \tag{246}$$

Developing da, db on ω_1, ω_2 via Eqs. (201 and 202), one obtains after rearrangement

$$\rho_n(s) = L_{22}\sin^2\Phi + L_{11}\cos^2\Phi + 2\sin\Phi\cos\Phi L_{12}. \tag{247}$$

The asymptotic line is defined by $\rho_n(s) = 0$, leading to

$$\tan\Phi = \frac{-L_{12} \pm \sqrt{L_{12}^2 - L_{11}L_{22}}}{L_{22}} \tag{248}$$

under the assumption that $L_{12}^2 - L_{11}L_{22} \geq 0$.

Curvature Lines. The relative torsion $\tau_r(s)$ is expressed by

$$I\tau_r(s) = \frac{d(OM)}{ds} \wedge v_3 \wedge \frac{dv_3}{ds} \tag{249}$$
$$= (v_1\cos\Phi + v_2\sin\Phi) \wedge v_3 \wedge \left(\frac{db}{ds}v_1 - \frac{da}{ds}v_2\right) \tag{250}$$
$$= \left(\cos\Phi\frac{da}{ds} + \frac{db}{ds}\sin\Phi\right) v_1 \wedge v_2 \wedge v_3. \tag{251}$$

Developing da, db on ω_1, ω_2 as above, we get

$$\tau_r(s) = \begin{bmatrix} \cos\Phi\,(L_{12}\cos\Phi + L_{22}\sin\Phi) \\ -\sin\Phi\,(L_{11}\cos\Phi + L_{12}\sin\Phi) \end{bmatrix} \tag{252}$$

$$= \left[L_{12}\cos 2\Phi + \frac{\sin 2\Phi}{2}\,(L_{22} - L_{11}) \right]. \tag{253}$$

The curvature lines are defined by $\tau_r(s) = 0$, hence

$$\tan 2\Phi = \frac{2L_{12}}{L_{11} - L_{22}} \tag{254}$$

as we have already obtained previously in Eq. (219).

Geodesics. The geodesic curvature $\rho_g(s)$ is given by

$$I\rho_g(s) = \frac{d(OM)}{ds} \wedge \frac{d^2(OM)}{ds^2} \wedge v_3 \tag{255}$$

$$= \frac{1}{ds}\,(v_1\cos\Phi + v_2\sin\Phi) \tag{256}$$

$$\wedge \begin{bmatrix} (d\Phi + dc)\,(-v_1\sin\Phi + v_2\cos\Phi) \\ +v_3\,(\sin\Phi\,da - \cos\Phi\,db) \end{bmatrix} \wedge v_3 \tag{257}$$

$$= \left(\frac{dc}{ds} + \frac{d\Phi}{ds} \right) I. \tag{258}$$

The geodesic lines correspond to $\rho_g(s) = 0$ and thus to $dc + d\Phi = 0$ or equivalently

$$c_1 du + c_2 dv + d\left(Arc\tan\frac{\omega_2}{\omega_1} \right) = 0. \tag{259}$$

Hence, the equation of the geodesic is given by [15, II, p. 414]

$$c_1 du + c_2 dv + d\left(Arc\tan\frac{A_2 du + B_2 dv}{A_1 du + B_1 dv} \right) = 0 \tag{260}$$

where c_1, c_2 are expressed by Eqs. (188 and 189).

8.7   Example

Consider as surface the sphere of radius r, $x(t) = (r\sin\theta\cos\varphi)Ii + (r\sin\theta\sin\varphi)Ij + (r\cos\theta)Ik$. One has $dx = \omega_1 v_1 + \omega_2 v_2$ with $\omega_1 = rd\theta, \omega_1 = r\sin\theta d\varphi$ $(A_1 = r, B_1 = 0, A_2 = 0, B_2 = r\sin\theta)$ and

$$v_1 = (\cos\theta\cos\varphi)\,Ii + (\cos\theta\sin\varphi)\,Ij - (\sin\theta)\,Ik \tag{261}$$

$$v_2 = -(\sin\varphi)\,Ii + (\cos\varphi)\,Ij \tag{262}$$

$$v_3 = (v_1 \wedge v_2)^* = (\sin\theta\cos\varphi)\,Ii + (\sin\theta\sin\varphi)\,Ij + (\cos\theta)\,Ik. \tag{263}$$

This frame $(v_i = re_i r_c)$ is obtained from the canonical basis e_i via a rotation r_1 of φ around e_3, followed by a rotation r_2 of θ around the axis $r_1 e_2 r_{1c}$, hence,

$$r = r_1 r_2 = e^{k\frac{\varphi}{2}} e^{j\frac{\theta}{2}} \tag{264}$$

$$= \cos\frac{\theta}{2}\cos\frac{\varphi}{2} - \left(\sin\frac{\theta}{2}\sin\frac{\varphi}{2}\right)i \tag{265}$$

$$+ \left(\sin\frac{\theta}{2}\cos\frac{\varphi}{2}\right)j + \left(\cos\frac{\theta}{2}\sin\frac{\varphi}{2}\right)k. \tag{266}$$

The affine connection bivector $d\iota = 2r_c dr$ is

$$d\iota = -\left(d\varphi\sin\theta\right)i + \left(d\theta\right)j + \left(d\varphi\cos\theta\right)k \tag{267}$$

$$= \left(da\right)i + \left(db\right)j + \left(dc\right)k, \tag{268}$$

leading to (with $da = a_1 d\theta + a_2 d\varphi$, etc.)

$$a_1 = 0, a_2 = \sin\theta, b_1 = 1, b_2 = 0 \tag{269}$$

$$c_1 = 0, c_2 = \cos\theta. \tag{270}$$

The Gaussian and mean curvature are respectively

$$K = \frac{a_1 b_2 - a_2 b_1}{A_1 B_2 - A_2 B_1} = \frac{1}{r^2} \tag{271}$$

$$H = \frac{-b_1 B_2 + b_2 A_2 + a_2 A_1 - a_1 B_1}{2\left(A_1 B_2 - A_2 B_1\right)} = -\frac{1}{r}. \tag{272}$$

9 Conclusion

The paper has presented a biquaternion calculus, having an associative exterior product, and shown how differential geometry can be expressed within this new algebraic framework. The method presented here can be extended to other spaces such as a pseudo-Euclidean 4D space. It is hoped that this paper will further interest in these new algebraic tools and provide new perspectives for geometric modeling.

Acknowledgments. This work was conducted in the framework of LabEx Celya and PRIMES (Physics Radiobiology Medical Imaging and Simulation).

A   Representation of Biquaternions by 4 × 4 Real Matrices

$$e_1 = Ii = \begin{bmatrix} 0 & 1 & 0 & 0 \\ 1 & 0 & 0 & 0 \\ 0 & 0 & 0 & -1 \\ 0 & 0 & -1 & 0 \end{bmatrix}, e_2 = Ij = \begin{bmatrix} -1 & 0 & 0 & 0 \\ 0 & 1 & 0 & 0 \\ 0 & 0 & -1 & 0 \\ 0 & 0 & 0 & 1 \end{bmatrix}$$

$$e_3 = Ik = \begin{bmatrix} 0 & 0 & 0 & -1 \\ 0 & 0 & -1 & 0 \\ 0 & -1 & 0 & 0 \\ -1 & 0 & 0 & 0 \end{bmatrix}, e_3 e_2 = i = \begin{bmatrix} 0 & 0 & 0 & -1 \\ 0 & 0 & 1 & 0 \\ 0 & -1 & 0 & 0 \\ 1 & 0 & 0 & 0 \end{bmatrix}$$

$$e_1 e_3 = j = \begin{bmatrix} 0 & 0 & -1 & 0 \\ 0 & 0 & 0 & -1 \\ 1 & 0 & 0 & 0 \\ 0 & 1 & 0 & 0 \end{bmatrix}, e_2 e_1 = k = \begin{bmatrix} 0 & -1 & 0 & 0 \\ 1 & 0 & 0 & 0 \\ 0 & 0 & 0 & 1 \\ 0 & 0 & -1 & 0 \end{bmatrix}$$

$$e_1 e_2 e_3 = I = \begin{bmatrix} 0 & 0 & -1 & 0 \\ 0 & 0 & 0 & 1 \\ 1 & 0 & 0 & 0 \\ 0 & -1 & 0 & 0 \end{bmatrix}, 1 = \begin{bmatrix} 1 & 0 & 0 & 0 \\ 0 & 1 & 0 & 0 \\ 0 & 0 & 1 & 0 \\ 0 & 0 & 0 & 1 \end{bmatrix}$$

B   Representation of Biquaternions by 2 × 2 Complex Pauli Matrices

(i': ordinary complex imaginary)

$$e_1 = Ii = \sigma_1 = \begin{bmatrix} 0 & 1 \\ 1 & 0 \end{bmatrix}, e_2 = Ij = \sigma_2 = \begin{bmatrix} 0 & -i' \\ i' & 0 \end{bmatrix}$$

$$e_3 = Ik = \sigma_1 = \begin{bmatrix} 1 & 0 \\ 0 & -1 \end{bmatrix}, e_3 e_2 = i = \begin{bmatrix} 0 & -i' \\ -i' & 0 \end{bmatrix}$$

$$e_1 e_3 = j = \begin{bmatrix} 0 & -1 \\ 1 & 0 \end{bmatrix}, e_2 e_1 = k = \begin{bmatrix} -i' & 0 \\ 0 & i' \end{bmatrix}$$

$$e_1 e_2 e_3 = I = \begin{bmatrix} i & 0 \\ 0 & i \end{bmatrix}, 1 = \begin{bmatrix} 1 & 0 \\ 0 & 1 \end{bmatrix}.$$

C   Work-Sheet: Biquaternions (Mathematica)

<<Quaternions '
    (*product of two biquaternions $a = a_1 + Ia_2, b = b_1 + Ib_2, a_i, b_i \in \mathbb{H}$; a double star ** means a quaternion product*)

$$CP[a\_, b\_] := \{(a[[1]] ** b[[1]]) - (a[[2]] ** b[[2]]),$$
$$(a[[2]] ** b[[1]]) + (a[[1]] ** b[[2]])\}$$

(*conjugate K*)

$$K[a\_] := \{Quaternion[a[[1,1]], -a[[1,2]], -a[[1,3]], -a[[1,4]]],$$
$$Quaternion[a[[2,1]], -a[[2,2]], -a[[2,3]], -a[[2,4]]]\}$$

(*sum and difference*)

$$csum[a\_, b\_] := \{a[[1]] + b[[1]], a[[2]] + b[[2]]\}$$
$$cdif[a\_, b\_] : = \{a[[1]] - b[[1]], a[[2]] - b[[2]]\}$$

(*multiplication by a scalar*)

$$fclif[f\_, a\_] \doteq \{f * a[[1]], f * a[[2]]\}$$

(*products $\frac{1}{2}(ab + ba), \frac{1}{2}(ab - ba)$*)

$$int[a\_, b\_] := \{fclif[1/2, csum[CP[a,b], CP[b,a]]]\}$$
$$ext[a\_, b\_] := \{fclif[1/2, cdif[CP[a,b], CP[b,a]]]\}$$

(*products $-\frac{1}{2}(ab + ba), -\frac{1}{2}(ab - ba)$*)

$$mint[a\_, b\_] := \{fclif[-1/2, csum[CP[a,b], CP[b,a]]]\}$$
$$mext[a\_, b\_] := \{fclif[-1/2, cdif[CP[a,b], CP[b,a]]]\}$$

(*example: product of two biquaternions A and B, $w = AB$*)

$$A = \{Quaternion[1, 3, 0, 4], Quaternion[2, 1, 5, 1]\}$$
$$B = \{Quaternion[1, 7, 8, 1], Quaternion[2, 1, 0, 1]\}$$
$$w = Simplify[CP[A, B]]$$

(*result*)

$$\{Quaternion[-26, -31, 23, 30], Quaternion[-51, 19, 28, -15]\}$$

References

1. Gürlebeck, K., Sprössig, W.: Quaternionic and Clifford Calculus for Physicists and Engineers. Wiley, New York (1997)
2. Vince, J.: Geometric Algebra for Computer Graphics. Springer, London (2008)
3. Snygg, J.: A New Approach to Differential Geometry Using Clifford's Geometric Algebra. Springer, New York (2012)
4. Casanova, G.: L'algèbre vectorielle. PUF, Paris (1976)
5. Girard, P.R., Pujol, R., Clarysse, P., Marion, A., Goutte, R., Delachartre, P.: Analytic video (2D+t) signals by Clifford Fourier Transforms in multi-quaternion Grassmann-Hamilton-Clifford algebras. In: Hitzer, E., Sangwine, S.J. (eds.) Quaternion and Clifford Fourier Transforms and Wavelets, pp. 197–219. Birkhäuser, Basel (2013)

6. Girard, P.R.: Quaternion Grassmann-Hamilton-Clifford-algebras: new mathematical tools for classical and relativistic modeling. In: Dössel, O., Schlegel, W.C. (eds.) World Congress 2009, IFMBE Proceedings, vol. 25/IV, pp. 65–68 (2009)

7. Girard, P.R.: Quaternions, Clifford Algebras and Relativistic Physics. Birkhäuser, Basel (2007)

8. Girard, P.R.: Quaternions, Algèbre de Clifford et Physique Relativiste. PPUR, Lausanne (2004)

9. Girard, P.R.: Quaternions, Clifford algebra and symmetry groups. In: Dorst, L., Doran, C., Lasenby, J. (eds.) Applications of Geometric Algebra in Computer Science and Engineering, pp. 307–315. Birkhäuser, Boston (2002)

10. Girard, P.R.: Einstein's equations and Clifford algebra. Adv. Appl. Clifford Algebras 9(2), 225–230 (1999)

11. Girard, P.R.: The quaternion group and modern physics. Eur. J. Phys. 5, 25–32 (1984)

12. Gsponer, A., Hurni, J.P.: Quaternions in mathematical physics (1): Alphabetical bibliography, math-ph, 0510059V4, (1430 references) (2008). http://www.arxiv.org/abs/arXiv:mathph/0510059

13. Gsponer, A., Hurni, J.P.: Quaternions in mathematical physics (2): Analytical bibliography, math-ph, 0511092V3, (1100 references) (2008). http://www.arxiv.org/abs/arXiv:mathphy/0511092

14. Hamilton, W.R.: The Mathematical Papers. Conway, A.W. (ed.) for the Royal Irish Academy. Cambridge University Press, Cambridge (1931–1967)

15. Hamilton, W.R.: Elements of Quaternions, 2 vols. (1899–1901). Reprinted 1969, Chelsea, New York (1969)

16. Hankins, T.L.: Sir William Rowan Hamilton. Johns Hopkins University Press, Baltimore (1980)

17. Crowe, M.J.: A History of Vector Analysis: The Evolution of the Idea of a Vectorial System. University of Notre Dame, Notre Dame, London (1967)

18. Grassmann, H.: Die lineale Ausdehungslehre: ein neuer Zweig der Mathematik, dargestellt und durch Anwendungen auf die übrigen Zweige der Mathematik, wie auch die Statik, Mechanik, die Lehre von Magnetismus und der Krystallonomie erläutert. Wigand, Leipzig (1844). Second ed. (1878)

19. Grassmann, H.: Mathematische und physikalische Werke, 3 vols. in 6 pts. Engel, F. (ed.) Leipzig (1894–1911)

20. Grassmann, H.: Der Ort der Hamilton'schen Quaternionen in der Ausdehnungslehre. Math. Ann. 12, 375–386 (1877)

21. Petsche, H.-J.: Grassmann. Birkhäuser, Basel (2006)

22. Clifford, W.K.: Applications of Grassmann's extensive algebra. J. Math. 1, 350–358 (1878)

23. Clifford, W.K.: Mathematical Papers. Tucker, R. (ed.) pp. 266–276 (1882). Reprinted Chelsea, New-York, (1968)

24. Lagally, M.: Vorlesungen über Vektorrechnung, 5th edn. Akademische Verlagsgesellschaft, Leipzig (1956)

25. Fehr, H.: Application de la Méthode Vectorielle de Grassmann à la Géométrie Infinitésimale. Carré et Naud, Paris (1899)

26. Guggenheimer, H.W.: Differential Geometry. Dover, New York (1976)

Ridgelet Methods for Linear Transport Equations

Philipp Grohs and Axel Obermeier[✉]

Seminar for Applied Mathematics, Mathematics Department,
Swiss Federal Institute of Technology, Zürich, Switzerland
axel.obermeier@sam.math.ethz.ch

Abstract. In this paper we present an overview of a novel method for the numerical solution of linear transport equations, which is based on ridgelets and has been introduced in [12,16]. Such equations arise for instance in radiative transfer or in phase contrast imaging. Due to the fact that ridgelet systems are well adapted to the structure of linear transport operators, it can be shown that our scheme operates in optimal complexity, even if line singularities are present in the solution. After presenting the basic algorithm, we prove that certain operators are compressible, which is the key to obtain unconditional convergence results. Finally, we show some applications in radiative transport.

1 Introduction

In the past two decades, a wide range of multiscale systems have been introduced with lasting impact in many different fields, starting with wavelets [9] and continuing with ridgelets [3], curvelets [4–6], shearlets [19,20], contourlets [11] etc. – the latter three of which fall into the framework of so-called "parabolic molecules" [15], while all of the mentioned systems are encompassed by the even broader framework of α-molecules [18].

These systems share the property that they are very well-adapted to representing certain classes of functions optimally (in the sense of the decay rate of the best N-term approximation) – functions with point singularities for wavelets, line singularities for ridgelets and curved singularities for parabolic molecules. Since these classes make up the fundamental phenomenological features of most images in an extremely diverse set of applications, it is perhaps not surprising, that many of the above-mentioned systems were originally investigated in view of their properties regarding image processing.

With a certain time-lag, it is becoming apparent that these systems are also very suitable for solving partial differential equations – again, wavelets were the first in this regard, for example leading to provably optimal solvers for elliptic equations [7]. For differential equations with strong directional features – such as transport equations – it is intuitively clear that optimal solvers will need to take these features into account, however, the development of solvers based on directional systems is still in its infancy.

J.-D. Boissonnat et al. (Eds.): Curves and Surfaces 2014, LNCS 9213, pp. 243–262, 2015.
DOI: 10.1007/978-3-319-22804-4_18

Following recent results [14], that ridgelets permit the construction of simple diagonal preconditioners for linear transport equations which arise in collocation-type discretization methods for kinetic transport equations (such as radiative transport), we consider the papers [12,16] to be a first step towards establishing directional representation systems as a useful tool for solving PDEs. The present paper proves results about the compressibility of certain operators that have been left open in [16].

Secondly, we present results from [12], where we introduced an **FFT**-based implementation of a ridgelet transform: FFRT – **F**ast **F**inite **R**idgelet **T**ransform. We use this for the numerical solution of kinetic transport equations arising in radiative transport. Using the preconditioner from [14] for linear transport equations together with a sparse discrete ordinates method similar to [13], we construct a solver which mitigates the curse of dimensionality and which results in uniformly well-conditioned linear systems which can be solved efficiently with CG.

1.1 Radiative Transport Equation

The motivation for this work is the numerical solution of the following model equation, described by the radiative transport equation (RTE),

$$Au := \boldsymbol{s} \cdot \nabla u + \kappa u = f + \int_{\mathbb{S}^{d-1}} \sigma u \, \mathrm{d}\boldsymbol{s}'.$$

It is a steady state continuity equation describing the conservation of radiative intensity in an absorbing, emitting and scattering medium, see e.g. [21]. We will, however, not treat the scattering operator in this paper, which can be incorporated through a variety of methods, one of which – the source iteration – we implemented in [12]. Let us assume that the following quantities are known at all locations $\boldsymbol{x} \in \Omega \subset \mathbb{R}^d$ and for all directions $\boldsymbol{s} \in \mathbb{S}^{d-1} := \{\boldsymbol{s} \in \mathbb{R}^d \colon \|\boldsymbol{s}\|_2 = 1\}$:

- absorption coefficient $\kappa(\boldsymbol{x}, \boldsymbol{s}) \geq \kappa_0 > 0$
- source term $f(\boldsymbol{x}, \boldsymbol{s}) \in \mathbb{R}$

Then, the above equation allows us to find the unknown radiative intensity u as a function $\Omega \times \mathbb{S}^{d-1} \to \mathbb{R}$.

Although the RTE looks simple, standard numerical techniques for solving it do not perform well for a number of reasons, mainly:

- The transport term $\boldsymbol{s} \cdot \nabla u$ leads to ill-conditioned systems of equations.
- Singularities in the input data may remain in the solution.
- With the dimension of the domain of u being 3 in 2-dimensional physical space and 5 in 3-dimensional space, the problem is fairly high-dimensional.

These issues make the accurate numerical solution of the RTE very costly or even impossible due to memory and compute power limitations of today's hardware.

1.2 Model Problem

We first simplify the problem by removing the scattering operator and fixing the transport direction s. However, we will come back to the full problem once we have developed a solver for this easier problem, and then use solve the problem with scattering using a collocation approach, which can also be combined with sparse tensor methods to alleviate the curse of dimensionality.

Therefore, our starting point is the differential operator

$$A: H^s(\mathbb{R}^d) \ni u \mapsto s \cdot \nabla u(x) + \kappa(x)u(x) \in L^2(\mathbb{R}^d) \tag{1}$$

with fixed $s \in \mathbb{S}^{d-1}$ and a function $\kappa \in L^\infty(\mathbb{R}^d)$ that satisfies $\kappa(x) \geq \gamma > 0$, $\forall x \in \mathbb{R}^d$. The space H^s is defined as follows.

Definition 1. *Let $s \in \mathbb{S}^{d-1}$, then we define the* anisotropic Sobolev space

$$H^s(\Omega) := \{f \in L^2(\Omega)(s \cdot \nabla)f \in L^2(\Omega)\} \quad \text{with norm } \|f\|^2_{H^s(\Omega)}$$
$$:= \|f\|^2_{L^2(\Omega)} + \|(s \cdot \nabla)f\|^2_{L^2(\Omega)}.$$

This space is more easily characterised on the Fourier side

$$H^s(\widehat{\Omega}) := \{\hat{f} \in L^2(\widehat{\Omega}) : \langle s \cdot \xi \rangle \hat{f}(\xi) \in L^2(\widehat{\Omega})\} \quad \text{with norm} \quad \|\hat{f}\|_{H^s(\widehat{\Omega})} := \|\langle s \cdot \xi \rangle \hat{f}\|_{L^2(\widehat{\Omega})},$$

where $\langle x \rangle := \sqrt{1 + |x|^2}$ is the regularised absolute value.

To solve the equation $Au = f \in L^2(\mathbb{R}^d)$, we look for solutions by minimising the L^2-residual,

$$u_0 = \underset{v \in H^s}{\operatorname{argmin}} \|Av - f\|_{L^2}. \tag{2}$$

With a variation-of-constants-argument (following [17]), it is not difficult to show the following.

Proposition 2 ([16]). *For every $\ell \in (H^s)'$ there exists a unique $u_0 \in H^s$ which solves (2). Moreover, the solution is characterised by the variational equation*

$$a(v, u_0) = \ell(v) \quad \text{for all } v \in H^s, \quad \text{where} \quad a(v, u) := \langle Av, Au \rangle_{L^2}. \tag{3}$$

In particular, well-definedness holds for

$$\ell_f(v) := \langle Av, f \rangle_{L^2} \quad \text{with } f \in L^2(\mathbb{R}^d).$$

2 Discretization

In our paper we aim to solve (2) via solving a discretization of the linear system (3). Several ingredients are needed to render this approach efficient:

(i) Uniform well-conditionedness of the resulting infinite discrete linear system
(ii) Fast approximate matrix-vector multiplication for the discrete operator matrix
(iii) Efficient approximation of typical solutions.

There exists several results which essentially state that, whenever (i), (ii) and (iii) are satisfied, then the linear system (3) can be solved in optimal computational complexity [7,8,23].

2.1 Gelfand Frames

Consider a Gelfand triple $(\mathcal{H}, L^2(\mathbb{R}^d), \mathcal{H}')$ – i.e. $\mathcal{H} \subseteq L^2(\mathbb{R}^d) \subseteq \mathcal{H}'$ with both inclusions being dense and continuous – and an operator $F : \mathcal{H} \to \mathcal{H}'$ which is bounded and boundedly invertible, induces a symmetric bilinear form $a(v, u) := \langle v, Fu \rangle_{\mathcal{H} \times \mathcal{H}'}$ and is elliptic in the sense that $a(v, v) \sim \|v\|_{\mathcal{H}}^2$.

We call a frame $\Phi = (\varphi_\lambda)_{\lambda \in \Lambda}$ for $\mathcal{H}$ (with dual frame $\tilde{\Phi}$) a *Gelfand frame* if $\Phi \subseteq \mathcal{H}$, and there exists a Gelfand triple $(\mathcal{H}_d, \ell^2(\Lambda), \mathcal{H}'_d)$ of sequence spaces such that the operators

$$G_\Phi^* : \begin{cases} \mathcal{H}_d \to \mathcal{H} \\ \mathbf{c} \mapsto \Phi\mathbf{c} \end{cases} \quad \text{and} \quad G_{\tilde{\Phi}} : \begin{cases} \mathcal{H} \to \mathcal{H}_d \\ f \mapsto \langle \tilde{\Phi}, f \rangle_{\mathcal{H}' \times \mathcal{H}} = \langle \tilde{\Phi}, f \rangle_{L^2} \end{cases}$$

are bounded. In addition, suppose that there exists an isomorphism $D_{\mathcal{H}} : \mathcal{H}_d \to \ell^2(\Lambda)$ such that its $\ell^2(\Lambda)$-adjoint $D_{\mathcal{H}}^* : \ell^2(\Lambda) \to \mathcal{H}'_d$ is also an isomorphism.

Then the equation $Fu = f$ can be discretised (and preconditioned) with the frame Φ to $\mathbf{F}\mathbf{u} = \mathbf{f}$,

$$\mathbf{F} = (D_{\mathcal{H}}^*)^{-1} G_\Phi F G_\Phi^* D_{\mathcal{H}}^{-1} \quad \text{and} \quad \mathbf{f} = (D_{\mathcal{H}}^*)^{-1} G_\Phi f,$$

such that the following result holds.

Lemma 3 ([8, Lemma 4.1]). *The operator* $\mathbf{F} : \ell^2(\Lambda) \to \ell^2(\Lambda)$ *is bounded and boundedly invertible on its range. In particular, the system* $\mathbf{F}\mathbf{u} = \mathbf{f}$ *is a uniformly well-conditioned infinite linear system.*

2.2 Numerical Solution of the Discrete System

If we were able to compute with infinite vectors, at this point we could simply solve

$$\mathbf{F}\mathbf{u} = \mathbf{f}, \tag{4}$$

by using a standard iterative solver such as a damped Richardson iteration

$$\mathbf{u}^{(j+1)} = \mathbf{u}^{(j)} - \alpha(\mathbf{F}\mathbf{u}^{(j)} - \mathbf{f}), \quad \mathbf{u}^{(0)} = \mathbf{0}.$$

Due to the well-conditionedness of the matrix $\mathbf{F}$ ensured by Lemma 3 and the fact that the iterates stay in $\mathrm{ran}(\mathbf{F})$ in each step, it is easy to show that for appropriate damping α the sequence $\mathbf{u}^{(j)}$ converges geometrically to the sought solution $\mathbf{u}$ in the $\ell^2(\Lambda)$-norm, i.e.

$$\|\mathbf{u} - \mathbf{u}^{(j)}\|_{\ell^2(\Lambda)} \lesssim \rho^j$$

for some $\rho < 1$, depending on the spectral properties of the operator $\mathbf{F}$.

To deal with the fact that we can only compute with finite vectors, we consider how this is done for the by-now classical wavelet discretisations of elliptic PDEs. The approximative evaluation of the Richardson iteration utilises the following three procedures:

- **RHS**$[\varepsilon, \mathbf{f}] \rightarrow \mathbf{f}_\varepsilon$: determines for $\mathbf{f} \in \ell^2(\Lambda)$ a finitely supported $\mathbf{f}_\varepsilon \in \ell^2(\Lambda)$ such that

$$\|\mathbf{f} - \mathbf{f}_\varepsilon\|_{\ell^2(\Lambda)} \leq \varepsilon;$$

- **APPLY**$[\varepsilon, \mathbf{A}, \mathbf{v}] \rightarrow \mathbf{v}_\varepsilon$: determines for $\mathbf{A} : \ell^2(\Lambda) \rightarrow \ell^2(\Lambda)$ and for a finitely supported $\mathbf{v} \in \ell^2(\Lambda)$ a finitely supported $\mathbf{v}_\varepsilon$ such that

$$\|\mathbf{A}\mathbf{v} - \mathbf{v}_\varepsilon\|_{\ell^2(\Lambda)} \leq \varepsilon;$$

- **COARSE**$[\varepsilon, \mathbf{u}] \rightarrow \mathbf{u}_\varepsilon$: determines for a finitely supported $\mathbf{u} \in \ell^2(\Lambda)$ a finitely supported $\mathbf{u}_\varepsilon \in \ell^2(\Lambda)$ with at most N nonzero coefficients, such that

$$\|\mathbf{u} - \mathbf{u}_\varepsilon\|_{\ell^2(\Lambda)} \leq \varepsilon. \tag{5}$$

Moreover, $N \lesssim N_{\min}$ holds, $N_{\min}$ being the minimal number of entries with (5).

We refer to [7,8,23] for information on the numerical realization of these routines.

2.3 The Problem with ker(F)

For the finite system, the iterates are no longer guaranteed to stay in $\mathrm{ran}(\mathbf{F})$ – in particular, errors in the kernel of the discretisation of $\mathbf{F}$ may accumulate. One possible remedy is to apply a projection $\mathbf{P}$ with $\mathrm{ker}(\mathbf{P}) = \mathrm{ker}(\mathbf{F})$ periodically, to eliminate these errors in the kernel.

Assuming the existence of numerical procedures as above and such a projection $\mathbf{P}$, we can formulate the numerical algorithm [23] to solve the discrete linear system (4) up to accuracy $\varepsilon > 0$, given as Algorithm 4 below.

Conditional on the three routines above, we have thus formulated a feasible algorithm for the approximate solution of (4).

Algorithm 4. Modified Inexact Damped Richardson Iteration

Data: $\varepsilon > 0$, $\mathbf{F}$, $\mathbf{f}$
Result: $\mathbf{u}_\varepsilon = \mathrm{modSOLVE}[\varepsilon, \mathbf{F}, \mathbf{P}, \mathbf{f}]$
Let $\theta < \frac{1}{3}$ and $K \in \mathbb{N}$ such that $3\rho^K \|\mathbf{P}\| < \theta$. $i := 0$, $\mathbf{u}^{(0)} := \mathbf{0}$,
$\varepsilon_0 := \|\mathbf{P}\| \|\mathbf{F}\|_{\mathrm{ran}(\mathbf{F})}^{-1} \|\mathbf{f}\|_{\ell^2(\Lambda)}$
while $\varepsilon_i > \varepsilon$ **do**
    $i := i + 1$;
    $\varepsilon_i := 3\rho^K \|\mathbf{P}\| \varepsilon_{i-1}/\theta$;
    $\mathbf{f}^{(i)} := \mathbf{RHS}[\theta\varepsilon_i/(6\alpha K\|\mathbf{P}\|), \mathbf{f}]$;
    $\mathbf{u}^{(i,0)} := \mathbf{u}^{(i-1)}$;
    for $j = 1, \ldots, K$ **do**
        $\mathbf{u}^{(i,j)} := \mathbf{u}^{(i,j-1)} - \alpha(\mathbf{APPLY}[\theta\varepsilon_i/(6\alpha K\|\mathbf{P}\|), \mathbf{F}, \mathbf{u}^{(i,j-1)}] - \mathbf{f}^{(i)})$;
    $\mathbf{z}^{(i)} := \mathbf{APPLY}[\theta\varepsilon_i/3, \mathbf{P}, \mathbf{u}^{(i,K)}]$;
    $\mathbf{u}^{(i)} := \mathbf{COARSE}[(1 - \theta)\varepsilon_i, \mathbf{z}^{(i)}]$;
$\mathbf{u}_\varepsilon := \mathbf{u}^{(i)}$;

2.4   Compressibility

To achieve optimal convergence rates for our problem through the techniques introduced in [7], a key ingredient is *compressibility* of the discretised operator equation. Such a property guarantees the existence of linear-time approximate matrix-vector multiplication algorithms **APPLY** which are used in the iterative solution of the operator equation, see [7,23] for more information.

Definition 5. *A matrix* **A** *is called* σ^*-*compressible if for every* $\sigma < \sigma^*$ *and* $k \in \mathbb{N}$ *there exists a matrix* $\mathbf{A}^{[k]}$ *such that*

(i) the matrix $\mathbf{A}^{[k]}$ *has at most* $\alpha_k 2^k$ *non-zero entries in each column,*
(ii) the following estimate holds

$$\|\mathbf{A} - \mathbf{A}^{[k]}\| \leq C_k,$$

in such a way that the sequences $(\alpha_k)_{k\in\mathbb{N}}$, $(C_k 2^{\sigma k})_{k\in\mathbb{N}}$ *are both summable.*

Furthermore, we require the well-known weak ℓ^p-spaces.

Definition 6. *For* $0 < p < \infty$, *set*

$$\ell^p_w := \left\{ (c_n)_{n\in\mathbb{N}} : c_n^* \lesssim n^{-\frac{1}{p}} \right\},$$

where c_n^* *is the decreasing rearrangement of* $(|c_n|)_{n\in\mathbb{N}}$.

Remark 7. It is easy to see that $\ell^p \subseteq \ell^p_w \subseteq \ell^{p+\varepsilon}$ for all $\varepsilon > 0$. More importantly, the ℓ^p_w-spaces are intimately related to best N-term approximations with a given (dual) frame, in the sense that if the coefficient sequence for a function lies in ℓ^p_w with $p = (\alpha + \frac{1}{2})^{-1}$ then the error of approximating said function with only N terms behaves like $\mathcal{O}(N^{-\alpha})$. For details, see e.g. [10].

Definition 8 ([23, Definition 3.9]). *A vector* $\mathbf{c} \in \ell^2$ *is called* σ^*-*optimal, when for a suitable routine* **RHS**, *for each* $\sigma \in (0, \sigma^*)$ *with* $p := (\sigma + \frac{1}{2})^{-1}$, *the following is valid for* $\mathbf{c}_\varepsilon = \mathbf{RHS}[\varepsilon, \mathbf{c}]$:

1. $\#\mathrm{supp}\, \mathbf{c}_\varepsilon \lesssim \varepsilon^{-1/\sigma} |\mathbf{c}|_{\ell^p_w}^{1/\sigma}$
2. *The number of arithmetic operations used to compute* $\mathbf{c}_\varepsilon$ *is at most a multiple of* $\varepsilon^{-1/\sigma} |\mathbf{c}|_{\ell^p_w}^{1/\sigma}$.

We can now formulate the main result of [23, Theorem 3.11].

Theorem 9 (Convergence of modSOLVE). *Assume that for some* $\sigma^* > 0$, *the matrices* **F** *and* **P** *are* σ^*-*compressible and that for some* $\sigma \in (0, \sigma^*)$ *and* $p := (\sigma + \frac{1}{2})^{-1}$, *the system* $\mathbf{Fu} = \mathbf{f}$ *has a solution* $\mathbf{u} \in \ell^p_w(\Lambda)$. *Moreover, assume that* $\mathbf{f}$ *is* σ^*-*optimal. Then for all* $\varepsilon > 0$, $\mathbf{u}_\varepsilon := \mathbf{modSOLVE}[\varepsilon, \mathbf{F}, \mathbf{P}, \mathbf{f}]$ *satisfies*

(I) $\#\mathrm{supp}\, \mathbf{u}_\varepsilon \lesssim \varepsilon^{-1/\sigma} |\mathbf{u}|_{\ell^p_w(\Lambda)}^{1/\sigma}$,
(II) the number of arithmetic operations to compute $\mathbf{u}_\varepsilon$ *is at most a multiple of* $\varepsilon^{-1/\sigma} |\mathbf{u}|_{\ell^p_w(\Lambda)}^{1/\sigma}$.

Furthermore, $\|\mathbf{P}\mathbf{u} - \mathbf{u}_\varepsilon\|_{\ell^2(\Lambda)} \leq \varepsilon$ and so, the recovered approximation $(D_\mathcal{H}^{-1}$ undoes the preconditioning, and G_Φ^* is the frame synthesis operator) satisfies $\|u - G_\Phi^* D_\mathcal{H}^{-1} \mathbf{u}_\varepsilon\|_\mathcal{H} \lesssim \varepsilon$.

One crucial assumption in Theorem 9 is that $\mathbf{P}$ be σ^*-compressible, because for the most obvious choice – an orthogonal projection – compressibility cannot be verified with current mathematical technology except in trivial cases [8,23]. However, as we will demonstrate for ridgelets below, other projections can be carefully constructed hat are in fact compressible.

The line of attack to solve the operator equation (2) is now clear: We have to construct a Gelfand frame Φ for the Gelfand triple $(H^s, L^2, (H^s)')$ and show that the resulting matrices $\mathbf{F}$ and $\mathbf{P}$ are compressible.

2.5   Ridgelet Frames

To this end, in [14], a Parseval frame $\Phi = (\varphi_\lambda)_{\lambda \in \Lambda}$ of ridgelets for $\mathcal{H} = H^s$ was constructed. The key to the construction is a certain set of functions $\psi_{j,\ell} \in L^2(\mathbb{R}^d)$, which form a partition of unity in the frequency domain, i.e.

$$(\psi_{j,\ell})_{j \in \mathbb{N}_0, \ell \in \{0,\dots,L_j\}} \quad \text{such that} \quad \sum_{j=0}^\infty \sum_{\ell=0}^{L_j} \hat{\psi}_{j,\ell}^2 = 1. \tag{6}$$

The support $P_{j,\ell} := \operatorname{supp} \hat{\psi}_{j,\ell}$ is explicitly given from the construction in (hyper-) spherical coordinates

$$P_{j,\ell} = \{(r, \boldsymbol{\theta}) : r \in [2^{j-1}, 2^{j+1}], \operatorname{dist}_{\mathbb{S}^{d-1}}(\boldsymbol{\theta}, \boldsymbol{s}_{j,\ell}) \leq \alpha_j := 2^{-j+1}\}, \tag{7}$$

where on each scale j, the set $(\boldsymbol{s}_{j,\ell})_{\ell=0}^{L_j}$ – see [1] for its construction – is a quasi-uniform sampling of points on the sphere in the sense that

$$\bigcup_{\ell=0}^{L_j} B_{\mathbb{S}^{d-1}}(\boldsymbol{s}_{j,\ell}, \alpha_j) = \mathbb{S}^{d-1}, \qquad \text{while}$$

$$B_{\mathbb{S}^{d-1}}\left(\boldsymbol{s}_{j,\ell}, \frac{\alpha_j}{3}\right) \cap B_{\mathbb{S}^{d-1}}\left(\boldsymbol{s}_{j,\ell'}, \frac{\alpha_j}{3}\right) = \emptyset \quad \text{for} \quad \ell \neq \ell'.$$

Definition 10. *Using* (6), *a Parseval frame for* $L^2(\mathbb{R}^d)$ *is defined by*

$$\varphi_{j,\ell,k} = 2^{-\frac{j}{2}} T_{U_{j,\ell}k} \psi_{j,\ell}, \quad j \in \mathbb{N}_0, \ell \in \{0,\dots,L_j\}, \boldsymbol{k} \in \mathbb{Z}^d,$$

with T *the translation operator,* $T_{\boldsymbol{y}} f(\cdot) := f(\cdot - \boldsymbol{y})$, *and* $U_{j,\ell} := R_{j,\ell}^{-1} D_{2^{-j}}$. *The rotation* $R_{j,\ell}$ *takes* $\boldsymbol{s}_{j,\ell}$ *into the first canonical unit vector* $\boldsymbol{e}_1$ *and* D_a *scales the first element by* a. *Whenever possible, we will subsume the indices of* φ *by* $\lambda = (j, \ell, \boldsymbol{k})$.

Remark 11. In $d \geq 3$ dimensions, rotations $R_{\boldsymbol{s}}$ turning $\boldsymbol{s}$ into $\boldsymbol{e}_1$ are no longer necessarily unique. However, for fixed $R_{\boldsymbol{s}}$, it is always possible [16, Lemma A.4] to choose $R_{\boldsymbol{s}'}$ such that the following Lipschitz conditions still holds,

$$\|R_{\boldsymbol{s}} - R_{\boldsymbol{s}'}\| \lesssim \operatorname{dist}_{\mathbb{S}^{d-1}}(\boldsymbol{s}, \boldsymbol{s}'). \tag{8}$$

To prove compressibility of **F** and **P**, we will need the following assumption on the partition-of-unity. This is not an undue restriction, as we show in [16, Lemma B.1] that this can be satisfied by constructing $\hat{\psi}_{j,\ell}$ in a certain way, but still leaving many possibilities to choose the window functions in question.

Assumption 12 *The $\hat{\psi}_{j,\ell}$ are constructed in such a way, that for any rotation $R_{j,\ell}$ (taking $\boldsymbol{s}_{j,\ell}$ to $\boldsymbol{e}_1$), the pullbacks under the transformation $U_{j,\ell}^{-\top} = R_{j,\ell}^{-1} D_{2^j}$, have bounded derivatives independently of j and ℓ, i.e.*

$$\text{for} \quad \hat{\psi}_{(j,\ell)}(\boldsymbol{\eta}) := \hat{\psi}_{j,\ell}(U_{j,\ell}^{-\top}\boldsymbol{\eta}): \quad \|\hat{\psi}_{(j,\ell)}\|_{C^n} \leq \beta_n, \qquad \forall n \leq N.$$

Essentially, $\hat{\psi}_{(j,\ell)}$ can be thought of as the transformation of $\hat{\psi}_{j,\ell}$ back to a reference element.

With the ridgelet frame Φ in hand we go on to show that Φ is indeed a Gelfand frame for the Gelfand triple $(H^s, L^2(\mathbb{R}^d), (H^s)')$. First, we need to find suitable sequence spaces $\mathcal{H}_d$. To this end we introduce the diagonal preconditioning matrix

$$\mathbf{W}_{\lambda,\lambda'} = \begin{cases} 0, \lambda \neq \lambda', \\ w(\lambda) := 1 + 2^j|\boldsymbol{s} \cdot \boldsymbol{s}_{j,\ell}|, \lambda = \lambda', \end{cases} \tag{9}$$

and define the weighted ℓ^2-spaces

$$\mathcal{H}_d := \ell_{\mathbf{W}}^2(\Lambda) := \{\mathbf{c} \in \ell^2(\Lambda) : \|\mathbf{W}\mathbf{c}\|_{\ell^2(\Lambda)} < \infty\}$$

and the corresponding isomorphisms

$$D_{\mathcal{H}_d}: \begin{cases} \mathcal{H}_d \to \ell^2(\Lambda), \\ \mathbf{c} \mapsto \mathbf{W}\mathbf{c}, \end{cases} \quad \text{and} \quad D_{\mathcal{H}_d}^*: \begin{cases} \ell^2(\Lambda) \to \mathcal{H}_d' = \ell_{\mathbf{W}^{-1}}^2(\Lambda), \\ \mathbf{c} \mapsto \mathbf{W}\mathbf{c}. \end{cases}$$

Finally, we have the frame operators

$$G_\Phi^*: \begin{cases} \mathcal{H}_d \to \mathcal{H}, \\ \mathbf{c} \mapsto \Phi\mathbf{c}, \end{cases} \quad \text{and} \quad G_\Phi: \begin{cases} \mathcal{H}' \to \mathcal{H}_d', \\ f \mapsto \langle \Phi, f \rangle_{\mathcal{H}' \times \mathcal{H}}, \end{cases}$$

which are bounded.

Then, as desired, we have the following result:

Theorem 13 ([16]). *The ridgelet frame Φ as constructed above constitutes a Gelfand frame for the Gelfand triple $(H^s, L^2(\mathbb{R}^d), (H^s)')$.*

Together with Lemma 3, this yields the following.

Theorem 14 ([16]). *With Φ the ridgelet system and A the differential operator defined above, consider the (infinite) matrix*

$$\mathbf{F} := \mathbf{W}^{-1}\langle A\Phi, A\Phi \rangle_{L^2}\mathbf{W}^{-1}.$$

Then the operator $\mathbf{F} : \ell^2(\Lambda) \to \ell^2(\Lambda)$ is bounded as well as boundedly invertible on its range $\operatorname{ran}(\mathbf{F}) = \operatorname{ran}(D_{\mathcal{H}_d}^{-1}G_\Phi)$. Furthermore, $\ker(\mathbf{F}) = \ker(G_\Phi^* D_{\mathcal{H}_d}^{-1})$.

Proposition 15. *The matrix*

$$\mathbf{P} := \mathbf{W} \langle \Phi, \Phi \rangle_{L^2} \mathbf{W}^{-1}$$

is a projection and satisfies $\ker(\mathbf{P}) = \ker(\mathbf{F})$.

Proof. We begin by writing

$$\mathbf{P} : \begin{cases} \mathcal{H} \to & \ell^2(\Lambda), \\ f \mapsto D_{\mathcal{H}_d} G_\Phi G_\Phi^* D_{\mathcal{H}_d}^{-1} f. \end{cases}$$

Then

$$\mathbf{PP}f = D_{\mathcal{H}_d} G_\Phi G_\Phi^* D_{\mathcal{H}_d}^{-1} D_{\mathcal{H}_d} G_\Phi G_\Phi^* D_{\mathcal{H}_d}^{-1} f = D_{\mathcal{H}_d} G_\Phi G_\Phi^* G_\Phi G_\Phi^* D_{\mathcal{H}_d}^{-1} f = \mathbf{P}f,$$

due to the fact that we have a Parseval frame – i.e. $G_\Phi^* G_\Phi g = g$. Finally, because $D_{\mathcal{H}_d} G_\Phi$ is injective due to the frame property, we have that $\ker(\mathbf{P}) = \ker(G_\Phi^* D_{\mathcal{H}_d}^{-1})$, which matches $\ker(\mathbf{F})$ by Theorem 14. $\qquad\square$

2.6   Compressibility of F

The main property left to verify is the compressibility of $\mathbf{F}$ and $\mathbf{P}$ – however, compressibility is difficult to verify directly in general. Instead we use the following notion of sparsity for a (possible bi-infinite) matrix $\mathbf{A}$:

Definition 16. *Let* $p > 0$. *A matrix* $\mathbf{A} = (a_{\lambda,\lambda'})_{\lambda \in \Lambda, \lambda' \in \Lambda'}$ *is called* p-sparse *if*

$$\|\mathbf{A}\|_{p\text{-sparse}} := \max \left(\sup_{\lambda' \in \Lambda'} \sum_{\lambda \in \Lambda} |a_{\lambda,\lambda'}|^p, \sup_{\lambda \in \Lambda} \sum_{\lambda' \in \Lambda'} |a_{\lambda,\lambda'}|^p \right)^{\frac{1}{p}} < \infty. \qquad (10)$$

Then, as a consequence of Schur's test, one can show:

Proposition 17 ([16, Proposition 5.2]). *Assume that* $\mathbf{A}$ *is* p-sparse *for* $0 < p < 1$. *Then* $\mathbf{A}$ *is* $\frac{1}{2}(\frac{1}{p} - 1)$-compressible.

Theorem 18 ([16, Thm.5.4]). *We consider the frame* $\Phi = (\varphi_\lambda)_{\lambda \in \Lambda}$, *satisfying Assumption 12 with* $N = 2n > d$ *for some* $n \in \mathbb{N}$, *and choose* $p \in \mathbb{R}$ *such that* $1 \geq p > \frac{d}{2n}$. *For the operator A from (1), we require that the absorption coefficient* κ *has a decomposition* $\kappa = \gamma + \kappa_0$ *with constant* $\gamma > 0$, *and* κ_0 *satisfying* $\kappa_0, \hat{\kappa}_0 \in L_\infty(\mathbb{R}^d)$. *Finally, we demand the existence of* $r_0, c_0 > 0$, *such that the decay condition*

$$|\hat{\kappa}_0(\boldsymbol{\xi})| \leq \frac{c_0}{|\boldsymbol{\xi}|^q} \qquad \forall \boldsymbol{\xi} \in \mathbb{R}^d : |\boldsymbol{\xi}| \geq r_0,$$

is fulfilled for a fixed $q > 2d + 2n + \frac{3}{2} + \frac{d-1}{p}$. *Then the stiffness matrix for the operator A (with appropriate preconditioning, see (9)) is* p-sparse *in this frame – in other words,*

$$\left\| \mathbf{W}^{-1} \langle A\Phi, A\Phi \rangle_{L^2} \mathbf{W}^{-1} \right\|_{p\text{-sparse}} < \infty. \qquad (11)$$

Remark 19. As we have seen in Proposition 17, the smaller p, the better the compressibility. The theorem is formulated in a way that p is chosen according to the restrictions imposed by d and n – however, since it is possible to construct window functions of arbitrary smoothness (and thus arbitrarily smooth $\hat{\psi}_{j,\ell}$), the limiting factor for p then becomes the decay rate of $\hat{\kappa}_0$. In the case that $\hat{\kappa}_0$ decays faster than any polynomial (say, exponentially), arbitrarily small p can be achieved (for infinitely smooth $\hat{\psi}_{j,\ell}$) – of course at the cost of exploding constants.

Instead of relying on unverified assumptions about the compressibility of the orthogonal projection, we can very sparsity (and thus compressibility) for the projection $\mathbf{P}$ defined above directly.

Theorem 20. *Again, let $\Phi = (\varphi_\lambda)_{\lambda \in \Lambda}$ satisfy Assumption 12 with $N = 2n > d$ for some $n \in \mathbb{N}$, and choose $p \in \mathbb{R}$ such that $1 \geq p > \frac{d}{2n}$. Then the projection $\mathbf{P}$ is p-sparse in this frame – in other words,*

$$\|\mathbf{P}\|_{p\text{-sparse}} = \|\mathbf{W}\langle \Phi, \Phi \rangle_{L^2} \mathbf{W}^{-1}\|_{p\text{-sparse}} < \infty.$$

Before we come to the proof, we collect two auxiliary technical results.

Proposition 21 ([16, Corollary C.3]). *For functions $f, g \in C^{2n}$, we have the following estimate for the Laplacian in d dimensions,*

$$|[\Delta^n(fg)](\boldsymbol{\eta})| \leq (4d)^n |f(\boldsymbol{\eta})|_{C^{2n}} |g(\boldsymbol{\eta})|_{C^{2n}} \leq (4d)^n \|f\|_{C^{2n}} \|g\|_{C^{2n}}, \quad (12)$$

where $|f(\boldsymbol{\eta})|_{C^{2n}} = \max_{|\alpha| \leq 2n} |\frac{\partial^{|\alpha|} f}{\partial \boldsymbol{\eta}^\alpha}(\boldsymbol{\eta})|$ is the maximum of all derivatives up to order $2n$ of f at $\boldsymbol{\eta}$.

Proposition 22 ([16, Proposition A.3]). *For fixed j' and ℓ', the number of $P_{j,\ell}$ on scale j that can intersect $P_{j',\ell'}$ is bounded as follows*

$$\#\{\ell \in \{0, \dots, L\} : P_{j,\ell} \cap P_{j',\ell'} \neq \emptyset\} \lesssim \begin{cases} 2^{|j-j'|(d-1)}, & |j - j'| \leq 1, \\ 0, & |j - j'| > 1. \end{cases} \quad (13)$$

Proof (Proof of Theorem 20). Due to symmetry, we are able to express (11) without taking the maximum (cf. (10)),

$$\|\mathbf{P}\|_{p\text{-sparse}}^p = \sup_{\lambda' \in \Lambda} \sum_{\lambda \in \Lambda} \left| \frac{w(\lambda)}{w(\lambda')} \langle \varphi_\lambda, \varphi_{\lambda'} \rangle_{L^2} \right|^p$$

$$= \sup_{\lambda' \in \Lambda} \sum_{j \in \mathbb{N}_0} \sum_{\ell=0}^{L_j} \sum_{\boldsymbol{k} \in \mathbb{Z}^d} \left| \frac{w(\lambda)}{w(\lambda')} \langle \varphi_{j,\ell,\boldsymbol{k}}, \varphi_{j',\ell',\boldsymbol{k}'} \rangle_{L^2} \right|^p < \infty. \quad (14)$$

Step 1 – Transforming the Integral: Recalling the definition of the φ_λ, we compute

$$\mathbf{P}_{\lambda,\lambda'} = \frac{w(\lambda)}{w(\lambda')} \langle \varphi_\lambda, \varphi_{\lambda'} \rangle_{L^2} = \frac{w(\lambda)}{w(\lambda')} \langle \hat{\varphi}_\lambda, \hat{\varphi}_{\lambda'} \rangle_{L^2}$$

$$= 2^{-\frac{i+j'}{2}} \frac{w(\lambda)}{w(\lambda')} \int \overline{\mathcal{F}(\psi_{j,\ell}(\boldsymbol{x} - U_{j,\ell}\boldsymbol{k}))} \mathcal{F}(\psi_{j',\ell'}(\boldsymbol{x} - U_{j',\ell'}\boldsymbol{k}')) \, \mathrm{d}\boldsymbol{\xi}$$

$$= c_{\mathcal{F}} 2^{-\frac{i+j'}{2}} \frac{w(\lambda)}{w(\lambda')} \int \overline{\hat{\psi}_{j,\ell}(\boldsymbol{\xi})} \hat{\psi}_{j',\ell'}(\boldsymbol{\xi}) \exp(2\pi \mathrm{i}\boldsymbol{\xi} \cdot (U_{j,\ell}\boldsymbol{k} - U_{j',\ell'}\boldsymbol{k}')) \, \mathrm{d}\boldsymbol{\xi}$$

where $c_{\mathcal{F}} = (2\pi \mathrm{i})^2$.

The transformation $U_{j,\ell}$ modifying $\boldsymbol{k}$ in the exponential function makes summing $\boldsymbol{k}$ difficult, and therefore, we will transform all the integral by $\boldsymbol{\xi} = U_{j,\ell}^{-\top}\boldsymbol{\eta}$ – introducing a factor 2^j from the determinant of the Jacobian and yielding the exponent

$$2\pi \mathrm{i}\boldsymbol{\eta} \cdot (\boldsymbol{k} - U_{j,\ell}^{-1}U_{j',\ell'}\boldsymbol{k}') = 2\pi \mathrm{i}\boldsymbol{\eta} \cdot (\boldsymbol{k} - U_{j',\ell'}^{j,\ell}\boldsymbol{k}'), \quad \text{where} \quad U_{j',\ell'}^{j,\ell} := U_{j,\ell}^{-1}U_{j',\ell'}.$$

We want to use the representation of $\hat{\psi}_{j,\ell}$ from Assumption 12, which holds for arbitrary rotations $\widetilde{R}_{j',\ell'}$ taking $\boldsymbol{s}_{j',\ell'}$ to $\boldsymbol{e}_1$. We choose $\widetilde{R}_{j',\ell'}$ in $\widetilde{U}_{j',\ell'} := \widetilde{R}_{j',\ell'}^{-1}D_{2^{-j}}$ in such a way that (8) holds for $\boldsymbol{s} = \boldsymbol{s}_{j,\ell}$ and $\boldsymbol{s}' = \boldsymbol{s}_{j',\ell'}$, and unsurprisingly, we set $\widetilde{U}_{j',\ell'}^{j,\ell} := U_{j,\ell}^{-1}\widetilde{U}_{j',\ell'}$. Thus,

$$\mathbf{P}_{\lambda,\lambda'} = c_{\mathcal{F}} 2^{\frac{j-j'}{2}} \frac{w(\lambda)}{w(\lambda')} \int \underbrace{\hat{\psi}_{(j,\ell)}(\boldsymbol{\eta}) \hat{\psi}_{(j',\ell')}(\widetilde{U}_{j',\ell'}^{\top}U_{j,\ell}^{-\top}\boldsymbol{\eta})}_{=:h_{\lambda,\lambda'}(\boldsymbol{\eta})}$$

$$\exp(2\pi \mathrm{i}\boldsymbol{\eta} \cdot (\boldsymbol{k} - U_{j',\ell'}^{j,\ell}\boldsymbol{k}')) \, \mathrm{d}\boldsymbol{\eta}. \tag{15}$$

It should be noted that h-terms does not depend on $\boldsymbol{k}, \boldsymbol{k}'$ – however, we have chosen this notation for reasons of notational brevity.

We now have to show that the sum of (15) over all parameters in (14) is finite – which we will do for $\boldsymbol{k}$ first, then for ℓ and finally for j.

Step 2 – Integration by Parts: Even though the exponent is purely imaginary, we cannot estimate the exponential function by one, as we would then sum constants in $\boldsymbol{k}$. However, a simple calculation shows $\Delta_{\boldsymbol{\eta}} \exp(2\pi \mathrm{i}\boldsymbol{\eta} \cdot \boldsymbol{y}) = -(2\pi)^2|\boldsymbol{y}|^2 \exp(2\pi \mathrm{i}\boldsymbol{\eta} \cdot \boldsymbol{y})$, which entails

$$\Delta_{\boldsymbol{\eta}} \exp(2\pi \mathrm{i}\boldsymbol{\eta} \cdot (\boldsymbol{k} - U_{j',\ell'}^{j,\ell}\boldsymbol{k}')) = -(2\pi)^2|\boldsymbol{k} - U_{j',\ell'}^{j,\ell}\boldsymbol{k}'|^2 \exp(2\pi \mathrm{i}\boldsymbol{\eta} \cdot (\boldsymbol{k} - U_{j',\ell'}^{j,\ell}\boldsymbol{k}')).$$

Applying Green's second identity iteratively (boundary terms disappear due to the compact support of $h_{\lambda,\lambda'}$), we will use this to generate a denominator of sufficient power to be summed over all $\boldsymbol{k} \in \mathbb{Z}^d$ – on the other hand, this forces us to estimate the derivatives of the remaining factors of the integrands. All differential operators will be with respect to $\boldsymbol{\eta}$, which we will not indicate anymore in the following.

Thus, for $\boldsymbol{k} \neq U_{j',\ell'}^{j,\ell}\boldsymbol{k}'$,

$$\mathbf{P}_{\lambda,\lambda'} = c_{\mathcal{F}} 2^{\frac{j-j'}{2}} \frac{w(\lambda)}{w(\lambda')} \frac{(-1)^n(2\pi)^{-2n}}{|\boldsymbol{k} - U_{j',\ell'}^{j,\ell}\boldsymbol{k}'|^{2n}} \int \Delta^n(h_{\lambda,\lambda'}(\boldsymbol{\eta}))$$

$$\exp(2\pi \mathrm{i}\boldsymbol{\eta} \cdot (\boldsymbol{k} - U_{j',\ell'}^{j,\ell}\boldsymbol{k}')) \, \mathrm{d}\boldsymbol{\eta}. \tag{16}$$

Step 3 – Estimating the Derivatives: Using (12) and Assumption 12, we see

$$\|\Delta^n h_{\lambda,\lambda'}\| \le (4d)^n \|\hat{\psi}_{(j,\ell)}\|_{\mathcal{C}^{2n}} \|\hat{\psi}_{(j',\ell')}((\tilde{U}^{j,\ell}_{j',\ell'})^\top \cdot)\|_{\mathcal{C}^{2n}} \le (4d)^n \beta_{2n}^2 \|\tilde{U}^{j,\ell}_{j',\ell'}\|^{2n}$$

which allows us to estimate (16), while remembering to keep the support information of terms we estimate away:

$$|\mathbf{P}_{\lambda,\lambda'}| \lesssim \frac{w(\lambda)}{w(\lambda')} \frac{2^{\frac{i-j'}{2}}}{|\mathbf{k} - U^{j,\ell}_{j',\ell'}\mathbf{k}'|^{2n}} \int_{U^\top_{j,\ell}(P_{j,\ell} \cap P_{j',\ell'})} \|\tilde{U}^{j,\ell}_{j',\ell'}\|^{2n}\, d\boldsymbol{\eta} \qquad (17)$$

Step 4 – Estimating the Transformation: We begin by considering the matrix $R_{j,\ell}\tilde{R}^{-1}_{j',\ell'}$. Denoting the identity by $\mathbb{I}$, we exploit the orthogonality of the $R_{j,\ell}$ and (8) to yield

$$\|R_{j,\ell}\tilde{R}^{-1}_{j',\ell'} - \mathbb{I}\| = \|\tilde{R}_{j',\ell'} - R_{j,\ell}\| \lesssim \mathrm{dist}_{\mathbb{S}^{d-1}}(\boldsymbol{s}_{j,\ell}, \boldsymbol{s}_{j',\ell'}).$$

Due the necessary proximity of $\boldsymbol{s}_{j,\ell}$ and $\boldsymbol{s}_{j',\ell'}$ (for the intersection $P_{j,\ell} \cap P_{j',\ell'}$ to be non-empty),

$$\mathrm{dist}_{\mathbb{S}^{d-1}}(\boldsymbol{s}_{j,\ell}, \boldsymbol{s}_{j',\ell'}) \le \alpha_j + \alpha_{j'} = 2^{-j+1} + 2^{-j'+1}, \qquad (18)$$

we can finish the estimate,

$$\|\tilde{U}^{j,\ell}_{j',\ell'}\| = \|D_{2^j} R_{j,\ell} \tilde{R}^{-1}_{j',\ell'} D_{2^{-j'}}\| = \|D_{2^{j-j'}} + D_{2^j}(R_{j,\ell}\tilde{R}^{-1}_{j',\ell'} - \mathbb{I})D_{2^{-j'}}\|$$
$$\lesssim \max(2^{j-j'}, 1) + 2^j \mathrm{dist}_{\mathbb{S}^{d-1}}(\boldsymbol{s}_{j,\ell}, \boldsymbol{s}_{j',\ell'}) \lesssim 2^{|j-j'|}.$$

Step 5 – Estimating the Weights: The factor $\frac{w(\lambda)}{w(\lambda')}$ can be easily estimated, again due to (18). Inserting the definition, we have

$$\frac{1 + 2^j |\boldsymbol{s} \cdot \boldsymbol{s}_{j,\ell}|}{1 + 2^{j'}|\boldsymbol{s} \cdot \boldsymbol{s}_{j',\ell'}|} \le \frac{1 + 2^j |\boldsymbol{s} \cdot \boldsymbol{s}_{j',\ell'}| + 2^j |\boldsymbol{s} \cdot (\boldsymbol{s}_{j,\ell} - \boldsymbol{s}_{j',\ell'})|}{1 + 2^{j'}|\boldsymbol{s} \cdot \boldsymbol{s}_{j',\ell'}|} \le 1$$
$$+ 2^{j-j'} + 2^j \mathrm{dist}_{\mathbb{S}^{d-1}}(\boldsymbol{s}_{j,\ell}, \boldsymbol{s}_{j',\ell'}) \lesssim 2^{|j-j'|},$$

because $|\boldsymbol{s} - \boldsymbol{s}'| \le \mathrm{dist}_{\mathbb{S}^{d-1}}(\boldsymbol{s}, \boldsymbol{s}')$ for $\boldsymbol{s}, \boldsymbol{s}' \in \mathbb{S}^{d-1}$.

Step 6 – Estimating the Integral: For $j \ge 1$, the transformation $U^\top_{j,\ell}$ takes the "frequency tiles" $P_{j,\ell}$ back into a bounded set around the origin,

$$U^\top_{j,\ell} P_{j,\ell} \subseteq \left[\frac{1}{2}\cos(\alpha_j), 2\right] \times \mathcal{P}_{(\mathrm{span}\{e_1\})^\perp}(B_{\mathbb{R}^d}(0,4)) \subseteq B_{\mathbb{R}^d}(0,5).$$

This can be seen from (7), as $P_{j,\ell}$ is contained in the intersection between a spherical shell (between radii 2^{j-1} and 2^{j+1}) and a cone around $\boldsymbol{s}_{j,\ell}$ with opening angle $\alpha_j = 2^{-j+1}$. The rotation in $U^\top_{j,\ell} = D_{2^{-j}} R_{j,\ell}$ brings the axis of this cone into $\boldsymbol{e}_1$. We see that the smallest value of η_1 for $\boldsymbol{\eta} \in U^\top_{j,\ell} P_{j,\ell}$ is $2^{-j}2^{j-1}\cos(\alpha_j) = \frac{1}{2}\cos(\alpha_j) > \frac{1}{4}$ since $\alpha_j = 2^{-j+1} \le 1 < \frac{\pi}{3}$ for $j \ge 1$.

The largest extent perpendicular to e_1 can be calculated as

$$2^{j+1} \cos \alpha_j \sin \alpha_j = 2^j \sin 2\alpha_j \leq 2^j \cdot 2\alpha_j = 4,$$

which is what we wanted. Therefore,

$$\int_{U_{j,\ell}^\top (P_{j,\ell} \cap P_{j',\ell'})} \mathrm{d}\eta \lesssim 1. \tag{19}$$

However, we will keep the integrals for now, as the support conveniently encodes the condition that j and j', resp. ℓ and ℓ' have to be close. Applying the other estimates to (17), we arrive at

$$|\mathbf{P}_{\lambda,\lambda'}| \lesssim 2^{|j-j'|(2n+\frac{3}{2})} |k - U_{j',\ell'}^{j,\ell} k'|^{-2n} \int_{U_{j,\ell}^\top (P_{j,\ell} \cap P_{j',\ell'})} \mathrm{d}\eta$$

Step 7 – Summing k: Thus far, we have omitted the case $k = U_{j',\ell'}^{j,\ell} k'$ – in fact, to sum over k, we need treat even more elements differently. In order to estimate the term $|k - U_{j',\ell'}^{j,\ell} k'|$, we choose $K_{j',\ell'}^{j,\ell} k' \in \mathbb{Z}^d$ as a (possibly non-unique) closest lattice element to $U_{j',\ell'}^{j,\ell} k'$ (for example by rounding every component to the nearest integer), which may be interpreted as a projection of $U_{j',\ell'}^{j,\ell} k'$ onto the lattice $\mathbb{Z}^d$. Then $|K_{j',\ell'}^{j,\ell} k' - U_{j',\ell'}^{j,\ell} k'| \leq \frac{\sqrt{d}}{2}$, and if we restrict $k \in \mathbb{Z}^d$ such that $|k - K_{j',\ell'}^{j,\ell} k'| \geq \sqrt{d}$, it holds that

$$|k - U_{j',\ell'}^{j,\ell} k'| \geq |k - K_{j',\ell'}^{j,\ell} k'| - \frac{\sqrt{d}}{2} \geq \frac{1}{2} |k - K_{j',\ell'}^{j,\ell} k'|. \tag{20}$$

For $k \in \mathbb{Z}^d$ such that $|k - K_{j',\ell'}^{j,\ell} k'| < \sqrt{d}$, we retrace the derivation of all above estimates without the partial integration, which, in effect, only eliminates the divisor $|k - U_{j',\ell'}^{j,\ell} k'|^{2n}$ (and reduces the implicit constants). Summing up, this means that

$$|\mathbf{P}_{\lambda,\lambda'}| \lesssim 2^{|j-j'|(2n+\frac{3}{2})} |k - U_{j',\ell'}^{j,\ell} k'|^{-2n} \int_{U_{j,\ell}^\top (P_{j,\ell} \cap P_{j',\ell'})} \mathrm{d}\eta =: Z_{\lambda,\lambda'}$$

for $|k - K_{j',\ell'}^{j,\ell} k'| \geq \sqrt{d}$, and similarly for $|k - K_{j',\ell'}^{j,\ell} k'| < \sqrt{d}$,

$$|\mathbf{P}_{\lambda,\lambda'}| \lesssim 2^{|j-j'|(2n+\frac{3}{2})} \int_{U_{j,\ell}^\top (P_{j,\ell} \cap P_{j',\ell'}')} \mathrm{d}\eta = Z_{\lambda,\lambda'}$$

Note that the different cases for k are incorporated in the definition of the Z-terms.

The intention now is to prove (14) by showing

$$\sup_{\lambda' \in \Lambda} \sum_{\lambda \in \Lambda} |\mathbf{P}_{\lambda,\lambda'}|^p \lesssim \sup_{\lambda' \in \Lambda} \sum_{\lambda \in \Lambda} (Z_{\lambda,\lambda'})^p < \infty.$$

We begin by summing $\boldsymbol{k}$, which crucially requires the condition $p > \frac{d}{2n}$,

$$\sum_{\substack{\boldsymbol{k}\in\mathbb{Z}^d \\ |\boldsymbol{k}-K_{j',\ell'}^{j,\ell}\boldsymbol{k}'|\geq\sqrt{d}}} \frac{1}{|\boldsymbol{k}-U_{j',\ell'}^{j,\ell}\boldsymbol{k}'|^{2np}} \overset{(2.6)}{\leq} \sum_{\substack{\boldsymbol{k}\in\mathbb{Z}^d \\ |\boldsymbol{k}-K_{j',\ell'}^{j,\ell}\boldsymbol{k}'|\geq\sqrt{d}}} \frac{2^{2np}}{|\boldsymbol{k}-K_{j',\ell'}^{j,\ell}\boldsymbol{k}'|^{2np}}$$

$$= \sum_{\substack{\boldsymbol{k}\in\mathbb{Z}^d \\ |\boldsymbol{k}|\geq\sqrt{d}}} \frac{2^{2np}}{|\boldsymbol{k}|^{2np}} =: G_{d,2np} < \infty.$$

The remaining sum over $\boldsymbol{k}\in\mathbb{Z}^d : |\boldsymbol{k}-K_{j',\ell'}^{j,\ell}\boldsymbol{k}'| < \sqrt{d}$ has at most $\mathcal{O}(\sqrt{d}^d)$ terms. Taken together, this implies

$$\sum_{\lambda\in\Lambda} |\mathbf{P}_{\lambda,\lambda'}|^p \lesssim \sum_{j\in\mathbb{N}_0} \sum_{\ell=0}^{L_j} 2^{|j-j'|(2n+\frac{3}{2})p} \left(\int_{U_{j,\ell}^\top(P_{j,\ell}\cap P_{j',\ell'})} \right)^p d\boldsymbol{\eta} \ .$$

Step 8 – Summing ℓ and j: From (13), we know how many intersections in ℓ are maximally possible on scale j. Therefore

$$\sum_{\lambda\in\Lambda} |\mathbf{P}_{\lambda,\lambda'}|^p \lesssim \sum_{j\in\mathbb{N}_0} 2^{|j-j'|(2np+\frac{3p}{2}+d-1)} \left(\int_{U_{j,\ell}^\top(P_{j,\ell}\cap P_{j',\ell'})} d\boldsymbol{\eta} \right)^p .$$

Finally, because $|j-j'| \leq 1$ (and using (19)), we have bounded the left-hand side independently of λ', and therefore taking the sup doesn't change anything and the proof is finished.

3   Main Results

The results so far allow us to formulate the following corollary to Theorem 13, which, in essence, states that the complexity of **modSOLVE** is *linear* with respect to the number of relevant coefficients of the discretisation.

Corollary 23. *Assume that* $\mathbf{f}$ *is* σ^*-*optimal (compare Definition 8) and that the system* $\mathbf{Fu} = \mathbf{f}$ *has a solution* $\mathbf{u}\in\ell_w^p(\Lambda)$ *for* $\sigma\in(0,\sigma^*)$ *and* $p := (\sigma+\frac{1}{2})^{-1}$. *Then the solution* $\mathbf{u}_\varepsilon := \mathbf{modSOLVE}[\varepsilon,\mathbf{F},\mathbf{P},\mathbf{f}]$ *of the ridgelet-based solver recovers this approximation rate – i.e.*

$$\#\mathrm{supp}\ \mathbf{u}_\varepsilon \lesssim \varepsilon^{-1/\sigma}|\mathbf{u}|_{\ell_w^p(\Lambda)}^{1/\sigma},$$

and the number of arithmetic operations is at most a multiple of $\varepsilon^{-1/\sigma}|\mathbf{u}|_{\ell_w^p(\Lambda)}^{1/\sigma}$.

Finally, the last assumption – that the discretisation of typical solutions are in $\ell_w^p(\Lambda)$ – is also satisfied by the ridgelet discretisation. The proof of this theorem is based on arguments of [2] and is the subject of an upcoming paper [22].

Theorem 24. *Let f be a function in $L^2(\mathbb{R}^d)$ such that $f \in H^t(\mathbb{R}^d)$ apart from discontinuities across a finite number of (affine) hyperplanes h_i, with corresponding (normalised) orthogonal vectors $\boldsymbol{n}_i$ and offsets t_i. What this means is that – with $H(y) := \mathbb{1}_{\mathbb{R}^+}(y)$ denoting the Heaviside step function – f is of the form*

$$f(\boldsymbol{x}) = f_0(\boldsymbol{x}) + \sum_{i=1}^{N} f_i(\boldsymbol{x}) H(\boldsymbol{x} \cdot \boldsymbol{n}_i - t_i),$$

where $f_0, f_1, \dots, f_N \in H^t(\mathbb{R}^d)$.

Furthermore, with $\mathcal{P}_{h_i}\boldsymbol{x} = \boldsymbol{x} - (\boldsymbol{x} \cdot \boldsymbol{n}_i - t_i) \cdot \boldsymbol{n}_i$ being the projection onto h_i, the f_i ($i \geq 1$) have to satisfy the decay condition

$$|f_i(\mathcal{P}_{h_i}\boldsymbol{x})| \lesssim \langle \mathcal{P}_{h_i}\boldsymbol{x} \rangle^{-2n},$$

where $\langle y \rangle = \sqrt{1 + |y|^2}$ is the regularised absolute value and $n \in \mathbb{N}$ is arbitrarily large[1] – which covers the cases of compact support and exponential decay (this decay is only necessary along the singularities, the rest of f just has to satisfy $f \in H^t(\mathbb{R}^d)$).

Then, for arbitrary $\delta > 0$, the solution u to (1) with right-hand side f satisfies $\mathbf{W}\langle \Phi, u \rangle_{L^2} \in \ell_w^{p+\delta}$, where $p = (\frac{t}{d} + \frac{1}{2})^{-1}$. Similarly $\langle \Phi, f \rangle_{L^2} \in \ell_w^{p+\delta}$ with the same p, which is the best possible for $f \in H^t(\mathbb{R}^d)$, even without singularities!

3.1   Conclusion

The bottom line is that the presented construction "sparsifies" *both* the system matrix as well as typical solutions of transport problems (in the sense of compressibility and N-term approximations, respectively), which makes it the ideal candidate for the development of fast algorithms, as underscored also by the results of Corollary 23.

4   Applications

4.1   Discrete Ordinates Method

As outlined in the introduction, we will now construct an algorithm to solve the full transport problem by collocation in different directions (where we can use our uni-directional solver)[2]. Therefore, we consider

$$\boldsymbol{s} \cdot \nabla u + \kappa u = f, \tag{21}$$

but this time we let $\boldsymbol{s} \in \mathbb{S}^1$ also be an independent variable such that $u, \kappa, f : \Omega \times \mathbb{S}^1 \to \mathbb{R}$. The *discrete ordinates method* (DOM) as outlined in [13, Section 2] solves this problem in the following way:

---

[1] A result for finitely fast decay has also been obtained in [22], with δ then depending on n.

[2] A detailed account can be found in [12].

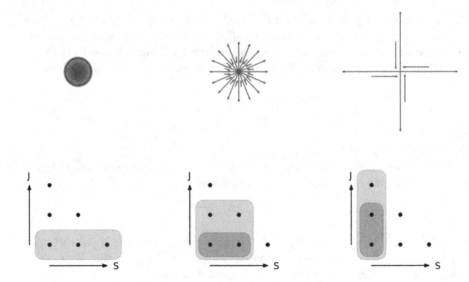

Fig. 1. Illustration of the SDOM

- Choose some directions $\{\boldsymbol{s}_i\}_{i=1}^{N_s} \subset \mathbb{S}^1, N_s \in \mathbb{N}$.
- Solve (21) for these fixed directions, yielding the one-directional solutions $u_i'(x, y)$.
- Interpolate the $(\boldsymbol{s}_i, u_i')$ to get a solution u' for the full domain $\Omega \times \mathbb{S}^1$.

The approximation error $\|u - u'\|_{L^2(\Omega \times \mathbb{S}^1)}$ depends on the quality of the interpolation – which is determined by N_s and the angular smoothness of u – on the one hand, and the error in the uni-directional solver on the other hand. Balancing these two errors gives a certain base b for choosing $N_s \sim b^J$, where, for example, $b = 2^{\frac{k}{3}}$ if the right-hand side $f \in H^k$ and linear interpolation is used (with u being sufficiently smooth in angle). The outlined discrete ordinates method – where the uni-directional solver goes up to scale J – has complexity $\mathcal{O}(N_s 4^J) = \mathcal{O}((4b)^J)$, which quickly becomes very expensive.

4.2   Sparse Discrete Ordinates Method

In order to mitigate the scaling problem of the (full) discrete ordinates method, the *sparse discrete ordinates method* (SDOM) was developed in [13, Section 4]. The main idea is that the highest resolution in radius does not need the highest resolution in angle, and vice versa. Under suitable smoothness assumptions, matching decreasing resolution in one with increasing resolution in the other allows to lower the complexity of the solver to $\mathcal{O}(\max(4, b)^J)$, while losing only a logarithmic factor in accuracy compared to the full DOM (with the highest radial and angular resolutions used for the SDOM), see [13, Lemma 4.3] or [12].

A graphical representation of the SDOM is given in Fig. 1 for $J = 3$ and $b = 4$ – the interpolations between the green arrows are added, while the interpolation

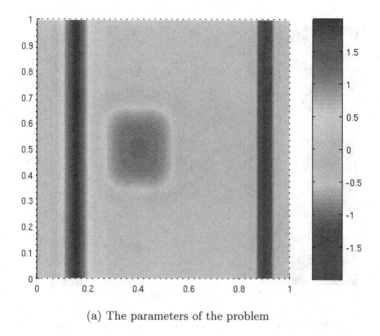

(a) The parameters of the problem

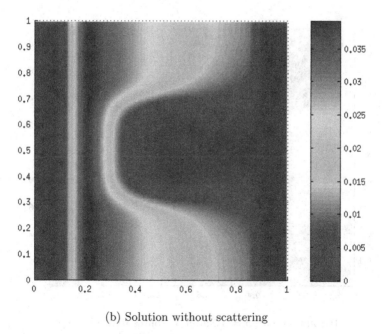

(b) Solution without scattering

Fig. 2. Scattering of radiation around an obstacle for different values of the scattering coefficient σ (Color figure online) (Continued on next page)

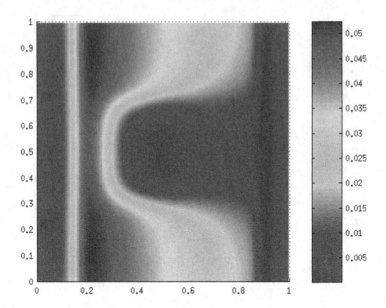

(c) Solution with $\sigma = 0.2$ after 10 source iteration steps

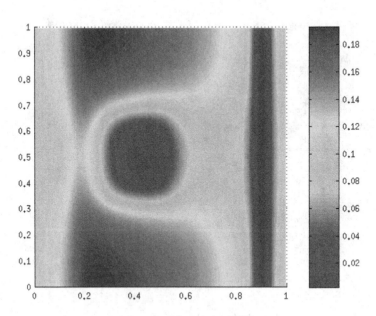

(d) Solution with $\sigma = 0.5$ after 40 source iteration steps

Fig. 2. (*continued*)

between the red arrows is subtracted. In the upper row, the lengths of the arrows represent the number of scales that were used whereas their number and directions indicate how many and which directions are used for angular interpolation. The bottom row shows in which detail spaces the functions obtained in this way live (the J-arrow denotes increasing frame size, the S-arrow denotes increasing number of angular interpolation points).

4.3 Source Iterations

Finally, we are able to tackle the complete RTE including the scattering term:

$$s \cdot \nabla u + \kappa u = f + \int_{\mathbb{S}^1} \sigma u \mathrm{d}s'$$

This problem can be solved using the *source iteration* method, which is:

- Set $u'^{(0)}(x, y, s) = 0$.
- For $t = 1, ..., T$, solve

$$s \cdot \nabla u'^{(t)} + \kappa u'^{(t)} = f + \int_{\mathbb{S}^1} \sigma u'^{(t-1)} \mathrm{d}s'$$

using e.g. the DOM or SDOM based on the basic ridgelet RTE solver.

The idea of the source iterations is that the $u^{(t)}$ will converge to the true solution u for large enough t – which is what we observe numerically.

Since three-dimensional functions are difficult to visualize, we will only look at the *incident radiation*

$$G[u](x, y) := \int_{\mathbb{S}^1} u(x, y, s) \, \mathrm{d}s.$$

Figure 2 shows the incident radiation of solutions of to the complete radiative transfer problem with sources and sinks as in (a) – details below. Subfigures (b), (c) and (d) show the Fincident radiation for three different value of the scattering coefficient σ (0, 0.2 and 0.5, respectively). In Fig. 2a, the f and κ are illustrated: The red[3] line on the left shows the shape of the source term

$$f(x, y, \varphi) = \mathrm{e}^{-500\,(x-0.15)^2 - 10\min\{\varphi, 2\pi-\varphi\}^2}$$

for some constant φ. The light blue area in the middle represents the obstacle which corresponds to the second term in

$$\kappa(x, y) = 2 + 18\,\mathrm{e}^{-2000\,(x-0.4)^4 - 1000\,(y-0.5)^4} + 98\,\mathrm{e}^{-900\,(x-0.9)^2}.$$

The last term in κ, shown in dark blue on the right in Fig. 2a, was introduced in order to avoid that radiation flows across the y-boundary.

Acknowledgements. The second author gratefully acknowledges support for this work by the Swiss National Science Foundation, Project 146356.

---

[3] For colour images, see online version.

References

1. Borup, L., Nielsen, M.: Frame decomposition of decomposition spaces. J. Fourier Anal. Appl. **13**(1), 39–70 (2007)
2. Candès, E.: Ridgelets and the representation of mutilated sobolev functions. SIAM J. Math. Anal. **33**(2), 347–368 (2001)
3. Candès, E.: Ridgelets: theory and applications. Ph.D Thesis. Stanford University (1998)
4. Candès, E., Donoho, D.L.: Continuous curvelet transform: I. resolution of the wavefront set. Appl. Comput. Harmon. Anal. **19**(2), 162–197 (2005)
5. Candès, E., Donoho, D.L.: Continuous curvelet transform: II. discretization and frames. Appl. Comput. Harmon. Anal. **19**(2), 198–222 (2005)
6. Candès, E., et al.: Fast discrete curvelet transforms. Mult. Model. Simul. **5**, 861–899 (2006)
7. Cohen, A., Dahmen, W., DeVore, R.: Adaptive wavelet methods for elliptic operator equations: convergence rates. Math. Comp. **70**(233), 27–75 (2001)
8. Dahlke, S., Fornasier, M., Raasch, T.: Adaptive frame methods for elliptic operator equations. Adv. Comput. Math. **27**(1), 27–63 (2007)
9. Daubechies, I.: Ten lectures on wavelets. CBMS-NSF Regional Conference Series in Applied Mathematics. In: Society for Industrial and Applied Mathematics (SIAM), vol. 61, pp. xx+357. Philadelphia, PA (1992)
10. DeVore, R.: Nonlinear approximation. In: Acta numerica, 1998. Acta Numer. vol. 7, pp. 51–150. Cambridge University Press, Cambridge (1998)
11. Do, M.N., Vetterli, M.: The contourlet transform: an efficient directional multiresolution image representation. IEEE Trans. Image Proc. **14**, 2091–2106 (2005)
12. Etter, S., Grohs, P., Obermeier, A.: FFRT - A Fast Finite Ridgelet Transform for Radiative Transport. In: submitted (2014)
13. Grella, K., Schwab, Ch.: Sparse discrete ordinates method in radiative transfer. Comput. Methods Appl. Math. **11**(3), 305–326 (2011)
14. Grohs, P.: Ridgelet-type frame decompositions for Sobolev spaces related to linear transport. J. Fourier Anal. Appl. **18**(2), 309–325 (2012)
15. Grohs, P., Kutyniok, G.: Parabolic Molecules. Found. Comput. Math. **14**(2), 299–337 (2014)
16. Grohs, P., Obermeier, A.: Optimal Adaptive Ridgelet Schemes for Linear Transport Equations. In: Submitted (2014)
17. Grohs, P., Schwab, Ch.: Sparse twisted tensor frame discretization of parametric transport operators. In: Preprint available as a SAM Report (2011), ETH Zürich (2011) http://www.sam.math.ethz.ch/sam_reports/index.php?id=2011-41
18. Grohs, P., et al.: α-Molecules. In: Submitted (2014). Preprint available as a SAM Report (2014), ETH Zürich, http://www.sam.math.ethz.ch/sam_reports/index.php?id=2014-16
19. Kutyniok, G., Labate, D.: Shearlets: multiscale analysis for multivariate data. In: Birkhäuser (ed.) Chapter Introduction to Shearlets, pp. 1–38 (2012)
20. Kutyniok, G., et al.: Sparse multidimensional representation using shearlets. In: Wavelets XI(San Diego, CA), Procedings of SPIE vol. 5914, pp. 254–262 (2005)
21. Modest, M.F.: Radiative heat transfer. Academic press, San Diego (2013)
22. Obermeier, A., Grohs, P.: On the approximation of functions with line singularities by ridgelets. In: In preparation (2015)
23. Stevenson, R.: Adaptive solution of operator equations using wavelet frames. SIAM J. Numer. Anal. **41**(3), 1074–1100 (2003)

Basis Functions for Scattered Data Quasi-Interpolation

Nira Gruberger[1]([⊠]) and David Levin[2]

[1] Ruppin Academic Center, Emek Hefer, Israel
nirag@ruppin.ac.il
[2] School of Mathematical Sciences, Tel-Aviv University,
Tel Aviv-yafo, Israel

Abstract. Given scattered data of a smooth function in $\mathbb{R}^d$, we consider quasi-interpolation operators for approximating the function. In order to use these operators for the derivation of useful schemes for PDE solvers, we would like the quasi-interpolation operators to be of compact support and of high approximation power. The quasi-interpolation operators are generated through known quasi-interpolation operators on uniform grids, and the resulting basis functions are represented by finite combinations of box-splines. A special attention is given to point-sets of varying density. We construct basis functions with support sizes and approximation power related to the local density of the data points. These basis functions can be used in Finite Elements and in Isogeometric Analysis in cases where a non-uniform mesh is required, the same as T-splines are being used as basis functions for introducing local refinements in flow problems.

Keywords: Scattered Data · Quasi-Interpolation · Box-splines

1 Preliminaries

Consider the approximation of a smooth function on a closed domain $\Omega \subset \mathbb{R}^d$, $f : \Omega \to \mathbb{R}$, given its values at scattered points $\{f(a)\}_{a \in A}$, $A \subset \Omega$. We look for a quasi-interpolation approximation operator of the form

$$Qf(x) = \sum_{a \in A} f(a) q_a(x), \tag{1}$$

where $\{q_a\}_{a \in A}$ are the quasi-interpolation basis functions with the following properties:

1. Polynomial reproduction,

$$\sum_{a \in A} p(a) q_a(x) = p(x), \quad \forall x \in \Omega, \quad p \in \Pi_m, \tag{2}$$

where Π_m is the space of polynomials of total degree $\leq m$ in $\mathbb{R}^d$.

© Springer International Publishing Switzerland 2015
J.-D. Boissonnat et al. (Eds.): Curves and Surfaces 2014, LNCS 9213, pp. 263–271, 2015.
DOI: 10.1007/978-3-319-22804-4_19

2. Compact support,

$$supp(q_a) \subset B(a, Rh). \tag{3}$$

Here $B(a, r)$ is the open ball of radius r centered at a, and h is the fill-distance of the point-set A, namely, the radius of the largest open ball centered in Ω which does not contain any point from A. R is independent of $a \in A$ and h.

Definition 1. *Bound on derivatives of total order k:*

$$M_{k,\Omega}(f) = sup_{x \in \Omega}\{|\partial_k f(x)| \; : \; \partial_k = \partial_{x_1^{i_1}} \cdots \partial_{x_d^{i_d}} f \; , \; \sum_{j=1}^{d} i_j = k\} \tag{4}$$

Proposition 1. *Let $f \in C^{m+1}(\Omega)$, i.e., all partial derivatives of f of total order $\leq m + 1$ are continuous on Ω. If Q is a quasi-interpolation operator satisfying properties (2) and (3), then*

$$\|f - Qf\|_{\infty, \Omega} \leq C \cdot M_{m+1,\Omega}(f)h^{m+1}. \tag{5}$$

The proof can be found, e.g., in [1]. The most familiar example of such a quasi-interpolation operator is for $m = 1$ in $\mathbb{R}^2$. For simplicity we assume that the set A includes points on the boundary $\partial\Omega$ of Ω, and that the fill-distance on $\partial\Omega$ is $O(h)$. Using a Delaunay triangulation of Ω we define, for each $a \in A$, the corresponding basis function q_a over the first ring of triangles around a, as the piecewise linear 'hat function' centered at a. The resulting quasi-interpolation approximations are C^0, and they provide an $O(h^2)$ approximation power.

For the approximation of smooth functions over rectangular domains, the simplest approach is to use a uniform square grid, and the tensor product splines over this grid. However, in many applications (e.g., for PDE solvers) the geometry of the domain Ω is not that simple, and the functions to be approximated behave differently at different sub-domains of Ω. To take care of the inhomogeneity of the function, it is required to use a mesh of different densities at different sub-domains. e.g. in solving Laplace equation in an L-shaped domain, a finer mesh is required near the corner. See Fig. 1.

The desirable approximation result for a quasi-interpolation operator in the inhomogeneous case would be as follows:

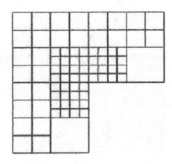

Fig. 1. Finite elements refined mesh near the corner.

Proposition 2. *Let $f \in C^{m+1}(\Omega)$, and $\Omega = \cup_{i \in I} \Omega_i$, and let h_i be the fill-distance of the data points A in Ω_i, $i = 1, ..., I$. Then, there exists a quasi-interpolation operator Q satisfying properties (2) and*

$$supp(q_a) \subset B(a, Rh_i), \quad a \in \Omega_i, \quad i = 1, ..., I. \tag{6}$$

Furthermore,

$$\|f - Qf\|_{\infty, \Omega_i} \leq C \cdot M_{m+1, \Omega_i}(f) h_i^{m+1}. \tag{7}$$

In the next section we describe the construction of quasi-interpolation operators for scattered data, of high smoothness and of high approximation power, following [1]. We continue by presenting the construction for the inhomogeneous case, in the spirit of Proposition 2. There we will introduce basis functions that can be used in Finite Elements and in Isogeometric Analysis in cases where a non-uniform mesh is required, the same as in [7], where T-splines are being used as basis functions for introducing local refinements in flow problems.

2 The Construction

We start with presenting the construction for uniform grids and then we will extend the theory to scattered data.

2.1 Construction of Quasi-interpolation Operators on Uniform Grids

The quasi-interpolation operator we will present, is based on Box-splines. In [5], Box-splines are also used for quasi-interpolation. We use a similar approach and bring different proofs to obtain our results. Box-splines [4] are well defined on a uniform grid $A = h\mathbb{Z}^d$, and for a specified degree m we can find a box-spline $M(x|X)$ such that

$$span\{M(\cdot/h - i|X)\}_{i \in \mathbb{Z}^d} = \Pi_m. \tag{8}$$

where X is the set of vectors $\{x_1, ..., x_n\}$, $x_i \in \mathbb{R}^d$, $\forall i = 1, ..., n$. Using these box-spline basis functions, $\{M(\cdot/h - i|X)\}_{i \in \mathbb{Z}^d}$ we would like to build a basis function Q_m of finite support. The shifts of Q_m, $\{Q_m(\cdot/h - i)\}_{i \in \mathbb{Z}^d}$, will serve as the basis functions for quasi-interpolation over $A = h\mathbb{Z}^d$:

$$Q^{h\mathbb{Z}^d} f(x) = \sum_{i \in \mathbb{Z}^d} f(ih) Q_m(x/h - i), \tag{9}$$

with the property

$$Q^{h\mathbb{Z}^d} p = p, \quad \forall p \in \Pi_m. \tag{10}$$

One way of constructing Q_m is described in [2]. In the following we present another approach for deriving Q_m.

Definition 2. *Support size of Q_m:*

$$\rho(Q_m) = sup_x\{|x| : Q_m(x) \neq 0\} \tag{11}$$

The following propositions can be found in [4].

Proposition 3. *The box-spline $M(x|X)$ is a piecewise polynomial that satisfies*

$$deg\{M(x|X)\} \leq n - d. \tag{12}$$

where n is the number of vectors $x_i \in X$ and d is the dimension of the space $(x_i \in \mathbb{R}^d)$.

Proposition 4. *The box-spline $M(x|X)$ is of class C^{s-1}, where $s + 1$ is the smallest of the cardinalities of the subsets $Y \subset X$, such that $X \setminus Y$ does not generate $\mathbb{R}^d$.*

Proposition 5. *Let D_v denote the differentiation operator in the direction of a vector $v \in X$ and suppose $X \setminus v$ generates $\mathbb{R}^d$. Then $D_v M(x|X) = \nabla_v M(x|X \setminus v)$ where ∇_v is the backwards difference operator in the direction of v.*

We define Schoenberg operator T^X acting on the space of smooth functions f, $f : \Omega \subset \mathbb{R}^d \to R$, where X is the set of vectors $\{x_1, ..., x_n\}$, $x_i \in \mathbb{R}^d, \forall i = 1, ..., n$:

$$T^X\{f\}(x) = \sum_{i \in \mathbb{Z}^d} f(i)M(x - i|X) \tag{13}$$

Suppose the box-splines $M(x|X)$ are of class C^{s-1} such that $s = m$ and $n - d = m$. We denote $\Pi_X = span\{M(\cdot - i|X)\}$. The following theorem describes a property of the operator T^X.

Theorem 1. *The operator T^X satisfies the following properties: $\forall p \in span\{M(\cdot - i|X)\}$, $p : \mathbb{R}^d \to R$:*

- *The operation $T^X p - p$, reduces the degree of p, i.e. $deg(T^X p - p) = deg(p) - 1$.*
- *$T^X p - p \in span\{M(\cdot - i|X \setminus v)\}, \forall v \in X$ such that $X \setminus v$ generates $\mathbb{R}^d$.*

Proof. We will prove the theorem for $d = 2$. The proof can be extended to higher values of d. Assume $v \in X$, $v = \delta_1 e_1 + \delta_2 e_2$ and apply the directional derivative operator D_v on (13). From Proposition 5 we will have: $D_v T^X\{f\}(x) = \sum_{i \in \mathbb{Z}^d} f(i) \nabla_v M(x - i|X \setminus v)$, where ∇ is the backwards difference operator: $\nabla_w f(x) = f(x) - f(x - w)$. It can be written as: $D_v T\{f\}(x) = \sum_{i \in \mathbb{Z}^d}[f(i) - f(i - (\delta_1, \delta_2))]M(x - i|X \setminus v)$. Hence, $D_v T^X\{1\}(x) = 0$ and consequently $T^X\{1\} = const$. But due to the partition of unity property of box-Splines, we get: $T^X\{1\} - 1 = 0$. The proof for $p \in \Pi_r, 0 < r \leq s$, is by induction on r as follows. Assume $T^X\{p\} - p \in span\{M(\cdot - i|X \setminus v)\}$, where $p \in span\{M(\cdot - i|X)\}$ and $deg(p) \leq r - 1 < s$. We will prove for $deg(p) \leq r \leq s$. Let $p = p(x, y) = x^k \cdot y^{r-k}$. Then, $T^X\{p\} = \sum_{i=(i_1,i_2)} i_1^k \cdot i_2^{r-k} M(x - (i_1, i_2)|X)$. Let $v \in X$ and assume $v = \delta_1 e_1 + \delta_2 e_2$. Applying the operator D_v on the last identity we get: $D_v T^X \{p\} = \sum_{i=(i_1,i_2)} g(i_1,i_2) M(x - (i_1,i_2)|X \setminus v)$, where $g(i_1,i_2) = i_1^k \cdot i_2^{r-k} - (i_1 - \delta_1)^k \cdot (i_2 - \delta_2)^{r-k}$. But it can be shown that

$$g(i_1,i_2) = (r-k) \cdot \delta_2 \cdot i_1^k \cdot i_2^{r-k-1} + k \cdot \delta_1 \cdot i_1^{k-1} \cdot i_2^{r-k} + p_{r-2} \qquad (14)$$

where p_{r-2} is a polynomial of total degree $r - 2$. Thus, $g(i_1,i_2)$ can be written as: $g(i_1,i_2) = p_{r-1} + p_{r-2}$ and is of total degree $r - 1$. Thus, we can write

$$D_v T^X \{p\} = \sum_{i=(i_1,i_2)} [p_{r-1}(i_1,i_2) + p_{r-2}(i_1,i_2)] M(x - (i_1,i_2)|X \setminus v). \qquad (15)$$

Also, p_{r-1} has the following form:

$$p_{r-1}(i_1,i_2) = \begin{cases} r \cdot \delta_2 \cdot i_2^{r-1} & \text{if } k = 0 \\ (r-k) \cdot \delta_2 \cdot i_1^k \cdot i_2^{r-k-1} + k \cdot \delta_1 \cdot i_2^{r-k} \cdot i_1^{k-1} & \text{if } 0 < k < r \\ k \cdot \delta_1 \cdot i_1^{r-1} & \text{if } k = r \end{cases} \qquad (16)$$

It can be shown that p_{r-1} in (16), satisfies the following:

$$p_{r-1}(x,y) = D_v p \qquad (17)$$

where $p = x^k \cdot y^{r-k}$. Now, from (15) and (16) and the definition of the operator T^X, we obtain: $D_v T^X \{x^k \cdot y^{r-k}\} = \sum_{i=(i_1,i_2)} p_{r-1} M(x - (i_1,i_2)|X \setminus v) + \sum_{i=(i_1,i_2)} p_{r-2} M(x - (i_1,i_2)|X \setminus v)$. Thus: $D_v T^X \{x^k \cdot y^{r-k}\} = T^{X \setminus v} p_{r-1} + T^{X \setminus v} p_{r-2}$. And by the induction assumption for both p_{r-1} and p_{r-2}, we have: $D_v T^X \{x^k \cdot y^{r-k}\} = p_{r-1} + \hat{p}_{r-2} + p_{r-2} + \hat{p}_{r-3}$. Hence, it can be written in the form: $D_v T^X \{x^k \cdot y^{r-k}\} = p_{r-1} + \tilde{p}_{r-2}$. But from (16), $p_{r-1} = D_v \{x^k \cdot y^{r-k}\}$ and hence $D_v [T^X \{x^k \cdot y^{r-k}\} - x^k \cdot y^{r-k}] = \tilde{p}_{r-2}$ and therefore: $T^X \{x^k \cdot y^{r-k}\} - x^k \cdot y^{r-k} = \tilde{p}_{r-1}$, where $\tilde{p}_{r-1} \in span \, M(\cdot - (i_1,i_2)|X \setminus v)$ and this ends the proof.

From the results of Theorem 1, we will introduce a method to construct a quasi-interpolation operator based on Box-splines. Employing the operator T^X defined in (13), we define another operator R as follows:

$$R = I - T^X. \qquad (18)$$

From Theorem 1, the operator R in (18) satisfies the following property:

$$R : \Pi_k \to \Pi_{k-1} \qquad (19)$$

and consequently, if $p \in \Pi_k$, we have $R^2 \{p\} \in \Pi_{k-2}, ..., R^{k+1}\{p\} = 0$. And next we introduce a new operator:

$$\hat{T} = I + R + ... + R^k. \qquad (20)$$

The operator $\hat{T}$ in (20) satisfies the following: $T^X \cdot \hat{T} = (I - R) \cdot (I + R + ... + R^k)$ and hence

$$T^X \cdot \hat{T} \{p\} = p, \text{ if } p \in \Pi_k. \qquad (21)$$

From (21) we realize that $T^X \cdot \widehat{T}$ is a quasi interpolation operator since it reproduces polynomials in Π_k and has a finite support. In order to evaluate the quasi interpolation operator for a given Box-spline and given d and m, we substitute integer values $j \in \mathbf{Z}^d$ in the operator T^X in (13) and obtain: $T^X\{f\}(j) = \sum_{i \in \mathbf{Z}^d} f(i) \cdot M(j - i|X)$ for $j \in \mathbf{Z}^d$. Using the box spline values at the integers we can evaluate: $R\{f\}(j) = (I - T^X)\{f\}(j) = \{f\}(j) - \sum_{i \in \mathbf{Z}^d} f(i) \cdot M(j - i|X)$. Then $\widehat{T}$ can be computed by (20) and the quasi interpolation operator is:
$T^X \cdot \widehat{T}\{f\}(x) = \sum_{i \in \mathbf{Z}^d} \widehat{T}\{f\}(i)M(x - i|X)$.

Example 1. Let $M(x|X)$ be the box-spline with $d = 2$, associated with the
Matrix: $X = \begin{pmatrix} 1\,1\,0\,0\,1\,1 \\ 0\,0\,1\,1\,1\,1 \end{pmatrix}$.

The nonzero values of the box-spline at the integers are: $M(0, 0) = \frac{1}{2}$, $M(-1, -1) = M(-1, 0) = M(0, 1) = M(1, 1) = M(1, 0) = M(0, -1) = \frac{1}{12}$.

Therefore,

$T^X\{f\}(i, j) = \frac{1}{2}f(i, j) + \frac{1}{12}\{f(i+1, j+1) + f(i+1, j) + f(i, j-1) + f(i-1, j-1) + f(i-1, j) + f(i, j+1)\}$

and R defined in (18) is:

$R\{f\}(i, j) = \frac{1}{2}f(i, j) - \frac{1}{12}\{f(i+1, j+1) + f(i+1, j) + f(i, j-1) + f(i-1, j-1) + f(i-1, j) + f(i, j+1)\}$

and the operator $\widehat{T}$ in (21) is:

$\widehat{T}\{f\}(i, j) = \frac{3}{2}f(i, j) - \frac{1}{12}\{f(i+1, j+1) + f(i+1, j) + f(i, j-1) + f(i-1, j-1) + f(i-1, j) + f(i, j+1)\}$,

since for $f \in \Pi_3, R^2\{f\}(x) = f^{(4)}(x) + O(f^{(5)})$

2.2   Moving from Uniform Grids to Scattered Data

In this section we derive the quasi-interpolation basis functions for the case of scattered point-set A. The construction follows the approach in [1]. The difference is in the assumption on the point-set A, which is less restrictive here. Another difference lies in the consideration of a bounded domain Ω, compared to $\Omega = \mathbb{R}^d$ treated in [1]. Let A be a point-set in Ω with a fill-distance h. For a given degree m, following [3], there exists an r_m, $r_m = O(m)$, such that for any $x \in \Omega$ such that $B(x, r_m h) \subset \Omega$, the following least-squares approximation is well defined:

$$p^* = argmin_p\{ \sum_{a \in B(x, r_m h)} [f(a) - p(a)]^2 \; : \; p \in \Pi_m\}. \tag{22}$$

Lemma 1. *(Following [3]) The linear operator* $L : \{f(a)\}_{a \in B(x, r_m h)} \to p^*$ *is uniformly bounded.*

Consider a subset of the uniform mesh $h\mathbb{Z}^d$,

$$E = \{e \in h\mathbb{Z}^d \ : \ dist(e, \Omega) < \rho(Q_m)h\}. \tag{23}$$

The following result is deduced.

Lemma 2. *The quasi-interpolation operator on E reproduces Π_m on Ω:*

$$\sum_{e \in E} p(e)Q_m(x/h - e/h) = p(x), \quad \forall x \in \Omega. \tag{24}$$

Definition 3. *Approximation sets $\{N_e\}_{e \in E}$:*

1. For $e \in E$, let $x_e \in \Omega$ be the closest point to e such that $B(x_e, r_m h) \subset \Omega$.
2. Denote by L_e the line segment connecting e to x_e.
3.

$$N_e = \{a \ : \ dist(a, L_e) < r_m h\}. \tag{25}$$

We are now ready to construct the basis function $\{q_a\}_{a \in A}$ for the quasi-interpolation operator.

For every $e \in E$, let us compute the approximation to $f(e)$ by the least-squares approximation to f from Π_m, using the data on N_e. Let $\{b_j\}_{j=1}^J$ be a basis for Π_m, and let $n_e = \#\{N_e\}$, the number of data points in N_e. Finding the coefficients of the least-squares polynomial is equivalent to the least-squares solution of the over determined system

$$A\bar{\beta} = \bar{f}. \tag{26}$$

Here A is a matrix of size $n_e \times J$, $A_{a,j} = b_j(a)$, $\bar{\beta}$ is the vector of coefficients of the required polynomial, $p^* = \sum_{j=1}^J \beta_j b_j$ and $\bar{f}$ is the vector of values indexed by the points $a \in N_e$, $\bar{f}_a = f(a)$. The solution is

$$\bar{\beta} = (A^t A)^{-1} A^t \bar{f}. \tag{27}$$

Evaluating p^* gives the approximation to $f(e)$, which turns to be

$$p^*(e) = \bar{b}^t (A^t A)^{-1} A^t \bar{f} \equiv \sum_{a \in N_e} w_{e,a} f(a), \tag{28}$$

where $\bar{b}_j = b_j(e)$, $j = 1, ..., J$. Evidently, the approximation is exact for $f \in \Pi_m$, which gives the desired coefficients $\{w_{e,a}\}_{a \in N_e}$ such that

$$\sum_{a \in N_e} w_{e,a} p(a) = p(e) , \quad \forall p \in \Pi_m. \tag{29}$$

Setting $w_{e,a} = 0$ for $a \notin N_e$, we define the new basis functions for quasi-interpolation $\{q_a\}_{a \in A}$:

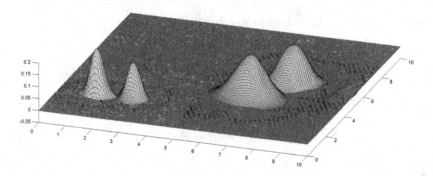

Fig. 2. Four of the quasi-interpolation basis functions in a scattered data situation. Notice that the support size of the basis functions matches the local fill distance of the data points.

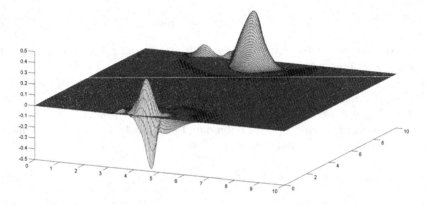

Fig. 3. Two of the quasi-interpolation basis functions near the boundaries of the domain in a scattered data situation.

Definition 4. *The quasi-interpolation basis functions* $\{q_a\}_{a \in A}$ *are:*

$$q_a = \sum_{e \in E} w_{e,a} Q_m(\cdot/h - e) \tag{30}$$

where $w_{e,a}$ *are evaluated by (29) and* $w_{e,a} = 0$ *for* $a \notin N_e$.

Theorem 2. *The functions* $\{q_a\}_{a \in A}$ *are spline functions on a uniform grid, with the following properties:*

1. *A bounded support size*

$$supp(q_a) \subset B(a, (\rho(Q_m) + r_m)h). \tag{31}$$

2. *Reproduction of* Π_m

$$\sum_{a \in A} p(a) q_a(x) = p(x), \quad \forall x \in \Omega, \quad p \in \Pi_m, \tag{32}$$

Proof.

$$\sum_{a \in A} p(a) q_a(x) = \sum_{a \in A} p(a) \sum_{e \in E} w_{e,a} Q_m(x/h - e)$$

$$= \sum_{e \in E} Q_m(x/h - e) \sum_{a \in A} w_{e,a} p(a) \qquad (33)$$

$$= \sum_{e \in E} Q_m(x/h - e) p(e) = p(x).$$

Figure 2 depicts basis functions built on scattered data with 2 different fill-distances, for each of the areas the figure shows 2 basis functions. Similarly, Fig. 3 depicts 2 basis functions near the boundaries of the domain built on scattered data with 2 different fill-distances. The least squares approximation was based on the formulae for the Moving Least Squares Method described in [6]. In Moving Least-Squares approximation (MLS) one has to form a linear system such as as (26) and to solve it for any point where the approximation is required and there is no convenient closed form for the approximation. In the quasi-interpolation approach, one solves once a linear system for any knot-mesh point of the spline and then one forms the quasi-interpolation basis functions. Since these basis functions are combinations of B-splines, they are easily evaluated and easily differentiated and integrated for applications in Finite Elements.

References

1. Buhmann, M.D., Dyn, N., Levin, D.: On quasi-interploation by radial basis functions with scattered centers. Constr. Approx. **11**(2), 239–254 (1995)
2. de Boor, C., Hollig, K.: B-splines from parallelepipeds. J. d'Analyse Math. **42**, 99–115 (1982–1983)
3. Wendland, H.: Scattered Data Approximation. Cambridge University Press, Cambridge (2005)
4. Risler, J.J.: Mathematical Methods for CAD. Cambridge University Press, Cambridge (1993)
5. de Boor, C., Hollig, K., Riemenschneider, S.: Box Splines. Springer, New York (1993)
6. Levin, D.: The approximation power of moving least-squares. Math. Comp. **67**(224), 1517–1531 (1998)
7. Bazilevs, Y., Calo, V.M., Cottrell, J.A., Evans, J.A., Hughes, T.J.R., Lipton, S., Scott, M.A., Sederberg, T.W.: Isogeometric analysis using T-splines. Comput. Meth. Appl. Mech. Eng. **199**(5), 229–263 (2010). Elsevier

Mass Smoothers in Geometric Multigrid for Isogeometric Analysis

Clemens Hofreither[✉] and Walter Zulehner

Institute of Computational Mathematics, Johannes Kepler University Linz,
Altenbergerstr. 69, 4040 Linz, Austria
chofreither@numa.uni-linz.ac.at

Abstract. We investigate geometric multigrid methods for solving the
large, sparse linear systems which arise in isogeometric discretizations of
elliptic partial differential equations. In particular, we study a smoother
which incorporates the inverse of the mass matrix as an iteration matrix,
and which we call mass-Richardson smoother. We perform a rigorous
analysis in a model setting and perform some numerical experiments to
confirm the theoretical results.

1 Introduction

Isogeometric analysis (IGA), a numerical technique for the solution of partial
differential equations first proposed in [12], has attracted considerable research
attention in recent years. The use of spline spaces both for representation of
the geometry and for approximation of the solution affords the method several
very interesting features, such as the possibility to use exactly the geometry
generated by CAD systems, refinement without further communication with the
CAD system, the possibility of using high-continuity trial functions, the use of
high-degree spaces with comparatively few degrees of freedom, and more. We
refer to [1,12] as well as the monograph [5] and the references therein for an
overview of the topic.

The efficient solution of the discretized systems arising in isogeometric analysis has been the topic of several publications in recent years, e.g., [3,4,6–9,13].
In the present paper, our interest lies in geometric multigrid methods for isogeometric analysis. It is known by now that the simple classical multigrid smoothers
do not result in multigrid solvers with convergence rates which are robust in the
spline degree of the IGA discretization. In this paper, motivated by promising
results of a preliminary local Fourier analysis, we propose a smoother based on
inverting the mass matrix. We perform a rigorous analysis in a model setting and
find that, due to boundary effects, the smoother does not achieve total robustness in the spline degree with a single smoothing step, but requires a number
of additional smoothing steps. We are able to quantify the needed number of
smoothing steps for robust convergence. To our knowledge, this is the first rigorous analysis of a multigrid method for IGA which takes the spline degree into
account explicitly.

© Springer International Publishing Switzerland 2015
J.-D. Boissonnat et al. (Eds.): Curves and Surfaces 2014, LNCS 9213, pp. 272–279, 2015.
DOI: 10.1007/978-3-319-22804-4_20

The remainder of the paper is structured as follows. In Sect. 2, we outline a simple model problem and a geometric multigrid solver for IGA. In Sect. 3, we perform the analysis of the two-grid method with our proposed mass-Richardson smoother. In Sect. 4, we report the results of some numerical experiments which confirm the theory.

2   Geometric Multigrid for Isogeometric Analysis

For a detailed description of the IGA methodology, see, e.g., [12]. For the sake of simplicity, we consider here only a simple model problem with a trivial geometry map. Previous numerical experiments indicate that nontrivial but well-behaved geometry maps do not significantly impact the convergence behaviour of multigrid methods for IGA.

Let $\mathcal{V}_h \subset H_0^1(\Omega)$ denote a tensor product B-spline space over $\Omega = (0,1)^d$. We use here spline spaces with open knot vectors which have the same spline degree p in each coordinate direction, and which have the same smoothness parameter k in each coordinate direction, where $k \in \{1,\ldots,p\}$ describes splines which are globally C^{k-1}. We also assume that the spline spaces are quasi-uniform in the sense that the minimum knot span in any direction can be bounded from below by some uniform constant times the maximum knot span.

We consider an IGA discretization of the Poisson equation with pure Dirichlet boundary conditions: find $u_h \in \mathcal{V}_h$ such that

$$a(u_h, v_h) = \langle F, v_h \rangle \qquad \forall v_h \in \mathcal{V}_h$$

with the bilinear form and linear functions, respectively,

$$a(u,v) = \int_\Omega \nabla u \cdot \nabla v \, dx, \qquad \langle F, v \rangle = \int_\Omega f v \, dx - a(\tilde{g}, v).$$

Here $\tilde{g} \in H^1(\Omega)$ is a suitable extension of the given Dirichlet data.

In the following, we outline the construction of a simple geometric multigrid scheme for this problem. Let $\mathcal{V}_0$ denote a coarse tensor product spline space over $(0,1)^d$. Performing uniform and global h-refinement by knot insertion, we obtain a sequence of refined spline spaces $\mathcal{V}_1, \mathcal{V}_2, \ldots$ Let $\mathcal{V}_H$ and $\mathcal{V}_h$ denote two successive spline spaces in this sequence, and let $P : \mathcal{V}_H \to \mathcal{V}_h$ denote the prolongation operator from the coarse to the fine space. One step of the two-grid iteration process is given by a pre-smoothing step, the coarse-grid correction, and a post-smoothing step; i.e., given $u_0 \in \mathcal{V}_h$, the next iterate u_1 is obtained by

$$u^{(1)} := u_0 + S^{-1}(f_h - A_h u_0),$$
$$u^{(2)} := u^{(1)} + PA_H^{-1}P^\top(f_h - A_h u^{(1)}),$$
$$u_1 := u^{(3)} := u^{(2)} + S^{-\top}(f_h - A_h u^{(2)}).$$

Here, S is a suitable smoother for the fine-space stiffness matrix A_h.

As usual, a multigrid scheme is obtained by considering a hierarchy of nested spline spaces and replacing the exact inverse A_H^{-1} in the above procedure recursively with the same procedure applied on the next coarser space, until V_0 is reached. We consider only the case of a single coarse-grid correction step, i.e., the V-cycle.

It is known that this multigrid algorithm with standard smoothers like the Richardson, Jacobi or Gauss-Seidel smoothers is robust in the mesh size h. However, the same is not true for the spline degree p: especially in higher space dimensions d, the iteration numbers for obtaining a desired accuracy increase strongly with p. This soon results in iteration numbers which are no longer practical.

It is therefore of interest to find smoothers which result in better or even completely robust iteration numbers with respect to p. An interesting choice is what we will call the mass-Richardson smoother,

$$S = \tau^{-1} M,$$

where M denotes the mass matrix and τ a real damping parameter. Local Fourier analysis suggests that this smoother should lead to convergence rates which are independent of p. However, as we will see, in actual boundary value problems a certain dependence on p remains due to boundary effects.

We note that applying this smoother requires inverting the isogeometric mass matrix in every iteration. Although the mass matrix is itself ill-conditioned for higher spline degrees p, an efficient approach for inverting it has been described in [10] by exploiting the tensor product structure of the spline spaces.

3   Analysis

We follow the ideas of the multigrid convergence theory as given by Hackbusch [11]. For simplicity, we restrict ourselves to the analysis of the two-grid method in either one or two space dimensions. From this, the convergence of the W-cycle follows by a perturbation argument, but the convergence of the V-cycle needs a different proof technique. Nevertheless, in practice we have observed that the V-cycle performs similarly to the two-grid method.

For any function u defined on the fine grid, we will write $\underline{u}$ for its coefficient vector in the fine-grid B-spline basis. In addition to the L_2-norm

$$\|u\|_0 := \|u\|_{L_2(\Omega)} = \|\underline{u}\|_M,$$

we will make use of the H^2-like norm

$$\|u\|_2 := \sup_{w \in V} \frac{a(u, w)}{\|w\|_0} = \sup_{\underline{w}} \frac{\underline{w}^\top A \underline{u}}{(\underline{w}^\top M \underline{w})^{1/2}} = \sup_{\underline{w}} \frac{\underline{w}^\top M^{-1/2} A \underline{u}}{(\underline{w}^\top \underline{w})^{1/2}}$$

$$= \sup_{\|\underline{w}\|=1} \underline{w}^\top M^{-1/2} A \underline{u} = \|M^{-1/2} A \underline{u}\| = \|A\underline{u}\|_{M^{-1}} =: \|\underline{u}\|_2.$$

By C, we will denote a generic constant which does not depend on the discretization parameters h, p and k.

For purposes of the analysis, we will always make the choice $\tau = 1/\lambda_{\max}(M^{-1}A)$ for the damping parameter.

We point out that we make use of the explicit spline approximation error estimates presented in [2]. So far, these results are limited to the case of relatively low-continuity splines, and this limitation therefore extends to our work.

3.1   Smoothing Property

We will make use of the following polynomial inverse inequalities or Markov-type inequalities (see, e.g., Schwab [14]).

Theorem 1. *Let $d = 1$, $I = (a, b)$ and $h = b - a$, then for any polynomial f of degree at most p, we have*

$$\|f'\|_{L_2(I)} \leq 2\sqrt{3}\frac{p^2}{h}\|f\|_{L_2(I)}$$

Let $d = 2$. For an arbitrary quadrilateral $I = (a, b) \times (c, d)$ with size $h_1 = b - a$, $h_2 = d - c$, setting $h := \max\{h_1, h_2\}$ and assuming $h \leq C\min\{h_1, h_2\}$ with a uniformly bounded constant C, we have

$$\|\nabla f\|_{L_2(I)} \leq C\frac{p^2}{h}\|f\|_{L_2(I)}. \tag{1}$$

Theorem 2. *After ν steps of mass-Richardson smoothing, the resulting error $\underline{e}^{(\nu)} = (I - \tau M^{-1}A)^\nu \underline{e}$ satisfies*

$$\|\underline{e}^{(\nu)}\|_2 \leq \frac{Cp^4}{h^2\nu}\|\underline{e}^{(0)}\|_M.$$

Proof. By iterating the statement of Theorem 1 over every non-empty knot span of the spline space, we obtain for any $f \in C(\Omega) \cap L_2(\Omega)$ which is piecewise a polynomial of maximum degree at most p, and in particular for $f \in \mathcal{V}_h$,

$$\|\nabla f\|_{L_2(\Omega)} \leq C\frac{p^2}{h}\|f\|_{L_2(\Omega)}.$$

It follows that $\langle A\underline{v}, \underline{v}\rangle \leq C\frac{p^4}{h^2}\langle M\underline{v}, \underline{v}\rangle$ and therefore $\lambda_{\max}(M^{-1}A) \leq C\frac{p^4}{h^2}$.

We have

$$
\begin{aligned}
\|A\underline{e}^{(\nu)}\|_{M^{-1}} &= \|A(I - \tau M^{-1}A)^\nu\underline{e}\|_{M^{-1}} \\
&= \|M^{-1}A(I - \tau M^{-1}A)^\nu\underline{e}\|_M \\
&= \tfrac{1}{\tau}\|\tau M^{-1}A(I - \tau M^{-1}A)^\nu\underline{e}\|_M \\
&= \tfrac{1}{\tau}\|\tau M^{-1/2}AM^{-1/2}(I - \tau M^{-1/2}AM^{-1/2})^\nu M^{1/2}\underline{e}\| \\
&\leq \tfrac{1}{\tau}\|X(I - X)^\nu\|\|\underline{e}\|_M \\
&= \tfrac{1}{\tau}\max\{\lambda(1 - \lambda)^\nu : \lambda \in \sigma(X)\}\|\underline{e}\|_M
\end{aligned}
$$

with the symmetric matrix $X = \tau M^{-1/2}AM^{-1/2}$ which has the same spectrum as $\tau M^{-1}A$. Thus, $\sigma(X) \subset [0,1]$, and with the estimate $\lambda(1-\lambda)^{\nu} \leq 1/(e\nu)$ for $\lambda \in [0,1]$, we obtain

$$\|A\underline{e}^{(\nu)}\|_{M^{-1}} \leq \frac{C\lambda_{\max}(M^{-1}A)}{\nu}\|\underline{e}\|_M \leq \frac{Cp^4}{h^2\nu}\|\underline{e}\|_M. \qquad \square$$

3.2 Approximation Property

We summarize some recent results from [2] on the approximation properties of spline spaces with explicit dependence on h, p and k.

Theorem 3. *Assume that, if $d = 1$, then $k \leq \sigma \leq p+1$, or if $d = 2$, then $2k \leq \sigma \leq p+1$. There exists a spline interpolation operator $\Pi : H^{\sigma}(\Omega) \to \mathcal{V}_h$ such that for all $v \in H^{\sigma}(\Omega)$ and $j = 0, \ldots, \sigma$, we have*

$$|v - \Pi v|_{H^{\ell}(\Omega)} \leq C(p-k+1)^{-(\sigma-\ell)}h^{\sigma-\ell}|v|_{H^{\sigma}(\Omega)}.$$

Theorem 4. *Under the assumptions of Theorem 3 and full H^2-regularity of the boundary value problem, the errors e before and e^{CGC} after the coarse-grid correction step satisfy*

$$\|\underline{e}^{CGC}\|_M \leq Ch^2(p-k+1)^{-2}\|\underline{e}\|_2.$$

Proof. With $u_0 = u - e$ being the iterate before the coarse-grid correction step, the correction $t_C \in V_C$ such that $e^{CGC} = e - t_C$ is given by

$$a(t_C, w_C) = \langle F, w_c \rangle - a(u_0, w_C) = a(e, w_C) \qquad \forall w_C \in V_C.$$

For an arbitrary bounded linear functional $F^* : L_2(\Omega) \to \mathbb{R}$, we introduce the solutions of the dual problems $\tilde{\xi} \in H_0^1(\Omega)$, $\xi \in V$, and $\xi_C \in V_C$, respectively, by

$$a(\tilde{x}, \tilde{\xi}) = \langle F^*, \tilde{x} \rangle \qquad \forall \tilde{x} \in H_0^1(\Omega),$$
$$a(x, \xi) = \langle F^*, x \rangle \qquad \forall x \in V,$$
$$a(x_C, \xi_C) = \langle F^*, x_C \rangle \qquad \forall x_C \in V_C.$$

We have the identity

$$\langle F^*, e - t_C \rangle = a(e - t_C, \xi) = a(e, \xi) - \langle F^*, t_C \rangle = a(e, \xi - \xi_C)$$

and thus

$$|\langle F^*, e - t_C \rangle| \leq \frac{|a(e, \xi - \xi_C)|}{\|\xi - \xi_C\|_0}\|\xi - \xi_C\|_0 \leq \sup_{w \in V}\frac{|a(e, w)|}{\|w\|_0}\|\xi - \xi_C\|_0.$$

Using Theorem 3 and a standard Nitsche duality argument under the assumption of full H^2-regularity, we find that

$$\|\xi - \xi_C\|_0 \leq \|\xi - \tilde{\xi}\|_0 + \|\tilde{\xi} - \xi_C\|_0 \leq Ch^2(p-k+1)^{-2}\|F^*\|_*.$$

From the above two inequalities, it follows that

$$\|e^{CGC}\|_0 = \|e - t_C\|_0 = \sup_{F^*}\frac{|\langle F^*, e - t_C \rangle|}{\|F^*\|_*} \leq Ch^2(p-k+1)^{-2}\sup_{w \in V}\frac{a(e, w)}{\|w\|_0}. \qquad \square$$

3.3   Two-Grid Convergence Result

Theorem 5. *Under the assumptions of Theorem 4, the errors $e^{(0)}$ and $e^{(1)}$ before and after one two-grid cycle with ν mass-Richardson presmoothing steps and no postsmoothing satisfy*

$$\|e^{(1)}\|_0 \leq \frac{Cp^4}{\nu(p-k+1)^2}\|e^{(0)}\|_0.$$

In particular, there exists a positive constant C_ν independent of h, p and k such that the choice

$$\nu \geq C_\nu p^2$$

for the number of smoothing steps guarantees a convergence rate $\sigma \in (0,1)$ such that $\|e^{(1)}\|_0 \leq \sigma\|e^{(0)}\|_0$ with σ independent of h, k and p.

Proof. The first estimate follows directly by combining the statements of Theorem 2 and Theorem 4. By the assumptions of Theorem 4, we can conclude $p - k + 1 \geq p/2$ and hence $p^4/(p-k+1)^2 \leq Cp^2$, which proves the second statement.   □

4   Numerical Experiments

We solve the described model Poisson problem with an exact solution $u(x) = \prod_{i=1}^d \sin(\pi(x_i + 0.5))$ using two-grid iteration. We start with a random starting vector and iterate until the initial residual is reduced by a factor of 10^{-8} in the ℓ_2-norm. The obtained iteration numbers are more or less independent of the mesh size, and we do therefore not report numbers for different h. The problem sizes were relatively small with at most a few thousand degrees of freedom.

Both in 1D (Table 1) and in 2D (Table 2), the iteration numbers for p^2 smoothing steps remain uniformly bounded as we increase p. This confirms the theory with a choice of the constant $C_\nu = 1$ from Theorem 5. For comparison, we have also included the choice $\nu = p$, which is not supported by theory but serves as an interesting comparison.

Table 1. Iteration numbers in 1D. From left to right: spline degree p, smoothness parameter k, damping parameter τ; iteration numbers for $\nu = p^2$, $\nu = p$ smoothing steps.

p	k	τ	ν	iter.	ν	iter.
1	1	0.13	1	34	1	34
2	2	0.16	4	9	2	17
3	2	0.16	9	5	3	13
4	2	0.15	16	4	4	11
5	2	0.14	25	4	5	10
6	2	0.13	36	3	6	13

Table 2. Iteration numbers in 2D. From left to right: spline degree p, smoothness parameter k, damping parameter τ; iteration numbers for $\nu = p^2$, $\nu = p$ smoothing steps.

p	k	τ	ν	iter.	ν	iter.
1	1	0.08	1	65	1	65
2	2	0.10	4	15	2	30
3	2	0.09	9	8	3	23
4	2	0.09	16	5	4	44
5	2	0.08	25	4	5	16

We do not include CPU times here as we have not implemented the optimal mass matrix inversion algorithm from [10] at present, and therefore the results would be skewed towards lower smoothing numbers. We also note that the iteration numbers can be reduced, often significantly so, by replacing the two-grid iteration with CG iteration preconditioned by one two-grid cycle.

Acknowledgments. This work was supported by the National Research Network "Geometry + Simulation" (NFN S117, 2012–2016), funded by the Austrian Science Fund (FWF).

References

1. Bazilevs, Y., Beirão da Veiga, L., Cottrell, J.A., Hughes, T.J.R., Sangalli, G.: Isogeometric analysis: approximation, stability and error estimates for h-refined meshes. Math. Models Methods Appl. Sci. **16**(07), 1031–1090 (2006)
2. Beirão da Veiga, L., Buffa, A., Rivas, J., Sangalli, G.: Some estimates for h-p-k-refinement in isogeometric analysis. Numer. Math. **118**(2), 271–305 (2011)
3. Buffa, A., Harbrecht, H., Kunoth, A., Sangalli, G.: BPX-preconditioning for isogeometric analysis. Comput. Methods Appl. Mech. Eng. **265**, 63–70 (2013)
4. Collier, N., Pardo, D., Dalcin, L., Paszynski, M., Calo, V.M.: The cost of continuity: a study of the performance of isogeometric finite elements using direct solvers. Comput. Methods Appl. Mech. Eng. **213–216**, 353–361 (2012)
5. Cottrell, J.A., Hughes, T.J.R., Bazilevs, Y.: Isogeometric Analysis: Toward Integration of CAD and FEA. Wiley, Chichester (2009)
6. Beirão da Veiga, L., Cho, D., Pavarino, L., Scacchi, S.: Overlapping Schwarz methods for isogeometric analysis. SIAM J. Num. Anal. **50**(3), 1394–1416 (2012)
7. Beirão da Veiga, L., Cho, D., Pavarino, L., Scacchi, S.: BDDC preconditioners for isogeometric analysis. Math. Models Methods Appl. Sci. **23**(6), 1099–1142 (2013)
8. Beirão da Veiga, L., Pavarino, L., Scacchi, S., Widlund, O., Zampini, S.: Isogeometric BDDC preconditioners with deluxe scaling. SIAM J. Sci. Comput. **36**(3), A1118–A1139 (2014)
9. Gahalaut, K.P.S., Kraus, J.K., Tomar, S.K.: Multigrid methods for isogeometric discretization. Comput. Methods Appl. Mech. Eng. **253**, 413–425 (2013)
10. Gao, L., Calo, V.M.: Fast isogeometric solvers for explicit dynamics. Comput. Methods Appl. Mech. Eng. **274**, 19–41 (2014)

11. Hackbusch, W.: Multi-Grid Methods and Applications, vol. 4. Springer Series in Computational Mathematics. Springer-Verlag, Heidelberg (2003)
12. Hughes, T.J.R., Cottrell, J.A., Bazilevs, Y.: Isogeometric analysis: CAD, finite elements, NURBS, exact geometry and mesh refinement. Comput. Methods Appl. Mech. Eng. **194**(39–41), 4135–4195 (2005)
13. Kleiss, S.K., Pechstein, C., Jüttler, B., Tomar, S.: IETI - Isogeometric tearing and interconnecting. Comput. Methods Appl. Mech. Eng. **247–248**, 201–215 (2012)
14. Schwab, C.: *p*- and *hp*-Finite Element Methods. Clarendon Press, Oxford (1998)

On the Set of Trajectories of the Control Systems with Limited Control Resources

Nesir Huseyin, Anar Huseyin, and Khalik G. Guseinov$^{(\boxtimes)}$

Mathematics Department, Anadolu University, 26470 Eskisehir, Turkey
{nhuseyin,ahuseyin,kguseynov}@anadolu.edu.tr

Abstract. The set of trajectories of the control system with limited control resources is studied. It is assumed that the behavior of the system is described by a Volterra type integral equation which is nonlinear with respect to the state vector and is affine with respect to the control vector. The closed ball of the space L_p ($p > 1$) with radius μ and centered at the origin, is chosen as the set of admissible control functions. It is proved that for each fixed p the set of trajectories is Lipschitz continuous with respect to μ and for each fixed μ is continuous with respect to p. The upper evaluation for the diameter of the set of trajectories is given.

Keywords: Integral equation · Control system · Integral constraint · Set of trajectories

1 Introduction

Control systems with limited control recourses arise in various problems of the theory and applications. In general, control systems with integral constraint on the controls naturally arise as control problems with bounded L_p norms on the controls, control problems with prescribed bounded total energy and finance and control problems with design uncertainties (see, e.g. [1–7] and references therein). For example, the dynamics of flying objects with variable mass is described in the form of control system, where the control function has an integral constraint (see, e.g. [4,5]). Control system with integral constraint on the controls whose behavior is described by a differential equation is studied in [1–6]. In [1–3] closedness, compactness, dependence of the parameters properties and numerical construction methods of the attainable sets and the set of trajectories of the control system with integral constraint on the controls are studied, where the behavior of the system is described by an affine ordinary differential equation.

The theory of linear and nonlinear integral equations plays an important role in the mathematics and its applications (see, e.g. [7–11] and references therein). They arise in many problems of contemporary physics and mechanics. In this paper the control system with integral constraint on the controls whose behavior is described by a Volterra type integral equation is considered. It is assumed that integral equation is affine, i.e. it is nonlinear with respect to the state vector and is affine with respect to the control vector. The closed ball of the space L_p ($p > 1$)

© Springer International Publishing Switzerland 2015
J.-D. Boissonnat et al. (Eds.): Curves and Surfaces 2014, LNCS 9213, pp. 280–288, 2015.
DOI: 10.1007/978-3-319-22804-4_21

with radius μ and centered at the origin, is chosen as the set of admissible control functions. Dependence of the set of trajectories on the system's parameters is studied. Under stronger conditions, a similar problem was considered in [12], where the system is nonlinear with respect to both the state and the control vector.

The paper is organized as follows. In Sect. 2 the conditions which satisfy the system are formulated (Conditions 2.A, 2.B and 2.C). In Sect. 3 it is proved that the set of trajectories of the system generated by all admissible control functions is Lipschitz continuous with respect to μ which characterizes the constraint on the control resource (Theorem 1). In Sect. 4 it is shown that the set of trajectories is continuous with respect to p which is parameter of the space L_p (Theorem 2). In Sect. 5 an upper estimation for the diameter of the set of trajectories is given (Theorem 3).

2   Preliminaries

Consider control system the behavior of which is described by a Volterra type integral equation

$$x(t) = g\left(t, x\left(t\right)\right) + \lambda \int_{t_0}^{t} \left[K_1\left(t, s, x\left(s\right)\right) + K_2\left(t, s, x\left(s\right)\right) u\left(s\right)\right] ds, \qquad (1)$$

where $x \in \mathbb{R}^n$ is the state vector, $u \in \mathbb{R}^m$ is the control vector, $\lambda \in \mathbb{R}$ and $t \in [t_0, \theta]$. The system (1) is affine with respect to the control vector, but it is nonlinear with respect to the state vector. Therefore the system (1) will be called an affine control system.

Let $p > 1$ and $\mu > 0$ be given numbers. The function $u(\cdot) \in L_p\left([t_0, \theta]; \mathbb{R}^m\right)$ such that

$$\|u(\cdot)\|_p \le \mu \qquad (2)$$

is said to be an admissible control function, where $\|u(\cdot)\|_p = \left(\int_{t_0}^{\theta} \|u(t)\|^p \, dt \right)^{\frac{1}{p}}$,

$\|\cdot\|$ denotes the Euclidean norm.

The set of all admissible control functions is denoted by symbol $U_{p,\mu}$. Thus

$$U_{p,\mu} = \left\{ u(\cdot) \in L_p\left([t_0, \theta]; \mathbb{R}^m\right) : \|u(\cdot)\|_p \le \mu \right\}. \qquad (3)$$

It is assumed that the following conditions are satisfied:

2.A the vector functions $g(\cdot, \cdot) : [t_0, \theta] \times \mathbb{R}^n \to \mathbb{R}^n$, $K_1(\cdot, \cdot, \cdot) : [t_0, \theta] \times [t_0, \theta] \times \mathbb{R}^n \to \mathbb{R}^n$ and matrix function $K_2(\cdot, \cdot, \cdot) : [t_0, \theta] \times [t_0, \theta] \times \mathbb{R}^n \to \mathbb{R}^{n \times m}$ are continuous;

2.B there exist $L_0 \in [0, 1)$, $L_1 \geq 0$ and $L_2 \geq 0$ such that

$$\|g(t, x_1) - g(t, x_2)\| \leq L_0 \|x_1 - x_2\|,$$

$$\|K_1(t, s, x_1) - K_1(t, s, x_2)\| \leq L_1 \|x_1 - x_2\|,$$

$$\|K_2(t, s, x_1) - K_2(t, s, x_2)\| \leq L_2 \|x_1 - x_2\|$$

for every $(t, s, x_1) \in [t_0, \theta] \times [t_0, \theta] \times \mathbb{R}^n$, $(t, s, x_2) \in [t_0, \theta] \times [t_0, \theta] \times \mathbb{R}^n$;
2.C for given $p_0 > 1$ and $\mu_0 > 0$ the inequality

$$0 \leq \lambda \left(L_1 (\theta - t_0) + L_2(\theta - t_0)^{\frac{p_0 - 1}{p_0}} \mu_0 \right) < 1 - L_0$$

holds.

We denote

$$L(\lambda; p, \mu) = L_0 + \lambda \left[L_1 (\theta - t_0) + L_2 (\theta - t_0)^{\frac{p-1}{p}} \mu \right]. \tag{4}$$

According to the condition 2.C we obtain that $0 \leq L(\lambda; p_0, \mu_0) < 1$. Then there exists $\alpha > 0$ such that

$$0 \leq L(\lambda; p, \mu) < 1 \tag{5}$$

for every $p \in [p_0 - \alpha, p_0 + \alpha]$ and $\mu \in [\mu_0 - \alpha, \mu_0 + \alpha]$, where $p_0 - \alpha > 1$ and $\mu_0 - \alpha > 0$. We set

$$L_*(\lambda) = \max \left\{ L(\lambda; p, \mu) : p \in [p_0 - \alpha, p_0 + \alpha], \ \mu \in [\mu_0 - \alpha, \mu_0 + \alpha] \right\}. \tag{6}$$

From now on, it will be assumed that $p \in (p_0 - \alpha, p_0 + \alpha)$ and $\mu \in (\mu_0 - \alpha, \mu_0 + \alpha)$.

Now, let us define a trajectory of the system (1) generated by an admissible control function. Let p and μ be fixed and $u_*(\cdot) \in U_{p,\mu}$. A continuous function $x_*(\cdot) : [t_0, \theta] \to \mathbb{R}^n$ satisfying the integral equation

$$x_*(t) = g(t, x_*(t)) + \lambda \int_{t_0}^{t} [K_1(t, s, x_*(s)) + K_2(t, s, x_*(s)) u_*(s)] \, ds$$

for every $t \in [t_0, \theta]$ is said to be a trajectory of the system (1) generated by the admissible control function $u_*(\cdot) \in U_{p,\mu}$. The trajectory of the system (1) generated by the control function $u(\cdot) \in U_{p,\mu}$ is denoted by $x(\cdot; u(\cdot))$ and we set

$$\mathbf{X}_{p,\mu} = \{x(\cdot; u(\cdot)) : u(\cdot) \in U_{p,\mu}\}.$$

$\mathbf{X}_{p,\mu}$ is called the set of trajectories of the system (1) with integral constraint (2). It is obvious that $\mathbf{X}_{p,\mu} \subset C([t_0, \theta]; \mathbb{R}^n)$, where $C([t_0, \theta]; \mathbb{R}^n)$ is the space of continuous functions $x(\cdot) : [t_0, \theta] \to \mathbb{R}^n$ with norm $\|x(\cdot)\|_C = \max\{\|x(t)\| : t \in [t_0, \theta]\}$.

For $t \in [t_0, \theta]$ we denote

$$\mathbf{X}_{p,\mu}(t) = \{x(t) : x(\cdot) \in \mathbf{X}_{p,\mu}\}.$$

The set $\mathbf{X}_{p,\mu}(t)$ is the section of the set of trajectories at the instant of t.

Note that the mathematical models of different controlled processes are given by Eq. (1). For example, the control system described by nonlinear integral Volterra equation

$$x(t) = f(t) + \int_{t_0}^{t} [\varphi(x(t-s))p(s) + b(t,s)\,u(s)]\,ds$$

can be used (1) as a model of controlled population growth rate depending on population size and a probability of death depending only on age; (2) in the study of the controlled spread of a disease for which recovery from the disease confers no immunity to reinfection; (3) in an controlled economic model, where $x(t)$ represents the total value of capital of time t, and the production of new capital within the economy depends on $x(\cdot)$ and control effort $u(\cdot)$, where $\|u(\cdot)\|_p \leq \mu$ and μ characterizes the stock of food in case (1), the stock of medicine in case (2), and the stock of reserved capital in the case (3) (see, [8]).

Now let us give propositions which characterize properties of the set of trajectories and will be used in following arguments.

Proposition 1. *[7] Every admissible control function $u_*(\cdot) \in U_{p,\mu}$ generates a unique trajectory $x(\cdot; u_*(\cdot))$ of the system (1).*

Proposition 2. *[7] The set of trajectories $\mathbf{X}_{p,\mu}$ is a compact subset of the space $C\left([t_0, \theta]; \mathbb{R}^n\right)$ and the set valued map $t \to \mathbf{X}_{p,\mu}(t)$, $t \in [t_0, \theta]$, is continuous.*

Proposition 3. *[7] There exists $r_* > 0$ such that*

$$\|x(\cdot)\|_C \leq r_* \tag{7}$$

for every $x(\cdot) \in \mathbf{X}_{p,\mu}$, $p \in [p_0 - \alpha, p_0 + \alpha]$ and $\mu \in [\mu_0 - \alpha, \mu_0 + \alpha]$, where $\alpha > 0$ is defined in (5).

Let (Y, d) be a metric space. The Hausdorff distance between the sets $F \subset Y$ and $E \subset Y$ is denoted by $h(F, E)$ and defined as

$$h(F, E) = \max\left\{\sup_{x \in F} d(x, E), \sup_{y \in E} d(y, F)\right\},$$

where $d(x, E) = \inf\{d(x, y) : y \in E\}$.

3   Continuity of the Set of Trajectories with Respect to μ

Denote

$$B_n(r_*) = \{x \in \mathbb{R}^n : \|x\| \leq r_*\},$$

$$\gamma_* = \max \left\{ (\theta - t_0)^{\frac{p-1}{p}} : p \in [p_0 - \alpha, p_0 + \alpha] \right\}, \tag{8}$$

$$M_2 = \max \{ \|K_2(t, s, x)\| : (t, s, x) \in [t_0, \theta] \times [t_0, \theta] \times B_n(r_*) \}, \tag{9}$$

$$R_* = \frac{\lambda}{1 - L_0} M_2 \gamma_* \cdot \exp \left[\frac{L_*(\lambda) - L_0}{1 - L_0} \right], \tag{10}$$

where $\alpha > 0$ is defined in (5), $L_*(\lambda)$ is defined by (6), r_* is given in (7).

Theorem 1. *Let* $p \in (p_0 - \alpha, p_0 + \alpha)$, $\mu \in (\mu_0 - \alpha, \mu_0 + \alpha)$, $\mu_* \in (\mu_0 - \alpha, \mu_0 + \alpha)$. *Then*

$$h(\mathbf{X}_{p,\mu}, \mathbf{X}_{p,\mu_*}) \leq R_* |\mu - \mu_*|$$

and consequently

$$h(\mathbf{X}_{p,\mu}(t), \mathbf{X}_{p,\mu_*}(t)) \leq R_* |\mu - \mu_*|$$

for every $t \in [t_0, \theta]$, *where* $R_* \geq 0$ *is defined by (10).*

Proof. Let us choose an arbitrary $x_*(\cdot) \in \mathbf{X}_{p,\mu_*}$ generated by the admissible control function $u_*(\cdot) \in U_{p,\mu_*}$. Since $u_*(\cdot) \in U_{p,\mu_*}$, then $\|u_*(\cdot)\|_p \leq \mu_*$. Define a function $u(\cdot) : [t_0, \theta] \rightarrow \mathbb{R}^m$ setting

$$u(t) = \frac{\mu}{\mu_*} u_*(t), \quad t \in [t_0, \theta]. \tag{11}$$

It is not difficult to verify that $u(\cdot) \in U_{p,\mu}$. Let $x(\cdot) : [t_0, \theta] \rightarrow \mathbb{R}^m$ be the trajectory of the system (1) generated by the control function $u(\cdot) \in U_{p,\mu}$. Then $x(\cdot) \in \mathbf{X}_{p,\mu}$ and (10), (11), Condition 2.B and Gronwall's inequality yield

$$\|x(t) - x_*(t)\| \leq \frac{\lambda}{1 - L_0} M_2 \gamma_* |\mu - \mu_*| \cdot \exp \left[\frac{\lambda}{1 - L_0} \int_{t_0}^{t} [L_1 + L_2 \|u(s)\|] \, ds \right]$$

$$\leq \frac{\lambda}{1 - L_0} M_2 \gamma_* |\mu - \mu_*| \cdot \exp \left[\frac{L(\lambda; p, \mu) - L_0}{1 - L_0} \right]$$

$$\leq \frac{\lambda}{1 - L_0} M_2 \gamma_* |\mu - \mu_*| \cdot \exp \left[\frac{L_*(\lambda) - L_0}{1 - L_0} \right] = R_* |\mu - \mu_*| \tag{12}$$

for every $t \in [t_0, \theta]$, where γ_* is defined by (8), M_2 is defined by (9). From (12) we conclude that for each $x_*(\cdot) \in \mathbf{X}_{p,\mu_*}$ there exists $x(\cdot) \in \mathbf{X}_{p,\mu}$ such that $\|x_*(\cdot) - x(\cdot)\|_C \leq R_* |\mu - \mu_*|$ and consequently

$$\sup_{x_*(\cdot) \in \mathbf{X}_{p,\mu_*}} d(x_*(\cdot), \mathbf{X}_{p,\mu}) \leq R_* |\mu - \mu_*|. \tag{13}$$

Analogously it is possible to show that

$$\sup_{x(\cdot) \in \mathbf{X}_{p,\mu}} d(x(\cdot), \mathbf{X}_{p,\mu_*}) \leq R_* |\mu - \mu_*|. \tag{14}$$

(13) and (14) complete the proof.

Corollary 1. *Let $p \in (p_0 - \alpha, p_0 + \alpha)$ be fixed. Then the set valued map $\mu \to \mathbf{X}_{p,\mu}$, $\mu \in (\mu_0 - \alpha, \mu_0 + \alpha)$, is Lipschitz continuous with Lipschits constant R_*, where $R_* \geq 0$ is defined by (10).*

4 Continuity of the Set of Trajectories with Respect to p

First of all let us define a distance between the subsets of the spaces $L_{p_1}([t_0, \theta]; \mathbb{R}^m)$ and $L_{p_2}([t_0, \theta]; \mathbb{R}^m)$, where $p_1 \in [1, \infty)$, $p_2 \in [1, \infty)$. Since $L_p([t_0, \theta]; \mathbb{R}^m) \subset L_1([t_0, \theta]; \mathbb{R}^m)$ for every $p \in [1, \infty)$, then the norm of the space $L_1([t_0, \theta]; \mathbb{R}^m)$ will be used for definition of the Hausdorff distance between subsets of the spaces $L_{p_1}([t_0, \theta]; \mathbb{R}^m)$ and $L_{p_2}([t_0, \theta]; \mathbb{R}^m)$, where $p_1 \in [1, \infty)$, $p_2 \in [1, \infty)$.

Let $G \subset L_{p_1}([t_0, \theta]; \mathbb{R}^m)$ and $W \subset L_{p_2}([t_0, \theta]; \mathbb{R}^m)$, where $1 \leq p_1 < +\infty$, $1 \leq p_2 < +\infty$. The Hausdorff distance between the sets G and W is denoted by $\hbar_1(G, W)$ and is defined by

$$\hbar_1(G, W) = \max \left\{ \sup_{x(\cdot) \in G} d_{L_1}(x(\cdot), W), \ \sup_{y(\cdot) \in W} d_{L_1}(y(\cdot), G)) \right\},$$

where

$$d_{L_1}(x(\cdot), W) = \inf_{y(\cdot) \in W} \|x(\cdot) - y(\cdot)\|_1, \quad \|z(\cdot)\|_1 = \int_{t_0}^{\theta} \|z(t)\| \, dt.$$

We set

$$\beta_* = \frac{\lambda M_2}{1 - L_0} \cdot \exp \left[\frac{L_*(\lambda) - L_0}{1 - L_0} \right], \tag{15}$$

where $L_*(\lambda)$ is defined by (6), M_2 is defined by (9).

Theorem 2. *Let $\mu \in (\mu_0 - \alpha, \mu_0 + \alpha)$ and $p_* \in (p_0 - \alpha, p_0 + \alpha)$ be fixed. Then for every $\varepsilon > 0$ there exists $\delta = \delta(\varepsilon, p_*, \mu) > 0$ such that for every $p \in (p_* - \delta, p_* + \delta)$ the inequality*

$$h(\mathbf{X}_{p,\mu}, \mathbf{X}_{p_*,\mu}) \leq \varepsilon$$

holds and consequently

$$h(\mathbf{X}_{p,\mu}(t), \mathbf{X}_{p_*,\mu}(t)) \le \varepsilon$$

for every $t \in [t_0, \theta]$, where $\alpha > 0$ is given in (5).

Proof. By virtue of Theorem 3.6 from [3], for given $\mu \in (\mu_0 - \alpha, \mu_0 + \alpha)$, $p_* \in (p_0 - \alpha, p_0 + \alpha)$ and $\dfrac{\varepsilon}{\beta_*}$ there exists $\delta = \delta(\varepsilon, p_*, \mu) \in (0, p_* - 1)$ such that

$$\hbar_1(U_{p,\mu}, U_{p_*,\mu}) < \frac{\varepsilon}{\beta_*} \tag{16}$$

for every $p \in (p_* - \delta, p_* + \delta)$.

Without loss of generality let us assume that

$$\delta = \delta(\varepsilon, p_*, \mu) < \min\{p_* - p_0 + \alpha, p_0 - p_* + \alpha\}.$$

Since $p_* \in (p_0 - \alpha, p_0 + \alpha)$, then we have

$$(p_* - \delta, p_* + \delta) \subset (p_0 - \alpha, p_0 + \alpha). \tag{17}$$

Now, let us choose arbitrary $p \in (p_* - \delta, p_* + \delta)$ and trajectory $x(\cdot) \in \mathbf{X}_{p,\mu}$ generated by the admissible control function $u(\cdot) \in U_{p,\mu}$. It follows from (16) that there exists $u_*(\cdot) \in U_{p_*,\mu}$ such that

$$\int_{t_0}^{\theta} \|u(s) - u_*(s)\| \, ds \le \frac{\varepsilon}{\beta_*}, \tag{18}$$

where $\beta_* > 0$ is defined by (15). Let $x_*(\cdot) : [t_0, \theta] \to \mathbb{R}^n$ be the trajectory of the system (1) generated by the admissible control function $u_*(\cdot) \in U_{p_*,\mu}$. Then $x_*(\cdot) \in \mathbf{X}_{p_*,\mu}$ and (6), (9), (15), (17), (18), Condition 2.B and Gronwall's inequality imply

$$\|x(t) - x_*(t)\| \le \frac{\lambda M_2}{\beta_*(1 - L_0)} \varepsilon \cdot \exp\left(\frac{\lambda}{1 - L_0} \int_{t_0}^{t} [L_1 + L_2 \|u(s)\|] \, d\tau\right)$$

$$\le \frac{\lambda M_2}{\beta_*(1 - L_0)} \varepsilon \cdot \exp\left[\frac{L_*(\lambda) - L_0}{1 - L_0}\right] = \varepsilon \tag{19}$$

for every $t \in [t_0, \theta]$. Thus, from (19) we get that for each $x(\cdot) \in \mathbf{X}_{p,\mu}$ there exists $x_*(\cdot) \in \mathbf{X}_{p_*,\mu}$ such that $\|x(\cdot) - x_*(\cdot)\|_C \le \varepsilon$ and hence

$$\sup_{x(\cdot) \in \mathbf{X}_{p,\mu}} d(x(\cdot), \mathbf{X}_{p_*,\mu}) \le \varepsilon. \tag{20}$$

Similarly one can prove that

$$\sup_{x_*(\cdot) \in \mathbf{X}_{p_*,\mu}} d(x_*(\cdot), \mathbf{X}_{p,\mu}) \le \varepsilon. \tag{21}$$

(20) and (21) imply the proof.

Corollary 2. *For each fixed $\mu \in (\mu_0 - \alpha, \mu_0 + \alpha)$ the set valued map $p \to \mathbf{X}_{p,\mu}$, $p \in (p_0 - \alpha, p_0 + \alpha)$, is continuous.*

5   Diameter of the Set of Trajectories

For given metric space (Y, d) and a set $E \subset Y$ the diameter of E is denoted by $diam\, E$ and is defined by

$$diam\, E = \sup \{d(x, y) : x \in E,\ y \in E\}.$$

For $\mu \in (\mu_0 - \alpha, \mu_0 + \alpha)$, $p \in (p_0 - \alpha, p_0 + \alpha)$ and $t \in [t_0, \theta]$ we set

$$k(t; p, \mu) = \frac{2\lambda M_2 \mu}{1 - L_0} (t - t_0)^{\frac{p-1}{p}}$$
$$\cdot \exp \left[\frac{\lambda}{1 - L_0} \left(L_1(t - t_0) + L_2 \mu \left(t - t_0\right)^{\frac{p-1}{p}} \right) \right], \tag{22}$$

where $\alpha > 0$ is given in (5), M_2 is defined by (9).

Theorem 3. *For each* $\mu \in (\mu_0 - \alpha, \mu_0 + \alpha)$, $p \in (p_0 - \alpha, p_0 + \alpha)$ *and* $t \in [t_0, \theta]$ *the inequalities*

$$diam\, \mathbf{X}_{p,\mu}(t) \leq k(t; p, \mu) \tag{23}$$

and

$$diam\, \mathbf{X}_{p,\mu} \leq k(\theta; p, \mu) \tag{24}$$

are satisfied.

Proof. Let us choose arbitrary $\mu \in (\mu_0 - \alpha, \mu_0 + \alpha)$, $p \in (p_0 - \alpha, p_0 + \alpha)$ and let $x(\cdot) \in \mathbf{X}_{p,\mu}$ and $y(\cdot) \in \mathbf{X}_{p,\mu}$ be arbitrarily chosen trajectories generated by the admissible control functions $u_1(\cdot) \in U_{p,\mu}$ and $u_2(\cdot) \in U_{p,\mu}$ respectively. From (22), Condition 2.B and Gronwall's inequality it follows

$$\|x(t) - y(t)\| \leq \frac{2\lambda M_2 \mu}{1 - L_0} (t - t_0)^{\frac{p-1}{p}} \cdot \exp \left[\frac{\lambda}{1 - L_0} \int_{t_0}^{t} (L_1 + L_2 \|u_1(s)\|)\, ds \right]$$
$$\leq \frac{2\lambda M_2 \mu}{1 - L_0} (t - t_0)^{\frac{p-1}{p}} \cdot \exp \left[\frac{\lambda}{1 - L_0} \left(L_1(t - t_0) + L_2 \mu \left(t - t_0\right)^{\frac{p-1}{p}} \right) \right]$$
$$= k(t; p, \mu) \tag{25}$$

for every $t \in [t_0, \theta]$. Since $x(\cdot) \in \mathbf{X}_{p,\mu}$ and $y(\cdot) \in \mathbf{X}_{p,\mu}$ are arbitrarily chosen, then (25) yields the validity of inequality (23).

Since the function $k(\cdot, p, \mu) : [t_0, \theta] \to [0, \infty)$ is monotone increasing for every fixed p and μ, then we have from (25) that

$$\|x(t) - y(t)\| \leq k(\theta; p, \mu)$$

for every $t \in [t_0, \theta]$ and consequently

$$\|x(\cdot) - y(\cdot)\|_C = \sup \{\|x(t) - y(t)\| : t \in [t_0, \theta]\} \leq k(\theta; p, \mu)$$

which implies the validity of the inequality (24).

References

1. Akyar, E.: Dependence on initial conditions of attainable sets of control systems with p-integrable controls. Nonlinear Anal. Model. Control. **12**, 293–306 (2007)
2. Guseinov, K.G., Ozer, O., Akyar, E., Ushakov, V.N.: The approximation of reachable sets of control systems with integral constraint on controls. Nonlinear Diff. Equat. Appl. (NoDEA) **14**, 57–73 (2007)
3. Guseinov, K.G., Nazlipinar, A.S.: On the continuity property of L_p balls and an application. J. Math. Anal. Appl. **335**, 1347–1359 (2007)
4. Krasovskii, N.N.: Theory of Control of Motion: Linear systems. Nauka, Moscow (1968). (In Russian)
5. Ukhobotov, V.I.: One Dimensional Projection Method in Linear Differential Games with Integral Constraints. Chelyabinsk University Press, Chelyabinsk (2005). (In Russian)
6. Ushakov, V.N.: Extremal strategies in differential games with integral constraints. J. Appl. Math. Mech. **36**, 12–19 (1972)
7. Huseyin, N., Huseyin, A.: Compactness of the set of trajectories of the controllable system described by an affine integral equation. Appl. Math. Comp. **219**, 8416–8424 (2013)
8. Brauer, F.: On a nonlinear integral equation for population growth problems. SIAM J. Math. Anal. **6**, 312–317 (1975)
9. Polyanin, A.D., Manzhirov, A.V.: Handbook of Integral Equation. CRC Press, Boca Raton (1998)
10. Väth, M.: Volterra and Integral Equations of Vector Functions. M. Deccer. Inc., New York (2000)
11. Krasnoselskii, M.A., Krein, S.G.: On the principle of averaging in nonlinear mechanics. Uspekhi Mat. Nauk. **10**, 147–153 (1955). (in Russian)
12. Huseyin, A., Huseyin, N.: Dependence on the parameters of the set of trajectories of the control system described by a nonlinear Volterra integral equation. Appl. Math. **59**, 303–317 (2014)

High Order Reconstruction from Cross-Sections

Yael Kagan$^{(\boxtimes)}$ and David Levin

School of Mathematical Sciences, Tel Aviv University, Tel Aviv, Israel
yaelkaga@gmail.com

Abstract. Given parallel cross-sections data of a smooth object in $I\!R^d$, one of the ways of generating approximation to the object is by interpolating the signed-distance functions corresponding to the cross-sections. This well-known method is useful in many applications, and yet its approximation properties are not fully established. The known result is that away from cross-sections that are parallel to the boundary of the object, this method gives high approximation order. However, near such tangent cross-sections the approximation order is drastically reduced. This is due to the singular behaviour of the signed-distance function near tangent cross-sections. In this paper we suggest a way to restore the high approximation order everywhere. The new method involves a recent development in the approximation of functions with singularities. We present the application of this approach to our case, analyze its approximation properties, and discuss the numerical issues involved.

Keywords: Set-valued approximation · Multivariate reconstruction · Singularity detection

1 Preliminaries

The problem of body reconstruction from cross-sections originates from applications in computerized tomography (CT). The main numerical issue in CT is to reconstruct cross-sections of an organ from its projections. The computed cross-sections are then to be used in the 3-D reconstruction of the organ, for the purpose of graphic 3-D display, and for medical analysis. Several methods have been suggested for solving the reconstruction problem, e.g., [1,3,6], who build the surface of the 3D object as a triangulated surface. Other ideas for 3D reconstruction from parallel cross-sections appear in [8,9,12]. Some works deal with the more general problem of reconstruction from general cross sections, e.g., [2,4]. The focus in most works is on the robustness of the method, rather than on its approximation power. The present work in confined to reconstruction from parallel cross-section, aiming at achieving high approximation order. The reconstruction framework presented in [10] suggests a high order interpolation method between cross-sections of objects of a most general topology. The method is based upon interpolation between the signed-distance functions representing the cross-sections. In fact, this method is quite general, and it can be used to reconstruct an d-dimensional object from a sequence of its $(d-1)$-dimensional

© Springer International Publishing Switzerland 2015
J.-D. Boissonnat et al. (Eds.): Curves and Surfaces 2014, LNCS 9213, pp. 289–303, 2015.
DOI: 10.1007/978-3-319-22804-4_22

cross-sections for any d. The method has been used in many applications, such as MRI reconstruction [7], 2D and 3D metamorphosis [5], and surface recovery from planar sectional contours.

Let us first recall the essential ingredients of the reconstruction method in [10], and recall its strong and its weak qualities.

1.1   Distance-Field Interpolation

Consider a bounded closed domain B in $I\!\!R^d$, also referred to as a d-dimensional object, and consider $(d-1)$-dimensional cross-sections of B. W.l.o.g., we assume the cross-sections are intersection of B with planes $P(t)$ of constant last coordinate (level) in $I\!\!R^d$, $t = x_d$, and we denote such a cross-section by $F(t)$:

$$F(t) = P(t) \cap B \subset I\!\!R^d. \tag{1}$$

Given a finite sequence of cross-sections, $\{F(t_i)\}_{i=1}^n$, one would like to approximate the object B. The boundary of each cross-section can be defined as the zero level set of a continuous function in $I\!\!R^d$, and in [10] it is suggested to use the signed-distance function for this purpose. We denote a point in the $(d-1)$ dimensional subspace of $I\!\!R^d$ by

$$p = (x_1, ..., x_{d-1}).$$

The signed-distance function of a cross-section $F(t)$ is defined as:

$$d(p,t) = \begin{cases} dist\big(p, F(t)^c\big) & p \in F(t), \\ -dist\big(p, F(t)\big) & p \notin F(t), \\ -D & F(t) = \emptyset \end{cases} \tag{2}$$

$F(t)^c$ denotes the complement of the set $F(t)$, and D is some large number. By construction, the set B is the set of all points with a non-negative $d(p,t)$:

$$B = \{(p,t) \mid d(p,t) \geq 0\}. \tag{3}$$

The Signed-Distance Reconstruction Procedure - SDR

Given $\{F(t_i)\}_{i=1}^n$, the method in [10] considers a fixed line L_p in $I\!\!R^d$, defined by a constant vector $p = (x_1, ..., x_{d-1})$, and approximates $d(p,t)$ by a univariate approximation (or interpolation) based upon the data values $d(p, t_i)$, $i = 1, 2, ..., n$. Denoting this univariate approximant by $\bar{d}(p,t)$, the approximation $\bar{B}_p$ of $B_p = B \cap L_p$ is defined as

$$\bar{B}_p = \{(p,t) \mid \bar{d}(p,t) \geq 0\}. \tag{4}$$

To achieve a high order approximation to B_p, $\bar{d}(p,t)$ should give a high order approximation to $d(p,t)$. Consider a kth order univariate approximation method of a finite support m, namely, the approximation in $[t_i, t_{i+1}]$ depends only upon $\{t_{i+j}\}_{j=-m}^{m+1}$. A kth order method gives approximation errors $O(h^k)$ to C^k functions, where $h = max_i\{|t_{i+1} - t_i|\}$ The following approximation result is given in [10] :

Theorem 1. *Let B be a closed domain in R^d whose boundary ΓB is composed of a finite number of C^k mutually disconnected surfaces, and let $\{s_\ell\}_{\ell=1}^r$ be the cross-section levels which are tangent to ΓB. Consider a level t which is $m \cdot h$ away from a tangency level where $h = max_i\{|t_{i+1} - t_i|\}$,*

$$min\{|t - s_\ell|\}_{\ell=1}^r \geq m \cdot h.$$

Using a kth order univariate approximation of support size m, if the SDR procedure gives a wrong result, i.e., if

$$\bar{d}(p,t) \cdot d(p,t) < 0,$$

then

$$dist\big((p,t), \Gamma B\big) = O(h^k) \quad as \ h \to 0. \tag{5}$$

Remark 1. The result in [10] is stated for $d = 3$, and for equidistant cross-sections' data, but it holds for any dimension d and for a general levels' distribution.

Remark 2. Theoretically, the approximation result in Theorem 1 holds for any level t which is not a tangency level. However, in practical applications h does not tend to zero, and the approximation near the tangency levels is quite poor.

1.2   The Behaviour Near Tangency Levels

The goal of this paper is to enhance the SDR procedure so that the resulting approximation would be of high order near tangency levels as well. For that we shall analyze the special behaviour near tangency levels, and then develop a special treatment using an adaptation of an approximation method suggested in [11].

Let B be a closed domain in $\mathbb{R}^d$ with a C^k boundary ΓB composed of a finite number of disconnected surfaces. If t^* is a tangency level, and $x^* = (x_1^*, ..., x_{d-1}^*, t^*) \in \Gamma B$ is a tangency point, then ΓB can be locally described as a C^k function $f(x_1, ..., x_{d-1})$. In a generic case, the second order partial derivatives of f would not vanish at $p^* = (x_1^*, ..., x_{d-1}^*)$. We later prove the following lemma:

Lemma 1. *Assume $p^* = (x_1^*, ..., x_{d-1}^*)$ is a point of local minima of the above function $f(x_1, ..., x_{d-1})$. For a fixed vector $p = (x_1, ..., x_{d-1})$ close to p^*, and for $t > t^*$,*

$$d(p,t) = C_p + O(|t - t^*|^{0.5}), \quad as \ t \to t^*. \tag{6}$$

The above behaviour of the signed-distance function is the reason for the failure of the SDR procedure near tangency levels. The first derivative $\partial_t d(p,t)$ is unbounded near t^*, and any standard univariate approximation method for approximating $d(p,t)$ (for a fixed p) would fail. The approximation is even more problematic since t^*, the tangency level, is unknown. The given data are only the cross-sections $\{F(t_i)\}$, and, in general, t^* falls in between two (unknown) cross-section levels. In Sect. 3 we suggest an approximation method which can handle the singular behavior near tangency levels and yields nice approximation results. This method is based upon the basic framework in [11] presented below in its more general form.

1.3   Approximating Piecewise-Smooth Functions

Consider a function of the following form:

$$f = g + \rho, \tag{7}$$

where $g \in PC^{m+1}[a, b]$ and

$$\rho(t) = \sum_{j=1}^{k} c_j u_j^s(t), \tag{8}$$

where $\{u_j^s\}_{j=1}^k$ are known functions, with some singularity at $s \in (a, b)$. The singularity location s and the coefficients $\{c_j\}$ being unknown. e.g., in [11],

$$u_j^s(t) = (t - s)_+^{j-1}. \tag{9}$$

Given data values $\{f(t_i)\}_{i=1}^n$, one would like to approximate f on $[a, b]$ with a high approximation order. A convenient tool for obtaining high order approximation to smooth functions is the quasi-interpolation operator

$$Qf(t) = \sum_{i=1}^{n} q_i(t) f(t_i), \tag{10}$$

where $\{q_i\}_{i=1}^n$ are compactly supported function, satisfying

$$Qp = p, \quad p \in \Pi_m[a, b], \tag{11}$$

where Π_m is the space of polynomials of total degree $\leq m$. Assuming $supp(q_i) = O(h)$, where $h = max_i\{|t_{i+1} - t_i|\}$, it follows that for $g \in PC^{m+1}[a, b]$

$$g(t) - Qg(t) = O(h^{m+1}), \quad as \ \ h \to 0, \ \ \forall t \in [a, b]. \tag{12}$$

Applying the quasi-interpolation operator on a function of the form (7) one gets good approximation away from the singular point s, and high errors near s. These significant errors play an important role in the singularity reconstruction described below.

The Piecewise-Smooth Approximation Procedure - PSA

The main idea in [11] is to exploit the known high errors at the data points $\{t_i\}$ near s in order to approximate the unknown singular point s and the known coefficients $\{c_j\}$. Recalling the small approximation errors $\{g(t_i) - Qg(t_i)\}$, the procedure for approximating the parameters in the function ρ is

$$\{s^*, \{c_j^*\}\} = \arg\min_{s, \{c_j\}} \sum_{i=1}^{n} [(f - \rho - Q(f - \rho))(t_i)]^2. \tag{13}$$

Using the approximate parameters of ρ we define

$$\rho^*(t) = \sum_{j=1}^{k} c_j^* u_j^{s^*}(t), \tag{14}$$

and the suggested corrected approximation to f is:

$$\bar{Q}(f) = Q(f) + \rho^* - Q(\rho^*). \tag{15}$$

The main result in [11] shows an $O(h^{m+1})$ approximation order by $\bar{Q}$ to functions of the form (9), and an $O(h^{m+1})$ approximation to s as well. The above approximation mehod is applied in Sect. 3 for improving the approximation of the signed distance function near tangency levels. In order to implement the method at its full capacity, first we have to identify the right singularity behaviour near a tangency level.

2   The Distance Function Near Tangency Levels

We have already stated in Lemma 1 that near a tangency level t^* the signed distance function demonstrates a singular behavior. We would like to find the full form of this singularity in order to use it within the method of Sect. 1.3. Let us first examine the case $d = 2$. In this case the local boundary ΓB can be locally approximated near a tangency point (x_1^*, t^*) as

$$f(x_1) = t^* + a_2(x_1 - x_1^*)^2 + a_3(x_1 - x_1^*)^3 + \dots + a_m(x_1 - x_1^*)^m. \tag{16}$$

W.l.o.g., we assume that $a_2 > 0$, and consider $t > t^*$. We note that $d(p,t) = p - \bar{p} \equiv x_1 - \bar{x}_1$ where $\bar{p} = \bar{x}_1$ is the point for which $f(\bar{x}_1) = t$, and we write this as

$$d(p,t) = p - x_1^* - (\bar{p} - x_1^*), \quad \text{such that } f(\bar{p}) = t. \tag{17}$$

i.e.,

$$t = t^* + a_2(\bar{p} - x_1^*)^2 + a_3(\bar{p} - x_1^*)^3 + \dots + a_m(\bar{p} - x_1^*)^m. \tag{18}$$

Denoting $\delta = \bar{p} - x_1^*$, we get

$$\delta^2 = \alpha(t - t^*) - \sum_{k=3}^{m} b_k \delta^k. \tag{19}$$

Taking the square root of both sides of (19), we get

$$\delta = \beta(t - t^*)^{0.5}(1 - \sum_{k=3}^{m} b_k'\left(\frac{\delta^k}{t - t^*}\right))^{0.5}. \tag{20}$$

Note that for a small δ, by (19) $\delta = O((t - t^*)^{0.5})$, which implies that

$$\frac{\delta^k}{t - t^*} = O((t - t^*)^{\frac{k}{2} - 1}). \tag{21}$$

Re-substituting the above expression for δ in the r.h.s. of (20), and using the binomial expansion for the second square root there, we obtain the following asymptotic expansion for δ:

$$\delta \sim \sum_{k=1}^{\infty} c_k (t - t^*)^{\frac{k}{2}}, \quad as \ \ t \to t^*. \tag{22}$$

Together with (17) this gives us the asymptotic expansion of $d(p, t)$ near a tangency level for $d = 2$:

$$d(p, t) \sim \sum_{k=0}^{\infty} c_k (t - t^*)^{\frac{k}{2}}, \quad as \ \ t \to t^*. \tag{23}$$

A similar expansion holds in higher dimension $d > 2$:

Proposition 1. *Let B be a closed domain in $\mathbb{R}^d$ with a C^k boundary ΓB composed of a finite number of disconnected surfaces. Let t^* be a tangency level, and $x^* = (x_1^*, ..., x_{d-1}^*, t^*) \in \Gamma B$ be the tangency point. For a fixed vector $p = (x_1, ..., x_{d-1})$ and for t close to t^*, $t \geq t^*$, assume that $dist((p, t), \Gamma B)$ is attained near x^*. Then*

$$d(p, t) \sim \sum_{k=0}^{\infty} c_k (t - t^*)^{\frac{k}{2}}, \quad as \ \ t \to t^*. \tag{24}$$

Proof. Let $(\bar{p}, t)$ the point on ΓB at which $dist((p, t), \Gamma B)$ is attained. Consider the two dimensional plane through (p^*, t^*) and $(\bar{p}, t)$:

$$P = \{(p^*, t^*) + \alpha((\bar{p}, t) - (p^*, t^*)) + \beta(0, ..., 0, 1) \mid (\alpha, \beta) \in \mathbb{R}^2\}. \tag{25}$$

On this plane we are back in the framework of the case $d = 2$ discussed above, and we have, by (22),

$$\|\bar{p} - p^*\| \sim \sum_{k=1}^{\infty} c_k (t - t^*)^{\frac{k}{2}}. \tag{26}$$

The coefficients $\{c_k\}$ in (26) depend on t since $\bar{p}$ depends on t. The dependence is due the fact that for different values of t, the plane P in (25) may be different. However, this dependent is smooth in t and the expansion (26) is still valid, with maybe different coefficients.

To evaluate $d(p, t)$, we note that $d(p, t) = \|p - \bar{p}\|$, and use the relation

$$p - \bar{p} = p - p^* - (\bar{p} - p^*). \tag{27}$$

Using the law of cosine

$$\|p - \bar{p}\|^2 = \|p - p^*\|^2 + \|\bar{p} - p^*\|^2 - 2\|p - p^*\|\|\bar{p} - p^*\|cos(\theta), \tag{28}$$

where θ is the angle between $p - p^*$ and $\bar{p} - p^*$. For $p \neq p^*$, $\theta \to 0$ as $t \to t^*$, and $cos(\theta) \to 1$ smoothly in $t - t^*$. Taking the square root of both sides of (28), recalling that $p - p^*$ is a constant vector here, and using (26), implies the expansion (24).

3   The Enhanced SDR Procedure

As discussed in [10], at levels which are not close to a tangency level $d(p, \cdot)$ is C^k smooth. Using the above expansion for $d(p, \cdot)$ near tangency levels we are now ready to present the enhanced SDR procedure. Note that for proving Proposition 1 we have assumed that at the tangency point ΓB is a local minima. The other two cases include local maxima and a local saddle-point. This gives us three possible choices for the singular basis functions $\{u_j^s\}$ to be used in the PSA procedure (Sect. 1.3) :

$$I. \quad u_j^s(t) = (t - s)_+^{\frac{j-1}{2}}, \qquad j = 1, ..., k.$$

$$II. \quad u_j^s(t) = (s - t)_+^{\frac{j-1}{2}}, \qquad j = 1, ..., k. \tag{29}$$

$$III. \quad u_j^s(t) = |t - s|^{\frac{j-1}{2}}, \qquad j = 1, ..., k.$$

For simplicity, we consider equidistant cross-sections levels,

$$t_i = ih, \qquad i = 1, ..., n.$$

The PSA procedure involves a quasi-interpolation operator, and here we suggest to use spline quasi-interpolants of degree m, reproducing Π_m, and having a support size $2mh$. As explained in [11], the quasi-interpolant of a function with a singular point (of some low order derivative) exhibits the location of the singularity by large approximation errors at the data points near the singularity. In the d-dimensional object reconstruction problem from cross-sections, one needs to apply approximation to $d(p, \cdot)$ along many lines L_p. In any case of a smooth bounded object, there are always some tangency levels. We assume there is a finite number of isolated tangency levels, and, for simplicity, we assume that each tangency level includes only one tangency point on B. Recalling that the PSA procedure involves a non-linear optimization step (13), it would be quite expensive to apply for many lines L_p. Hence, we suggest the following preprocessing step, in which one first identify the tangency levels of the object, and approximate them with a high accuracy. Having these approximations of the tangency levels, the approximation of the signed-distance functions $\{d(p, \cdot)\}$ becomes a linear algorithm. The following enhanced form of the SDR procedure can be used within most generalizations of the basic SDR procedure, such as [5,7].

3.1   Preprocessing: Approximating the Tangency Levels

Applying the quasi-interpolation operator on a function of the form (7)−(8), with $\{u_j^s\}$ of the forms in (29), one gets good approximation away from the singular point s, and high errors near s, as shown in Fig. 1. The errors are even higher than those obtained for a singularity of the form (9). The large variation in the error pattern for close values of s, as demonstrated in Fig. 1, implies the possibility of accurately pinpointing s using such error data. This property is analysed in Theorem 2.

To approximate the tangency levels, we suggests the following algorithm:

I. Define a dense enough array of lines $\{L_p\}$.

II. For each line compute the distances $\{d(p, t_i)\}_{i=1}^n$.

III. For each line apply the quasi-interpolation operator Q, and compute the approximation errors

$$E_{p,i} = d(p, t_i) - \sum_{j=1}^n q_j(t) d(p, t_j), \quad i = 1, ..., n. \qquad (30)$$

IV. Locate lines with high approximation errors, and identify the intervals (in t) with high errors.

V. Choose a subset of the lines $\{L_{p_\ell}\}_{\ell=1}^r$ with distinctly different high errors intervals.

VI. Define $\rho(t) = \sum_{j=1}^k c_j u_j^s(t)$, for each of the three choices of singular basis functions in (29).

VII. For $1 \leq \ell \leq r$ apply the singularity approximation procedure from Sect. 1.3, namely, find, for each of the three choices of singular basis functions in (29),

$$\{s^*, \{c_j^*\}\} = \arg\min_{s, \{c_j\}} \sum_{i \in I_{p_\ell}} [(d(p_\ell, \cdot) - \rho - Q(d(p_\ell, \cdot) - \rho))(t_i)]^2, \qquad (31)$$

where I_{p_ℓ} is the set of indices of high approximation errors along L_{p_ℓ}.

VIII. For $1 \leq \ell \leq r$ choose out of the three models of singular basis functions in (29) the one which gives the minimal value to the sum of squares in (31), define the approximation to the tangency level $s_\ell = s^*$, and mark the corresponding singularity model.

The levels $\{s_\ell\}_{\ell=1}^r$ are the approximate tangency levels.

Remark 3. Note that one line may include several intervals with high errors. Each of these intervals should be treated separately in step VII above.

Remark 4. The sum of squares in (31) may be replaced by a weighted sum, ·with a compact smoothly varying weight function, in order to achieve a smooth approximation.

3.2   Reconstruction

The approximate tangency levels, together with their corresponding singularity models, are now being used in the reconstruction algorithm, as follows:

A. For any line L_p, apply steps II-IV of the previous algorithm.

B. Denote by I_p a set of indices of high approximation errors. Match to I_p one of the pre-computed tangency levels, s_ℓ, and its corresponding singularity model.

C. Find the coefficients of the model as

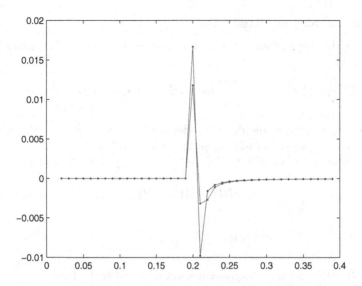

Fig. 1. A cubic spline quasi-interpolation approximation is applied to the function $f(x) = sin(x) + (x-s)_+^{0.5}$. The function values are given at equidistant points $t_i = 0.01i$, and the errors $f(t_i) - Qf(t_i)$ are displayed for $s = 0.2$ (blue) and for $s = 0.205$ (red) (Color figure online).

$$\{c_j^*\} = \arg\min_{\{c_j\}} \sum_{i \in I_p} [(d(p, \cdot) - \rho - Q(d(p, \cdot) - \rho))(t_i)]^2, \qquad (32)$$

where

$$\rho(t) = \sum_{j=1}^{k} c_j u_j^{s_\ell}(t). \qquad (33)$$

Note that this is a linear problem.

D. Define the corrected approximation (using (15))

$$d^*(p, \cdot) = Q(d(p, \cdot) + \rho^* - Q(\rho^*). \qquad (34)$$

where

$$\rho^*(t) = \sum_{j=1}^{k} c_j^* u_j^{s_\ell}(t). \qquad (35)$$

E. Define the approximation $\bar{B}_p$ to $B_p = L_p \cap B$ as in (4):

$$\bar{B}_p = \{(p, t) \mid d^*(p, t) \geq 0\}. \qquad (36)$$

Remark 5. The reconstruction step is a local linear process, and it may also be replaced by an interpolatory scheme.

Remark 6. The choice of the index set I_p in (31) and of the approximation order k in (33) is explained and specified in the next section.

3.3 Approximation Rates

To analyze the approximation $\bar{B}_p$ to B_p we have to study the minimization problem (31)

$$\{s^*, \{c_j^*\}\} = \operatorname*{arg\,min}_{s, \{c_j\}} \sum_{i \in I_{p_\ell}} [(d(p_\ell, \cdot) - \rho - Q(d(p_\ell, \cdot) - \rho))(t_i)]^2. \tag{37}$$

We discuss here the first singularity model in (29-I), the other models are treated the same. For the model (29-I) the problem is as follows:

Let us assume we are given values at $\{t_i\} = \{ih\}$ of a function f of the form

$$f(t) = g(t) + \bar{\rho}(t), \tag{38}$$

where

$$\bar{\rho}(t) = \sum_{j=1}^{k} \bar{c}_j (t - \bar{s})_+^{\frac{j-1}{2}}, \tag{39}$$

and $g \in PC^{m+1}$. Define the approximation to s and to $\{c_j\}$ as

$$\{s^*, \{c_j^*\}\} = \operatorname*{arg\,min}_{s, \{c_j\}} F(s, \{c_j\}), \tag{40}$$

where

$$F(s, \{c_j\}) = \sum_{i \in I} [(f - \rho - Q(f - \rho))(t_i)]^2, \tag{41}$$

and

$$\rho(t) = \sum_{j=1}^{k} c_j (t - s)_+^{\frac{j-1}{2}}. \tag{42}$$

In (40) I is a set including $2(m + \kappa)$ data points, s is in the two mid-intervals, and Q is a quasi-interpolation operator of a finite support $2m$, reproducing Π_m.

Before presenting the approximation theorem, let us consider the case where $g \equiv 0$ in (38). Of course, the functional $F(s, \{c_j\})$ attains its minimum with $s^* = \bar{s}$ and $\{c_j^*\} = \{\bar{c}_j\}$. Yet, it is not clear that this is the unique minimizer in the space of the free parameters s and $\{c_j\}$. Since we are using $m + \kappa$ data points to the right of $\bar{s}$, and these are the only points which carry information about $\bar{s}$ and $\{\bar{c}_j\}$, we can expect to have a unique solution only if $k < m + \kappa$. In order to match the approximation order to the singularity in $d(p, \cdot)$ to that of the quasi-interpolation operator we shall use $k = 2m + 2$. That implies $\kappa > m + 2$, and we take $\kappa = m + 3$.

The optimization problem is non-linear in s, and linear in $\{c_j\}_{j=1}^{2m+2}$ and it can be shown to be scale independent. i.e., showing that the solution is unique and stable for $h = 1$, implies the same for any h. We have performed this check numerically, and from now we shall assume it holds:

Assumption A. The optimization problem (40) with $g \equiv 0$, with $k = 2m+2$ and $\#I = 2(2m+3)$ has a unique solution, and it is stable. i.e., a small perturbation yields small variations in the solution.

The following theorem follows the main theorem in [11].

Theorem 2. *Let s^* and $\{c_j^*\}$ be defined as above, and let*

$$\bar{Q}f = Qf + \rho^* - Q(\rho^*) \tag{43}$$

where

$$\rho^*(t) = \sum_{j=1}^{2m+2} c_j^*(t - s^*)_+^{\frac{j-1}{2}}. \tag{44}$$

Then,

$$|\bar{s} - s^*| = O(h^{m+2}), \tag{45}$$

and

$$|f(t) - \bar{Q}f(t)| = O(h^{m+1}), \quad \forall t. \tag{46}$$

Proof. The proof starts as in [11]. By the approximation properties of Q, together with the asymptotic expansion (29-I), we have

$$F(\bar{s}, \{\bar{c}_j\}) = O(h^{m+1}). \tag{47}$$

Hence, the minimal value of $F(s, \{c_j\})$ is also $O(h^{m+1})$, and since there is a finite number of terms in (41), each of them os also $O(h^{m+1})$. By the definition of the corrected approximation $\bar{Q}$ in (43) it follows that the errors in the approximation are $O(h^{m+1})$ at the data points near $\bar{s}$:

$$|f(t_i) - \bar{Q}f(t_i)| = O(h^{m+1}), \quad i \in I. \tag{48}$$

Using this observation, together with Assumption A, we would like to show that the above approximation error result holds for all t.

Using (48) and (43) it follows that

$$|(\bar{\rho} - \rho^* - Q(\bar{\rho} - \rho^*))(t_i)| = O(h^{m+1}), \quad i \in I. \tag{49}$$

After applying Q and regathering the coefficients of the $\{(t_i - \bar{s})^{\frac{j-1}{2}}\}$ and of $\{(t_i - s^*)^{\frac{j-1}{2}}\}$ we get,

$$\left| \sum_{j=1}^{2m+2} \bar{a}_j(t_i - \bar{s})_+^{\frac{j-1}{2}} - \sum_{j=1}^{2m+2} a_j^*(t_i - s^*)_+^{\frac{j-1}{2}} \right| = O(h^{m+1}), \quad i \in I, \tag{50}$$

where $\{\bar{a}_j\}$ and $\{a_j^*\}$ are uniquely determined by $\{\bar{c}_j\}$ and $\{c_j^*\}$. Let us assume that $s^* \in [0, h)$, $I = \{-2m - 2, ..., 2m + 3\}$, and $t_i = ih$. The relations (50) imply that s^* and $\{a_j^*\}$ satisfy the system

$$\sum_{j=1}^{2m+2} a_j^*(t_i - s^*)_+^{\frac{j-1}{2}} = \sum_{j=1}^{2m+2} \bar{a}_j(t_i - \bar{s})_+^{\frac{j-1}{2}} + O(h^{m+1}), \quad i \in I_+, \tag{51}$$

where $I_+ = \{1, 2, ..., 2m + 3\}$. Defining $b_j^* = a_j^* h^{\frac{j-1}{2}}$, $\bar{b}_j = \bar{a}_j h^{\frac{j-1}{2}}$, $\bar{\sigma} = \bar{s}/h$ and $\sigma^* = s^*/h$, we can write (51) as

$$A(\sigma^*)v^* = A(\bar{\sigma})\bar{v} + O(h^{m+1}), \tag{52}$$

where $\bar{v} = (\bar{b}_1, ..., \bar{b}_{2m+2})$ and $v^* = (b_1^*, ..., b_{2m+2}^*)$. Note that $A(s)$ is a Vandermonde matrix, independent of h. Omitting the equation for $i = 1$ in (51) we are left with a square linear system for v^* (given σ^*):

$$A'(\sigma^*)v^* = A'(\bar{\sigma})\bar{v} + \bar{e}, \tag{53}$$

where $\bar{e} = O(h^{m+1})$ and $A'(\sigma^*)$ is an invertible matrix. The omitted equation is

$$\sum_{j=1}^{2m+2} b_j^*(1 - \sigma^*)_+^{\frac{j-1}{2}} = \sum_{j=1}^{2m+2} \bar{b}_j(1 - \bar{\sigma})_+^{\frac{j-1}{2}} + O(h^{m+1}). \tag{54}$$

The existence of the solution is assured by Assumption A. In order to learn about the approximation properties of this solution, let us consider a fictitious iterative procedure for computing σ^*. Starting with $\sigma^{(0)} = \bar{\sigma}$, and $v^{(0)} = \bar{v}$, we use (53) to define $v^{(n+1)} = (b_1^{(n+1)}, ..., b_{2m+2}^{(n+1)})$:

$$v^{(n+1)} = \left(A'(\sigma^{(n)})\right)^{-1}\left(A'(\bar{\sigma})\bar{v} + \bar{e}\right). \tag{55}$$

Now we use $v^{(n+1)}$ in (54) to define $\sigma^{(n+1)}$ as the solution of

$$\sum_{j=1}^{2m+2} b_j^{(n+1)}(1 - \sigma^{(n+1)})_+^{\frac{j-1}{2}} = \sum_{j=1}^{2m+2} \bar{b}_j(1 - \bar{\sigma})_+^{\frac{j-1}{2}} + O(h^{m+1}). \tag{56}$$

It follows from (55) that for $n > 1$

$$\|v^{(n)} - \bar{v}\| = O(h^{m+1}), \tag{57}$$

and by (56) that

$$|\sigma^{(n)} - \bar{\sigma}| = O(h^{m+1}). \tag{58}$$

In order to ensure convergence of the above iterative procedure we have to add another assumption, saying that $|\bar{b}_2|$ is large enough. The convergence of the procedure implies that the limit values v^* and σ^* satisfy

$$\|v^* - \bar{v}\| = O(h^{m+1}), \tag{59}$$

$$|\sigma^* - \bar{\sigma}| = O(h^{m+1}). \tag{60}$$

Using the relations $b_j^* = a_j^* h^{\frac{j-1}{2}}, \bar{b}_j = \bar{a}_j h^{\frac{j-1}{2}}, \bar{\sigma} = \bar{s}/h$ and $\sigma^* = s^*/h$, we obtain

$$\|a_j^* - \bar{a}_j\| = O(h^{\frac{2m+3-j}{2}}), \quad j = 1, ..., 2m + 2, \tag{61}$$

$$|s^* - \bar{s}| = O(h^{m+2}). \tag{62}$$

By (61) and (62) it follows that

$$|\rho^*(t) - \bar{\rho}(t)| = O(h^{m+1}), \tag{63}$$

which yields the main result (46).

4 Numerical Examples

To illustrate the proposed method and its performance we suggest an illuminating example of reconstructing a 2D object from its 1D parallel cross sections. In Fig. 2 we have a 2D object with a smooth boundary, for which the topology of the cross-sections is changed at a single level. We also observe the square-root type singularity of the signed-distance function near this level.

For the same data which appears in Fig. 2, we now apply the procedure for computing the correction terms. On Fig. 3 the small crosses mark the errors in the quasi-interpolation approximation to the signed-distance function at the given levels. The continuous graph describes the fitting of $\rho - Q\rho$ to this errors, using the optimal s^* and $\{c_j^*\}$ computed by (13).

The reconstruction method is applied along several vertical cross-sections, achieving high accuracy also near the problematic area of topology change, as seen in Fig. 4.

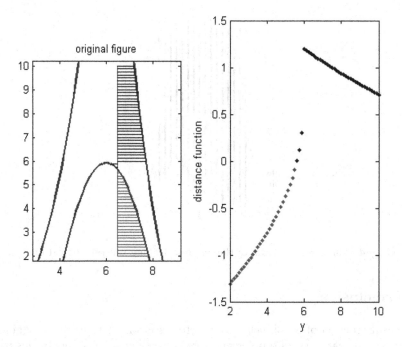

Fig. 2. On the left graph we see the boundaries of the object (in blue), and the distances to the boundary from a fixed vertical line for the different cross-sections. On the right graph we depict the values of the signed-distance function. One clearly observe the singular behavior near level $y = 6$ (Color figure online).

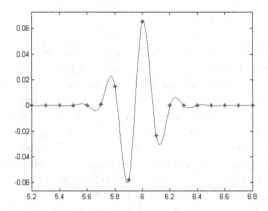

Fig. 3. The crosses mark the errors in the quasi-interpolation approximation to the signed-distance function at the given levels. The continuous graph describes the fitting of $\rho - Q\rho$ to this errors, using the optimal s^* and $\{c_j^*\}$.

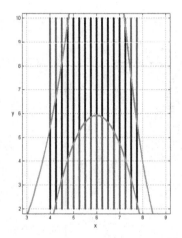

Fig. 4. The reconstruction of the 2D object along several vertical lines.

5   Conclusions

The reconstruction of an object from its cross-sections data exhibits significant accuracy lose near areas of topology change between cross-sections. In this paper we analyse the singularity behavior of the sign-distance function of the cross-sections near such areas. Then, we use this special singularity behaviour within an efficient method for approximating functions with singularities. It is proved that the new enhanced method overcomes the problem of accuracy lose in the object reconstruction algorithm.

References

1. Bajaj, C.L., Coyle, E.J., Lin, K.N.: Arbitrary topology shape reconstruction from planar cross sections. Graph. Models Image Process. **58**(6), 524–543 (1996)
2. Bermano, A., Vaxman, A., Gotsman, C.: Online reconstruction of 3D objects from arbitrary cross-sections. ACM Trans. Grap. (TOG) **30**(5), 113 (2011)
3. Boissonnat, J.-D.: Shape reconstruction from planar cross sections. Comput. Vis. Graph. Image Process. **44**(1), 1–29 (1988)
4. Boissonnat, J.-D., Memari, P.: Shape reconstruction from unorganized cross-sections. In: Symposium on Geometry Processing, pp. 89–98 (2007)
5. Cohen-Or, D., Solomovic, A., Levin, D.: Three-dimensional distance field meta-morphosis. ACM Trans. Graph. (TOG) **17**(2), 116–141 (1998)
6. Fuchs, H., Kedem, Z.M., Uselton, S.P.: Optimal surface reconstruction from planar contours. Commun. ACM **20**(10), 693–702 (1977)
7. Herman, G.T., Zheng, J., Bucholtz, C.A.: Shape-based interpolation. IEEE Comput. Graph. Appl. **12**(3), 69–79 (1992)
8. Kels, S., Dyn, N.: Reconstruction of 3D objects from 2D cross-sections with the 4-point subdivision scheme adapted to sets. Comput. Graph. **35**(3), 741–746 (2011)
9. Kels, S., Dyn, N.: Subdivision schemes of sets and the approximation of set-valued functions in the symmetric difference metric. Found. Comput. Math. **13**(5), 835–865 (2013)
10. Levin, D.: Multidimensional reconstruction by set-valued approximations. IMA J. Numer. Anal. **6**(2), 173–184 (1986)
11. Lipman, Y., Levin, D.: Approximating piecewise-smooth functions. IMA J. Numer. Anal. **30**(4), 1159–1183 (2010)
12. Nurzynska, K.: 3D object reconstruction from parallel cross-sections. In: Bolc, L., Kulikowski, J.L., Wojciechowski, K. (eds.) ICCVG 2008. LNCS, vol. 5337, pp. 111–122. Springer, Heidelberg (2009)

Adaptive Atlas of Connectivity Maps

Ali Mahdavi-Amiri$^{(\boxtimes)}$ and Faramarz Samavati

University of Calgary, Calgary, Canada
{amahdavi,samavati}@ucalgary.ca

Abstract. The Atlas of Connectivity Maps (ACM) is a data structure designed for semiregular meshes. These meshes can be divided into regular, grid-like patches, with vertex positions stored in a 2D array associated with each patch. Although the patches start at the same resolution, modeling objects with a variable level of detail requires adapting the patches to different resolutions and levels of detail. In this paper, we describe how to extend the ACM to support this type of adaptive subdivision. The new proposed structure for the ACM accepts patches at different resolutions connected through one-to-many attachments at the boundaries. These one-to-many attachments are handled by a linear interpolation between the boundaries or by forming a transitional quadrangulation/triangulation, which we call a *zipper*. This new structure for the ACM enables us to make the ACM more efficient by dividing the initial mesh into larger patches.

Keywords: Atlas of connectivity maps · Adaptive subdivision · Data structure

1 Introduction

Semiregular models are very common in computer graphics, appearing in subdivision and multiresolution surfaces, Digital Earth representations and many parametrization techniques. Semiregular models are made of a set of regular patches that are attached to each other and the extraordinary vertices may only be located at the corner of patches [27,28]. These models result from applying repetitive refinements on a model with arbitrary connectivity (see Fig. 1). In a semiregular model, most vertices and faces are regular, allowing one to design an efficient data structure called the Atlas of Connectivity Maps (ACM) [1]. In the ACM, the connectivity queries within each patch are handled by simple algebraic operations and, to transit from one patch to another, simple pre-calculated transformations are used. In addition, there exists a hierarchy between the coarse and refined faces of each patch that factors into the design of the ACM. Consider Fig. 1, which shows patches before and after refinement. The ACM also provides simple algebraic relations to support such a hierarchical correspondence between faces and vertices, which proves to be advantageous in applications such as mesh editing and multiresolution.

Although ACM is efficient for supporting semiregular models in which all the patches have the same resolution (Fig. 1(b)), we wish to support adaptive

© Springer International Publishing Switzerland 2015
J.-D. Boissonnat et al. (Eds.): Curves and Surfaces 2014, LNCS 9213, pp. 304–320, 2015.
DOI: 10.1007/978-3-319-22804-4_23

refinement, where different regions of the mesh can have different resolutions satisfying a geometric or a user specified condition (e.g. smoothness). The original description of the ACM considers a connectivity map for each patch of a given semiregular model and it is useful for meshes that are uniformly refined. It is not designed for adaptive scenarios. In this paper, we extend the ACM to support adaptive refinement by modifying its structure to support patches at different resolutions. To adaptively refine a connectivity map, it is split into an optimized number of new connectivity maps. Necessary transformations for traversing between the connectivity maps are also calculated for each new connectivity map. In the original description of ACM, each connectivity map is connected to four other connectivity maps and transformations between them are pre-calculated and encoded into four integer numbers. However in the adaptive case, connectivity maps can be connected to an arbitrary number of other connectivity maps and transformations are not restricted to only four forms. We extend the ACM in such a way that it can support connections to multiple connectivity maps by maintaining a list of each connectivity map's neighbors. For each neighbor, we also store additional adjacency information, from which transformations are calculated using simple algebraic relationships. Gaps between different resolutions are resolved either by a set of transitioning triangles called zippers (see Fig. 1(c)) or by placing the vertices of the higher resolution patch on the edges of the lower resolution patch using a linear interpolation.

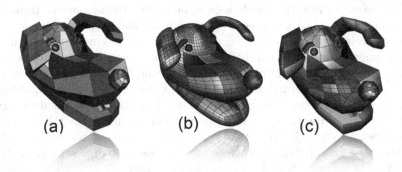

(a)   (b)   (c)

Fig. 1. (a) Coarse model. (b) A semiregular model obtained by uniformly subdividing the coarse model in (a). (c) Adaptively subdividing regions of interest. Patches at different resolutions are connected by zippers highlighted with orange faces (Color figure online).

With our proposed structure for the ACM, we can achieve a more compact representation of a mesh by forming larger initial connectivity maps. To do so, we provide an algorithm that can find regular quadrilateral patches. We also show that this algorithm significantly improves the performance of the ACM.

The paper is organized as follows: Related work is presented in Sect. 2. An overview of the ACM is described in Sect. 3. The extension of the ACM for adaptive subdivision is discussed in Sect. 4, followed by algorithms and discussions for improving the ACM by considering larger patches in Sect. 5. In Sect. 6, results and discussion are presented and we finally conclude in Sect. 7.

2 Related Work

Many data structures have been proposed to work with polygonal meshes [3]. Among the proposed data structures, edge-based data structures and their variations such as the half-edge structure are very common for applications that make liberal use of connectivity queries (e.g. subdivision) [11,12]. As a result, the half-edge structure is professionally implemented in many libraries such as CGAL with the ability to support many types of subdivision schemes [4]. They are also used to support adaptive subdivision, as they can handle very local connectivity queries. In addition to subdivision, half-edges have also been extended in supporting multiresolution surfaces [13]. Although half-edges are very useful to support meshes with arbitrary connectivity, they do not benefit from the regularity of meshes and the hierarchy between faces that result after applying subdivision.

To benefit from the regularity and hierarchy of patches obtained from subdivision, some hierarchical data structures have been proposed. Quadtrees are one of the most common data structures used to support subdivision, particularly when the factor of refinement is four (e.g. Loop and Catmull-Clark subdivision) [9,10,24,25]. Variations on quadtrees, such as balanced quadtrees, have been proposed for adaptive subdivision in order to establish smooth transitions between patches [29]. A quadtree is balanced if any two neighboring nodes differ at most one in depth. However, since quadtrees generally require many hierarchical connections to maintain the hierarchy of faces, other data structures have been proposed that are specifically designed for subdivision. Some of these data structures are based on indexing methods specific to the type of subdivision. Shiue et al. use a 1D spiral indexing for subdividing quadrilateral and triangular meshes in [6] and use the same indexing method to employ the GPU in subdividing meshes with arbitrary topology [5]. Some patch-based methods have been also proposed for subdivision schemes in which each patch is separately stored in a 2D array [7,8]. These data structures are very efficient when employing subdivision. However, they often restrict the initial connectivity of the mesh or specific type of subdivision.

The ACM is a data structure that supports connectivity queries on semi-regular models. Similar to a patch-based data structure, the ACM stores the geometric information of each patch in a 2D array. A unique aspect of the ACM is its ability to handle global and inter-patch connectivity queries using inter-patch transformations. In addition, the ACM comprehensively handles all existing refinements used in subdivision surfaces [1,2,27]. However, adaptive subdivision is not supported by the ACM since the primary assumption in its design is that all patches are at the same resolution.

Several adaptive subdivision methods have been proposed. In these methods, the geometry and the connectivity of the original subdivision scheme are modified to maintain the adaptivity. Geometric modifications to vertices in a locally subdivided region are handled differently [14–16,23]. Moreover, since cracks may be created in the transition from a coarse region to a smooth region, the connectivity has to be modified. While directly connecting the newly inserted vertices to existing vertices is a solution [15], more sophisticated approaches have been taken, such as red-green and incremental adaptive subdivision. In red-green algorithm, faces with one crack are bisected (green triangulation) while faces with more than one crack per edge are split into four (red triangulation) [21]. Under the incremental adaptive subdivision, a one ring neighborhood is introduced to transition from a smooth region to a coarse region, and in the process avoids high valence vertices and skinny triangles [18–20]. In addition, Pannozo and Puppo designed a method for adaptive Catmull-Clark subdivision by limiting the transitional polygons to pentagons and triangles [22]. The main challenge to creating an efficient data structure for adaptive subdivision is how to handle the change in connectivity. We provide two solutions to handle connectivity modifications by providing transitional domains (zippers) between coarse and fine connectivity maps, or linearly interpolating the boundary edge between a high and low resolution connectivity map.

3   Overview of the ACM

In this section, we provide an overview of the ACM [1]. As mentioned earlier, the ACM is designed for semiregular meshes, which are made of connected regular patches resulting from a regular refinement (Fig. 2). Each patch i is assigned to a 2D domain for which a 2D coordinate system is considered. This 2D domain along with its connectivity information is called a *connectivity map* $(CM(i))$. In an ACM, an array of connectivity maps CM is stored for a semiregular model with M patches in which $CM(i)$ $(0 \leq i < M)$ refers to the ith patch of the model. The coordinate system of $CM(i)$ is used to index vertices within a 2D location array that records the 3D positions of the vertices through the resolutions. Connectivity queries for internal vertices of $CM(i)$ are handled by neighborhood vectors that are added to the index of a vertex (Fig. 2(c)). As refinements are applied, the connectivity information of the model continues to be maintained by the connectivity maps. Based on the type of refinement, hierarchical relationships are defined for vertices and faces that are useful in applications such as mesh editing and multiresolution.

The connectivity information between $CM(i)$ and its neighboring connectivity maps, denoted by $CM(N_j(i))$, should be recorded as well in order to support connectivity queries outside of each patch. As a result, the of neighbors of $CM(i)$ are stored in an array, say *neighbors*. Inter-patch queries between $CM(i)$ and $CM(N_j(i))$ are handled by a set of transformations that map the coordinate system of $CM(i)$ to the coordinate system of $CM(N_j(i))$. To simplify these transformations, they are encoded as integer numbers (Fig. 3(b)). Consequently, $CM(i)$ has a 2D array of 3D points storing the locations of vertices, a 1D array recording $CM(N_j(i))$ and an array storing the inter-patch transformation codes (see Fig. 3(a)). Since the corner vertices of $CM(i)$ may be

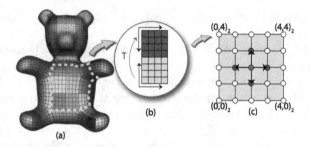

(b)

(a)

Fig. 2. (a) A semiregular mesh. (b) For each patch, a 2D domain (connectivity map) with a coordinate system is assigned. To move from one patch to another, a transformation between the coordinate systems of two adjacent connectivity maps can be used. (c) To index vertices, the coordinate system defined for each patch is used. The subscript of the index refers to the resolution of the refinement. Neighborhood vectors are used to obtain neighbors of internal vertices.

irregular/extraordinary, a separate structure is used to store corner neighbors of $CM(i)$. This structure - a list for each corner - stores the connectivity maps that are attached to each corner of $CM(i)$.

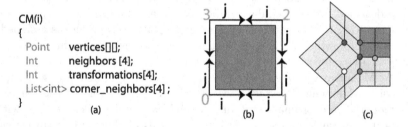

Fig. 3. (a) The elements that are stored for each connectivity map. (b) Inter-patch transformations are encoded as integer numbers. (c) A corner (red vertex) can be extraordinary. Connectivity maps attached to corners (in this case, two of them) are saved in corner_neighbors (Color figure online)

The ACM can support a variety of refinements for applications such as subdivision surfaces, Digital Earth frameworks and multiresolution [1,2].

4   Adaptive ACM

Although the ACM is an efficient data structure for semiregular models, adaptive subdivision cannot be supported within the original formulation for the ACM. One simple approach to make the ACM usable for adaptive subdivision is to consider the possibility of allowing patches to exist at different resolutions (see Fig. 4(a)). To do so, we only need to add an integer value recording the resolution

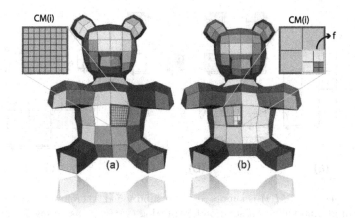

Fig. 4. (a) Adaptive subdivision with patches at different resolutions. (b) Adaptive subdivision with faces in a patch at different resolutions.

of each patch. However, this does not support the adaptive refinement of faces within a patch (see Fig. 4(b)).

To support such a case, we need to enable the possibility of breaking a given connectivity map $CM(i)$ into a set of smaller connectivity maps. Let $f \in CM(i)$ be the face selected for adaptive subdivision (see Fig. 4(b)). A new connectivity map with a new coordinate systems should be assigned to f (see Fig. 5). The question is what should happen to the other faces of $CM(i)$. As illustrated in Fig. 5(b), one simple solution is to generate individual connectivity maps for every face in $CM(i)$. However, it is more efficient to divide these faces between larger connectivity map blocks (see Fig. 5(c)). To benefit from the regularity of $CM(i)$, we can categorize all possible cases of this division based on the position of $f \in CM(i)$. As shown in Fig. 6, three possible cases exist when $CM(i)$ is divided to blocks. When f is located at the corner of, the boundary of, or internal to $CM(i)$, we split $CM(i)$ into three, four, or five connectivity maps respectively. Notice that while the dividing patterns are not unique, the number of blocks in each case is minimal. Coordinate systems aligned with that of $CM(i)$ are assigned to each new connectivity map (Fig. 6).

Dividing a connectivity map into these blocks requires that we support one-to-many attachments between connectivity maps (see Fig. 7(a)). These one-to-many attachments are not supported in the original ACM, as connectivity maps are only connected to one neighbor along each boundary edge. Here, we show how to extend the ACM to support such one-to-many attachments at boundary edges.

To store the 3D locations of vertices, we still use a 2D array. A list recording the connectivity information at the corners is also still sufficient. Hence, we only need to change the neighbors array, which we accomplish by changing it to an array of lists of neighbors (Fig. 7). In the original ACM, all possible transformations are encoded in four integer numbers. However, due to the existence of

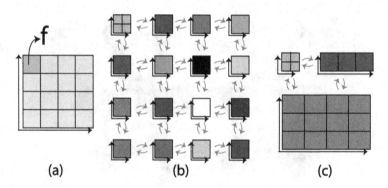

Fig. 5. (a) Face $f \in CM(i)$ is supposed to be subdivided, therefore $CM(i)$ has to be split. (b) Simple and inefficient solution for splitting $CM(i)$ by assigning a connectivity map to each face. (c) A more efficient split of $CM(i)$ that benefits from the regularity of $CM(i)$.

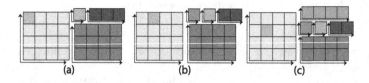

Fig. 6. $CM(i)$ is split into three, four, or five connectivity maps when f is located at (a) a corner of, (b) a boundary of, or (c) internal to $CM(i)$.

one-to-many attachments, it may be the case that more complex transformations map a connectivity map to one of its side neighbors. As a result, a new method of storing transformations has to be used that is flexible enough to include any type of transformation. To do so, the range of vertices connected to $CM(i)$ along a boundary edge is stored. To capture this, we store the indices of each neighbors first and last vertices along the shared boundary edge. For example, in Fig. 7, $CM(i)$ is connected to its neighbor (N_0) at indices $(e, f)_{\tilde{r}}$ and $(g, h)_{\tilde{r}}$ (Fig. 7(b)). These two indices are both stored for $CM(i)$. Using these two indices and the relative positions of these two connectivity maps, transformation T_0 that maps the coordinate system of $CM(i)$ to N_0 is calculated. Similarly, $(s, t)_{\tilde{r}}$ and $(u, v)_{\tilde{r}}$ are used to calculate T_1 (Fig. 7(c)). Consequently, we do not need to explicitly store transformations as they can be found from the indices. Therefore, our adaptive ACM has five components, as listed in Fig. 7(d).

When two neighboring patches have different resolutions, this creates an ambiguity in the definition of a vertex's neighborhood. For instance, the green vertex in Fig. 8(a) has two different sets of neighbors depending on whether we consider it from the high or low resolution patch. Hence, this ambiguity in the neighborhood definition should be somehow addressed based on the needs of the application. Using the adaptive ACM, connectivity information between patches is accessible even when they are at two different resolutions. Therefore,

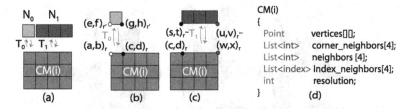

Fig. 7. (a) $CM(i)$ is connected to two connectivity maps N_0 and N_1. Transformations T_0 and T_1 are necessary to traverse these neighbors. (b), (c) The transformations that map a vertex in $CM(i)$ to its neighbors are determined by storing necessary indices from the neighboring connectivity maps. (d) The new structure of the ACM.

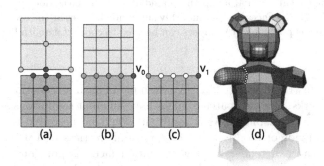

Fig. 8. (a) There exists ambiguity in determining the neighbors of a vertex when it is shared between a coarse and fine connectivity map. (b) The geometry of the blue vertices is obtained using regular refinement filters when two patches have the same resolution. (c) Two connectivity maps at two different resolutions. The white vertices at the boundary are linearly interpolated. (d) An example of refinement where vertices along the boundary between two connectivity maps at different resolutions are linearly interpolated (Color figure online).

any adjacency queries or geometric modifications on the vertices needed to perform adaptive subdivision remain possible. For example, a possible method for performing adaptive subdivision is to smoothly subdivide patches to a desired resolution (Fig. 8(b)) and linearly interpolate the vertices on boundary edges shared between high and low resolution patches to avoid cracks (Fig. 8(c)). In this case, the geometry of the ith boundary vertex on the high resolution patch is calculated by $\frac{i}{n}(v_1 - v_0) + v_0$ where n is the number of vertices on the high resolution patch and v_0 and v_1 are vertices on the low resolution patch. Figure 8(d) illustrates an example where a mesh is subdivided using Catmull-Clark subdivision and the edges between two patches at two different resolutions are linearly interpolated.

Although linearly subdividing patches is useful to insert more vertices and create low scale features (see Fig. 15), in some cases it may produce unwanted

Fig. 9. (a) Linear subdivision along the boundaries of the faces does not smooth the hand of the Teddy. (b) Faces can be smoothly subdivided. Faces at the coarse resolution can be rendered as polygons. (c) Zippers can be used to connect smooth faces to coarse faces.

artifacts, as illustrated in Fig. 9(a). An example of such a scenario is in Digital Earth frameworks, in which patches might have different resolutions and vertices are projected to the sphere. As a result, it would be desirable to be able to smoothly subdivide or modify the geometry of all vertices in a patch. To avoid artifacts, we can can render the low resolution faces as polygons rather than quads, with the edges of the polygons aligned with the vertices of the high resolution quads (see Fig. 9(b)). To obtain a mesh with higher quality, the high resolution patches can be connected to low resolution patches using transitional quadrangulations or triangulations called *zippers* (see Fig. 9(c)) [26].

These zippers are applicable for triangles and quads. Consider two connectivity maps $CM(i)$ and $CM(j)$ with m and n faces ($m > n$), respectively, along the common boundary. To connect $CM(i)$ to $CM(j)$, one simple solution is to insert a vertex in the adjacent triangle (or quad) strip of the lower resolution connectivity map or in the gap between these two connectivity maps. We can then simply connect all vertices of $CM(i)$ to $CM(j)$ (Fig. 10(a)). However, this is not a good solution due to the existence of high valence vertices and skinny triangles. A better solution is to insert $\frac{m}{n}$ vertices that connect $\frac{m}{n}$ number of faces in $CM(i)$ to a face in $CM(j)$ (Fig. 10(b)). We can achieve an even better transition by inserting extra vertices and requadrangulating or retriangulating the zipper as shown in Figs. 9(c) and 10(c). Figure 11 illustrates an example of zippers on the surface of the Earth. Note that these zippers can be formed on the fly using a simple algorithm, therefore an additional data structure is not needed.

5   Compact ACM

In the original ACM, a connectivity map is assigned to each quad at the coarsest resolution (i.e. control mesh) where each connectivity map has four neighbors

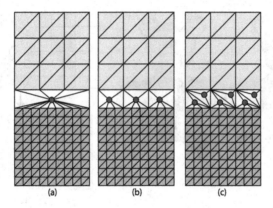

Fig. 10. (a) A zipper with one high valence vertex produces poor triangles (b) Adding more vertices to the zippers produces better triangulations. (c) It is possible to add vertices and retriangulate the zipper in (b) to achieve a better transition from a high resolution connectivity map to a low resolution one.

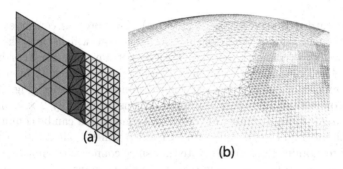

Fig. 11. (a) Zipper between two triangular connectivity maps. (b) Connectivity maps and zippers on a portion of the Earth.

(See Fig. 12(a)). The boundary vertices of each connectivity map are duplicated at each neighbor to obtain a separate, simple quadrilateral domain for each connectivity map. By merging two connectivity maps into one, such repetitions can be discarded and a more efficient ACM is obtained (See Fig. 12(c)). Using the adaptive ACM, connectivity maps of any size ($m \times n$) are acceptable, and they can have multiple neighbors (See Fig. 12(b)). Therefore, in an adaptive ACM, it is possible to combine several initial connectivity maps into one connectivity map as long as they form a quadrilateral domain.

We systematize our approach to this problem by proposing a simple algorithm that has three main functions: Pair, Union, and CleanUp (see Algorithm 1). Our algorithm takes a list of connectivity maps (CM) as input and outputs a list of combined connectivity maps denoted by l. In the Pair function, two neighboring connectivity maps $CM(i), CM(j) \in CM$ are combined and added

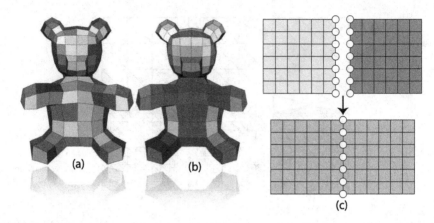

Fig. 12. (a) The original ACM, where a connectivity map is assigned to each individual quad at the first resolution. (b) Adaptive ACM supports larger initial connectivity maps. (c) When two connectivity maps are combined, duplicated vertices are discarded.

to l and discarded from CM (see Fig. 13(a)). This process continues until no connectivity map with a neighbor in CM exists, at which time the isolated connectivity maps (connectivity maps that do not have any neighbor in Q) are added to l. The Union function combines neighboring connectivity maps in l that share the same number of vertices at their common boundaries (Fig. 13(b)). In CleanUp, connectivity maps with a small dimension (usually 2×2, $m \times 1$, or $1 \times m$) are divided into a number of connectivity maps that can be combined with their neighbors (Fig. 13(c)). Union and CleanUp will be called interchangeably until no more modifications occur. Note that since connectivity maps each need a coordinate system, after combining two connectivity maps, a coordinate system aligned with the coordinate system of one of the connectivity maps is chosen for the new connectivity map.

Data: Mesh M given by a list of connectivity maps denoted by CM
Result: List l of combined Connectivity Maps.
Pair();
while *There are modifications* **do**
| Union();
| CleanUp();
end

Algorithm 1. An algorithm to create the ACM by combining patches into connectivity map blocks.

Table 1 lists the number of connectivity maps for several models when the algorithm is and is not applied. As apparent, this algorithm significantly reduces the number of connectivity maps. As a result, there are fewer repetitive vertices and less redundancy in the resulting ACM. For instance, after three applications

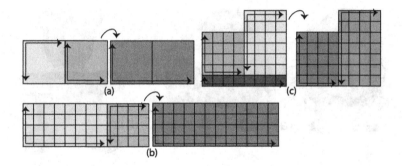

Fig. 13. (a) Pair, (b) Union, and (c) CleanUp.

Table 1. Number of connectivity maps in the original ACM and adaptive ACM. The reduction in the number of connectivity maps is apparent.

Models	ACM	AACM
Cube	6	3
Teddy	272	45
Big Guy	1450	191
Monster Frog	1292	187

Fig. 14. (a) Teddy, (b) Monster Frog, (c) Big Guy stored in an adaptive ACM.

of Catmull-Clark subdivision on the Big Guy model, about 30000 redundant vertices are removed if the connectivity maps are combined using the proposed algorithm. Figure 14 illustrates the models that are used in Table 1. Consequently, we can conclude that, using the adaptive ACM, the performance of the ACM is significantly improved and adaptive subdivision is also supported.

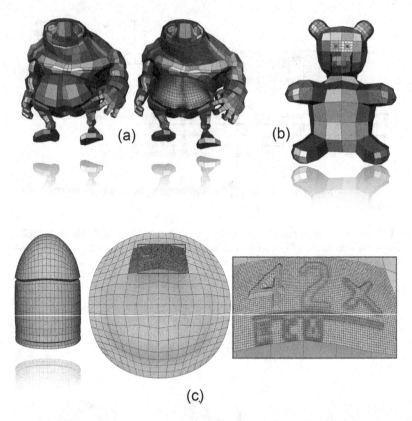

Fig. 15. (a) The Big Guy is adaptively subdivided on its belly and neck. Fine faces resulting from adaptively subdividing two different connectivity maps on its neck (purple and blue) are merged into a single connectivity map (red). (b) The hands and legs of Teddy are adaptively smoothed and some low scale details are added to the lips and the eyes of using adaptive refinement and local vertex manipulation. (c) The caliber of the Bullet is engraved on its back using adaptive refinement (Color figure online).

6   Comparisons and Results

Using the adaptive ACM, we no longer need to uniformly refine an entire model to add further details, and can efficiently model a wider variety of objects. For example, as illustrated in Fig. 15, we can add local low scale details such as engravings. Using the algorithm for compacting the connectivity maps, we can also join together connectivity maps created from splitting the connectivity maps during adaptive refinement. For instance, as illustrated in Fig. 15(a), adaptively subdivided faces on the neck of the Big Guy from two different connectivity maps (blue and purple) are joined into a single connectivity map (red).

The ACM performs efficiently in comparison to other data structures employed for adaptive subdivision. Consider the case illustrated in Fig. 5, in which face $f \in CM(i)$ is subdivided. $CM(i)$ is an $N \times N$ patch obtained from

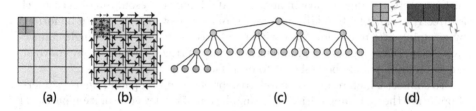

Fig. 16. (a) $f \in CM(i)$ is subdivided. (b) Half-edge pointers needed for edges of the patch in (a). (c) Pointers needed to store the patch in (a). (d) Adaptive ACM and its essential connectivity information (Color figure online).

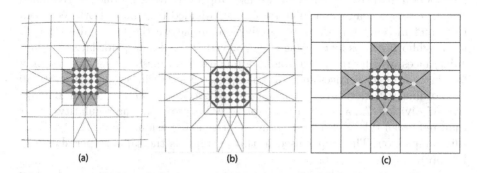

Fig. 17. (a) Red-green rule. (b) Incremental adaptive subdivision. (c) Zippers in the adaptive ACM. Images (a) and (b) are taken from [19,20] (Color figure online).

subdividing a coarse face C at the first resolution. We compare the memory consumption of the adaptive ACM with a half-edge structure and a quadtree; two standard data structures used to support adaptive subdivision. We also discuss the efficiency of zippers in comparison to red-green rules and incremental adaptive subdivision.

In the case illustrated in Fig. 16, three connectivity maps are stored with dimensions $((N+1) \times N)$, $(2 \times N)$, and (3×3). The connections between these connectivity maps are also stored in an adaptive ACM through 12 additional indices. By contrast, a half-edge data structure needs to store $(\approx (5 \times N \times N) + 12)$ for half-edge objects as well as $((N+1) \times (N+1) + 5)$ vertices and $(N \times N + 3)$ faces. In adaptive ACM, $(N+2)$ more vertices are needed, though we avoid $(\approx (5 \times N \times N) + 12)$ pointers for edges and $(N \times N + 3)$ faces. Hence, by respecting the regularity of the patches in an adaptive ACM, a large amount of memory can be saved.

It is possible to use quadtrees when a 1-to-4 refinement is used. As a result, a quadtree is often used to store $N \times N$ patch when $N = 2^M$. In this case, $4 \left(\frac{1-4^M}{1-4} \right)$ pointers are needed to identify the resolutions of the patches. For instance, for the patch illustrated in Fig. 16, 24 pointers are needed. However, such pointers are unnecessary in adaptive ACM and the resolution of each patch is directly accessible. In addition, a more compact adaptive ACM can be formed by grouping first resolution faces with regular connectivity into a single connectivity map, removing some redundant boundary vertices. Quadtrees, by contrast, require a unique node be assigned to each face of the first resolution.

To connect a high resolution patch to a lower resolution patch, we use zippers. Zippers are the quadrangulation or triangulation of the lower resolution patch (or the gap between them) to avoid cracks. Red-green rules and incremental adaptive subdivision are two other methods that may be used to connect high resolution patches to lower ones and establish a progressive resolution change among faces. Although we can connect the connectivity maps using red-green rules or incremental adaptive subdivision, we use zippers to avoid producing irregularities. Consider a coarse quadrilateral face subdivided two times. Both red-green rules and incremental adaptive subdivision propagate the face splits through the neighboring faces by creating irregular vertices and non-quadrilateral (triangular) faces, which are undesirable in a quadrilateral mesh. Using zippers, only four extraordinary vertices and four triangles are formed, in comparison to the 40 extraordinary vertices and 36 triangles of red-green rules and incremental adaptive subdivision. As a result, the adaptive ACM preserves the regularity better than red-green rules and incremental adaptive subdivision. Figure 17 illustrates this comparison. The zipper used in this example is the same zipper shown in Fig. 9(c).

Consequently, the adaptive ACM performs well when adaptively subdividing models in which the patches are mostly regular. However, if the subdivision is very local and the object exhibits irregular behaviors in the connectivity, the ACM does not provide a significant advantage over data structures that efficiently support arbitrary local modifications, such as the half-edge structure. As a result, we do not claim that adaptive ACM is the best data structure for adaptive subdivision in all cases, but that it is advantageous for applications in which there exists a relatively strong notion of regularity in the connectivity of the object.

7   Conclusion and Future Work

By extending the ACM, adaptive subdivision is supported and a more efficient ACM is achieved. We provide two solutions to support adaptive subdivision: one that linearly interpolates the edges at the boundary and one that uses transitional patches (zippers) to connect connectivity maps at two different resolutions. An algorithm is also proposed to obtain a compact set of regular patches from a given mesh. This algorithm performs fairly well and can discard redundant vertices found on the boundaries of adjacent patches. Although the algorithm provided in this paper significantly improves the ACM, future work may look into discovering a more optimized algorithm as our proposed algorithm does not guarantee the best possible set of regular patches in a given mesh. The adaptive ACM can be extended and improved in several other possible directions. For example, one possible problem is to determine the optimal zippers for a specific type of face, refinement, or application.

References

1. Mahdavi-Amiri, A., Samavati, F.: Atlas of connectivity maps. J. Comput. Graph. **39**, 1–11 (2014)
2. Mahdavi-Amiri, A., Harrison, E., Samavati, F.: Hexagonal connectivity maps for Digital Earth. Int. J. Digit. Earth 1–11 (2014)
3. De Floriani, L., Magillo, P.: Multiresolution Mesh Representation: Models and Data Structures. In: Iske, A., Quak, E., Floater, M.S. (eds.) Tutorials on Multiresolution in Geometric Modelling, pp. 363–418. Springer, Heidelberg (2002)
4. Kettner, L.: Halfedge data structures. In: CGAL User and Reference Manual (2013)
5. Shiue, L.-J., Peters, J.: A pattern-based data structure for manipulating meshes with regular regions. In: Proceedings of Graphics Interface 2005, GI 2005, pp. 153–160. Canadian Human-Computer Communications Society, Victoria, British Columbia (2005)
6. Shiue, L., Peters, J.: A realtime GPU subdivision kernel. In: Graphics Interfaces 2005, pp. 153–160 (2005)
7. Peters, J.: Patching Catmull-Clark meshes. In: SIGGRAPH 2000: Proceedings of the 27th Annual Conference on Computer Graphics and Interactive Techniques, pp. 255–258 (2000)
8. Bunnell, M.: Adaptive tessellation of subdivision surfaces with displacement mapping. In: Pharr, M., Fernando, R. (eds.) GPU Gems 2. Addison-Wesley Professional, Boston (2005)
9. Samet, H.: Foundations of Multidimensional and Metric Data Structures. Morgan Kaufmann Publishers Inc., San Francisco (2005)
10. Zorin, D., Schröder, P., Sweldens, W.: Interactive multiresolution mesh editing. In: SIGGRAPH 1997: Proceedings of the 24th Annual Conference on Computer Graphics and Interactive Techniques, pp. 259–268 (1997)
11. Weiler, K.: Edge-based data structures for solid modeling in curved-surface environments. IEEE Comput. Graph. Appl. **5**, 21–40 (1985)
12. Kettner, L.: Designing a data structure for polyhedral surfaces. In: SCG 1998: Proceedings of the Fourteenth Annual Symposium on Computational Geometry, pp. 146–154 (1998)
13. Kraemer, P., Cazier, D., Bechmann, D.: Extension of half-edges for the representation of multiresolution subdivision surfaces. Vis. Comput. **25**, 149–163 (2009)
14. Nießner, M., Loop, C., Meyer, M., Derose, T.: Feature-adaptive GPU rendering of Catmull-Clark subdivision surfaces. ACM Trans. Graph. **31**, 6:1–6:11 (2012)
15. Müller, H., Jaeschke, R.: Adaptive subdivision curves and surfaces. In: Proceedings of the Computer Graphics International 1998 (CGI 1998), pp. 6:1–6:11 (1998)
16. Kobbelt, L.: $\sqrt{3}$ subdivision. In: Proceedings of the 27th Annual Conference on Computer Graphics and Interactive Techniques, pp. 103–112 (2000)
17. Guiqing, L., Weiyin, M., Hujun, B.: $\sqrt{2}$ subdivision for quadrilateral meshes. Vis. Comput. **20**, 180–198 (2004)
18. Pakdel, H., Samavati, F.: Incremental subdivision for triangle meshes. Int. J. Comput. Sci. Eng. **3**, 80–92 (2007)
19. Pakdel, H., Samavati, F.: Incremental Catmull-Clark subdivision. In: Fifth International Conference on 3-D Digital Imaging and Modeling 2005 (3DIM 2005), pp. 95–102 (2005)

20. Pakdel, H.-R., Samavati, F.F.: Incremental adaptive loop subdivision. In: Laganá, A., Gavrilova, M.L., Kumar, V., Mun, Y., Tan, C.J.K., Gervasi, O. (eds.) ICCSA 2004. LNCS, vol. 3045, pp. 237–246. Springer, Heidelberg (2004)
21. Andrew, R.B., Sherman, A.H., Weiser, A.: Some refinement algorithms and data structures for regular local mesh refinement. In: Stepleman, R., et al. (eds.) Scientific Computing, pp. 3–17. IMACS/North-Holland, Amsterdam (1983)
22. Panozzo, D., Puppo, E.: Implicit hierarchical quad-dominant meshes. Comput. Graph. Forum. **30**, 1617–1629 (2011)
23. Li, G., Ma, W., Bao, H.: $\sqrt{2}$ subdivision for quadrilateral meshes. Vis. Comput. **20**, 180–198 (2004)
24. Loop, C.: Smooth subdivision surfaces based on triangles. Masters thesis, Department of Mathematics, University of Utah (1987)
25. Catmull, E., Clark, J.: Recursively generated B-spline surfaces on arbitrary topological meshes. Comput.-Aided Des. **10**, 350–355 (1978)
26. Heredia, V.M., Urrutia, J.: On convex quadrangulations of point sets on the plane. In: Akiyama, J., Chen, W.Y.C., Kano, M., Li, X., Yu, Q. (eds.) CJCDGCGT 2005. LNCS, vol. 4381, pp. 38–46. Springer, Heidelberg (2007)
27. Mahdavi-Amiri, A.: ACM: atlas of connectivity maps. Ph.D. thesis, Department of Computer Science, University of Calgary (2015)
28. Bommes, D., Lévy, B., Pietroni, N., Puppo, E., Silva, C., Tarini, M., Zorin, D.: State of the art in quad meshing. In: Eurographics STARS (2012)
29. Har-Peled, S.: Geometric Approximation Algorithms, vol. 173. American Mathematical Society, Providence (2011)

Matrix Generation in Isogeometric Analysis by Low Rank Tensor Approximation

Angelos Mantzaflaris[1], Bert Jüttler[1,2][✉], B.N. Khoromskij[3], and Ulrich Langer[1,2]

[1] RICAM, Austrian Academy of Sciences, Linz, Austria
[2] Johannes Kepler University, Linz, Austria
bert.juettler@jku.at
[3] MPI for Mathematics in the Sciences, Leipzig, Germany

Abstract. It has been observed that the task of matrix assembly in Isogeometric Analysis (IGA) is more challenging than in the case of traditional finite element methods. The additional difficulties associated with IGA are caused by the increased degree and the larger supports of the functions that occur in the integrals defining the matrix elements. Recently we introduced an interpolation-based approach that approximately transforms the integrands into piecewise polynomials and uses look-up tables to evaluate their integrals [17,18]. The present paper relies on this earlier work and proposes to use tensor methods to accelerate the assembly process further. More precisely, we show how to represent the matrices that occur in IGA as sums of a small number of Kronecker products of auxiliary matrices that are defined by univariate integrals. This representation, which is based on a low-rank tensor approximation of certain parts of the integrands, makes it possible to achieve a significant speedup of the assembly process without compromising the overall accuracy of the simulation.

Keywords: Isogeometric analysis · Matrix assembly · Tensor decomposition · Low rank tensor approximation · Numerical integration · Quadrature

1 Introduction

Isogeometric Analysis (IGA) has been conceived as a new approach to reconcile the conflicting approaches that are used in design and analysis [5,8]. This is achieved by using NURBS (non-uniform rational B-splines) for defining the discretization spaces that provide the basis of a numerical simulation. IGA has led to several new challenges, one of which is to find efficient methods for assembly, i.e., for evaluating the elements of the matrices and vectors that appear in the linear systems arising in isogeometric simulations.

The standard approach to perform the assembly task consists in using Gaussian quadrature, but this does not give optimal runtimes in IGA, due to its high computational cost. On the one hand, this has motivated the use of GPU

© Springer International Publishing Switzerland 2015
J.-D. Boissonnat et al. (Eds.): Curves and Surfaces 2014, LNCS 9213, pp. 321–340, 2015.
DOI: 10.1007/978-3-319-22804-4_24

programming [11,12] for accelerating the assembly process. On the other hand, several alternatives to Gauss quadrature have been explored. Special quadrature rules for spline functions [2,4,9,19] have been derived, but these rules are potentially difficult to compute (since they require solving a non-linear system of equations) and provide only modest improvements. Reduced Bézier element quadrature rules for low degree discretizations have been investigated recently [21]. Another approach, which is based on spline projection and exact integration via look-up tables, has been presented in [17,18]. The asymptotic time complexity of the method is shown to be $\mathcal{O}(np^{2d})$, where n is the number of elements, p is the spline degree and d the spatial dimension. While this approach provides a significant speedup when using larger polynomial degrees (i.e. degree more than three), it did not exhibit strong improvement with respect to using Gauss quadrature for the low degree case. Similarly, the sum-factorization technique, applied to IGA in [1], provides a significant improvement to the cost of Gauss quadrature, by reducing it to $\mathcal{O}(np^{2d+1})$. Again, the benchmarks presented in this work indicate that this advantage becomes significant for higher polynomial degrees. Finally, it has been proposed to derive discretizations via collocation [20], but the theoretical foundations of this mathematical technology are less well understood than in the case of Galerkin projection.

Tensor-product splines are a very popular approach to construct multivariate representations in IGA. In fact, bi- and tri-variate parameterizations in IGA rely almost exclusively on tensor-product representations. Tensor methods are also a well-known approach to deal with high-dimensional problems and to address the "curse of dimensionality" for simulations of high-dimensional problems.

A major breakthrough concerning the problem of data-sparse representation of multivariate functions and operators on large fine grids in high dimensions was made possible by employing the principle of separation of variables. Modern grid-based tensor methods [14,15] achieve linear memory costs $\mathcal{O}(dn)$ with respect to dimension d and grid size n. The novel method of quantized tensor approximation is proven to provide a logarithmic data-compression for a wide class of discrete functions and operators [13]. It allows to discretize and to solve multi-dimensional steady-state and dynamical problems with a logarithmic complexity in the volume size of the computational grid.

Even though the problems considered in isogeometric analysis are defined almost exclusively on computational domains of dimension ≤ 3, the use of tensor methods is a promising approach to reduce the computational costs that are associated with isogeometric discretizations, due to the tensor-product structure of the spline spaces that are used for the discretization.

In fact, the complexity of any quadrature-based assembly is bounded from below by the size of the matrix which is to be generated. A further improvement is possible only when considering alternative representations of this matrix, and such representations can be constructed by using tensor decomposition methods. We will show that the use of low rank tensor approximations provides significant gains and is feasible for non-trivial geometries.

In the present paper, we obtain a representation of the matrix as a sum of a few Kronecker products of lower-dimensional matrices. Since linear algebra operations can be performed efficiently on the Kronecker product structure [7], this representation is directly useful for solving the PDE, e.g., when using iterative solvers that are based on matrix-vector multiplications. Nevertheless, even computing the (sparse) Kronecker product as a final step of the assembly has optimal computational complexity, simply because it can be performed in time asymptotically equal to the size of its output.

The remainder of the paper is organized as follows. Section 2 describes the problem of matrix assembly in IGA. The following section recalls the spline projection of the integrands, which was presented in detail in [18]. Section 4 introduces the concept of tensor decomposition and Sect. 5 shows how this can be exploited for an efficient evaluation of the matrix elements. The efficiency of the overall approach is demonstrated by numerical results, which are described in Sect. 6. Finally, we conclude the paper in Sect. 7.

2   Preliminaries

The spaces used for the discretization of partial differential equations on two- and three-dimensional domains in isogeometric analysis are constructed almost exclusively with the help of (polynomial or rational) tensor-product splines. In order to keep the presentation simple, we restrict the presentation to polynomial splines and to the bivariate case. The case of rational splines (non-uniform rational B-splines – NURBS) can be dealt with similarly.

We consider two *univariate spline bases*

$$\boldsymbol{S}(s) = (S_1(s), \ldots, S_m(s))^T \text{ and } \boldsymbol{T}(t) = (T_1(t), \ldots, T_n(t))^T$$

of degree p, which are obtained from the knot vectors

$$(\sigma_1, \ldots, \sigma_{p+m+1}) \text{ and } (\tau_1, \ldots, \tau_{p+n+1}),$$

respectively. The B-splines $S_i(s)$ and $T_j(t)$ are defined by the well-known recurrence formulas, see [6]. We use column vectors $\boldsymbol{S}$ and $\boldsymbol{T}$ to collect the B-splines forming the two spline bases.

More precisely, the boundary knots appear with multiplicity $p+1$, the inner knots have multiplicity at most p, and the knot vectors are non-decreasing. Moreover we choose $\sigma_1 = \tau_1 = 0$, $\sigma_{p+m+1} = \tau_{p+n+1} = 1$.

The *tensor-product spline space*

$$\mathbb{S} = \text{span } \boldsymbol{S} \otimes \text{span } \boldsymbol{T} = \text{span } \{N_{\boldsymbol{i}} : (1,1) \leq \boldsymbol{i} \leq (m,n)\}$$

is spanned by the products of pairs of univariate B-splines,

$$N_{\boldsymbol{i}}(s,t) = S_{i_1}(s)T_{i_2}(t),$$

which are identified by double-indices $\boldsymbol{i} = (i_1, i_2)$.

The isogeometric approach is based on a parameterization of the computational domain Ω. This domain is represented by a suitable *geometry mapping* $\boldsymbol{G} = (G^{(1)}, G^{(2)})$,

$$\boldsymbol{G} : [0,1]^2 \to \Omega \subset \mathbb{R}^2.$$

The domain parameterization is assumed to be regular, i.e., $\det(\nabla \boldsymbol{G})$ does not change its sign in $[0,1]^2$. It is described by the two univariate spline bases (one for each coordinate)

$$G^{(k)}(s,t) = \boldsymbol{S}(s)^T \boldsymbol{D}^{(k)} \boldsymbol{T}(t), \quad k = 1, 2.$$

The coefficients of the two coordinate functions form the $m \times n$-matrices

$$\boldsymbol{D}^{(k)} = (D_{\boldsymbol{i}}^{(k)})_{(1,1) \leq \boldsymbol{i} \leq (m,n)}.$$

Equivalently, this mapping can be written as

$$\boldsymbol{G}(s,t) = \sum_{(1,1) \leq \boldsymbol{i} \leq (m,n)} N_{\boldsymbol{i}}(s,t) \, \boldsymbol{d_i}$$

with the control points

$$\boldsymbol{d_i} = (D_{\boldsymbol{i}}^{(1)}, D_{\boldsymbol{i}}^{(2)}),$$

whose coordinates are the elements of the matrices $\boldsymbol{D}^{(k)}$, $k = 1, 2$.

The discretization space used in isogeometric analysis is spanned by the functions

$$\beta_{\boldsymbol{i}} = N_{\boldsymbol{i}} \circ \boldsymbol{G}^{-1} : \ \beta_{\boldsymbol{i}}(\boldsymbol{G}(s,t)) = \boldsymbol{N_i}(s,t). \tag{1}$$

In many situations, the coefficient matrix of the resulting system of linear equations can be constructed from the mass and the stiffness matrices that are associated with the functions (1). We do not discuss the construction of the right-hand side vectors in this paper, since it is very similar.

The elements of the *mass matrix* take the form

$$a_{ij} = \iint_{\Omega} \beta_{\boldsymbol{i}}(\boldsymbol{x}) \beta_{\boldsymbol{j}}(\boldsymbol{x}) \, \mathrm{d}\boldsymbol{x}$$

$$= \int_0^1 \int_0^1 S_{i_1}(s) \, T_{i_2}(t) \, S_{j_1}(s) \, T_{j_2}(t) \, \underbrace{|\det((\nabla \boldsymbol{G})(s,t))|}_{\star} \, \mathrm{d}s \, \mathrm{d}t.$$

The integrand is a product of two terms. The second one, which has been marked by an asterisk, is independent of the indices $\boldsymbol{i}$ and $\boldsymbol{j}$ and therefore shared by all matrix elements.

The *stiffness matrix* contains the elements

$$b_{ij} = \iint\limits_{\Omega} (\nabla \beta_i)(x) \cdot (\nabla \beta_j)(x) \, dx \qquad (2)$$

$$= \sum_{k=1}^{2} \sum_{\ell=1}^{2} \int_0^1 \int_0^1 (\partial^{\delta_1^k} S_{i_1})(s) \, (\partial^{\delta_1^\ell} T_{i_2})(t)$$

$$(\partial^{\delta_2^k} S_{j_1})(s) \, (\partial^{\delta_2^\ell} T_{j_2})(t) \, \underbrace{q_{k,\ell}(s,t)}_{\star} \, ds \, dt,$$

where the weight functions $q_{k,\ell}$ form the 2×2-matrix

$$Q = |\det(\nabla G)| (\nabla G)^{-1} K (\nabla G)^{-T}.$$

Each of the four integrands is again a product of two terms. Once more, the second one (marked by an asterisk) is independent of the indices i and j and therefore shared by all matrix elements.

The matrix (or matrix-valued function) K contains material coefficients (or is the identity matrix in the simplest case). The symbol ∂ denotes the differentiation operator for univariate functions, i.e., $\partial f = f'$. The Kronecker symbol δ specifies whether a B-spline is differentiated or not.

The stiffness matrix (2) arises when considering the weak form of the diffusion equation, i.e., when solving the following elliptic problem:

$$\text{Find } u \in H^1(\Omega) \text{ such that } \begin{cases} -\nabla \cdot (K \nabla u) = f \text{ in } \Omega \\ u = 0 \text{ on } \partial\Omega \end{cases} \qquad (3)$$

with the right-hand side $f : \Omega \to \mathbb{R}$, homogeneous boundary conditions, and the (possibly non-constant) symmetric and uniformly positive definite matrix $K : \Omega \to \mathbb{R}^{2 \times 2}$. This problem was also considered in [18] when introducing the method of *integration by interpolation and look-up (IIL)*.

Our approach (introduced in [18]) to the computation of the stiffness and mass matrix elements – which we will refer to as *matrix assembly* – relies on an exact or approximate *projection* of the integrands into suitable spline spaces. More precisely, it is performed in two steps.

1. The terms in the integrals that have been marked by the asterisk $\star$ are independent of the indices i, j and therefore shared by all matrix elements. These terms, which will be called the *weight functions*, are replaced by suitable tensor-product spline functions.
2. The resulting integrals of piecewise polynomial functions, are evaluated exactly, either using look-up tables (as described in [18]) or Gauss quadrature.

In [18], the first step was performed solely by interpolation or quasi-interpolation. The present paper improves the first step by employing *low-rank tensor approximations* for constructing sparse representations of the weight functions. This also implies substantial modifications on the second step, since the integration can be reduced to the evaluation of a small number of univariate integrals.

3   Spline Projection

The first step of the computation uses spline spaces $\bar{\mathbb{S}}$, which are (potentially) different from those used to define the geometry mapping and the isogeometric discretizations. We consider univariate spline bases

$$\bar{S}(s) = (\bar{S}_1(s), \ldots, \bar{S}_{\bar{m}}(s))^T \text{ and } \bar{T}(t) = (\bar{T}_1(t), \ldots, \bar{T}_{\bar{n}}(t))^T$$

consisting of B-splines of degree $\bar{p}$, which are defined by knot vectors

$$(\bar{\sigma}_1, \ldots, \bar{\sigma}_{\bar{p}+\bar{m}+1}) \text{ and } (\bar{\tau}_1, \ldots, \bar{\tau}_{\bar{p}+\bar{n}+1}),$$

respectively. Once again, the boundary knots possess multiplicity $\bar{p}+1$, the inner knots have multiplicity less or equal to $\bar{p}$, the knot vectors are non-decreasing, $\sigma_1 = \tau_1 = 0$, and $\sigma_{\bar{p}+\bar{m}+1} = \tau_{\bar{p}+\bar{n}+1} = 1$. In practice, these new knots will be either equal to the original knots of the coarsest isogeometric discretization (determined by the geometry mapping) or a superset of those. For convenience, in this section we consider equidistant (i.e., uniform) inner knots. In this situation, the element size (which quantifies the fineness of the discretization) satisfies $h = \mathcal{O}(\max\{1/m, 1/n\})$ as $m, n \to \infty$. Also, we consider the same degree $\bar{p}$ in both directions only for the sake of simplicity.

The products $\bar{N}_i(s, t) = \bar{S}_{i_1}(s)\bar{T}_{i_2}(t)$ of pairs of univariate B-splines span another tensor-product spline space

$$\bar{\mathbb{S}} = \text{span } \bar{S} \otimes \text{span } \bar{T} = \text{span } \{\bar{N}_i \ : \ (1,1) \leq i \leq (\bar{m}, \bar{n})\}.$$

We are interested in the approximate evaluation of integrals of the form

$$C_{ij}^{d}(w) = \int_0^1 \int_0^1 (\partial^{d_1} S_{i_1})(s) \, (\partial^{d_3} T_{i_2})(t) \, (\partial^{d_2} S_{j_1})(s) \, (\partial^{d_4} T_{j_2})(t) \, w(s, t) \, ds \, dt. \tag{4}$$

These integrals depend on the *weight function* w and on the three multi-indices $i, j \in \mathbb{Z}^2$ and $d \in \{0,1\}^4$. The first two indices identify the corresponding matrix element. The remaining one specifies the order of differentiation of the univariate B-splines.

Indeed, the elements of the mass and stiffness matrix can be rewritten as

$$a_{ij} = C_{ij}^{(0,0,0,0)}(|\det(\nabla G)|) \text{ and } b_{ij} = \sum_{k=1}^{2} \sum_{\ell=1}^{2} C_{ij}^{(\delta_1^k, \delta_2^k, \delta_1^\ell, \delta_2^\ell)}(q_{k\ell}).$$

The *first step* of the matrix assembly is to project the weight function into the spline space $\bar{\mathbb{S}}$. It is realized by using a *spline projection operator*

$$\Pi : \mathcal{C}([0,1]^2) \to \bar{\mathbb{S}},$$

e.g., interpolation or quasi-interpolation. A detailed discussion is given in [18, Section 4.3].

More precisely, for each weight function w we create a tensor-product spline approximation $\Pi w \in \bar{\mathbb{S}}$. It is represented in the tensor-product B-spline basis,

$$(\Pi w)(s,t) = \bar{\boldsymbol{S}}(s)^T \, \boldsymbol{W} \, \bar{\boldsymbol{T}}(t)^T \tag{5}$$

with a $\bar{m} \times \bar{n}$ coefficient matrix $\boldsymbol{W}$. Replacing the weight functions (4) with their spline approximations then leads to the *approximate integrals*

$$C_{ij}^d(\Pi w). \tag{6}$$

The choice of the spline space $\bar{\mathbb{S}}$ depends on the integrand that is to be computed (i.e., mass or stiffness matrix).

For evaluating the *mass matrix* we use degree $\bar{p} = 2p$ and choose the multiplicity of the knots such that smoothness of the spline functions is equal to the smoothness of $\nabla \boldsymbol{G}$. More precisely, if the geometry mapping possesses an inner knot σ_i of multiplicity μ, then the corresponding knot $\bar{\sigma}$ of $\bar{\mathbb{S}}$ has multiplicity $\bar{\mu} = p + \mu + 1$, and similar for the knots τ_j. It should be noted that this applies only to the knots that are present in the actual geometry mapping, but not to the knots which are created later to obtain a finer isogeometric discretization (i.e. by h-refinement). This choice of knots allows to represent the weight function $|\det(\nabla \boldsymbol{G})|$ exactly in $\bar{\mathbb{S}}$. Thus, no error is introduced for the mass matrix.

For assembling the *stiffness matrix* we use $\bar{p} = p$ as discussed in [18]. Moreover, the knots which are already present in the geometry mapping are kept and their multiplicity is increased by one. Additional single knots can be inserted in order to improve the accuracy of the spline projection.

In the *second step* of matrix assembly, the approximations of the integrals are evaluated exactly, using either Gauss quadrature or suitable look-up tables. This is possible since the integrands are simply piecewise polynomial functions.

For the stiffness matrix, the error ε_Π introduced by the spline projection

$$\varepsilon_\Pi = \max_{k,\ell=1,2} \|q_{k,\ell} - \Pi q_{k,\ell}\|_{\infty,[0,1]^2}$$

is closely related to the overall error of the isogeometric simulation. We consider the problem (3), which leads to the stiffness matrix (2). Combining [18, Corollary 12] with Strang's first lemma (cited as [18, Lemma 2]) and using results of [3] (cited as [18, Eq. 31]) gives the bound

$$\|u - u^\star\|_{1,\Omega} \le C_1 h^p + C_2 \varepsilon_\Pi$$

for the H^1 (semi-) norm of the difference between the exact solution u and the result $u^\star$ of the isogeometric simulation, which is based on the approximate integrals (6). The first term of the right-hand side is the *discretization error*, which is caused by choosing $u^\star$ from the finite-dimensional space of isogeometric functions. The second term represents the *consistency error* which is due to the approximate evaluation of the integrals.

In order to keep the presentation simple we do not consider the right-hand side explicitly. Instead we will simply assume that its effect is also included in the error bound for the consistency error of the stiffness matrix.

Consequently, under suitable assumptions about the error introduced by the spline projection, the matrix assembly preserves the overall order of approximation. Since we use projection into a spline space $\bar{\mathbb{S}}$ of degree $\bar{p}$, we may expect that the difference between the weight functions and their spline projections is bounded by

$$\varepsilon_{\Pi} \leq C_3 h^{\bar{p}+1}, \tag{7}$$

where the constant C_3 depends only on the problem (geometry map, material properties, right-hand side) but not on the isogeometric discretization (i.e., on the knots which are introduced in addition to the knots of the geometry mapping). This has been formulated in [18, Assumption 5], and a possibility to find a spline projection providing theoretical guarantees (based on quasi-interpolation operators) is also described there.

Choosing $\bar{p} = p - 1$ was proved to be sufficient to preserve the overall order of approximation with respect to the H^1 semi-norm (according to Strang's first lemma). However, this choice was experimentally found to give sub-optimal results for the convergence with respect to the L^2 norm for even degrees p, while $\bar{p} = p$ gave optimal results with respect to both norms in all cases. We will therefore assume that assumption (7) is satisfied with $\bar{p} = p$ in the remainder of this paper.

Once a spline projection has been computed, the actual error ε_{Π} can be estimated by sampling.

4   Tensor Decomposition

We consider a singular value decomposition

$$W = U^T \Sigma V$$

of the coefficient matrix in (5). It consists of the diagonal $m \times n$ matrix of singular values $\Sigma = \mathrm{diag}(\sigma_1, \ldots, \sigma_{\min(m,n)})$, and of the two orthogonal matrices

$$U^T = (u_1, \ldots, u_m) \text{ and } V^T = (v_1, \ldots, v_n),$$

which are represented by their row vectors $u_i \in \mathbb{R}^m$ and $v_j \in \mathbb{R}^n$. More precisely, the *rows* of these two matrices are given by the *column vectors* u_i and v_j. Without loss of generality we assume that the singular values are ordered, $\sigma_r \geq \sigma_{r+1}$.

Consequently, we may rewrite the spline projection of the weight function (5) as a sum of products of univariate spline functions, as follows:

$$(\Pi w)(s,t) = \bar{S}(s)^T W \bar{T}(t) = \sum_{r=1}^{\min(m,n)} \left(\bar{S}(s)^T u_r \sqrt{\sigma_i} \right) \left(\sqrt{\sigma_i} v_r{}^T \bar{T}(t) \right).$$

As we will see in the next section, using this representation and truncated versions thereof makes the numerical quadrature very efficient.

While the representation in (5) is a sum of mn products of univariate spline functions (one for each tensor-product B-spline), using singular value decomposition allows to reduce the number of terms in the sum to $\min(m, n)$.

The spline projection Πw of the weight function can be approximated by omitting the terms that correspond to smaller singular values. We choose a positive integer $R \leq \min(m, n)$, which specifies the number of terms to be retained by the approximation, and define the *rank R tensor approximation operator*

$$\Lambda_R : \mathcal{C}([0, 1]^2) \to \bar{\mathbb{S}}.$$

Given a weight function w, we project into the spline space $\bar{\mathbb{S}}$, generate a singular value decomposition of the coefficient matrix and keep only the contributions of the largest R singular values,

$$(\Lambda_R w)(s, t) = \sum_{r=1}^{R} \left(\bar{S}(s)^T \sqrt{\sigma_r} u_r \right) \left(\bar{T}(t)^T \sqrt{\sigma_r} v_r \right). \tag{8}$$

The function is generated by this operator is called a *low rank* (more precisely: rank R) *tensor approximation of the weight function* w, and R is called its *rank*.

For typical geometry mappings, using a small rank R is sufficient and gives accurate results. The Frobenius norm of the difference between the coefficient matrices can be bounded as follows:

$$\| \sum_{r=R+1}^{\min(m,n)} \sigma_r u_r v_r^T \|_F \leq \sqrt{\sum_{r=R+1}^{\min(m,n)} \sigma_r^2}.$$

Since the Frobenius norm of the matrix is an upper bound of the 2-norm, which in turn is an upper bound of the element-wise maximum norm, we can use the convex hull property of tensor-product splines to conclude that

$$\| \Pi w - \Lambda_R w \|_{\infty, [0,1]^2} \leq \sqrt{\sum_{i=R+1}^{\min(m,n)} \sigma_r^2}.$$

Once the singular value decomposition has been computed, the error can be controlled by adjusting the rank R.

In particular, performing the singular value decomposition for the four coefficient matrices of the spline projections $\Pi q_{k\ell}$ ($k, \ell = 1, 2$) gives four sets of singular values $(\sigma_{r,k\ell})_{r=1,\ldots,\min(m,n)}$. Each of the four low-rank tensor approximations satisfies the error bound

$$\max_{k,\ell=1,2} \| \Pi q_{k\ell} - \Lambda_R q_{k\ell} \|_{\infty, [0,1]^2} \leq \varepsilon_\Lambda = \max_{k,\ell=1,2} \sqrt{\sum_{r=R+1}^{\min(m,n)} \sigma_{r,k\ell}^2}.$$

The contribution of the low rank tensor approximation to the overall error can be analyzed as in the previous section. We consider again the problem (3), which leads to the stiffness matrix (2). When using the low rank tensor approximation for assembling the stiffness matrix, the H^1 (semi-) norm of the difference between the exact solution u and the result u^* of the isogeometric discretization, which is based on the approximate integrals

$$C_{ij}^{d}(\Lambda_R w)$$

is bounded by (6)

$$\|u - u^*\|_{1,\Omega} \le C_1 h^p + C_2(\varepsilon_\Pi + \varepsilon_\Lambda).$$

Consequently, when choosing the tensor approximation error ε_Λ in the same order as the spline projection error ε_Π – which can be estimated by sampling – then the order of accuracy of the method remains unchanged.

5   Evaluation of Matrix Elements

Using a low rank tensor approximation (8) of the weight function w allows to rewrite the integrals (4) as sums of products

$$C_{ij}^{d}(\Lambda_R w) = \sum_{r=1}^{R} X_{r,i_1 j_1}^{d}(\Lambda_R w) Y_{r,i_2 j_2}^{d}(\Lambda_R w)$$

of R pairs of univariate integrals

$$X_{r,ij}^{d}(\Lambda_R w) = \int_0^1 (\partial^{d_1} S_i)(s)\, (\partial^{d_2} S_j)(s)\, (\bar{S}(s)^T \sqrt{\sigma_r} u_r)\, \mathrm{d}s \text{ and}$$

$$Y_{r,ij}^{d}(\Lambda_R w) = \int_0^1 (\partial^{d_3} T_i)(t)\, (\partial^{d_4} T_j)(t)\, (\bar{T}(t)^T \sqrt{\sigma_r} v_r)\, \mathrm{d}t.$$

For each multi-index index d, the integrals (4) form a $mn \times mn$ matrix

$$C^{d}(\Lambda_R w) = (C_{ij}^{d}(\Lambda_R w))_{(1,1) \le i,j \le (m,n)},$$

and the univariate integrals form R $m \times m$ and R $n \times n$ matrices

$$X_r^{d}(\Lambda_R w) = (X_{r,ij}^{d}(\Lambda_R w))_{1 \le i,j \le m} \text{ and } Y_r^{d}(\Lambda_R w) = (Y_{r,ij}^{d}(\Lambda_R w))_{1 \le i,j \le n}.$$

The first matrix is the sum of R Kronecker products of matrices obtained from the univariate matrices,

$$C^{d}(\Lambda_R w) = \sum_{r=1}^{R} X_r^{d} \otimes Y_r^{d}. \tag{9}$$

This structure suggests itself for designing an efficient evaluation procedure: In the first step, the matrices $X_r^{d}(\Lambda_R w)$ and $Y_r^{d}(\Lambda_R w)$ which are defined by the univariate integrals are evaluated, e.g., by using Gaussian quadrature. Then, in a second step, the matrix is assembled by summing up the Kronecker products in (9).

We conclude this section with a brief analysis of the computational complexity of the overall procedure, where we assume $m \cong n$ and $\bar{p} = p$. Moreover we assume that the knot vectors of the univariate spline bases S and $\bar{S}$ have the same knots (possibly with different multiplicities) and hence $\mathcal{O}(m)$ knot spans. A similar assumption is made concerning T and $\bar{T}$.

We do not include the singular value decomposition (SVD) into the complexity analysis. In practice one will prefer approximate SVD methods with lower computational costs. A detailed discussion of such methods is beyond the scope of the present paper.

We analyze the computational costs for assembling the $2R$ matrices $X_r^d(\Lambda_R w)$ and $Y_r^d(\Lambda_R w)$, which possess $\mathcal{O}(mp)$ non-zero entries each.

When using Gaussian quadrature, we need to evaluate the non-zero B-spline basis functions (and their derivatives) at $\mathcal{O}(p)$ Gauss nodes for each of the $\mathcal{O}(m)$ knot spans at costs of $\mathcal{O}(p^2)$ for each point. In addition we evaluate $2R$ spline functions using de Boor's algorithm. The total complexity of this step is $\mathcal{O}(Rmp^3)$.

We then continue with the assembly step, where each Gauss node contributes to $\mathcal{O}(p^2)$ matrix entries of each of the $2R$ matrices. Thus, the effort for each Gauss node equals $\mathcal{O}(Rp^2)$. The total complexity of this step amounts to $\mathcal{O}(Rmp^3)$.

The matrix (9) has $\mathcal{O}(m^2p^2)$ non-zero entries, and each is evaluated in $\mathcal{O}(R)$ operations from the univariate integrals. Since $m \gg p$, the overall effort is dominated by this step and equals

$$\mathcal{O}(Rm^2p^2).$$

Clearly, the efficiency of the method depends on the rank R which needs to be considered in the tensor approximation. The IIL method described in [18] has complexity of $\mathcal{O}(m^2p^4)$, and this was found to compare well with the Gauss quadrature, where the complexity was found to be $\mathcal{O}(m^2p^6)$. Consequently, if R is small and does not grow with m (i.e., as $h \to 0$), then matrix assembly by tensor decomposition has a clear advantage.

We will demonstrate in the next section that using a low rank tensor approximation is sufficient in most cases and leads to a substantial speedup.

6   Numerical Results

In this section we present numerical examples to demonstrate the power of assembly by tensor decomposition and low rank representations. As benchmark examples, we choose a quarter annulus and a multipatch domain consisting of several patches. The annulus is particularly well suited for our method, since it is a "rank one domain" in the sense that using $R = 1$ is sufficient. The multipatch "yeti footprint" domain, which was taken from [16], is a harder benchmark, since the patches do not have any symmetries. We assemble the mass and stiffness of the underlying tensor B-spline basis. In practice, small polynomial degrees are used for discretization; we use the same (h−refined) quadratic B-spline functions of the domain as discrete space.

As a baseline algorithm to compare with we use Gauss quadrature assembly with $p + 1$ nodes per parametric direction (FG = "full" Gauss quadrature). The assembly based on tensor decomposition is abbreviated as TD and the computation of the Kronecker product as KP.

We test the computational effort and memory requirements for computing the mass and stiffness matrices. Since these matrices are symmetric, a straight-forward saving is to compute only the lower triangular part, and this was done in all examples.

The singular value decompositions were computed by a Jacobi method. Even though we are interested in the highest magnitude singular values and vectors only, our current implementation computes the full SVD. More sophisticated methods exist that compute reduced SVDs, namely the decomposition up to a given tolerance of the singular values. This will explored in more detail in our future work. For this reason, in our experiments we do not measure the time required for the computation of a (reduced) SVD. The time depends on the dimension of the interpolation space used and the required tolerance. Indeed, this space is independent on the refinement level of the isogeometric discretization, since we always approximated the weight functions with a very small tolerance (such as machine precision).

All experiments were conducted on a laptop with Intel Core i7 @ 2.70GHz processor having two cores and 6 GB of RAM, running Linux. The method is implemented in C++ using the G+SMO library for isogeometric analysis[1].

6.1   Quarter Annulus

A quarter annulus can be regarded as a line segment swept along a circular arc. Therefore we expect that the tensor rank is low. The SVD computation confirms that the tensor rank is exactly one, and all sub-dominant singular values are equal to zero. Consequently, the assembly of mass and stiffness on this domain using TD is almost as efficient as if we would treat a square domain, by hard-coding the identity matrix in place of det ∇G.

The domain is represented by linear components in the s direction and by quadratic polynomials in the t direction. Note that we assumed to have the same degree in both directions in the previous section. This can easily be generalized to non-uniform degrees, as in the present example. Therefore we have used 2 and 3 Gauss points respectively, i.e. a total of 6 points per element. For our experiments, we apply uniform h−refinement to obtain problem sizes from $6.6 \cdot 10^4$ up to $4.3 \cdot 10^9$ degrees of freedom (dofs).

First we consider the assembly of the mass matrix. The Jacobian determinant of the geometry mapping has rank one and is exactly represented in a spline basis of bidegree (4,2) on the (input) coarse mesh. Therefore no approximation is needed in the SVD step. This implies that the spline space in which the

---

[1] *Geometry + Simulation Modules*, see www.gs.jku.at/gismo, also [10].

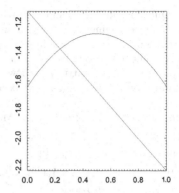

Fig. 1. Left: A B-spline quarter annulus domain. The color field is the value of the Jacobian determinant. Right: "Skeleton functions" of the Jacobian determinant. The linear function is associated with the first direction, while the curved one describes the determinant along the second parametric direction.

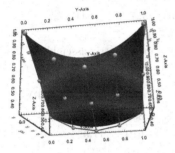
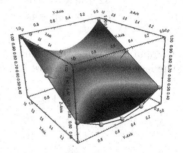

Fig. 2. Two views of the graph of the Jacobian determinant and its control grid over the parameter domain $[0, 1]^2$. The degree is 2 in the first direction and 4 in the second one.

determinant lies has dimension just 15, which is also the size of the matrix that we shall apply SVD computation to. Figure 1 shows the domain and the control grid, as well as the two "skeleton functions" of the rank one tensor representation of the Jacobian determinant. The product of these components is the Jacobian determinant functional, which is shown in Fig. 2.

We demonstrate the reduction of the required memory in Table 1. The required memory shrinks from $\mathcal{O}(n^2)$ to $\mathcal{O}(n)$. In our examples, this translates into using some megabytes of memory instead of several gigabytes.

Table 2 shows the computing times and the speedup with respect to using Gauss quadrature for the mass matrix assembly. Some entries of the table (for high numbers of degrees of freedom) are left blank, since our program aborted, due to insufficient memory; the testing machine (a laptop) has 6 GB of memory, which will not fit a matrix with 1G of elements, since this requires 8 GB of memory, using double arithmetic. It should be noted that TD was applicable up to 16 levels of dyadic refinement (about $4.3 \cdot 10^9$ dof), in only a fifth of a

Table 1. Quarter annulus: memory statistics for the (rank one) mass matrix for high levels of dyadic h−refinement. The second column shows the number of degrees of freedom per coordinate (tensor) direction. The third column contains the number of non-zero (nz) entries in each factor matrix of the Kronecker product in (9). The non-zero entries of the mass matrix are shown in the last column. The factor matrices in the decomposition of the stiffness matrix use four times as many non-zero elements.

h−ref. level	No. of dof	Mass matrix TD nz	Mass matrix nz
8	257×258	$769 + 1,284$	$987\,\mathrm{K}$
9	513×514	$1,537 + 2,564$	$3.9\,\mathrm{M}$
10	$1,025 \times 1,026$	$3,073 + 5,124$	$15.7\,\mathrm{M}$
11	$2,049 \times 2,050$	$6,145 + 10,244$	$62.9\,\mathrm{M}$
12	$4,097 \times 4,098$	$12,289 + 20,484$	$251.7\,\mathrm{M}$
13	$8,193 \times 8,194$	$24,577 + 40,964$	$1\,\mathrm{G}$
14	$16,385 \times 16,386$	$49,153 + 81,924$	$4\,\mathrm{G}$
15	$32,769 \times 32,770$	$98,305 + 163,844$	$16\,\mathrm{G}$
16	$65,537 \times 65,538$	$196,609 + 327,684$	$64\,\mathrm{G}$

Table 2. Computation times (in seconds) for the mass matrix assembly of the quarter annulus. The last two columns report the speedup (time ratio).

h−ref. level	FG	TD	KP	FG/TD	FG/(TD+KP)
8	0.30	$9 \cdot 10^{-4}$	0.017	339	16.87
9	1.22	$2 \cdot 10^{-3}$	0.07	610	17.9
10	4.82	$5 \cdot 10^{-3}$	0.23	947	20.9
11	18.94	$7 \cdot 10^{-3}$	0.79	2,600	23.6
12	76.21	$1.5 \cdot 10^{-2}$	3.23	5,154	23.4
13	-	$4.1 \cdot 10^{-2}$	-	-	-
14	-	$6.5 \cdot 10^{-2}$	-	-	-
15	-	0.1	-	-	-
16	-	0.18	-	-	-

second. Certainly, at this point we are unable to compute the Kronecker product of the skeleton matrices, but given enough memory and processors this is a task that could be performed easily in parallel.

We now turn our attention to the stiffness matrix. We find that the weight functions $q_{k,\ell}$ are also represented by rank one tensor approximations. However, a sufficiently good approximation of these weight functions requires some steps of h−refinement. In the experiments of Fig. 3, we used a 266K (515×517) grid for the interpolation of these weight functions. This was sufficient to approximate the four functions $q_{\ell,k}$ up to machine precision. Overall, the stiffness matrix has rank four.

Table 3. Computation times (in seconds) and speedup for the stiffness matrix assembly of the quarter annulus.

h−ref. level	FG	TD	KP	FG/TD	FG/(TD+KP)
8	0.36	0.019	0.016	19.02	10.2
9	1.45	0.083	0.072	17.40	9.3
10	5.84	0.085	0.28	68.86	15.8
11	23.59	0.089	1.09	264	19.9
12	96.50	0.099	4.56	979	20.7
13	-	0.11	-	-	-
14	-	0.14	-	-	-
15	-	0.21	-	-	-
16	-	0.34	-	-	-

Table 3 shows the computational time and reports the speedup for the stiffness matrix assembly. Note that at (dyadic) refinement level 13 we are dealing with around 67.1 million degrees of freedom, and the number of non-zeros in the global matrix reaches 1 billion. When using double precision, this requires at least 8 GB of memory, and was not feasible on our test machine (laptop). Consequently the columns FG and KP are filled up to refinement level 12.

One may observe that computing times are mostly determined by the memory required for the problem. In particular, note that computing e.g. the mass matrix using FG with 8 levels of refinement implies the computation of a matrix of around 526 K elements (taking into account symmetry). This took 0.3 s, which is quite similar to using TD with 16 levels of refinement for computing (in 0.18 seconds) a decomposition of a (much bigger) matrix with comparable amount of non-zero elements.

The annulus example has also been used as a benchmark for the IIL method. As observed in [18, Sect. 7.3], the speedup of the stiffness matrix assembly with respect to the baseline FG method is independent of h, that is, the computation time for both FG and IIL is linear with respect to the number of degrees of freedom. We did not include the comparison with IIL in Table 3, since its performance is quite similar to FG for low degrees (e.g., it is approximately twice as fast as FG for bi-quadratic discretizations, cf. [18, Fig. 10]).

6.2   Yeti Footprint

We now turn to a more challenging benchmark. The domain is a "footprint" domain parameterized by 21 patches (Fig. 3) introduced in [16]. All the patches have a number of interior knots per direction, ranging from one to three.

One question is whether the rank of each patch (i.e., the rank of the tensor approximation) stabilizes to a small value. To test this, we set a tolerance of 10^{-10} and compute the rank of det ∇G and $q_{k,\ell}$ for all mappings G of the different

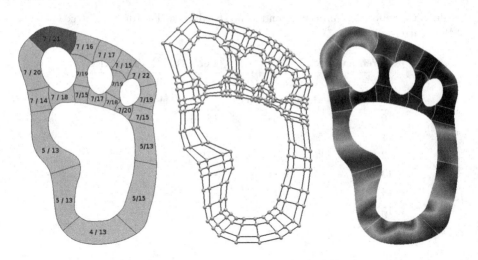

Fig. 3. The yeti footprint domain, parameterized by 21 quadratic B-spline patches. Patch segmentation (left), control grids (middle) and the (absolute) Jacobian determinant are shown. On the left picture rank(det ∇G) and max(rank($q_{k,\ell}$), $k, \ell \in \{1, 2\}$) are given.

patches and several levels of h−refinement. In all cases, the rank stabilized (i.e., it stayed constant under h−refinement) to a value between 4 and 7 for det ∇G and to values ranging from 14 to 22 for $q_{k,\ell}$. The exact numbers are shown in Fig. 3(left).

Clearly, the full multi-patch (mass or) stiffness matrix is a block-structured matrix where the blocks are the patch-wise matrices. Therefore, we now choose one patch (marked in Fig. 3 left) in order to perform our benchmark computation. Figure 4 provides a closer look at the selected patch and its Jacobian determinant.

In Table 4 we report computing times and memory requirements for the mass matrix assembly for the selected patch. The last column shows the memory savings that can be obtained by using a tensor representation, instead of assembling the global mass matrix.

We note that the mass matrix on this patch has rank 7, therefore memory requirements are seven times more than the annulus case, but the percentage shows that this is still minor compared to the size of the full matrix. The timings confirm that also in this case a substantial speedup is possible. In particular, in the last row we assembled a (tensor decomposed) mass matrix for 17.2 billion degrees of freedom, and this took one second. This suggests that in less than half a minute, the full mass matrix of all 21 patches can be computed with the same refinement level.

Table 5 reports on the stiffness matrix assembly for the same patch. Each weight $q_{k,\ell}$ has a rank 21 tensor approximation, leading to a total rank of 84 for the tensor form of the stiffness matrix. This increase with respect to the

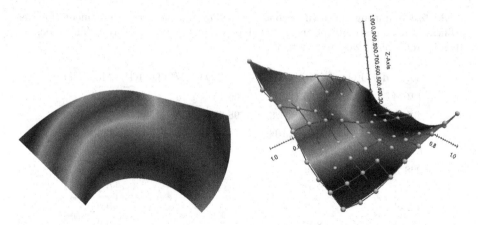

Fig. 4. The Jacobian determinant values of the selected patch (Fig. 3) on the domain (left). The graph and control net in parameter domain (right).

Table 4. Computing times (in seconds), speedup and memory requirements for the mass matrix assembly of the selected patch of the yeti footprint. The levels of (dyadic) refinement vary from 9 to 16.

no. of dof	FG	TD	KP	FG/TD	FG/(TD+KP)	Mem.(%)
1.0 M	7.96	0.015	0.60	544	13.0	0.273
4.2 M	32.50	0.025	2.34	1289.14	13.7	0.137
16.8 M	129.83	0.053	11.33	2461.57	11.4	0.068
67.1 M	-	0.091	-	-	-	0.034
268 M	-	0.148	-	-	-	0.017
1.1 G	-	0.262	-	-	-	0.009
4.2 G	-	0.496	-	-	-	0.004
17.2 G	-	0.997	-	-	-	0.002

mass matrix is reflected in the computation time and the number of non-zero elements, i.e., in the memory that is required. Nevertheless, the percentage of memory required is still around 3 % of the full matrix for one million degrees of freedom, and is reduced to half at every refinement level. The Kronecker product computation becomes slower since now we need to sum up $R = 84$ contributions to compute one entry in the Kronecker product. Nevertheless, we still get a speedup factor 1.52 compared to FG. The speedup observed if we refrain from computing the KP (sixth column) is still quite important and scales nicely with refinement. Finally, from the last row of the table we may estimate that about three minutes' time would suffice for assembling a (decomposed) giant stiffness matrix of 361 billion degrees of freedom on a laptop.

Table 5. Computing times (in seconds), speedup and memory requirements for the stiffness matrix assembly of the selected patch of the yeti footprint. The levels of (dyadic) refinement vary from 9 to 16.

no. of dof	FG	TD	KP	FG/TD	FG/(TD+KP)	Mem.(%)
1.0 M	9.94	0.073	7.25	136.75	1.36	3.33
4.2 M	40.45	0.14	29.1	298.97	1.39	1.66
16.8 M	164.4	0.27	108.1	614.38	1.52	0.83
67.1 M	-	0.53	-	-	-	0.41
268 M	-	1.06	-	-	-	0.21
1.1 G	-	2.17	-	-	-	0.10
4.2 G	-	4.46	-	-	-	0.05
17.2 G	-	8.85	-	-	-	0.02

7   Conclusion

We explored the use of low rank tensor approximation for performing the assembly task of the isogeometric mass and stiffness matrices. Clearly, the method is general and applicable to other types of isogeometric matrices, arising in PDE discretizations.

Our results show that the use of tensor methods in IGA possesses a great potential. The main advantage is the substantial memory reduction that is obtained by creating low rank approximations of large isogeometric matrices. This observation can be further exploited if one adopts the Kronecker representation both for storing matrices and for applying iterative solvers or other operations. Indeed, our benchmarks show that impressive speedups are possible if one avoids the evaluation of the full matrices.

As another advantage, using tensor methods allows to fully exploit the tensor-product structure of the B-spline basis to apply efficient quadrature, also for low polynomial degrees. This has been a long-standing challenge in isogeometric analysis, since the increased support and degree of the basis functions result in a rapidly growing quadrature grid, and affects dramatically the computation times. By reducing the problem to 1D components, the number of evaluations and quadrature points increase logarithmically with the number of degrees of freedom.

Future research will be devoted to the use of approximate SVD computation and to the extension to the trivariate case.

Acknowledgement. This research was supported by the National Research Network "Geometry + Simulation" (NFN S117), funded by the Austrian Science Fund (FWF).

References

1. Antolin, P., Buffa, A., Calabrò, F., Martinelli, M., Sangalli, G.: Efficient matrix computation for tensor-product isogeometric analysis: the use of sum factorization. Comp. Meth. Appl. Mech. Engrg. **285**, 817–828 (2015)
2. Auricchio, F., Calabrò, F., Hughes, T., Reali, A., Sangalli, G.: A simple algorithm for obtaining nearly optimal quadrature rules for NURBS-based isogeometric analysis. Comp. Meth. Appl. Mech. Engrg. **249–252**, 15–27 (2012)
3. Bazilevs, Y., Beirão da Veiga, L., Cottrell, J.A., Hughes, T.J.R., Sangalli, G.: Isogeometric analysis: approximation, stability and error estimates for h-refined meshes. Math. Models Methods Appl. Sci. **16**(7), 1031–1090 (2006)
4. Calabrò, F., Manni, C., Pitolli, F.: Computation of quadrature rules for integration with respect to refinable functions on assigned nodes. Appl. Numer. Math. **90**, 168–189 (2015)
5. Cottrell, J.A., Hughes, T.J.R., Bazilevs, Y.: Isogeometric Analysis: Toward Integration of CAD and FEA. Wiley, Chichester (2009)
6. De Boor, C.: A Practical Guide to Splines. Applied Mathematical Sciences. Springer, Berlin (2001)
7. Hackbusch, W.: Tensor Spaces and Numerical Tensor Calculus. Springer, Berlin (2012)
8. Hughes, T., Cottrell, J., Bazilevs, Y.: Isogeometric analysis: CAD, finite elements, NURBS, exact geometry and mesh refinement. Comp. Meth. Appl. Mech. Engrg. **194**(39–41), 4135–4195 (2005)
9. Hughes, T., Reali, A., Sangalli, G.: Efficient quadrature for NURBS-based isogeometric analysis. Comp. Meth. Appl. Mech. Engrg. **199**(5–8), 301–313 (2010)
10. Jüttler, B., Langer, U., Mantzaflaris, A., Moore, S.E., Zulehner, W.: Geometry + simulation modules: implementing isogeometric analysis. PAMM **14**(1), 961–962 (2014)
11. Karatarakis, A., Karakitsios, P., Papadrakakis, M.: Computation of the isogeometric analysis stiffness matrix on GPU. In: Papadrakakis, M., Kojic, M., Tuncer, I., (eds.) Proceedings of the 3rd South-East European Conference on Computational Mechanics (SEECCM) (2013). www.eccomasproceedings.org/cs2013
12. Karatarakis, A., Karakitsios, P., Papadrakakis, M.: GPU accelerated computation of the isogeometric analysis stiffness matrix. Comp. Meth. Appl. Mech. Engrg. **269**, 334–355 (2014)
13. Khoromskij, B.N.: $\mathcal{O}(d\log n)$-quantics approximation of N-d tensors in high-dimensional numerical modeling. Constr. Appr. **34**(2), 257–280 (2011)
14. Khoromskij, B.N.: Tensor-structured numerical methods in scientific computing: survey on recent advances. Chemometr. Intell. Lab. Syst. **110**(1), 1–19 (2012)
15. Khoromskij, B.N.: Tensor numerical methods for multidimensional PDEs: theoretical analysis and initial applications. In: Proceedings of ESAIM, pp. 1–28 (2014)
16. Kleiss, S., Pechstein, C., Jüttler, B., Tomar, S.: IETI - isogeometric tearing and interconnecting. Comp. Meth. Appl. Mech. Engrg. **247–248**, 201–215 (2012)
17. Mantzaflaris, A., Jüttler, B.: Exploring matrix generation strategies in isogeometric analysis. In: Floater, M., Lyche, T., Mazure, M.-L., Mørken, K., Schumaker, L.L. (eds.) MMCS 2012. LNCS, vol. 8177, pp. 364–382. Springer, Heidelberg (2014)
18. Mantzaflaris, A., Jüttler, B.: Integration by interpolation and look-up for Galerkin-based isogeometric analysis. Comp. Methods Appl. Mech. Engrg. **284**, 373–400 (2015). Isogeometric Analysis Special Issue

19. Patzák, B., Rypl, D.: Study of computational efficiency of numerical quadrature schemes in the isogeometric analysis. In: Proceedings of the 18 th International Conference on Engineering Mechanics, EM 2012, pp. 1135–1143 (2012)
20. Schillinger, D., Evans, J., Reali, A., Scott, M., Hughes, T.: Isogeometric collocation: cost comparison with Galerkin methods and extension to adaptive hierarchical NURBS discretizations. Comp. Meth. Appl. Mech. Engrg. **267**, 170–232 (2013)
21. Schillinger, D., Hossain, S., Hughes, T.: Reduced Bézier element quadrature rules for quadratic and cubic splines in isogeometric analysis. Comp. Meth. Appl. Mech. Engrg. **277**, 1–45 (2014)

Combination of Piecewise-Geodesic Curves for Interactive Image Segmentation

Julien Mille[1]([✉]), Sébastien Bougleux[2], and Laurent D. Cohen[3]

[1] CNRS Université Lyon 1, LIRIS UMR5202, Université de Lyon,
69622 Lyon, France
julien.mille@liris.cnrs.fr
http://liris.cnrs.fr/jmille
[2] CNRS, GREYC, UMR6072, Université de Caen-Basse Normandie,
14050 Caen, France
https://bougleux.users.greyc.fr
[3] CNRS, CEREMADE, UMR7534, Université Paris Dauphine,
75016 Paris, France
https://www.ceremade.dauphine.fr/cohen

Abstract. Boundary-based interactive image segmentation methods aim to build a closed contour, very often using paths linking a set of user-provided landmark points, ordered along the contour of the object of interest. Among these methods, the geodesically-linked active contour (GLAC) model generates a *piecewise-geodesic* curve, by linking each pair of successive landmark points by a geodesic curve. As an important shortcoming, the geodesically linked active contour model in its initial formulation does not guarantee the curve to be simple. It may have multiple points, creating self-tangencies and self-intersections, which is inconsistent with respect to the purpose of segmentation. To overcome this issue, we study some properties of non-simple closed curves and introduce a novel energy term to quantity the amount of non-simplicity. We propose to extract a relevant contour from a set of possible paths, such that the resulting structure fits the image data and is simple. We develop a local search method to choose the best combination among possible paths, integrating the novel energy term.

1 Introduction

Consider the extraction of an object of interest in an image, with prior interaction allowed. Typically, the user provides a set of landmark points along the boundary of interest. Then, minimum cost paths are computed between these points in order to build a closed contour. In most cases, a minimum cost path can be efficiently found as the global solution of the corresponding minimization problem, like Dijkstra's algorithm for shortest paths in directed graphs with no negative cycle. Interactive segmentation methods based on a graph modeling of the image, making them discrete by nature, include the intelligent scissors (or *live-wire*) [15] and their *on-the-fly* extension [8], as well as the riverbed approach [14] based on the image foresting transform [7].

© Springer International Publishing Switzerland 2015
J.-D. Boissonnat et al. (Eds.): Curves and Surfaces 2014, LNCS 9213, pp. 341–356, 2015.
DOI: 10.1007/978-3-319-22804-4_25

As regards continous models, the minimal path method [6] achieved global minimization of the geodesic active contour functional [4], with fixed user-provided endpoints, using the Fast Marching method [17,18]. An interactive extension was later proposed by [9]. Since the control points are fixed and must be located on the target contour, the minimal path approach does not represent a curve which deforms its shape. Moreover, due to the restricted number of these points, the geodesic may fail to capture a relevant contour if the image is too noisy, not contrasted enough, or if the target contour is too lengthy. While several methods concentrate on avoiding this second drawback [2,3,10], the geodesically linked active contour model of [13] allows to overcome the first one, as it can be evolved thanks to a local search method. This latter model combines the advantages of geodesics with the ones of region-based energies, such as the minimal variance term proposed by [5]. Despite its robustness to local minima, this model can fail to construct a simple curve, i.e. without double point, from the initialization step to the end of the evolution.

To overcome this drawback, we propose to generate several relevant paths between landmark points and to select the combination of paths generating the best closed contour. In this extent, we introduce an energy functional, combining contour and region terms with a novel term favoring the simplicity of the curve, penalizing self-tangencies and self-intersections. Our three contributions, i.e. the simplicity energy, the construction of admissible paths as well as the selection of the optimal combination of paths, are described in details.

2 Defects of Non-simple Curves

To outline the boundary of an object, one of the desirable properties of a closed curve $\mathcal{C}$ is that it should be simple, i.e. with no multiple point. Instead of imposing simplicity as a hard constraint, we propose the quantify the amount of non-simplicity, which will be used as a additional term in the energy of the contour. Dealing with the geometrical and topological properties of the obtained curve, we wish to measure to what extent the curve is not simple. We assume that $\mathcal{C}$ has a number of *multiple points* which may be of two kinds: self-tangencies and self-intersections. The proposed simplicity term is based on a distinction between these two aspects.

2.1   Self-Tangency

Let $(u, v) \in [0, 1]^2$ s.t. $u \neq v$ be the pair of curve positions identifying a double point: $\mathcal{C}(u) = \mathcal{C}(v)$. If (u, v) corresponds to a point of self-tangency, velocity vectors $\mathcal{C}'(u)$ and $\mathcal{C}'(v)$ are colinear:

$$\left| \frac{\mathcal{C}'(u) \cdot \mathcal{C}'(v)}{\|\mathcal{C}'(u)\| \, \|\mathcal{C}'(u)\|} \right| = 1.$$

On the other hand, if (u, v) corresponds to a self-intersection - also known as an *ordinary double point* or *crunode* - velocity vectors point towards different directions, making the curve cross itself. This distinction allows to address separately two different defects on curves, which are not necessarily related. A curve with points of self-tangencies will exhibit self-overlapping segments, as depicted in Fig. 1(b), whereas a curve with self-intersections shown in Fig. 1(c) will split the image domain into more than two connected regions.

As regards the first kind of double points, the amount of self-tangency is quantified by measuring the length of overlapping curve segments. Considering function ϕ_C measuring the squared Euclidean distance between two points on curve C,

$$\phi_C(u,v) = \|C(u) - C(v)\|^2,$$

the zero level set of ϕ_C, $\mathcal{Z}_C = \{(u,v) \mid C(u) = C(v)\}$, is the set of pairs of positions giving equal points. Trivially, this set is never empty, since it contains at least all pairs (u,u). The length of $\mathcal{Z}_C$ in the (u,v)-space corresponds to the total length of the overlapping segments:

$$
\begin{aligned}
|\mathcal{Z}_C| &= \int_0^1 \int_0^1 \|\nabla H(\phi_C(u,v))\| \, du dv \\
&= \int_0^1 \int_0^1 \delta(\phi_C(u,v)) \, \|\nabla \phi_C(u,v)\| \, du dv,
\end{aligned}
\tag{1}
$$

where ∇ is the gradient operator in the $(u,v)-$ space, H is the Heaviside step function and δ is the Dirac delta distribution.

It can be proven that for any simple regular curve C, the amount of self-tangency $|\mathcal{Z}_C|$ is $\sqrt{2}$ (a detailed proof with a more general distance function and parameterization is given in Appendix A). It can be intuitively understood by observing the plots of ϕ in the curve parameter space in Fig. 1, where dark lines correspond to self-overlapping segments. If the curve is non-simple, $|\mathcal{Z}_C| \geq \sqrt{2}$. Moreover, this term is advantageously intrinsic, i.e. independent of parameterization, and invariant to scaling. Note that intersection points may be viewed as overlapping segments of length zero, and thus have no contribution in $|\mathcal{Z}_C|$.

2.2   Twisting

While tangent double points are used to measure self-tangency, ordinary double points will serve as a basis for measuring the amount of *twisting* of the curve. Self-crossings are pairs of curve positions (u,v) where tangent vectors are not colinear:

$$\left| \frac{C'(u) \cdot C'(v)}{\|C'(u)\| \, \|C'(v)\|} \right| \neq 1.$$

Pairs of positions corresponding to crossing points are ordered, so that u is the position where the curve is *intersected* and v is the position where the curve is *intersecting*. Whether a position on the curve is intersected or intersecting only depends on the order in which these positions are met while one travels along the curve, so it follows that $u < v$. Whitney [19] studied crossing points of closed regular curves and divided them into two categories with respect to their orientation. Let us recall the orientation of curves, set up in the standard computer left-handed Cartesian coordinate system. In this setting, we choose the natural orientation of curves as the clockwise one, such that when one travels along the curve, looking forward, the interior of the curve is on the right. In such case, $C'^{\perp}$ is the inward normal. The crossing will be *positive* if the intersecting part of the curve $C(v)$ goes from right to left, or *negative* if it goes from left to right. In a positive (resp. negative) crossing, the intersecting section arrives from the interior (resp. exterior) of the intersected section. It follows that:

$$C'(u)^{\perp} \cdot C'(v) < 0 \text{ for a positive crossing}$$
$$C'(u)^{\perp} \cdot C'(v) > 0 \text{ for a negative crossing.}$$

Consider the self-intersecting curve shown in Fig. 2(a). It splits the image domain into disjoints subdomains, some of which are demarcated by *inverted segments*, i.e. portions of curves along which the normal vector points outward. If one decomposes the curve using *uncrossing moves* - replacing each couple of cross-

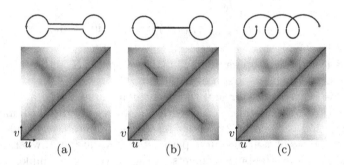

Fig. 1. Distances ϕ_C plotted in (u, v)-space for several types of curve: (a) On a simple closed curve, ϕ_C vanishes only on the graph diagonal. (b) On a curve with a section of self-tangency, ϕ_C additionally exhibits two symmetrical zero lines. (c) On a curve with self-intersections, ϕ_C additionally exhibits isolated zeros.

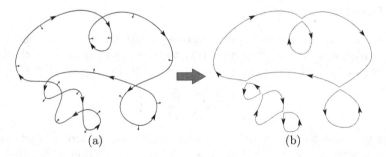

Fig. 2. Self-intersecting curve and inverted segments: (a) A nonsimple loop with ordinary double points is transformed into a set of (b) disjoint simple loops (Seifert circles) using a sequence of *uncrossing moves*. Resulting badly-oriented (counterclockwise) simple loops are drawn in red (Color figure online).

ing sections by two new non-crossing sections - one obtains a collection of disjoint simple loops, as in Fig. 2(b), also known as *Seifert circles* [1, p. 94]. While some simple loops are well oriented (clockwise), others are inverted (counterclockwise). We propose to quantify the twisting of $\mathcal{C}$ as the proportion of area demarcated by inverted segments to the total area of $\mathcal{C}$, or equivalently, the area inside the counter-clockwise oriented loops one would obtain when decomposing the curve. To some extent, this proportion is related to the energy that one should exert to untwist the curve.

A simple loop is described by a single inverted segment and involves a single intersection, whereas a double loop consists of two segments and involves two intersections. The two types of crossing points, that were previously presented, are used to detect such loops. A simple inverted loop will be detected thanks to a negative crossing point (u, v) such that there is no other crossing from u to v. A double inverted loop will be detected thanks to a negative crossing (u, v) such that the path from u to v contains a positive crossing. Let $\mathrm{SL}(\mathcal{C}) \subset [0, 1]^2$ be the set of ordered pairs of curve positions (u, v) s.t. $u < v$, describing intersections involved in single inverted loops, and $\mathrm{DL}(\mathcal{C}) \subset [0, 1]^2 \times [0, 1]^2$ be the set of double ordered pairs $((u_1, v_1), (u_2, v_2))$ s.t. $u_1 < u_2$ and $v_1 > v_2$ describing the couples of intersections involved in double inverted loops. Sets SL and DL can be extracted in linear time with respect to the curve length. When a portion of curve $\mathcal{C}$ from s to t, denoted $\mathcal{C}|_{s \to t}$, is closed and simple, the signed area inside $\mathcal{C}|_{s \to t}$ can be expressed using Green's theorem:

$$\int_{\Omega_{\mathrm{in}}(\mathcal{C}|_{s \to t})} \mathrm{d}\boldsymbol{x} = \frac{1}{2} \int_s^t x(u)y'(u) - x'(u)y(u)\mathrm{d}u = \int_s^t \frac{\mathcal{C}(u)^\perp \cdot \mathcal{C}'(u)}{2} \mathrm{d}u. \qquad (2)$$

Eventually, the total area of inverted loops of $\mathcal{C}$, denoted by $\mathcal{I}[\mathcal{C}]$, is expressed by considering all simple and double inverted loops in SL and DL:

$$\mathcal{I}[\mathcal{C}] = -\sum_{(s,t) \in \mathrm{SL}(\mathcal{C})} \int_s^t \frac{\mathcal{C}^\perp \cdot \mathcal{C}'}{2} \mathrm{d}u - \sum_{\substack{((s_1,t_1),(s_2,t_2)) \\ \in \mathrm{DL}(\mathcal{C})}} \int_{s_1}^{s_2} \frac{\mathcal{C}^\perp \cdot \mathcal{C}'}{2} \mathrm{d}u + \int_{t_2}^{t_1} \frac{\mathcal{C}^\perp \cdot \mathcal{C}'}{2} \mathrm{d}u. \qquad (3)$$

Note that $\mathcal{I}[\mathcal{C}]$ is positive, as the signed area of every inverted loop is negative.

3   Geodesic and Piecewise-Geodesic Curves

To extract structures in a given image $\boldsymbol{I} : \mathcal{D} \to \mathbb{R}^d$, [6] proposed to find curves of minimal length according to an heterogeneous isotropic metric defined from a potential $P : \mathcal{D} \to \mathbb{R}^{*+}$. In the context of contour extraction, curves should be located along edges. The potential is thus defined as $P = g + w$, where $g : \mathcal{D} \to \mathbb{R}^+$ is a decreasing function of the gradient magnitude of $\boldsymbol{I}$ at some scale s:

$$g(\boldsymbol{x}) = \frac{1}{1 + \|\nabla G_s * \boldsymbol{I}(\boldsymbol{x})\|} \qquad (4)$$

where G_s is a Gaussian kernel and $w \in \mathbb{R}^{*+}$ is a regularizing constant. Given two points $\boldsymbol{a}$ and $\boldsymbol{b}$, the curve of minimal length, or *geodesic path* between them,

$$\gamma_{\boldsymbol{a},\boldsymbol{b}} = \operatorname*{argmin}_{\mathcal{C} \subset \mathcal{D}} \left\{ L[\mathcal{C}] = \int_0^1 P(\mathcal{C}(u)) \, \|\mathcal{C}'(u)\| \, du \right\} \quad \text{s.t.} \quad \begin{cases} \mathcal{C}(0) = \boldsymbol{a} \\ \mathcal{C}(1) = \boldsymbol{b} \end{cases} \quad (5)$$

can be obtained by considering the geodesic distance map, also referred to as the *minimal action map*, $U_{\boldsymbol{a}} : \mathcal{D} \to \mathbb{R}^+$ which assigns, to each point $\boldsymbol{x} \in \mathcal{D}$, the length of the minimal path connecting $\boldsymbol{a}$ to $\boldsymbol{x}$. This map is the unique viscosity solution of the Eikonal equation, which makes it efficiently computable by the Fast Marching (FM) method [17,18] in $O(N \log N)$ operations, where N is the number of grid points. Once the distance map has been numerically computed, the minimal path between $\boldsymbol{a}$ and $\boldsymbol{b}$ can be extracted by a gradient descent on $U_{\boldsymbol{a}}$, starting from $\boldsymbol{b}$ until $\boldsymbol{a}$ is reached. The minimal path approach can fail to extract the desired curve, as depicted in Fig. 3(c). This happens for instance when P is too noisy or not contrasted enough, when the length of the target curve is too important, or when the regularization constant w is too high.

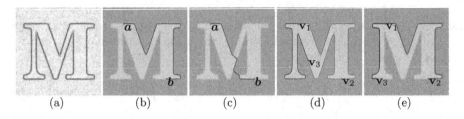

(a)   (b)   (c)   (d)   (e)

Fig. 3. (a) Potential P. (b) With a sufficiently low regularization weight w, the geodesic between two given points follows the object contour. (c) Due to an excessive regularization weight w, the geodesic makes an undesirable shortcut. (d) Relevant piecewise-geodesic curve with evenly spaced vertices. (e) Undesirable overlapping with unevenly spaced vertices

To address this issue, several approaches aim to find a *piecewise-geodesic*, i.e. a set of geodesics connecting pairs of successive landmark points or *vertices* (see Fig. 3(d)). Among them, the geodesically linked active contour (GLAC) model [13] is generated by joining end-to-end geodesic paths built from a set of vertices $\mathcal{V} = \{\mathbf{v}_i\}_{1 \le i \le n}$. Given a curve concatenation operator $\uplus$,

$$(\mathcal{C}_1 \uplus \mathcal{C}_2)(u) = \begin{cases} \mathcal{C}_1(2u) & \text{if } 0 \le u \le 1/2 \\ \mathcal{C}_2(2u - 1) & \text{if } 1/2 < u \le 1 \end{cases}, \quad (6)$$

which is valid only if $\mathcal{C}_1(1) = \mathcal{C}_2(0)$, the closed curve resulting from the assembly of successive geodesics is $\Gamma = \gamma_{\mathbf{v}_1,\mathbf{v}_2} \uplus \gamma_{\mathbf{v}_2,\mathbf{v}_3} \uplus \dots \uplus \gamma_{\mathbf{v}_{n-1},\mathbf{v}_n} \uplus \gamma_{\mathbf{v}_n,\mathbf{v}_1}$. It is important to keep in mind that a concatenation of geodesics is not a geodesic itself, thus it is relevant to refer to it as *piecewise-geodesic*. The GLAC approach consists in finding the sequence of n vertices which generates a piecewise-geodesic curve minimizing energy functional E, a weighted sum of an edge term and a region term (see [13] for more details). Numerical minimization of E with respect to every $\mathbf{v}_i$ is performed thanks to a greedy algorithm similar in principle to the one proposed in [20]. Considering a local discrete window $\mathcal{W}_i$ centered at each vertex $\mathbf{v}_i$, greedy evolution is performed, in several passes, by moving vertex $\mathbf{v}_i$ to the position in the window leading to the GLAC having the smallest energy:

$$\mathbf{v}_i^{(t+1)} = \underset{\tilde{\mathbf{v}}_i \in \mathcal{W}_i}{\operatorname{argmin}} E\left[\gamma_1 \uplus \ldots \gamma_{i-2} \uplus \tilde{\gamma}_{i-1} \uplus \tilde{\gamma}_i \uplus \gamma_{i+1} \uplus \ldots \uplus \gamma_n\right],$$

where $\tilde{\gamma}$ are the geodesics, according to Eq. (5), between the current position in the window and the adjacent vertices of $\mathbf{v}_i$, and γ_i is a shortened form for $\gamma_{\mathbf{v}_i, \mathbf{v}_{i+1}}$. While this model allows to blend the benefits of minimal paths and region-based terms, it turns out to have a significant drawback, as its initial state is not necessarily a simple closed curve. As depicted in Fig. 3(e), this can occur when the initial vertices are unevenly distributed around the target boundary. In this case, geodesics gather on particular sides of the target boundary, as $\gamma_{\mathbf{v}_3, \mathbf{v}_1}$ overlaps $\gamma_{\mathbf{v}_1, \mathbf{v}_2}$ and $\gamma_{\mathbf{v}_2, \mathbf{v}_3}$. The reason is that each geodesic is generated independently from the others, such that the obtained piecewise-geodesic curve does not depend on the visiting order of pairs of adjacent vertices. This undesirable phenomenon may occur either as soon as the geodesically linked contour is initialized, or after several evolution steps on a previously well initialized contour. One could consider imposing hard constraints on the overlapping between paths, or penalizing paths enclosing a region with excessively small area, but the independent construction of paths, which allows parallel implementation, prevents such constraints to be implemented. We address this shortcoming in what follows.

4   Combination of Admissible Paths

We study a more relevant contour construction preserving the advantages of piecewise geodesic curves. Assuming that a set of several possible relevant paths is available for each pair of successive vertices, we may select a single path from each set and combine these paths in order to build the best boundary curve. The relevancy of the generated contour is measured by an energy functional, combining existing contour and region terms with a novel term penalizing self-tangencies and self-intersections, ensuring the contour to be simple.

4.1   Sets of Admissible Paths

Given an ordered pair of successive vertices $(\mathbf{v}_i, \mathbf{v}_{i+1})$, let $\mathcal{A}_i$ be a set of K_i *admissible paths* linking the two vertices,

$$\mathcal{A}_i = \{\gamma_{i,j}\}_{1 \leq j \leq K_i},$$

which we refer to as *admissible set*. To generate these, we propose an approach based on the extraction of *saddle points*. As explained in Sect. 3, the geodesic

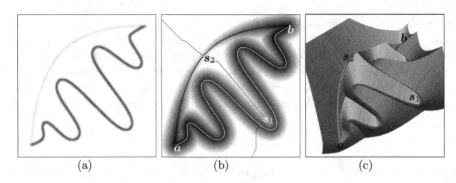

(a)   (b)   (c)

Fig. 4. Saddle points for determining relevant admissible paths between a and b: (a) Potential highlights two possible distinctive paths, (b) Action map with two admissible paths with their respective saddle points located halfway and (c) corresponding 3D plot

linking two points a and b is extracted by propagating the minimal action from a, stop when b is reached, and perform gradient descent from b. The opposite may also be done, provided that the path is reversed afterwards. A third way to compute $\gamma_{a,b}$ consists in propagating **simultaneously** from a and b, stopping at the first location where the two fronts meet, performing two gradient descents both sides apart from the meeting location, and assembling the two obtained paths adequately. This principle serves a basis for the generation of multiple paths. When propagation is performed from two source points, yielding the *combined action map* $U_{a,b} = \min(U_a, U_b)$, the two propagation fronts meet at the **saddle points** of $U_{a,b}$ (see Fig. 4(a) and (b)). If one intuitively imagine the action as the height in a mountainous area, the saddle points are the mountain passes on the different roads travelling from a to b, as depicted in Fig. 4(c). These roads can share common sections, but each road lies in the bottom of a particular valley.

Let $m_{a,b} : [0, 1] \longrightarrow \mathcal{D}$ be the *medial curve*, assumed to be of class C^2, sweeping along the points geodesically equidistant to a and b, i.e. $\{x \mid U_a(x) = U_b(x)\}$. It forms a crest on the combined action map and contains only critical points, put another way $\nabla U_{a,b}$ is not defined over $m_{a,b}$. However, $U_{a,b}$ may be differentiated **along** the medial curve, which allows to define the saddle points as the local minima of $U_{a,b}$ along $m_{a,b}$.

In order to extract robust local minima in spite of the effect of discretization, values of the action map are smoothed along $m_{a,b}$. Not all saddle points are kept as starting points for path construction. In practice, we limit the number of admissible paths per set below a fixed threshold $K_{\max}$. As can be seen in Fig. 5(b), many saddle points are intentionally ignored because of their action too high or their proximity to lower saddle points.

4.2   Combination of Admissible Paths Using the Simplicity Energy

The computation of an admissible closed contour can be formulated as determining the sequence of labels $\{x_1, x_2, \ldots, x_n\}$ minimizing

$E\left[\gamma_{1,x_1} \uplus \gamma_{2,x_2} \uplus \ldots \uplus \gamma_{n-1,x_{n-1}} \uplus \gamma_{n,x_n}\right]$, where label x_i corresponds to the chosen path in set $\mathcal{A}_i$. Energy E extends the energy involved in the GLAC. It is designed to penalize contours exhibiting strongly overlapping sections or self-intersections, poorly fitting to edges, or splitting regions with indistinct color statistics:

$$E[\Gamma] = E_{\text{simplicity}}[\Gamma] + \omega_{\text{edge}}E_{\text{edge}}[\Gamma] + \omega_{\text{region}}E_{\text{region}}[\Gamma]. \tag{7}$$

Weights ω_{edge} and ω_{region} are user-defined parameters controlling the relative significance of the edge and region terms over the simplicity term. This last one involves the normalized self-overlapping and twisting measures defined in Eqs. (1) and (3),

$$E_{\text{simplicity}}[\Gamma] = \frac{|\mathcal{Z}_\Gamma| - \sqrt{2}}{\sqrt{2}} + \frac{1}{|\Omega_{\text{in}}(\Gamma)|}\mathcal{I}[\Gamma]$$

whereas the edge term is defined as in [12]. For the region term, instead of using the piecewise-constant model as in the GLAC, which limits the segmentation to relatively homogeneous objects and backgrounds, we use the Bhattacharyya coefficient between the color probability distributions inside and outside Γ, following [11]:

$$E_{\text{region}}[\Gamma] = \int_C \sqrt{\mathrm{p}_{\text{in}}(\Gamma, \alpha)\mathrm{p}_{\text{out}}(\Gamma, \alpha)}\,\mathrm{d}\alpha \tag{8}$$

Probability distribution functions (PDF), for a given color α, are estimated using a Gaussian kernel-based histogram:

$$\mathrm{p}_{\text{in}}(\Gamma, \alpha) = \frac{1}{|\Omega_{\text{in}}(\Gamma)|} \int_{\Omega_{\text{in}}(\Gamma)} G_\sigma(\alpha - \boldsymbol{I}(\boldsymbol{x}))\mathrm{d}\boldsymbol{x}$$

$$\mathrm{p}_{\text{out}}(\Gamma, \alpha) = \frac{1}{|\Omega_{\text{out}}(\Gamma)|} \int_{\Omega_{\text{out}}(\Gamma)} G_\sigma(\alpha - \boldsymbol{I}(\boldsymbol{x}))\mathrm{d}\boldsymbol{x}$$

where Ω_{in} and Ω_{out} are the regions inside and outside Γ, respectively. To determine the best sequence of labels $\{x_1, \ldots, x_n\}$, a brute-force search would yield an exponential complexity of $O(K_{\max}^n)$. To avoid testing all possible configurations, we propose a greedy search in $O(n^2 K_{\max})$ relying on a specific ordering of paths. In each admissible set $\mathcal{A}_i$, paths are sorted according to increasing *exteriority* $\mathcal{X}$, i.e. the signed area, calculated with Green's theorem, formed by a given path C and the line segment from $C(1)$ returning to $C(0)$[1]:

$$\mathcal{X}[C] = \frac{1}{2} \int_0^1 C^\perp \cdot C'\mathrm{d}u + \frac{1}{2}C(1)^\perp \cdot C(0)$$

If the straight line from $C(0)$ to $C(1)$ is taken as a reference horizontal axis, the exteriority is negative (resp. positive) if C is predominantly below (resp. above) the axis. The vertices being located clockwise, admissible paths are sorted

---

[1] See detailed proof in appendix B.

Fig. 5. (a) Potential P (b) Medial curve (black) and saddle points (green) with corresponding paths drawn over combined action map (c-f) Admissible set for each pair of successive vertices with $n = 4$. Paths are sorted (blue to red) according to their exteriority (Color figure online)

from the innermost to the outermost (see Fig. 5(c–f)). Starting from an initial labelling corresponding to the most interior configuration, i.e. $\{1, \ldots, 1\}$, labels are changed according to a local search, by iteratively testing candidate labellings. At each iteration, given a current sequence of labels $\mathcal{S}$, candidate sequences are tested that differ from a single label from $\mathcal{S}$, by increasing labels only. For instance, if $\{2, 3, 1\}$ is the current sequence, candidate sequences will be $\{3,3,1\}$, $\{2,4,1\}$ and $\{2,3,2\}$. The candidate sequence leading to the smallest energy, regardless of the current energy, is chosen as the base sequence for the next iteration, while the globally minimal sequence is kept along the iterations. Generating candidate sequences by solely increasing labels makes the contour grow monotonically.

5   Experiments and Discussion

We demonstrate the ability of the model to recover closed boundaries of objects in natural color images, given a variable number of user-provided points along the target boundary. Experiments were carried out on the Grabcut dataset [16] in comparison with the original GLAC method [13] (without deformation, i.e. the piecewise-geodesic curve only), in order to show the benefits brought by the use of admissible paths and additional energy terms. To keep a critical look at our contribution, Fig. 6 depicts typical cases where the current approach provides strong global improvements over the GLAC, as well as slight localized improvements, for the same given sets of vertices. As regards the selection of parameters, both methods were evaluated in the most favorable configuration, i.e. parameters such as the regularization weight w - for both methods - or the energy weights ω_{edge} and ω_{region} - specifically for the proposed algorithm -, were tuned separately each time, in order to achieve the most relevant segmentation. The appropriate color space was also chosen for each image, i.e. RGB or the more perceptually uniform Lab, which affects potential P for both methods and the color PDFs involved in the region energy (8). Color components of I being normalized in range $[0, 1]$, regularization weight w varies between 0.01 and 0.1. Energy weights ω_{edge} and ω_{region} were both tuned between 0.5 and 2.

The BANANA image (row 1) depicts a situation where our approach does not improve segmentation over the GLAC, for this particular configuration of initial points. Despite the complex background and object containing many inner edges, boundaries are well defined and vertices are evenly distributed along the boundary so that the original GLAC manages to extract the object. The FLOWER (row 2) and DOLL (row 3) images are cases where the GLAC exhibits strong overlapping between geodesics when few vertices are provided, although the vertices are reasonably well distributed along the object boundary. Since it does not have any non-overlapping constraint, the GLAC systematically favors portions of contours with the lowest potential. Hence, smooth boundary segments are ignored, not because of their length, but because they may contain sparsely weak edges making the potential increase in small parts of the contour. Conversely, these boundary segments, despite from not being part of the minimal path, make valleys in the distance map and are very likely to be considered as parts of admissible paths by our approach. This proves the proposed method to be inherently less sensitive to weak edges. The CERAMIC (row 4) and TEDDY (row 5) images depicts situations where the GLAC makes *shortcuts* through the object, due to the presence of inner edges stronger than the actual boundaries. Both simplicity and region energies contribute to solve this issue in our algorithm. The former prevents overlapping while the latter favors high discrepancy between inner and outer color distributions, hence avoiding to select the undesirable shortcuts, which would yield to less distinct color histograms than the actual boundaries would. Finally, the MUSHROOM (row 6) combines the issues of inner shortcut and strong self-overlapping.

6 Conclusion

By searching the best paths configuration among sets of admissible paths, given an energy functional combining edge and region data terms with a novel term favoring the simplicity of the curve, we aimed at solving some important issues arising in geodesic-based segmentation. The histogram-based region term helps to avoid paths making shortcuts through the object, whereas the novel simplicity energy prevents self-overlapping or self-intersecting contours. Comparison against the geodesically linked active contour model, which has similar input and purpose, demonstrated the advantages of the approach.

Appendix

A   Self-Overlap Term

Let C be a regular curve parameterized over $[0, L]$. Let ϕ be a C^1 function defined over $[0, L]^2$ representing the distance between two positions on the curve:

$$\phi_C(u, v) = \|C(u) - C(v)\|^p$$

where p is an arbitrary positive real exponent. The length of the zero level set of ϕ_C,

$$|\mathcal{Z}_C| = \int_0^L \int_0^L \delta(\phi(u, v)) \, \|\nabla\phi(u, v)\| \, dudv, \tag{9}$$

quantifies the self-overlap of C.

Proposition: If C is simple, i.e. without self-intersection and self-tangency, then $|\mathcal{Z}_C| = L\sqrt{2}$

Proof: As a preliminary calculation, let us express the gradient of ϕ (partial derivatives are written using the indexed notation):

$$\nabla\phi(u, v) = [\phi_u(u, v) \quad \phi_v(u, v)]^T$$
$$= p \, \|C(u) - C(v)\|^{p-2} \begin{bmatrix} C'(u) \cdot (C(u) - C(v)) \\ -C'(v) \cdot (C(u) - C(v)) \end{bmatrix}$$

If C is regular and simple, varying with respect to u in range $[0, L]$, $\phi(u, v)$ is nowhere zero except when $u = v$. Hence, for a fixed v, we have:

$$\delta(\phi(u, v)) = \frac{\delta(u - v)}{|\phi_u(v, v)|} \tag{10}$$

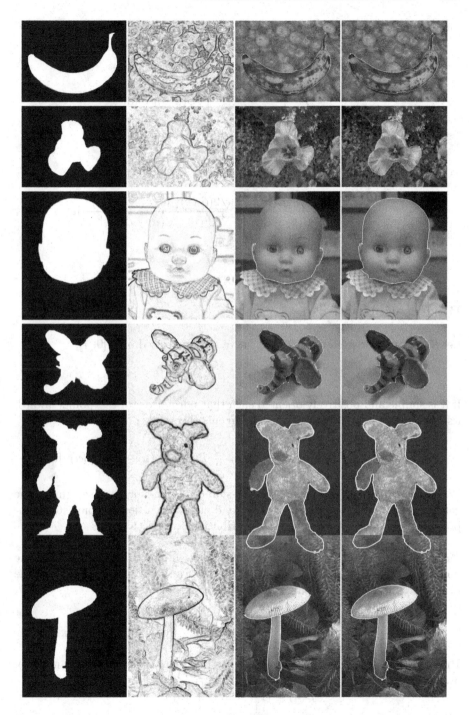

Fig. 6. Qualitative comparison with the GLAC on a sample of the Grabcut dataset. Column 1: ground truth segmentation, Column 2: potential, Column 3: GLAC without deformation (piecewise-geodesic curve), Column 4: combination of paths.

Integrating (10) into (9) and applying the definition of measure δ:

$$
\begin{aligned}
|\mathcal{Z}_C| &= \int_0^L \int_0^L \delta(\phi(u,v)) \, \|\nabla\phi(u,v)\| \, du\,dv \\
&= \int_0^L \int_0^L \frac{\delta(u-v)}{|\phi_u(v,v)|} \, \|\nabla\phi(u,v)\| \, du\,dv \\
&= \int_0^L \frac{\|\nabla\phi(v,v)\|}{|\phi_u(v,v)|} \, dv
\end{aligned}
$$

Trivially, $\phi(v,v) = 0$. However expanding the gradient gives:

$$
\begin{aligned}
|\mathcal{Z}_C| &= \int_0^L \frac{p\,\|\mathcal{C}(v) - \mathcal{C}(v)\|^{p-2} \sqrt{2(\mathcal{C}'(v)\cdot(\mathcal{C}(v)-\mathcal{C}(v)))^2}}{p\,\|\mathcal{C}(v)-\mathcal{C}(v)\|^{p-2}\,|\mathcal{C}'(v)\cdot(\mathcal{C}(v)-\mathcal{C}(v))|} dv \\
&= \int_0^L \sqrt{2}\, dv \\
&= L\sqrt{2}
\end{aligned}
$$

B   Exteriority Term

Let $\mathcal{C}$ be a piecewise-smooth regular curve parameterized over $[0,1]$,

$$
\mathcal{C} : u \longmapsto \mathcal{C}(u) = [x(u)\ y(u)]^T.
$$

If it is simple and positively oriented such that normal vector $\mathcal{C}'^{\perp}$ points inward, its inner area may be expressed using Green's theorem:

$$
|\Omega_{\text{in}}(\mathcal{C})| = \frac{1}{2}\int_0^1 \mathcal{C}^{\perp}(u)\cdot\mathcal{C}'(u)\,du = \frac{1}{2}\int_0^1 x(u)y'(u) - x'(u)y(u)\,du
$$

When one calculates the previous expression on a non-simple closed curve, one gets the signed area, in which positively and negatively oriented connected components have positive and negative contributions, respectively.

Proposition: The signed area formed by an open curve $\mathcal{C}$ over $[0,1]$ and the line segment from $\mathcal{C}(1)$ returning to $\mathcal{C}(0)$ (see Fig. 7), which we use to as the *exteriority* measure in the paper, may be expressed as:

$$
\mathcal{X}[\mathcal{C}] = \frac{1}{2}\int_0^1 \mathcal{C}^{\perp}\cdot\mathcal{C}'du + \frac{1}{2}\mathcal{C}^{\perp}(1)\cdot\mathcal{C}(0)
$$

Proof: Let S be the parametrization of the line segment joining $\mathcal{C}(1)$ and $\mathcal{C}(0)$, over $[0,1]$:

$$
S(u) = (1-u)\mathcal{C}(1) + u\mathcal{C}(0)
$$

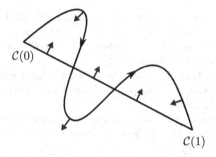

$\mathcal{C}(0)$

$\mathcal{C}(1)$

Fig. 7. The exteriority of an open curve is measured as the signed area of the multiple connected region that it forms with the line segment joining its two endpoints

The signed area is then obtained by applying Green's theorem on a piecewise basis:

$$
\begin{aligned}
\mathcal{X}[\mathcal{C}] &= \frac{1}{2}\int_0^1 \mathcal{C}^\perp \cdot \mathcal{C}'\,\mathrm{d}u + \frac{1}{2}\int_0^1 S^\perp \cdot S'\,\mathrm{d}u \\
&= \frac{1}{2}\int_0^1 \mathcal{C}^\perp \cdot \mathcal{C}'\,\mathrm{d}u + \frac{1}{2}\int_0^1 ((1-u)\mathcal{C}^\perp(1) + u\mathcal{C}^\perp(0)) \cdot (\mathcal{C}(0) - \mathcal{C}(1))\,\mathrm{d}u \\
&= \frac{1}{2}\int_0^1 \mathcal{C}^\perp \cdot \mathcal{C}'\,\mathrm{d}u + \frac{1}{2}\int_0^1 (1-u)\mathcal{C}^\perp(1) \cdot \mathcal{C}(0) + u\mathcal{C}^\perp(1) \cdot \mathcal{C}(0)\,\mathrm{d}u \\
&= \frac{1}{2}\int_0^1 \mathcal{C}^\perp \cdot \mathcal{C}'\,\mathrm{d}u + \frac{1}{2}\mathcal{C}^\perp(1) \cdot \mathcal{C}(0)
\end{aligned}
$$

References

1. Adams, C.: The Knot Book: An Elementary Introduction to the Mathematical Theory of Knots. American Mathematical Society, Williamstown (2004)
2. Appia, V., Yezzi, A.: Active geodesics: region-based active contour segmentation with a global edge-based constraint. In: IEEE International Conference on Computer Vision (ICCV), Barcelona, Spain, pp. 1975–1980 (2011)
3. Benmansour, F., Cohen, L.: Fast object segmentation by growing minimal paths from a single point on 2D or 3D images. J. Math. Imaging Vis. **33**(2), 209–221 (2009)
4. Caselles, V., Kimmel, R., Sapiro, G.: Geodesic active contours. Int. J. Comput. Vis. **22**(1), 61–79 (1997)
5. Chan, T., Vese, L.: Active contours without edges. IEEE Trans. Image Process. **10**(2), 266–277 (2001)
6. Cohen, L., Kimmel, R.: Global minimum for active contour models: a minimal path approach. Int. J. Comput. Vis. **24**(1), 57–78 (1997)
7. Falcão, A., Stolfi, J., Lotufo, R.: The image foresting transform: theory, algorithms, and applications. IEEE Trans. Pattern Anal. Mach. Intell. **26**(1), 19–29 (2004)
8. Falcão, A., Udupa, J., Miyazawa, F.: An ultra-fast user-steered image segmentation paradigm: live wire on the fly. IEEE Trans. Med. Imaging **19**(1), 55–62 (2000)

9. Gérard, O., Deschamps, T., Greff, M., Cohen, L.D.: Real-time interactive path extraction with on-the-fly adaptation of the external forces. In: Heyden, A., Sparr, G., Nielsen, M., Johansen, P. (eds.) ECCV 2002, Part III. LNCS, vol. 2352, pp. 807–821. Springer, Heidelberg (2002)

10. Kaul, V., Yezzi, A., Tsai, Y.: Detection of curves with unknown endpoints and arbitrary topology using minimal paths. IEEE Trans. Pattern Anal. Mach. Intell. 34(10), 1952–1965 (2012)

11. Michailovich, O., Rathi, Y., Tannenbaum, A.: Image segmentation using active contours driven by the Bhattacharyya gradient flow. IEEE Trans. Image Process. 16(11), 2787–2801 (2007)

12. Mille, J.: Narrow band region-based active contours and surfaces for 2D and 3D segmentation. Comput. Vis. Image Underst. 113(9), 946–965 (2009)

13. Mille, J., Cohen, L.D.: Geodesically linked active contours: evolution strategy based on minimal paths. In: Tai, X.-C., Mørken, K., Lysaker, M., Lie, K.-A. (eds.) SSVM 2009. LNCS, vol. 5567, pp. 163–174. Springer, Heidelberg (2009)

14. Miranda, P., Falcão, A., Spina, T.: Riverbed: a novel user-steered image segmentation method based on optimum boundary tracking. IEEE Trans. Image Process. 21(6), 3042–3052 (2012)

15. Mortensen, E., Barrett, W.: Interactive segmentation with intelligent scissors. Graph. Model. Image Process. 60(5), 349–384 (1998)

16. Rother, C., Kolmogorov, V., Blake, A.: Grabcut: interactive foreground extraction using iterated graph cuts. ACM Trans. Graph. 23(3), 309–314 (2004)

17. Sethian, J.: A fast marching level set method for monotonically advancing fronts. Proc. Natl. Acad. Sci. 93(4), 1591–1595 (1996)

18. Tsitsiklis, J.: Efficient algorithms for globally optimal trajectories. IEEE Trans. Autom. Contr. 40(9), 1528–1538 (1995)

19. Whitney, H.: On regular closed curves in the plane. Compos. Math. 4, 276–284 (1937)

20. Williams, D., Shah, M.: A fast algorithm for active contours and curvature estimation. Comput. Vis. Graph. Image Process. Image Underst. 55(1), 14–26 (1992)

A Fully-Nested Interpolatory Quadrature Based on Fejér's Second Rule

Jacques Peter$^{(\boxtimes)}$

ONERA, The French Aerospace Lab, CFD Department,
29 Av de la Division Leclerc, 92322 Chatillon Cedex, France
jacques.peter@onera.fr
http://www.onera.fr

Abstract. Our goal is to alleviate the constraint of the classical 1D interpolatory nested quadratures than one should go from a set of n to a set of $(2n + 1)$ points (for Fejér second rule [5]) or $(2n - 1)$ points (for Clenshaw-Curtis rule [1]) to benefit from the nesting property. In this work a sequence of recursively included quadrature sets for all odd number of quadrature points is proposed to define interpolatory rules. These sets are confounded with the one of Fejér's second rule when the cardinal is a power of two minus one and different if not. The weights of the corresponding interpolatory rule are studied. This rule is efficient for calculating integrals of very regular functions with control of accuracy via application of successive formulas of increasing order.

Keywords: Quadrature · Nested rule · Interpolatory rule

Nomenclature

$L_n^k(x)$	Lagrangian polynomial for x_n^k among the set $\mathcal{S}_n$
$EL_n^k(x)$	Even part of $L_n^k(x)$
$\mathcal{F}_n$	Set of nodes of Fejer n-point second rule
M	Largest odd integer lower than n such that $M = 2^Q - 1$
n	Number of points of the rule
$\mathcal{S}_n$	Set of nodes of the proposed n-point rule
$T_l(x)$	First-kind Chebychev polynomial of degree l
$U_l(x)$	Second-kind Chebychev polynomial of degree l
w_n^k	Weight of x_n^k and x_n^{n+1-k} in the rule
x_n^k	Abscissae of the n-point rule (decreasing order) equal to $\cos(\alpha_n^k)$
α_n^k	Set of increasing angles in $]0, \pi[$, $\cos(\alpha_n^k)$ equal to x_n^k
$\Delta_n^k(\theta)$	Sine polynomial with odd terms corresponding to $EL_n^k(x)$
γ_n^k	$\cos(\gamma_n^k)$, abscissae of Fejer second rule (decreasing order)
φ_j^m	Integral over $[0, \pi]$ of $\Phi_p^m(\theta)$
$\Phi_j^m(\theta)$	Sine polynomial with odd terms for $\Delta_n^k(\theta)$ analysis

When discussing the recursive relations between the successive Lagrange polynomial (or even part of Lagrange polynomial or trigonometric polynomials),

© Springer International Publishing Switzerland 2015
J.-D. Boissonnat et al. (Eds.): Curves and Surfaces 2014, LNCS 9213, pp. 357–383, 2015.
DOI: 10.1007/978-3-319-22804-4_26

$L_n^k(x)$ (or $EL_n^k(x)$ or Δ_n^k), associated to a specific abscissa, it is easier to use as exponent the polar angle α associated to the point as it does not change when enriching the set of points rather than the index k that may change from one set to another. Thus, in this context, the notations $L_n^\alpha(x)$, $EL_n^\alpha(x)$, $\Delta_n^\alpha(\theta)$ and w_n^α are prefered.

1   Introduction

1.1   Interest of Nested Quadrature Rules

A very large number of scientific problems require the calculation of integrals and this need has led to the development and study of various types of quadrature formulae [3]. In the framework of interpolatory quadrature rules, the so-called nested formulae are those in which the points of a coarse set are included in larger sets defining more accurate formulae. This property is hightly desirable when the evaluation of the interest function is expensive, *e.g.* when it comes from CPU-expensive numerical simulations.

The aim of this work is to release the constraint of the classical 1D interpolatory nested quadratures than one should go from a set of n points to a set of $(2n + 1)$ (for Fejér second rule [5]) or $(2n - 1)$ (for Clenshaw-Curtis rule [1]) to benefit from the nesting property. Subsects. 1.2 and 1.3 are remainders about the interpolatory 1D rules whereas Subsect. 1.4 defines the proposed rule and Subsect. 1.5 presents the outline of this document.

1.2   Interpolatory Quadrature Rules

The framework of this study is the calculation of 1D definite integrals by interpolatory quadrature rules [2,3]. By a linear change of variable a finite summation interval can be transformed to $[-1, 1]$ so that, without restrictions, integrals like

$$I[f] = \int_{-1}^{1} f(x)dx$$

can be considered. An interpolatory rule is of the form

$$I[f] \simeq \sum_{i=1}^{n} w_i f(x_i) \tag{1}$$

where $x_1 < x_2 < ... < x_n$ are distinct nodes and $w_1, w_2...w_n$ the corresponding weights. Due to the expression in the right-hand side, a Riemann integral has to be considered in the left hand-side; as concerning the function f, it is always supposed to be at least continuous on $[-1, 1]$.

The accuracy of formula (1) is classically defined by the highest degree of the polynomials that this formula can exactly integrate. In case the nodes are defined *a priori*, the weights are calculated to exactly integrate the canonic polynomial basis $1, x, x^2 ... x^{n-1}$. They are the solution of the following linear system

$$
\begin{cases}
w_1 & + & w_2 & + ... + & w_n & = & 2 \\
w_1\,x_1 & + & w_2\,x_2 & + ... + & w_n\,x_n & = & 0 \\
... & & ... & ... & ... & & ... \\
w_1\,x_1^{n-1} & + & w_2\,x_2^{n-1} & + ... + & w_n\,x_n^{n-1} & = & \frac{1-(-1)^n}{n}
\end{cases}
\tag{2}
$$

The Vandermonde matrix determinant is non zero as the nodes are distinct. The latter guarantees the existence and unicity of the solution. The corresponding weights define an exact quadrature for polynomials of degree up to $(n - 1)$. In case n is odd and the nodes are symmetricallydistributed in $[-1, +1]$, it is easy to check that the rule is also exact for polynomial of degree n. The weights are fully defined by Eq. (2) but they are often calculated by applying the interpolatory quadrature rule itself to the Lagrangian polynomials associated to the nodes: let $L^i(x)$ be the Lagrangian polynomial associated to x_i among the set of nodes $(x_1, x_2 ... x_n)$

$$
L^i(x) = \prod_{l=1; l \neq i}^{l=n} \frac{(x - x_l)}{(x_i - x_l)}
$$

L^i is a polynomial of degree $(n - 1)$. Hence its integral is calculated exactly by rule (1) and, as it is null at all points of the rule except x_i,

$$
w_i = \int_{-1}^{1} L^i(x)dx
\tag{3}
$$

It can be proved that the interpolatory rule based on $(x_1, x_2, ..., x_n)$ calculates the exact sum over $[-1, +1]$ of the Lagragian polynomial that interpolates f in $(x_1, x_2, ..., x_n)$ [3]. This second point of view clarifies the use of "interpolatory" to describe this type of rule.

The position of the nodes can be discussed using the following estimation of the error $R(f)$ of rule (1) for a function of class C^n [2]

$$
R(f) = \frac{1}{n!} \int_{-1}^{1} f^{(n)}(\xi_x) \prod_{l=1}^{l=n} (x - x_l)dx \qquad \xi_x \in [-1, 1]
$$

This leads in particular to the straightforward majoration

$$
|R(f)| \leq \frac{2}{n!} \, \|f^{(n)}\|_\infty \, \left\| \prod_{l=1}^{l=n} (x - x_l) \right\|_\infty
\tag{4}
$$

The degree of the polynomial $\prod_{l=1}^{l=n} (x - x_l)$ is n and its leading coefficient is 1. It is known that the lowest supremum norm for a polynomial of degree n over $[-1, 1]$ with leading coefficients 1 is $1/2^{n-1}$ and is reached by $T_n(x)/2^{n-1}$ ($T_n(x)$, the first-kind Chebychev polynomial of degree n). A low value of $(n + 1)/2^n$ is obtained for corresponding second-kind Chebychev polynomial $U_n(x)$ divided by

2^n. In general, the presence of the third term in the right hand-side of Eq. (4) suggests to consider a disposition of the roots similar to those of the Chebychev polynomials of first and second kind, that is cosine of regularly spread angles in $[0, \pi]$.

Finally, the quality of an interpolatory rule can also be inferred (with the low hypothesis of a continuous functions f) from the following majoration

$$|I[f] - \sum_{i=1}^{n} w_i f(x_i)| \leq 2E_{n-1} + \sum_{i=1}^{i=n} |w_i| E_{n-1}, \qquad (5)$$

where E_{n-1} is the supremum norm of the difference between f and its best approximation in this sense by a polynomial of degree $(n-1)$ that will be denoted $b_{n-1}(x)$ in the following lines. This property is easily proved noting that $b_{n-1}(x)$ is exactly integrated by rule (1) as polynomial of degree $(n-1)$ so that the left-hand side of previous equation can be rewritten

$$\left| \int_{-1}^{1} (f(x) - b_{n-1}(x))dx - \sum_{i=1}^{n} w_i(f(x_i) - b_{n-1}(x_i)) \right|.$$

As a consequence, when defining a quadrature rule, it is highly desirable that the sum of the absolute values of the weights has a constant bound. As the first line of system (2) states that the sum of the weights is equal to 2., the best possible bound is 2. This is obtained if and only if all weights are positive. In this case, Eq. (5) yields

$$|I[f] - \sum_{i=1}^{n} w_i f(x_i)| \leq 4E_{n-1}. \qquad (6)$$

1.3 Nested Rules

In some cases, a very reliable value of $I[f]$ is needed so that formulae of increasing accuracy should be used up to obtaining a converged valuation. If the evaluation of $f(x_i)$ is expensive, then it is highly desirable that all or some of the nodes of the n-point rule are also involved in some of the further $(n+p)$-point rules. If so, the sets of nodes are said to be nested.

This property is rather rare. It is not shared by the basic versions of Gauss quadratures (see [3] Subsect. 2.7.1, for possible adaption, in particular the methods of Kronrod and Patterson). The two general and classical rules that have this property are the second rule of Fejér [5] and the rule of Clenshaw and Curtis [1]. As it is the basis of this work, the second rule of Fejér is described hereafter in some details (As the first rule of Fejér, which nodes are not nested, does not appear in this work, the adjective "second" is sometimes omitted when refering to the former.) The nodes of the n-point rule are the roots of the second-kind Chebychev polynomial $U_n(x)$

$$\gamma_n^k = \frac{k\,\pi}{n+1} \qquad x_n^k = \cos(\gamma_n^k) \quad k \in \{1, ..., n\}.$$

The weights ensuring exact quadrature for polynomial up to degree $(n - 1)$ were calculated by Fejér using the property recalled in equation (3), a change of variable $(x = \cos(\gamma))$ and properties of trigonometric polynomials [5]. These weights can also be calculated using Christoffel-Darboux formula for U_n [3]. Their simplest expression is

$$w_n^k = \frac{4\sin(\gamma_n^k)}{n+1} \sum_{l=1}^{[(n+1)/2]} \frac{\sin((2l-1)\gamma_n^k)}{2l-1}. \tag{7}$$

Under this form, the positivity of the w_n^k is not clear. Using the following substitution

$$2\sin(\gamma_n^k)\sin((2l-1)\gamma_n^k) = \cos((2l-2)\gamma_n^k) - \cos((2l)\gamma_n^k),$$

and gathering the cosine terms of same argument [5], the positivity appears. These weights are studied in more details in Sect. 4 to derive a lower bound.

All the nodes of the n-th rule are also node of the $(2n+1)$ rule so that getting the estimation of the latter requires only $(n + 1)$ evaluations of f if the first has already been computed.

It has been noted that the x_k^n are the roots of the second kind Chebychev polynomial U_n. Besides, the supremum norm of U_n over $[-1, +1]$ is $n+1$. Another important property that will be used several times is

$$U_n(\cos(\theta)) = \frac{\sin((n+1)\theta)}{\sin(\theta)}. \tag{8}$$

Clenshaw and Curtis defined the second classical nested rule [1]. Their $(n + 1)$-rule has the same abscissa as the $(n - 1)$-rule of Fejér plus the two extrema -1 and $+1$. In their article [1], they evaluate the integral of interest from a spectral expansion on basis of the first-kind Chebychev polynomials. Later on, Imhof put their estimation under the form of Eq. (1) and proved that the corresponding weights are positive [2–4].

1.4   Proposed Rule

The goal of this work is to study a sequel of Fejér second rule that improves the nesting of the sets of nodes. The set of the nodes of the n-point Fejér rule is denoted $\mathcal{F}_n$ and the one of the rule to be defined, $\mathcal{S}_n$.

Definition. The interpolatory rule of interest is defined, for all odd numbers of point, n, by the set (denoted $\mathcal{S}_n$) of its nodes symmetrically dispached in $[-1, 1]$ and the requirement that it is exact for all polynomials up to degree $(n - 1)$. This set of nodes is defined as follows (Q being a strictly positive integer):

– if $n = 2^Q - 1$, $\mathcal{S}_n = \mathcal{F}_n$ and the rule is identical to Fejér's second rule;
– if $M = 2^Q - 1 < n < 2^{Q+1} - 1 = 2M + 1$ we define m such that $n = M + 2m$, then $\mathcal{D}_{2M+1}^{2m} = (x_{2M+1}^1, x_{2M+1}^3 ... x_{2M+1}^{2m-1}, x_{2M+1}^{2M+3-2m} ..., x_{2M+1}^{2M+1})$ and $\mathcal{S}_n = \mathcal{F}_M \cup \mathcal{D}_{2M+1}^{2m}$

The points of $\mathcal{S}_n$ are denoted $(x_n^1, x_n^2 ... x_n^{n-1}, x_n^n)$ (in decreasing order). The arccos of these abscissae are denoted (in corresponding increasing order) $(\theta_n^1, \theta_n^2 ... \theta_n^{n-1}, \theta_n^n)$. For example the set of nodes of the 9-point rule is $\mathcal{S}_9$ instead of $\mathcal{F}_9$, the one of the corresponding Fejér rule:

$$\mathcal{S}_9 = (\cos(\pi/16), \cos(\pi/8), \cos(2\pi/8), \cos(3\pi/8), \cos(4\pi/8),$$
$$\cos(5\pi/8), \cos(6\pi/8), \cos(7\pi/8), \cos(15\pi/16))$$
$$\mathcal{F}_9 = (\cos(\pi/10), \cos(2\pi/10), \cos(3\pi/10), \cos(4\pi/10), \cos(5\pi/10),$$
$$\cos(6\pi/10), \cos(7\pi/10), \cos(8\pi/10), \cos(9\pi/10))$$

It is obtained by adding $\cos(\pi/16)$ and $\cos(15\pi/16))$ to $\mathcal{S}_7$, the set of nodes of the standard 7-point quadrature of Fejér. $\mathcal{S}_{11}$ is obtained by adding $(\cos(3\pi/16), \cos(13\pi/16))$ to $\mathcal{S}_9$ and so on up to $\mathcal{S}_{15}$ that is equal to $\mathcal{F}_{15}$. At this stage, it is not clear why the points of $\mathcal{F}_{2M+1}$ missing in $\mathcal{F}_M$ should be progressively included always adding the largest and lowest missing values to define the intermediate sets of points. This point is addressed in Sect. 4 and in Appendix A.1. Besides, as an interpolatory rule calculates the exact sum over $[-1, +1]$ of the Lagragian polynomial that interpolates f in its set, a relevant connex matter is the values is the behaviour of the Lebesgue constants of the sets $\mathcal{S}_n$. This is discussed in Appendix A.4.

1.5   Outline

Our aim is to study the proposed fully-nested rule from the point of view of (A) Accuracy for very regular functions according to Eq. (4); (B) Accuracy for continuous functions according to Eq. (5).

(A) is a simple matter that is briefly discussed in Sect. 2, whereas it is not easy to prove that the rule has positive weights so that it satisfies the optimal error bound (6) for continuous functions. This point will be the object of Sect. 3 presenting the possible approaches, Sect. 4 presenting formulas for the weights, and Sect. 5 presenting a three-step demonstration. Finally, Sect. 6 describes numerical tests comparing the proposed quadrature rule with two classical rules for test functions.

2   Accuracy for Very Regular Functions

Let M be a power of 2 minus one so that the rule of interest is counfounded with Fejér's. When moving from this to Fejér $(M + 2m)$-point rule for a sufficiently regular function, the bound of the residual (4) gets

$$|R(f)| \leq \frac{2}{(M + 2m)!} \, ||f^{(M+2m)}||_\infty \, \left\| \prod_{j=1}^{j=M+2m} (x - x_j) \right\|_\infty,$$

or using the supremum norm of second-kind Chebychev polynomial

$$|R(f)| \leq \frac{2}{(M + 2m)!} \, ||f^{(M+2m)}||_\infty \, \frac{M + 2m + 1}{2^{M+2m}} \tag{9}$$

As concerning the proposed rule,

$$\mathcal{D}_{2M+1}^{2m} = (x_{2M+1}^1, x_{2M+1}^3 ... x_{2M+1}^{2m-1}, x_{2M+1}^{2M+3-2m} ..., x_{2M+1}^{2M+1})$$

is the set of nodes added to $\mathcal{S}_M$ to obtain $\mathcal{S}_{M+2m}$. We get

$$|R(f)| \leq \frac{2}{(M+2m)!} \ ||f^{(M+2m)}||_\infty \ \left\| \prod_{x_j \in \mathcal{F}_n} (x - x_j) \prod_{x_k \in \mathcal{D}_{2M+1}^{2m}} (x - x_k) \right\|_\infty.$$

A simple way to bound the second supremum norm is to bound separately the two products. The terms of $\mathcal{D}_{2M+1}^{2p}$ are introduced two by two by pairs of opposite abscissae and

$$\left\| \left(x - \cos(\frac{k\pi}{2M+2}) \right) \left(x - \cos(\frac{(2M+2-k)\pi}{2M+2}) \right) \right\|_\infty =$$

$$\max \left(\cos^2(\frac{k\pi}{2M+2}), \sin^2(\frac{k\pi}{2M+2}) \right),$$

so that for the proposed rule

$$|R(f)| \leq \frac{2}{(M+2m)!} \ ||f^{M+2m}||_\infty \ \frac{M+1}{2^M} \times$$

$$\prod_{k=1, k \ odd}^{k=2m-1} \max \left(\cos^2(\frac{k\pi}{2M+2}), \sin^2(\frac{k\pi}{2M+2}) \right) \qquad (10)$$

This bound is not as good as the one of Eq. (9), but it keeps the decreasing inverse of factorial term and does not involve any increasing term.

The remaining part of the manuscript is devoted to the study of the weights.

3   Different Approaches for Weights Calculation

Four approaches for weights' calculation are presented. The focus has been put on the last three methods that provide the deepest insight into the sequence of weights. They are used in Sect. 4 to derive weights' formulae and give a partial proof of weights' positiveness.

3.1   Solving the Linear System for the Weights

As the $\mathcal{S}_n$ are symmetric sets of nodes in $]-1, +1[$, symmetric sets of weights are obtained and the exact quadrature of all x^p, is ensured for odd p values. The system of linear equations (2) can hence be reduced to a system of $(n+1)/2$ unknowns and $(n+1)/2$ equations expressing the exact quadrature of $(1, x^2...x^{n-1})$. It seems difficult to obtain formulas for the weights and to prove their positivity in this approach.

3.2   Integrating the Even Part of Lagrange Polynomials

Let $(x_n^1, x_n^2 ... x_n^{n-1}, x_n^n)$ denote the abscissae of the n-point rule stored in decreasing order. The corresponding increasing angles in $]0, \pi[$ obtained by application of the arccos are denoted $(\alpha_n^1, \alpha_n^2 ... \alpha_n^{n-1}, \alpha_n^n)$. The calculation of weights based on Eq. (3) can be done only for the first $(n+1)/2$ abscissae since the definition of $\mathcal{S}_n$ ensures $x_n^{n+1-k} = -x_n^k$, $\alpha_n^{n+1-k} = \pi - \alpha_n^k$ that yields $w_n^{n+1-k} = w_n^k$ for all k in $\{1, ..., n-1\}$.

The Lagrange polynomial associated to $x_n^k = \cos(\alpha_n^k)$ where $k \in \{1, ..., (n+1)/2)\}$ is denoted L_n^k. Its even part is denoted EL_n^k. Nevertheless, when dealing with the recursive relations between the successive polynomials associated to a specific abscissa, it seems easier to use the angle α associated to the abscissa instead of the index in the set of points as exponent for L and EL, as the angle does not change from one rule to another whereas the index may change. L_n^k is the following polynomial of degree $(n-1)$

$$L_n^k(x) = \prod_{j=1; j \neq k}^{n} \frac{(x - x_n^j)}{(x_n^k - x_n^j)}$$

For the medium index values $(n+1)/2$, the abscissa and angle values are respectively $x_n^{(n+1)/2} = 0$ and $\alpha_n^{(n+1)/2} = \pi/2$. The corresponding polynomial is even and can be rewritten as

$$L_n^{(n+1)/2}(x) = EL_n^{(n+1)/2}(x) = \prod_{j=1}^{(n-1)/2} \frac{(x^2 - (x_n^j)^2)}{(x_n^j)^2}.$$

If k is not equal to the medium index values, L_n^α is not even. This polynomial and its even part are expressed as

$$L_n^k(x) = \frac{x(x - x_n^{n-k})}{2(x_n^k)^2} \prod_{j=1; j \neq k}^{(n-1)/2} \frac{(x^2 - (x_n^j)^2)}{((x_n^k)^2 - (x_n^j)^2)} \tag{11}$$

$$EL_n^k(x) = \frac{x^2}{2(x_n^k)^2} \prod_{j=1; j \neq k}^{(n-1)/2} \frac{(x^2 - (x_n^j)^2)}{((x_n^k)^2 - (x_n^j)^2)}, \tag{12}$$

where the identity $x_n^{n-k} = -x_n^k$ has been used to get an abridged expression of $x_n^k(x_n^k - x_n^{n-k})$. Let us denote $\pm \cos(\xi)$ ($\xi \in]0, \pi/2[$) the two abscissae added to $\mathcal{S}_n$ to get $\mathcal{S}_{n+2}$. If $\xi < \alpha_n^k$ then $\alpha_n^k = \alpha_{n+2}^{k+1}$ and $x_n^k = x_{n+2}^{k+1}$. On the contrary, if $\xi > \alpha_n^k$ then $\alpha_n^k = \alpha_{n+2}^k$ and $x_n^k = x_{n+2}^k$. In any case, the new Lagrange polynomial associated to an abscissa $\cos(\alpha)$ of $\mathcal{S}_n$ and its even part are:

$$L_{n+2}^\alpha(x) = L_n^\alpha(x) \frac{x^2 - \cos^2(\xi)}{\cos^2(\alpha) - \cos^2(\xi)} \qquad EL_{n+2}^\alpha(x) = EL_n^\alpha(x) \frac{x^2 - \cos^2(\xi)}{\cos^2(\alpha) - \cos^2(\xi)}. \tag{13}$$

3.3   Integrating the Even Part of Lagrangian Polynomials Using Polar Angle

A change of variable $x = \cos(\theta)$ is done in the expression of the weights according to Eq. (3):

$$w_n^\alpha = \int_{-1}^{1} EL_n^\alpha(x)dx = \int_0^\pi EL_n^\alpha(\cos(\theta))\sin(\theta)d\theta = \int_0^\pi \Delta_n^\alpha(\theta)d\theta$$

where Δ_n^α is defined by

$$\Delta_n^\alpha(\theta) = EL_n^\alpha(\cos(\theta))\sin(\theta), \tag{14}$$

The exponent X in the notation Δ_n^X, hereafter, is either the range of the point in $\mathcal{S}_n$ or the value of its arccos. The introduction of two new oppositive abscissae $\pm\cos(\xi)$ in $\mathcal{S}_n$ to define $\mathcal{S}_{n+2}$ yields the change of the Lagrangian polynomial L^α and its even part EL^α defined by Eq. (13). According to Eq. (14), the corresponding relation for functions Δ_{n+2}^α and Δ_n^α is

$$\Delta_{n+2}^\alpha(\theta) = \Delta_n^\alpha(\theta)\frac{\cos^2(\theta) - \cos^2(\xi)}{\cos^2(\alpha) - \cos^2(\xi)} \tag{15}$$

The Δ_n^α can be expanded in sine polynomial with only odd integer terms like $\sin(\theta)$, $\sin(3\theta)$... and whenever it is possible, a Chebychev second-kind polynomial should be involved in the expression.

3.4   Integrating the Even-Part of Lagrange Polynomials Using Fejér Rule

If the rule of interest has two points less than a standard n-point Fejér rule, then the weights of the points of $\mathcal{S}_{n-2}$ in the two rules can be easily compared. Let $\cos(\alpha)$ where $\alpha \in]0, \pi/2]$ denote a point of $\mathcal{S}_{n-2}$ and $\pm\cos(\beta)$ where $\beta \in]0, \pi/2[$ denote the points added to $\mathcal{S}_{n-2}$ to obtain $\mathcal{S}_n$ – from the construction of the sets $\beta = \pi/2 - \pi/(n+1)$. The degree of Lagrange polynomial L_{n-2}^α is $n - 3$. It is hence exactly integrated by the n-point rule that is exact up to degree $n - 1$ (and in practice up to degree n as n is odd). The only points of the n-point rule where L_{n-2}^α does not vanish are $\cos(\alpha)$ and $\pm\cos(\beta)$ so that

$$w_{n-2}^\alpha = \int_{-1}^{1} L_{n-2}^\alpha(x)dx = w_n^\alpha L_{n-2}^\alpha(\cos(\alpha)) + w_n^\beta L_{n-2}^\alpha(\cos(\beta)) + w_n^{\pi-\beta} L_{n-2}^\alpha(-\cos(\beta))$$

or

$$w_{n-2}^\alpha = w_n^\alpha + 2w_n^\beta EL_{n-2}^\alpha(\cos(\beta)) \tag{16}$$

From Eq. (12), the sign of $EL^\alpha(\cos(\beta))$ is discussed hereafter. This point is also illustrated by Fig. 1, where the dots represent the weights and, in particular, the positive weights of Fejér's second rule appear as red dots. The sign of the numerator of $EL_{n-2}^\alpha(\cos(\beta))$ is (-1) to the number of positive abscissae larger than $\cos(\beta)$ and different from $\cos(\alpha)$ (number of circles above the empty square

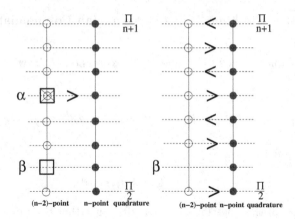

Fig. 1. Comparison of $(n-2)$-point rule weights vs n-point standard Fejér rule weights using $w_{n-2}^\alpha = w_n^\alpha + 2w_n^\beta EL^\alpha(\cos(\beta))$

in the left part of Fig. 1. The sign of the denominator of $EL_{n-2}^\alpha(\cos(\beta))$ is (-1) to the number of positive abscissae larger than $\cos(\alpha)$ (number of circles strictly above the α line in the left part of Fig. 1. This allows to order all weights of the $(n-2)$ rule with their counterparts in the n-point rule. It is easily checked that

$$w_{n-2}^\alpha > w_n^\alpha > 0 \quad \text{if} \quad \alpha = j\pi/(n+1) \quad j \quad \text{even} \in \{1, ...n\}$$

$$w_{n-2}^\alpha < w_n^\alpha \quad \text{if} \quad \alpha = j\pi/(n+1) \quad j \quad \text{odd} \in \{1, ...n\} - \{(n-1)/2, (n+3)/2\}$$

(this restriction excludes polar angle and abscissae of points that do not belong to $\mathcal{S}_{n-2}$). In Sect. 5.2, similar inequalities obtained in a more general context contribute to the study of weights positivity. All four methods were applied to the calculation of the weights of the 5-point rule. These weights are:

$$(w_5^{\pi/8}, w_5^{\pi/4}, w_5^{\pi/2}) = (\frac{4\sqrt{2}-4}{15}, \frac{10-2\sqrt{2}}{15}, \frac{18-4\sqrt{2}}{15})$$
$$(w_5^{7\pi/8}, w_5^{3\pi/4}) = (w_5^{\pi/8}, w_5^{\pi/4}).$$

4   Weight Formulas

The goal of this section is to provide simpler formulas for the weights than those derived from the sets of points $\mathcal{S}_n$ and Eq. (3). As in the definition of the rule in Sect. 1.3, the odd number of points of the rule, n, is bounded by two numbers M and $2M + 1$:

$$M = 2^Q - 1 < n < 2^{Q+1} - 1 = 2M + 1 \quad n = M + 2m$$

For the bounding values M and $2M + 1$, the interpolatory rules are the same as those of Fejér so that the weight formula is known. For all the $(M - 1)/2$

intermediate rules based on $\mathcal{S}_{M+2}$, $\mathcal{S}_{M+4}$... $\mathcal{S}_{2M-1}$, on the contrary, it is useful to explicit the weights. The study is based on the recursive calculation of Δ_n^k trigonometric polynomials presented in Sect. 3.3. The set of points $\mathcal{S}_n$ includes all points of $\mathcal{S}_M$,

$$\mathcal{S}_M = \{\cos(\frac{p\,\pi}{M+1})\ \ p \in \{1, ..., M\}\},$$

called hereafter "old" points (weights studied in Sect. 4.3) plus the $2m$ largest/ lowest points of $\mathcal{S}_{2M+1}$ that do not belong to $\mathcal{S}_M$,

$$\mathcal{D}_{2M+1}^{2m} = \{\cos(\frac{\pi}{2M+2}), \cos(\frac{3\pi}{2M+2})... \cos(\frac{(2m-1)\pi}{2M+2}),$$
$$\cos(\frac{(2M+3-2m)\pi}{2M+2}), ..., \cos(\frac{(2M-1)\pi}{2M+2}), \cos(\frac{(2M+1)\pi}{2M+2})\},$$

called hereafter "new" points (weights studied in Sect. 4.2). The final formulas for w_{M+2m} involve polynomial coefficient calculations but for degree m polynomials and not for degree $M + 2m - 1$ as in the direct application of Lagrange polynomial method for weights estimation; this indicates the benefit of the study. For the sake of conciseness, the derivation of the weight formulas is only presented extensively for (a) w_{M+2m}^{2m-1} the first occurrence of the new point $\pm\cos(\frac{(2m-1)\,\pi}{2M+2})$ introduced in the $(M+2m)$-point rule ($m \in \{1, 2...(M-1)/2\}$); (b) w_{M+2}^{2p}, w_{M+4}^{2p} the weights of the old point $\pm\cos(\frac{2p\,\pi}{2M+2}) = \pm\cos(\frac{p\,\pi}{M+1})$ in the first two intermediate rules involving $(M+2)$ and $(M+4)$ points. The derivation of the presented general formulas only requires complementary algebraic calculations.

4.1 Preliminary Calculations

The $\Delta_{M+2m}(\theta)$ function calculations to come may be simplified when isolating terms like

$$\Phi_M^j(\theta) \equiv \left(\frac{\sin(M\theta) + \sin((M+2)\theta)}{4}\right)\left(\frac{\cos(2\theta) - 1}{2}\right)^{j-1}$$

The weights expressions w_{M+2m} can be expressed based on the following integrals:

$$\varphi_M^j \equiv \int_0^\pi \Phi_M^j(\theta)d\theta$$

An explicit expression can be derived for φ_M^j:

$$\varphi_M^j = \frac{(2j-2)!\,(M+1)}{\prod_{l=1}^{l=j}((M+1)^2 - (2l-1)^2)} \tag{17}$$

The demonstration is given in Appendix A.2. It will appear that w_{M+2m} (for new points) and $w_{M+2m} - w_M$ (for old points) are linear combinations of $(\varphi_M^1, \varphi_M^2...\varphi_M^m)$. The largest value of interest for m is $(M-1)/2$. It is to be noted that the expressions of w_{M+2m} (new points) or $w_{M+2m} - w_M$ (old points) then only involve positive φ_M^j values.

4.2   Weights of the New Points $\pm\cos(\frac{k\,\pi}{2M+2})$, k odd

The abscissae $\cos(\frac{k\pi}{2M+2})$ and their opposites are introduced two by two in the set of quadrature points with odd k values increasing from 1 to (M-2) to successively define $S_{M+2}, S_{M+4},..., S_{2M-1}$. The set S_{M+2m} involves the new points $\pm\cos(\frac{\pi}{2M+2})$, $\pm\cos(\frac{3\pi}{2M+2})$ ... to $\pm\cos(\frac{(2m-1)\pi}{2M+2})$

Calculation of the w_{M+2m}^{2m-1} $m \in \{1, 2, ..., (M-1)/2\}$. The weight w_{M+2}^1 (also denoted $w_{M+2}^{\pi/(2M+2)}$) is calculated as follows:

$$w_{M+2}^1 = \int_0^\pi \frac{\cos(\theta) \times U_M(\cos(\theta))}{2\cos(\frac{\pi}{2M+2}) \times U_M(\cos(\frac{\pi}{2M+2}))} \sin(\theta)d\theta$$

$$w_{M+2}^1 = \int_0^\pi \frac{\cos(\theta)\sin(\theta) \times \sin((M+1)\theta)\sin(\frac{\pi}{2M+2})}{2\cos(\frac{\pi}{2M+2}) \times \sin(\theta)\sin(\frac{(M+1)\pi}{2M+2})}d\theta,$$

$$w_{M+2}^1 = 1/2\ \tan(\frac{\pi}{2M+2})\int_0^\pi \sin((M+1)\theta)\ \cos(\theta)\ d\theta.$$

Since $\sin((M+1)\theta)\ \cos(\theta) = 1/2(\sin(M\theta)+\sin((M+2)\theta)) = 2\Phi_M^1(\theta)$ and $\alpha_{M+2}^1 = \pi/(2M+2)$, we get

$$\Delta_{M+2}^1(\theta) = \tan(\alpha_{M+2}^1)\ \Phi_M^1(\theta),\quad w_{M+2}^1 = \tan(\alpha_{M+2}^1)\varphi_M^1.$$

w_{M+2m}^{2m-1} is calculated similarly to w_{M+2}^1 except that the sign of the terms arising from the application of Eq. (8) needs to be identified and that additional terms appear due to the inclusion of $\cos(\frac{\pi}{2M+2})$ to $\cos(\frac{(2m-3)\pi}{2M+2})$ (and their opposites) before $\cos(\frac{(2m-1)\pi}{2M+2})$ to define S_{M+2m}. The polynomial EL_{M+2m}^{2m-1} and the trigonometric polynomial Δ_{M+2m}^{2m-1} are written in a form that let the second-kind Chebychev polynomial U_M appear:

$$EL_{M+2m}^{2m-1}(x) = \frac{x\left(\prod_{l=1}^{(m-1)}(x^2 - \cos^2(\alpha_{M+2m}^{2l-1}))\right) \times U_M(x)}{2\cos(\alpha_{M+2m}^{2m-1})\left(\prod_{l=1}^{(m-1)}(\cos^2(\alpha_{M+2m}^{2m-1})^2 - \cos^2(\alpha_{M+2m}^{2l-1}))\right) \times U_M(\cos(\alpha_{M+2m}^{2m-1}))},$$

$$\Delta_{M+2m}^{2m-1}(\theta) = \frac{\cos(\theta)\left(\prod_{l=1}^{(m-1)}(\cos^2(\theta) - \cos^2(\alpha_{M+2m}^{2l-1}))\right) \times U_M(\cos(\theta))}{2\cos(\alpha_{M+2m}^{2m-1})\left(\prod_{l=1}^{(m-1)}(\cos^2(\alpha_{M+2m}^{2m-1})^2 - \cos^2(\alpha_{M+2m}^{2l-1}))\right) \times U_M(\cos(\alpha_{M+2m}^{2m-1}))}\sin(\theta).$$

Since

$$\frac{U_M(\cos(\theta))}{U_M(\cos(\alpha_{M+2m}^{2m-1}))} = \frac{\sin((M+1)\theta)}{\sin(\theta)}\frac{\sin(\alpha_{M+2m}^{2m-1})}{\sin((M+1)\alpha_{M+2m}^{2m-1})},$$

$$\Delta_{M+2m}^{2m-1}(\theta) = \frac{\sin(\alpha_{M+2m}^{2m-1})\cos(\theta)\sin((M+1)\theta)\prod_{l=1}^{(m-1)}(\cos^2(\theta) - \cos^2(\alpha_{M+2m}^{2l-1}))}{2\cos(\alpha_{M+2m}^{2m-1})\sin((M+1)\alpha_{M+2m}^{2m-1})\prod_{l=1}^{(m-1)}(\cos^2(\alpha_{M+2m}^{2m-1})^2 - \cos^2(\alpha_{M+2m}^{2l-1}))}.$$

Thanks to basic trigonometric identities,

$$\cos(\theta)\,\sin((M+1)\theta) = 0.5(\sin(M\theta) + \sin((M+2)\theta)) = 2\Phi_M^1(\theta)$$

$$\cos^2(\theta) - \cos^2(\alpha_{M+2m}^{2l-1}) = \frac{\cos(2\theta) - 1}{2} + \sin^2(\alpha_{M+2m}^{2l-1}),$$

$$\cos^2(\alpha_{M+2m}^{2m-1})^2 - \cos^2(\alpha_{M+2m}^{2l-1}) = \sin^2(\alpha_{M+2m}^{2l-1}) - \sin^2(\alpha_{M+2m}^{2m-1})$$

$$\sin((M+1)\alpha_{M+2m}^{2m-1}) = \sin(\frac{(M+1)(2m-1)\pi}{2M+2}) = \sin(m\pi - \frac{\pi}{2}) = (-1)^{(m-1)}$$

the final expressions of $\Delta_{M+2m}^{2m-1}(\theta)$ and w_{M+2m}^{2m-1} are then obtained (using the $(-1)^{(m-1)}$ term to reverse the sign of $(m-1)$ terms of the denominator)

$$\Delta_{M+2m}^{2m-1}(\theta) = \frac{\tan(\alpha_{M+2m}^{2m-1})\Phi_M^1(\theta) \prod\limits_{l=1}^{(m-1)} \left(\dfrac{\cos(2\theta) - 1}{2} + \sin^2(\alpha_{M+2m}^{2l-1})\right)}{\prod\limits_{l=1}^{(m-1)} \left(\sin^2(\alpha_{M+2m}^{2m-1}) - \sin^2(\alpha_{M+2m}^{2l-1})\right)}. \qquad (18)$$

$$w_{M+2m}^{2m-1} = \frac{\tan(\alpha_{M+2m}^{2m-1}) \displaystyle\int_0^\pi \Phi_M^1(\theta) \prod\limits_{l=1}^{(m-1)} \left(\dfrac{\cos(2\theta) - 1}{2} + \sin^2(\alpha_{M+2m}^{2l-1})\right) d\theta}{\prod\limits_{l=1}^{(m-1)} \left(\sin^2(\alpha_{M+2m}^{2m-1}) - \sin^2(\alpha_{M+2m}^{2l-1})\right)}.$$

$$(19)$$

The integrand of the numerator of Eq. (19) can be expanded in a series of $\Phi_M^j(\theta)$ $j \in \{1,..,m\}$ to obtain a computable expression thanks to the result of Sect. 4.1.

Expression of "new" Point Weights. For the sake of conciseness, only the resulting expressions of $\Delta_{M+2m}^k(\theta)$ w_{M+2m}^k (k odd $\in \{1,3,...,2m-1\}$ $m \in \{1,2,...,(M-1)/2\}$) are presented. The derivation of these expressions is done as above. The numerator and denominator of the fraction can be rewritten:

$$\Delta_{M+2m}^k(\theta) = \frac{\tan(\alpha_{M+2m}^k)\,\Phi_M^1(\theta) \left(\prod\limits_{l=1;l\neq(k+1)/2}^{m} (\dfrac{\cos(2\theta) - 1}{2} + \sin^2(\alpha_{M+2m}^{2l-1}))\right)}{\left|\prod\limits_{l=1;l<>(k+1)/2}^{m} (\sin^2(\alpha_{M+2m}^k) - \sin^2(\alpha_{M+2m}^{2l-1}))\right|}.$$

$$(20)$$

and the explicit final expression of the weight is:

$$w_{M+2m}^k = \frac{\tan(\alpha_{M+2m}^k)\left(\sum\limits_{l=1}^{l=m} \nu_{k,M+2m}^l \varphi_M^l\right)}{\left|\prod\limits_{l=1;l<>(k+1)/2}^{m} (\sin^2(\alpha_{M+2m}^k) - \sin^2(\alpha_{M+2m}^{2l-1}))\right|}, \qquad (21)$$

where the coefficients $\nu^l_{k,M+2m}$ are defined by the polynomial identity

$$X \left(\prod_{l=1; l\neq(k+1)/2}^{m} (X + \sin^2(\alpha^{2l-1}_{M+2m})) \right) = \sum_{l=1}^{m} \nu^l_{k,M+2m} X^l. \tag{22}$$

It is to be noted that formula (19) is obtained from formula (21) in the case where $k = 2m - 1$. Formulae (21) (22) hold for all new points weights and are the basis of a straightforward demonstration of their positivity.

4.3   Weights of the "old" Points $\pm\cos(\frac{p\,\pi}{M+1})$

Expression of Δ^p_M. The expression of the Δ^p_M functions has been derived by Fejér in the article presenting his two rules [5].

$$\Delta^p_M(\theta) = \frac{2\sin(\alpha^p_M)}{M+1} \sum_{l=1}^{(M+1)/2} \sin((2l-1)\theta)\,\sin((2l-1)\alpha^p_M) \tag{23}$$

Expression of $\Delta^p_M(\theta)\,(\cos^2(\theta) - \cos^2(\alpha^p_M))$. The calculations of this subsection aim at obtaining the expression of $\Delta^{\alpha^p_M}_{M+2}$ that is the base of the recursive calculation of $\Delta^{\alpha^p_M}_{M+2m}$ where $m \in \{1, 2, ..., (M-1)/2\}$. The points added to $\mathcal{S}_M$ to define $\mathcal{S}_{M+2}$ are $\pm\cos(\pi/(2M+2))$, that is, one smaller and one larger abscissae than all those of $\mathcal{S}_M$. The indices of all points of $\mathcal{S}_M$ are hence increased by one in $\mathcal{S}_{M+2}$ but, for one of the angles α^p_M, the following recurrence relation holds:

$$\begin{aligned}
\Delta^{\alpha^p_M}_{M+2}(\theta) &= \Delta^{\alpha^p_M}_M(\theta)\frac{\cos^2(\theta) - \cos^2(\frac{\pi}{2M+2})}{\cos^2(\alpha^p_M) - \cos^2(\frac{\pi}{2M+2})} \\
&= \Delta^{\alpha^p_M}_M(\theta) + \Delta^{\alpha^p_M}_M(\theta)\frac{\cos^2(\theta) - \cos^2(\alpha^p_M)}{\cos^2(\alpha^p_M) - \cos^2(\frac{\pi}{2M+2})}
\end{aligned} \tag{24}$$

A part of this expression can be simplified using Eq. (23) and trigonometric identities. $\Delta^p_M(\theta)(\cos^2(\theta) - \cos^2(\alpha^p_M))$, can be expressed without a sum over $(M+1)/2$ terms:

$$\begin{aligned}
\Delta^p_M(\theta)(\cos^2(\theta) - \cos^2(\alpha^p_M)) &= \frac{(-1)^{p+1}\,\sin^2(\alpha^p_M)}{2(M+1)}(\sin(M\theta) + \sin((M+2)\theta)) \\
&= \frac{2\,(-1)^{p+1}}{(M+1)}\sin^2(\alpha^p_M)\,\Phi^1_M(\theta) = \kappa^p_M\sin^2(\alpha^p_M)\,\Phi^1_M(\theta).
\end{aligned}$$

where the constant term, $2\,(-1)^{p+1}/(M+1)$, has been denoted κ^p_M

Expression of Old Point Weights. From the previous calculation, the expressions of the first trigonometric polynomial for the calculation of $w_{M+2}^{p\pi/(M+1)}$ is

$$
\Delta_{M+2}^{p\pi/(M+1)}(\theta) = \Delta_M^{p\pi/(M+1)}(\theta) + \kappa_M^p \frac{\sin^2(\frac{p\pi}{M+1})}{\sin^2(\frac{\pi}{2M+2}) - \sin^2(\frac{p\pi}{M+1})} \Phi_M^1(\theta)
$$

The multiplicative factor of Eq. (15) is written once again as in Eq. (24) and the old point weight trigonometric polynomial for the $(M+4)$ rule is

$$
\Delta_{M+4}^{p\pi/(M+1)}(\theta) = \Delta_M^{p\pi/(M+1)}(\theta) + \kappa_M^p \frac{\sin^2(\frac{p\pi}{M+1})}{\sin^2(\frac{3\pi}{2M+2}) - \sin^2(\frac{p\pi}{M+1})} \Phi_M^1(\theta) +
$$

$$
\kappa_M^p \frac{\sin^2(\frac{p\pi}{M+1})}{\left(\sin^2(\frac{\pi}{2M+2}) - \sin^2(\frac{p\pi}{M+1})\right)\left(\sin^2(\frac{3\pi}{2M+2}) - \sin^2(\frac{p\pi}{M+1})\right)} \left(\Phi_M^2(\theta) + \sin^2(\frac{3\pi}{2M+2})\Phi_M^1(\theta)\right).
$$

It is noted that the polar angles corresponding to the points successively introduced in the quadrature sets from S_M to S_{2M+1} are $\pi/(2M+2), 3\pi/(2M+2)\ldots$ $M\pi/(2M+2)$ (and π minus these angles). As we want to shorten the notation of these angles without refering to a specific rule, their position in the S_{2M+1} set is used. Using these shortened notations, the expression for $\Delta_{M+2m}^{p\pi/(M+1)}$, induced from first expressions, is defined and easily verified:

$$
\Delta_{M+2m}^{p\pi/(M+1)}(\theta) = \Delta_M^{p\pi/(M+1)}(\theta) + \kappa_M^p \sin^2(\frac{p\pi}{M+1}) \sum_{k=1}^{m} \frac{\left(\sum_{l=1}^{l=k} \psi_{k,m}^l \Phi_M^l(\theta)\right)}{\prod_{l=1}^{k}(\sin^2(\alpha_{2M+1}^{2m+1-2l}) - \sin^2(\alpha_{2M+1}^{2p}))} \tag{25}
$$

where the $\psi_{k,m}^l$ coefficients are defined by the identity

$$
X \prod_{j=1}^{k-1}\left(X + \sin^2(\frac{(2m+1-2j)\pi}{2M+2})\right) = \sum_{l=1}^{k} \psi_{k,m}^l X^l. \tag{26}
$$

Of course the equation for the weights corresponding to (25) is

$$
w_{M+2m}^{p\pi/(M+1)} = w_M^{p\pi/(M+1)} + \kappa_M^p \sin^2(\frac{p\pi}{M+1}) \sum_{k=1}^{m} \frac{\left(\sum_{l=1}^{l=k} \psi_{k,m}^l \varphi_M^l\right)}{\prod_{l=1}^{k}(\sin^2(\alpha_{2M+1}^{2m+1-2l}) - \sin^2(\alpha_{2M+1}^{2p}))} \tag{27}
$$

5 Weights Positivity

The new points weights are easily proved to be positive. This is the subject of Sect. 5.1. The ordering of weights as described in Sect. 3.4, then provides a quick demonstration of the positivity of part of the old points weights (Sect. 5.2). The points covered by the results for the $(M+2m)$-point rule are the $\pm \cos(\frac{p\pi}{M+1})$ with p lower equal than $(m + 1)$. Then more calculations are needed to demonstrate the positivity of the remaining old point weights for the $M + 2$ to $M + 12$ rules (Sects. 5.3 and 5.4).

5.1   New Points Weights Positivity

The equation for the new points weights (21) of the $M+2m$ rule is a linear combination of $(\varphi_M^1, ..., \varphi_M^m)$ with only positive coefficients from Eq. (22). Besides, as m is lower or equal to $(M-1)/2$ all the involved φ values are positive. The new points weights are thus positive.

5.2   Ordering of Old Point Weights in the Successive Rules

The method exposed in Sect. 3.4 is used to compare the weights of the old points, $\pm \cos(p\,\pi/(M+1))$ with $p \in \{1, .., M\}$, in all the rules from M to $(2M+1)$ points. In the presentation of Sect. 3.4, the rule with the larger number of points was a standard Fejér rule whereas the rules based on $\mathcal{S}_{M+2}$ to $\mathcal{S}_{2M-1}$ have another family of quadrature points. Nevertheless, their weights are fully defined by the linear algebra point of view. Besides, the weights of the two abscissae of the $(n + 2)$-point rule missing in the n-point rule have a known sign (they are all positive from Subsect. 4.2) so that technique of Sect. 3.4 can be applied. Its use for the comparison of the weights of the $(M + k - 1)$- and $(M + k + 1)$-point rule yields (k odd $k \in \{1, 3, ..., M\}$)

$$w_{M+k-1}^{p\pi/(M+1)} = w_{M+k+1}^{p\pi/(M+1)} + 2w_{M+k+1}^{k\pi/(2M+2)} EL_{M+k-1}^{p\pi/(M+1)}(\cos(\frac{k\pi}{2M+2}))$$

The sign of $EL^{p\pi/(M+1)}$ for $\cos(k\pi/(2M+2))$ is discussed below. It can be checked with the help of Fig. 2 as indicated in the beginning of Sect. 3.4, by counting the number of abscissae greater than $\cos(\frac{k\pi}{2M+2})$ and $\cos(\frac{p\pi}{M+1})$ in the $M + k - 1$ column:

1 – if $p \leq (k - 1)/2$ the numerator and the denominator of the EL value are negative. Hence

$$w_{M+k-1}^{p\pi/(M+1)} \geq w_{M+k+1}^{p\pi/(M+1)} \quad \text{if } p \leq (k-1)/2$$

2 – If $p = (k + 1)/2$ the numerator and the denominator of the EL value are positive. Hence

$$w_{M+k-1}^{p\pi/(M+1)} \geq w_{M+k+1}^{p\pi/(M+1)} \quad \text{if } p = (k+1)/2$$

3 – if $p = (k + 3)/2$, the numerator of the EL evaluation is positive and the denominator is negative. For increasing p values up to $(M+1)/2$, the sign of the EL evaluation is alternatively positive and negative (positive numerator, denominator changing sign) and the weights of the old points in the $(M+k-1)$- and $(M+k+1)$-point rule are alternately larger and lower.

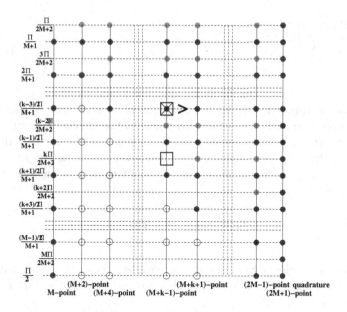

Fig. 2. Comparison of weights proving the positivity of the old points weights marked as full dark blue points in the figure. The α_n^k angles are reported on the left of the figure (shifted to left for old points, shifted to right for new points). The total number of points of the quadrature is reported at the bottom of the figure.

This yields the following series of inequalities

$$w_{M+2p-2}^{p\pi/(M+1)} > w_{M+2p}^{p\pi/(M+1)} > \ldots > w_{2M-1}^{p\pi/(M+1)} > w_{2M+1}^{p\pi/(M+1)} > 0$$

and then the following property.

Property. The weights of the old point $\cos(p\pi/(M+1))$ and its opposite is positive in the all rules from $(M+2p-2)$ points to $(2M-1)$ points. For a fixed rule, with $(M+2m)$ points, all $\cos(p\pi/(M+1))$ and opposite abscissae with $p \leq m+1$ have a positive weight.

If particular, all points of the $(2M-1)$-rule have a positive weight and the two abscissae $\cos(\frac{\pi}{2M+2})$ and $\cos(\frac{3\pi}{2M+2})$ have positive weights in all intermediate rules based on the sets $\mathcal{S}_{M+2}$ to $\mathcal{S}_{2M-1}$.

The points that have a positive weight by the ordering argument are marked with a full blue point in Fig. 2 (where the new point weights appear as green full points)

5.3 Positivity of Old Point Weights for $m \in \{1, \ldots 6\}$ – Preliminary Calculations

For the sake of conciseness, only final statements are given in this section and the next one. More details about this subject can be obtained from the author.

In Sect. 5.2 the positivity of

$$\frac{(M+1)(M-1)}{4} - \frac{(M-3)^2}{8} - \frac{(M-3)}{4} \quad \text{among} \quad \frac{(M+1)(M-1)}{4}$$

old point weights of the intermediate rules based on $\mathcal{S}_{M+2}$ to $\mathcal{S}_{2M-1}$, has been proved without using the algebraic expression of these weights. Unfortunately, this results concerns only part of the old points of the intermediate rules (except $\mathcal{S}_{2M-1}$). Additional work is devoted in this section and the section after to the positivity of the weights of the six first rules after a standard Fejer rule. Note that the symmetry of the weights and the result of Sect. 5.2 allows to restrict the considered points to

$$x_M^p = \cos(\frac{p\,\pi}{M+1}) \quad m+2 \le p \le \frac{M+1}{2}$$

Minoration of Fejér's Second Rule Weights. Fejér's M-point second rule weights are defined in [5]. Their expression has been recalled in Eq. (7)

$$\gamma_M^p = \frac{p\,\pi}{M+1} \quad x_M^p = \cos(\gamma_M^p) \quad w_M^p = \frac{4\sin(\gamma_M^p)}{M+1} \sum_{l=1}^{[(M+1)/2]} \frac{\sin((2l-1)\gamma_M^p)}{2l-1}.$$

with $p \in \{1, ..., M\}$. It is easily checked that

$$S_M(f) = \frac{4}{\pi} \sum_{l=1}^{[(M+1)/2]} \frac{\sin((2l-1)\gamma)}{2l-1}$$

is the $(M+1)/2$-term partial Fourier series of a 2π-periodic crenel function f defined on $[-\pi, \pi[$ by:

if $\gamma \in]-\pi, 0.[\quad f(\gamma) = -1$ if $\gamma \in]0., \pi[\quad f(\gamma) = +1 \quad f(-\pi) = f(0) = 0$

In order to find a lower bound for w_M^k, the convergence of $S_M(f)$ towards f is studied. As f is C^1 by part and all the angles of interest are located in the C^1 region $]0., \pi/2]$ (thanks to the symmetry of the $\mathcal{S}_n$ and equality of corresponding weights), the pointwise convergence in ensured. Conversely, as f is discontinuous in 0, the convergence of the series $S_M(f)$ towards f is not uniform on $]0., \pi/2]$ for both, the standard series with the number of terms increasing one by one and the specific series with $(M+1)/2 = 2^{Q-1}$ terms that needs to be considered in this study (on the contrary, f provides a classical example of Gibbs phenomenon and non-uniform convergence). A Dirichlet kernel is used to write difference between $S_M(f)$ and f.

$$S_M(f)(\gamma_M^p) - f(\gamma_M^p) = -\frac{1}{\pi} \int_{\gamma_M^p}^{\pi+\gamma_M^p} \frac{\sin(Mu + u/2)}{\sin(u/2)} du \ .$$

Integrating twice by parts the right integral leads to the following expression

$$S_M(f)(\gamma_M^p) - f(\gamma_M^p) = \frac{2(-1)^{p+1}}{\pi \sin(\gamma_M^p)} \left[\frac{1}{M} - \frac{1}{M^2} \right] + \frac{1}{2M^2\pi} \int_{\gamma_M^p}^{\pi+\gamma_M^p} \frac{\sin(Mu)\cos(u/2)}{\sin^3(u/2)} du \tag{28}$$

For M greater or equal to 7, the sign of the difference is given by the one $(-1)^{p+1}$ as appears in graphically in many courses about Fourier series (in other words, the right-hand side of Eq. (28) has the sign of his first term). For γ_M^p in $]0, \pi/2]$, the following bound is then obtained

$$|S_M(f)(\gamma_M^p) - f(\gamma_M^p)| \leq \frac{2}{M^2\pi \sin(\gamma_M^p)} \left[M - 1 + \frac{1}{\sin(\gamma_M^p)} \right] \tag{29}$$

For γ_M^p in $]0, \pi/2]$, $\sin(\gamma_M^p) \geq \frac{2}{\pi} \frac{p\pi}{M+1} \geq \frac{2p}{M+1}$, so that

$$|S_M(f)(\gamma_M^p) - f(\gamma_M^p)| \leq \frac{M+1}{M^2\,p\,\pi} \left[M - 1 + \frac{M+1}{2p} \right] \leq \frac{1}{p\,\pi} \left[1 + \frac{1}{2p}(1 + \frac{2}{M} + \frac{1}{M^2}) \right] \tag{30}$$

It is to be noted that the upper bounds of Eqs. (29) (30) tends towards 0 for fixed γ_M^p (that is fixed p/M) and $M \to \infty$ and do not for fixed p and $M \to \infty$ as the difference they bound (due to Gibbs phenomenom). Those results lead to the following numerical bounds for Fejér second rule weights

$$\text{for all } \; M = 2^Q - 1 \; \text{ for an odd } p \quad w_M^p \geq \frac{4\sin(\gamma_M^p)}{M+1} \frac{\pi}{4} = \frac{\pi \sin(\gamma_M^p)}{M+1}$$

$$\text{for all } \; M = 2^Q - 1 \quad \text{ for an even } p$$

$$\text{for } M \geq 7 \;\; p \geq 4 \;\; w_M^p \geq \frac{4\sin(\gamma_M^p)}{M+1} \frac{0.90\;\pi}{4} = \frac{0.90\;\pi \sin(\gamma_M^p)}{M+1}$$

$$\text{for } M \geq 15 \;\; p \geq 6 \;\; w_M^p \geq \frac{4\sin(\gamma_M^p)}{M+1} \frac{0.94\;\pi}{4} = \frac{0.94\;\pi \sin(\gamma_M^p)}{M+1}$$

$$\text{for } M \geq 15 \;\; p \geq 8 \;\; w_M^p \geq \frac{4\sin(\gamma_M^p)}{M+1} \frac{0.95\;\pi}{4} = \frac{0.95\;\pi \sin(\gamma_M^p)}{M+1}$$

Bounds for $\varphi_M^j \times (M+1)^{2j-1}$ (M Positive and Even, j in $\{1, ..., (M-1)/2\}$). From Eq. (17), we derive

$$\varphi_M^j(M+1)^{2j-1} = \frac{(2j-2)!\,(M+1)^{2j}}{\displaystyle\prod_{l=1}^{l=j}((M+1)^2 - (2l-1)^2)} = \frac{(2j-2)!}{\displaystyle\prod_{l=1}^{l=j}\left(1 - \frac{(2l-1)^2}{(M+1)^2}\right)}$$

The index j is assumed to be lower than $(M+1)/2$ so that all the corresponding φ values are positive. Bounds for $\varphi_M^j(M+1)^{2j-1}$ are then searched for a fixed j, for all M greater than a minimum M_{min} value. Previous equation yields:

$$(2j-2)! \leq \varphi_M^j(M+1)^{2j-1} \leq \frac{(2j-2)!}{\displaystyle\prod_{l=1}^{l=j}\left(1 - \frac{(2l-1)^2}{(M_{min}+1)^2}\right)} \quad \text{for } M \geq M_{min} \;\; j \leq \frac{M-1}{2}$$

Bounds of the Denominators of Old Point Weight Formula. A bound is searched for a quantity denoted $T^{p,k}_{M,m}$, derived from denominator of the k-th term in the discrete sum appearing in Eq. (27). $T^{p,k}_{M,m}$ is obtained by taking the absolute value of this factor (which is the same as taking the opposite of all terms of the product as $p \geq m + 2$) and multiplying it by $\sin^{2k}(\dfrac{p\pi}{M+1})$

$$
T^{p,k}_{M,m} = \frac{\sin^{2k}(\dfrac{p\pi}{M+1})}{\displaystyle\prod_{l=1}^{l=k} \left(\sin^2(\dfrac{p\pi}{M+1}) - \sin^2(\dfrac{(2m+1-2l)\pi}{2M+2}) \right)}
$$

M_{min} being the lowest value of interest for M (for a fixed m the smallest integer of the form $2^Q - 1$ such that $M \geq 2m+1$), p_{min} being the lowest value of interest for p (for a fixed m $p_{min} = m + 2$ thanks to the property presented in Sect. 5.2) the bound for $T^{p,k}_{M,m}$ is simply

$$
3 \leq m + 2 \leq p \leq \frac{(M+1)}{2} \quad 1 \leq k \leq m \quad T^{p,k}_{M,m} \leq T^{p_{min},k}_{M_{min},m} \tag{31}
$$

5.4 Positivity of Old Point Weights for $M \in \{1...6\}$

Lower bounds for the old point weights expression, Eq. (27), are searched for. All bounds obtained in the previous subsection are involved. For the sake of conciseness, only the final results are presented here:

$$
w^{p\pi/(M+1)}_{M+2} \geq \frac{1.80p\pi}{(M+1)^2} - \frac{2.04}{(M+1)^2} - \frac{0.0949}{(M+1)^2} \qquad (p \geq 3 \quad M \geq 7)
$$

$$
w^{p\pi/(M+1)}_{M+4} \geq \frac{1,80p\pi}{(M+1)^2} - \frac{2.04}{(M+1)^2} - \frac{0.0359}{(M+1)^2} - \frac{1.78}{p^2\,(M+1)^2} \qquad (p \geq 4 \quad M \geq 7)
$$

$$
w^{p\pi/(M+1)}_{M+6} \geq \frac{1,88p\pi}{(M+1)^2} - \frac{2.01}{(M+1)^2} - \frac{1.87\ 10^{-3}}{(M+1)^2} - \frac{1.75}{p^2\,(M+1)}
$$
$$
- \frac{43.1}{p^4\,(M+1)^2} \qquad (p \geq 5 \quad M \geq 15)
$$

$$
w^{p\pi/(M+1)}_{M+8} \geq \frac{1,88p\pi}{(M+1)^2} - \frac{2.01}{(M+1)^2} - \frac{7.87\ 10^{-4}}{p^2\,(M+1)^2} - \frac{2.67}{p^2\,(M+1)} - \frac{145.}{p^4\,(M+1)^2}
$$
$$
- \frac{2.75\ 10^3}{p^6\,(M+1)^2} \qquad (p \geq 6 \quad M \geq 15)
$$

$$
w^{p\pi/(M+1)}_{M+10} \geq \frac{1,88p\pi}{(M+1)^2} - \frac{2.01}{(M+1)^2} - \frac{6.90\ 10^{-4}}{(M+1)} - \frac{4.73}{p^2\,(M+1)} - \frac{5.13\ 10^2}{p^4\,(M+1)^2}
$$
$$
- \frac{2.09\ 10^4}{p^6\,(M+1)^2} - \frac{3.62\ 10^5}{p^8\,(M+1)^2} \qquad (p \geq 7 \quad M \geq 15)
$$

$$
w^{p\pi/(M+1)}_{M+12} \geq \frac{1,90p\pi}{(M+1)^2} - \frac{2.01}{(M+1)^2} - \frac{1.80\ 10^{-3}}{(M+1)^2} - \frac{11.7}{p^2\,(M+1)} - \frac{250.\ 10^3}{p^4\,(M+1)^2}
$$
$$
- \frac{2.02\ 10^5}{p^6\,(M+1)^2} - \frac{7.31\ 10^6}{p^8\,(M+1)^2} - \frac{1.05\ 10^8}{p^{10}\,(M+1)^2} \qquad (p \geq 8 \quad M \geq 15)
$$

For each minoration, the negative terms and the positive term of the right hand side are calculated for the lowest possible value of p. The comparison of the factors multiplying the inverse of $(M + 1)^2$ then proves the positivity of the weights based on the increase with p of these lower bounds.

6   Numerical Tests

The coefficients of the proposed quadrature are computed (see Appendix A.3) and it is compared with Gauss-Legendre and Fejér second rule for the six test functions of a classical article by Trefethen [6]. The results are presented in Fig. 3. The proposed quadrature appears to converge as well as Fejér's standard rule for the polynomial and the two entire functions. It is then more efficient than Fejér's rule if rules of increasing odd number of points are used as all its sets of points are successively included. It can even be slightly more efficient than Gauss-Legendre quadrature in this perspective of successive estimations by larger rules as appears in Fig. 4.

Unfortunately the proposed rule behaves badly for the three last functions (the analytic, the C^∞ and the C^2 function): for these three functions a threshold in the convergence is observed from $n = 2^Q + 1$ points to $n = 2^{Q+1} - 1$; the level of error being equal or higher than the one obtained for $2^Q - 1$ points, when the rule is counfounded with the one of Fejér. This may be interpreted from the small values of new points weights and the fact that the old points weights do not change very much from $n = 2^Q - 1$ to $n = 2^{Q+1} - 3$ before being approximately halved for $n = 2^{Q+1} - 1$. We assume that the small corrections brought by the new points are efficient for very regular functions but not for less regular functions.

7   Conclusion

A 1D interpolatory quadrature derived from Fejér's second rule has been proposed. The nesting is improved with respect to standard nested quadratures in the sense that any set of points (of odd cardinal) includes all the smaller sets, whereas the standard accuracy is maintained (order $(n - 1)$ polynomial exactly integrated for the n point rule). This stronger nesting is an interesting property when the evaluation of the function of interest is costly.

The calculated weights (as long as the numerical values satisfy proper numerical checks) are positive that is a desirable property for interpolatory quadratures. Explicit formulae have been derived for the weights and an effort has been devoted in demonstrating their positivity. It has been established for the new points weights and set of old point weights. These results are in particular sufficient to establish the positivity of the weights for all rules from 3-point to 43-point that is more than needed for many engineering problems. As concerning this property, the reader is also referred to Appendix A.1. It indicates that, for small sets of points, the chosen procedure for point addition is the only one leading to positive weights.

The proposed quadrature was applied to six test functions of a classical study by Trefethen. When calculating integrals by quadrature formulas of increasing number of points for accuracy checking, the proposed formula appeared to be more efficient than Fejér's second rule and possibly more than Gauss-Legendre formulas for very regular functions. Of course, Gauss-Legendre formulas are the best possible in the sense of the accuracy for polynomials integration and for a fixed number of points; the advantage of the proposed method then only comes from the fact that its sets of point are fully nested.

Unfortunately, the behavior of the proposed quadrature is by far less satisfactory for less regular functions. Future work will be devoted to this issue.

Acknowledgments. The author is very grateful to professor Albert Cohen for his explanations and recommendations, in particular about the comparison of a function and its Fourier series. He thanks the reviewer for his useful remarks. He warmly thanks Andrea Resmini for his careful check of this document.

A   Appendices

A.1   Properties of the Non-ordered Rules

In the definition of the proposed rule, in Sect. 1.4, the way to successively add the points of $\mathcal{F}_{2M+1}$ that do not belong to $\mathcal{F}_M$, to define the intermediate sets of points $\mathcal{S}_{M+2}$ to $\mathcal{S}_{2M-1}$ has been explicitly prescribed: the opposite largest and lowest missing values are successively added. This procedure (qualified here as "ordered" as in Sect. 1.3) may seem arbitrary and the reader may wonder whether there are other ways of defining a fully nested rule with positive weights. The possible other sets (qualified as "non-ordered" as in Sect. 1.3.) of 5 points (resp. 9 11 and 13 points) are considered in this appendix. In order to avoid confusions between formulae and numerical values for ordered and non-ordered rules, an underline is added to the notation of the sets, the weights and the delta functions related to non-ordered rules.

The first proposed rule distinct from the ones of Fejér is the five-point rule. Instead of $\pm\cos(\pi/8)$, $\pm\cos(3\pi/8)$ could be added to $\mathcal{S}_3$ to build $\underline{\mathcal{S}_5}$ while preserving the property of successive inclusion of the sets. The corresponding weights are

$$(\underline{w}_5^{\pi/4}, \underline{w}_5^{3\pi/8}, \underline{w}_5^{\pi/2}) = (\frac{10+2\sqrt{2}}{15}, \frac{-4-4\sqrt{2}}{15}, \frac{18+4\sqrt{2}}{15})$$

$$(\underline{w}_5^{3\pi/4}, \underline{w}_5^{5\pi/8}) = (\underline{w}_5^{\pi/4}, \underline{w}_5^{3\pi/8})$$

A weight is negative; that is not satisfactory for the reason discussed in Sect. 1.1 and in the conclusion. The definition of sequentially included 9-point, 11-point and 13-point sets between $\mathcal{F}_7$ and $\mathcal{F}_{15}$ is discussed. The numerical values of the weights have been calculated using LAPACK. For systems of corresponding sizes with known solution, an accuracy of eight to nine digits is obtained. The weights of the non-ordered rules are printed below with four digits for the sake of readability.

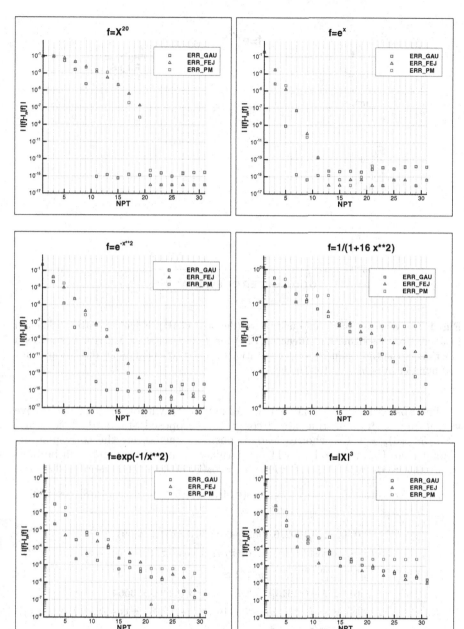

Fig. 3. Convergence of Gauss quadrature (denoted GAU), Fejer quadrature (denoted FEJ) and proposed method (denoted PM) for the six test functions proposed by L.N. Trefethen [6]

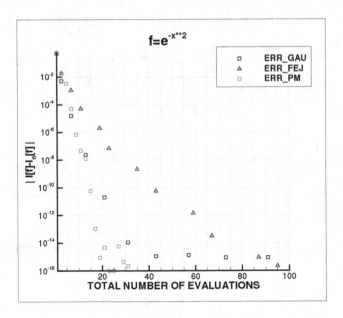

Fig. 4. Convergence of Gauss quadrature (denoted GAU), Fejer quadrature (denoted FEJ) and proposed method (denoted PM) for $f(x) = exp(-x^2)$. Error plotted as a function of the total number of f evaluations needed to calculate all quadrature with odd number of points up to the current.

The abscissae of $\mathcal{F}_{15}$ that can be added to $\mathcal{F}_7$ without having a negative weight in the 9-point rule are $\pm \cos(\pi/16)$ (proposed standard choice) and $\pm \cos(5\pi/16)$. The two choices lead to positive weights for the corresponding 9-point rule. Unfortunately, no 11-point rule with positive weights can be built by adding $\pm \cos(\pi/16)$, $\pm \cos(3\pi/16)$ or $\pm \cos(7\pi/16)$ to the non-ordered set $\underline{\mathcal{S}}_9$:

If $\underline{\mathcal{S}}_{11} = (\pm \cos(\pi/16), \pm \cos(\pi/8), \pm \cos(\pi/4), \pm \cos(5\pi/16), \pm \cos(3\pi/8), \cos(\pi/2))$
$(\underline{w}_{11}^{\pi/16}, \underline{w}_{11}^{\pi/8}, \underline{w}_{11}^{\pi/4}, \underline{w}_{11}^{5\pi/16}, \underline{w}_{11}^{3\pi/8}, \underline{w}_{11}^{\pi/2}) \simeq$
$$(0.0281, 0.1292, 0.2953, -0.0216, 0.3754, 0.3869)$$

If $\underline{\mathcal{S}}_{11} = (\pm \cos(\pi/8), \pm \cos(3\pi/16), \pm \cos(\pi/4), \pm \cos(5\pi/16), \pm \cos(3\pi/8), \cos(\pi/2))$
$(\underline{w}_{11}^{\pi/8}, \underline{w}_{11}^{3\pi/16}, \underline{w}_{11}^{\pi/4}, \underline{w}_{11}^{5\pi/16}, \underline{w}_{11}^{3\pi/8}, \underline{w}_{11}^{\pi/2}) \simeq$
$$(0.2247, -0.1613, 0.4801, -0.1713, 0.4495, 0.3564)$$

If $\underline{\mathcal{S}}_{11} = (\pm \cos(\pi/8), \pm \cos(\pi/4), \pm \cos(5\pi/16), \pm \cos(3\pi/8), \pm \cos(7\pi/16), \cos(\pi/2))$
$(\underline{w}_{11}^{\pi/8}, \underline{w}_{11}^{\pi/4}, \underline{w}_{11}^{5\pi/16}, \underline{w}_{11}^{3\pi/8}, \underline{w}_{11}^{7\pi/16}, \underline{w}_{11}^{\pi/2}) \simeq$
$$(0.1941, 0.0339, 0.7011, -0.8951, 1.7167, -1.5017)$$

A sequence of sets corresponding to positive weights is then searched based on $\mathcal{S}_9$. Adding $\pm \cos(3\pi/16)$ (as in the 11-point ordered rule) or $\pm \cos(7\pi/16)$ leads to a set of positive weights. If the second choice is retained, neither $\pm \cos(3\pi/16)$ nor $\pm \cos(5\pi/16)$ can be added to the non-ordered 11-point set while still obtaining positive weights.

If the ordered set $\mathcal{S}_{11}$ is considered, then the last four abscissae $\pm \cos(5\pi/16)$, $\pm \cos(7\pi/16)$ can only be added to the set in this order to build the 13- and then 15-point rule (if not, the 13-point rule has a negative weight).

Hence, for small numbers of points, it is impossible (set of 5 points) or only partly possible (set of 9, 11, 13 points) to define fully nested successive sets of quadrature points between two standard sets of Fejér quadrature that lead to interpolatory quadratures with only positive weights.

A.2   Calculation of φ_M^j

Let us first define $\overline{\Phi}_s^j(\theta)$ (s being an even integer) as

$$\overline{\Phi}_s^1(\theta) = \frac{\sin((s-1)\theta) + \sin((s+1)\theta)}{4}, \qquad \overline{\Phi}_s^{j+1}(\theta) = \overline{\Phi}_s^j(\theta)\,\frac{\cos(2\theta) - 1}{2}. \quad (32)$$

Using the trigonometric identity for $\sin(a)\cos(b)$ the $\overline{\Phi}_s^j$ can be expanded in a series of sine functions with odd-integer-times-θ arguments. This series is "symmetric" about $\sin(s\theta)$. Let $\overline{\varphi}_s^j$ denote the integral of $\overline{\Phi}_s^j$ over $[0, \pi]$. Another consequence of $\sin(a)\cos(b)$ identity is that $\overline{\Phi}_s^{j+1}(\theta)$, $\overline{\Phi}_{s-2}^j(\theta)$, $\overline{\Phi}_s^j(\theta)$, and $\overline{\Phi}_{s+2}^j(\theta)$ satisfy a simple recurrence relation:

$$\overline{\Phi}_s^{j+1}(\theta) = \frac{1}{4}\overline{\Phi}_s^j(\theta)(2\cos(2\theta) - 2) = \frac{1}{4}(2\overline{\Phi}_s^j(\theta)\cos(2\theta) - 2\overline{\Phi}_s^j(\theta))$$

When applying the trigonometric identity $\sin(a)\cos(b)$ with $B = 2\theta$ for all $\sin(k\theta)$ terms of $\overline{\Phi}_s^j(\theta)$, the sum of two series with the same coefficients $\overline{\Phi}_s^j$ but shifted factors $\sin((k-2)\theta)$ and $\sin((k+2)\theta)$ is obtained so that

$$\overline{\Phi}_s^{j+1}(\theta) = \frac{1}{4}(\overline{\Phi}_{s-2}^j(\theta) - 2\overline{\Phi}_s^j(\theta) + \overline{\Phi}_{s+2}^j(\theta)),$$

and

$$\overline{\varphi}_s^{j+1} = \frac{1}{4}(\overline{\varphi}_{s-2}^j - 2\overline{\varphi}_s^j + \overline{\varphi}_{s+2}^j). \quad (33)$$

From the first values of $\overline{\varphi}_s^j$, the general following form is assumed for these integrals:

$$\overline{\varphi}_s^j = \frac{a_j\ s}{(s^2 - 1)(s^2 - 3^3)...(s^2 - (2j-1)^2)} \quad (34)$$

Plugging this expression in the right-hand side of Eq. (33), a similar expression for $\overline{\varphi}_{s+1}^j$ with $a_{j+1} = (4j^2 - 2j)a_j$ is obtained. The latter and the expression for $\overline{\varphi}_s^1$ validate Eq. (34). Finally, $\overline{\varphi}_s^j$ takes the following form:

$$\overline{\varphi}_s^j = \frac{(2j-2)!\ s}{\prod_{l=1}^{l=j}(s^2 - (2l-1)^2)}.$$

The convention for the index s of $\overline{\Phi}$ and $\overline{\varphi}$ leads to symmetric expressions about s and makes the derivation of $\overline{\varphi}_s^j$ not do difficult. On the contrary, for the derivation of Δ_{M+2m}, w_{M+2m} expressions in Sect. 4, its is more practical to introduce

$$\Phi_M^j(\theta) = \overline{\Phi}_{M+1}^j(\theta) \qquad \varphi_M^j = \overline{\varphi}_{M+1}^j$$

$$\varphi_M^j = \frac{(2j-2)!\,(M+1)}{\prod_{l=1}^{l=j}((M+1)^2 - (2l-1)^2)} \tag{35}$$

A.3 Computational Weight Calculations

Among the four methods presented in Sect. 3, those of Subsects. 3.1, 3.2 and 3.3 were used for weight calculation and the values were checked with respect to the theoretical values obtained by Fejér (Eq. (7)) whenever $\mathcal{S}_n$ and $\mathcal{F}_n$ are the same. The resolution of a linear system for the weights (Eq. (2)) and the polynomial method (Eq. (13)) appeared to be more prone to numerical errors than the trigonometric polynomial method (Eq. (15)).

The focus was then put on the last method. Eq. (15) was used to calculate the successive trigonometric polynomials of the points already involved in the quadrature starting from Eq. (23) for M=3.

$$\Delta_3^{\pi/4}(\theta) = 0.5\sin(\pi/4) * (\sin(\pi/4)\sin(\theta) + \sin(3\pi/4)\sin(3\theta)) = \frac{\sqrt{2}}{4}(\sin(\theta) + \sin(3\theta))$$

$$\Delta_3^{\pi/2}(\theta) = 0.5\sin(\pi/2) * (\sin(\pi/2)\sin(\theta) + \sin(3\pi/2)\sin(3\theta)) = \frac{1}{2}(\sin(\theta) - \sin(3\theta))$$

The trigonometric polynomial of a new abscissa in $[0,1]$, $\cos(\xi)$, is calculated for the one corresponding to the null abscissa times $(\cos(2\theta)+1)$ plus a proper renormalisation of the coefficients:

$$\Delta_{n+2}^{\xi}(\theta) \sim (\cos(2\theta)+1)\Delta_n^{\pi/2}(\theta) \quad \Delta_{n+2}^{\xi}(\xi) = 1/2 \;\sin(\xi)$$

This non-intuitive normalisation is simply the counterpart of:

$$EL_{n+2}^{\xi}(x) \sim x^2 \; EL_n^{\pi/2}(x) \quad EL_{n+2}^{\xi}(\cos(\xi)) = 1/2$$

For this method, the Δ_n^{α} functions and the weights have been calculated for increasing n. The numerical error was checked (a) for the standard Fejér rule for $n = 7, 15, 31...$ based on Eqs. (7) and (23); (b) with respect to analytical expressions obtained in Sect. 4. The difference between theoretical coefficients of Fejér up to the 31-point rule and the computed values appeared to be less than $8.\,10^{-14}$ and the residual of the first four equations of the linear system for the weights (exact quadrature of $0, x^2, x^4, x^6$) appeared to be less than $5.\,10^{-13}$. The author can provide the calculated weights for odd n from 3 to 31 (not printed here for the sake of conciseness) and/or the program used to calculate them.

A.4 Lebesgue Constants of Sets $\mathcal{S}_n$

As an interpolatory rule calculates the exact sum over $[-1,+1]$ of the Lagragian polynomial that interpolates the considered function in its set, it is interesting to check the Lebesgue constants of the sets $\mathcal{S}_n$ proposed for quadrature. Their values are plotted in Fig. 5. They unfortunately strongly grow between two successive $2^Q - 1$ integers. The lowest values for large sets are Λ_{2^Q-1} that is found numerically to be equal to the cardinal of the set, $2^Q - 1$, and Λ_{2^Q+1} that is

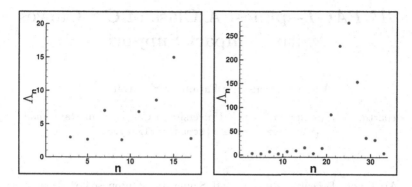

Fig. 5. Lebesgue constants for the sets $\mathcal{S}_n$

found numerically to be close to $ln(2^Q + 1)$.

The corresponding Gauss-Lobatto sets $\mathcal{L}_n = \{-1, 1\} \cup \mathcal{S}_{n-2}$ were considered. They do not exhibit significantly better or worse Lebesgue constants.

References

1. Clenshaw, C.W., Curtis, A.R.: A method for numerical integration on an automatic computer. Numer. Math. **2**, 197–205 (1960)
2. Dahlquist, G., Björck, A.: Numerical Methods in Scientific Computing. SIAM, New York (2008)
3. Davis, P.J., Rabinowitz, P.: Methods of Numerical Integration. Academic press, New York, San Francisco, London (1975)
4. Imhof, J.P.: On the method for numerical integration of Clenshaw and Curtis. Numer. Math. **5**, 138–141 (1963)
5. Fejér, L.: Mechanische Quadraturen mit positiven Cotesschen Zahlen. Math. Z. **37**, 287–309 (1933)
6. Trefethen, L.N.: Is Gauss quadrature better than Clenshaw-Curtis? SIAM Rev. **50**(1), 67–87 (2008)

CINPACT-splines: A Class of C^∞ Curves with Compact Support

Adam Runions and Faramarz Samavati$^{(\boxtimes)}$

Department of Computer Science, University of Calgary, Alberta, Canada
{runionsa,samavati}@cpsc.ucalgary.ca

Abstract. Recently, Runions and Samavati [7] proposed Partion of Unity Parametrics (PUPs), a generalization of NURBS which replaces B-spline basis functions with arbitrary Weight-Functions (WFs) while preserving affine invariance. A key problem identified by Runions and Samavati was the identification of classes of weight-functions which are well-suited to geometric modeling. In this paper, we propose a class of WF based on *bump*-functions, which arise in the study of smooth, non-analytic manifolds. These give rise to a class of C^∞ curves with compact-support, which we call *CINPACT* splines. The WFs are similar in form to B-spline basis functions, and are parameterized by a *degree-like* shape parameter. We examine the approximating and interpolating curves created using the proposed class of WF. Furthermore, we propose and demonstrate a method to specify the tangents and higher order derivatives of the curve at control points for CINPACT and PUPs curves.

Keywords: Parametric curves · Interpolating curves · Approximating curves · PUPs · B-spline · NURBS · Bump functions

1 Introduction

Parametric curves are fundamental to geometric modeling, and an important primitive for Computer-Aided-Design (CAD). In CAD applications, parametric curves are typically generated from a set of control points using splines. In this setting, an important distinction arises between *interpolating* and *approximating* curves, with the former passing through control points and the latter only passing nearby the control points.

Recently, Runions and Samavati [7] proposed Partition of Unity Parametrics (PUPs), a generalization of NURBS which replaces basis functions with arbitrary Weight-Functions (WFs) while preserving affine invariance. In the PUPs framework, each control point is associated with a WF, and a parametric curve is produced by summing the weighted control points. The resulting class of curves is a super set of those generated by other NURBS generalizations, such as T-Splines [8] and G-NURBS [9]. A question raised in the work was the identification of WFs that would take advantage of the generality offered by PUPs. This question motivates the work in this paper, where we use the work of Runions

© Springer International Publishing Switzerland 2015
J.-D. Boissonnat et al. (Eds.): Curves and Surfaces 2014, LNCS 9213, pp. 384–398, 2015.
DOI: 10.1007/978-3-319-22804-4_27

and Samavati [7] and Zhang and Ma [10] to identify a class of C^∞ curves with compact support.

In [7], a method for interpolating control points was proposed. In this method, interpolation was achieved through the appropriate choice of weight function. In particular, it employed a normalized sinc function, to interpolate control points, multiplied by a compactly supported kernel, which localized the contribution of each control point to the curve. A related interpolation method was proposed by Zhang and Ma [10], which employed a different choice of compactly supported kernel.

For their kernel, Zhang and Ma used a Gaussian, which generated C^∞ curves but required a tradeoff between compact support and affine invariance. In contrast, Runions and Samavati employed Partition of Unity Parametrics (PUPs), which guarantee affine invariance, but their construction relied on B-spline basis functions. Consequently, their method offers compact support and affine invariance, but only C^k continuity (for a specified integer k).

Drawing inspiration from these works, we propose an interpolation scheme which exhibits the beneficial properties of both. It generates high-quality interpolating curves which guarantee affine invariance, compact support and C^∞ continuity. The interpolation scheme generates a class of PUPs curves, and the weight-functions used are based on the *bump-functions* employed in the analysis of smooth, non-analytic manifolds [5]. The proposed weight-functions have a simple exponential form, which is used to localize the effects of the sinc function. A common requirement for interpolating curves is the interpolation of tangents as well as control points. To address this requirement, we outline a method for interpolating a specified set of tangents and higher order derivatives at control points. This problem, in particular, was not addressed by the curve schemes of Runions and Samavati [7], or Zhang and Ma [10], and the method we propose is general in the sense that it is applicable to any PUPs curve.

A limitation of the approach of Zhang and Ma [10] is that the scheme they propose does not readily give rise to a class of approximating curves with similar properties. Consequently, it is difficult to use their method as the basis for a comprehensive curve modeling system-which must support both approximating and interpolating curves. In contrast, the method we propose is derived from PUPs, permitting approximating curves to be generated in a straightforward manner. Specifically, by directly using the proposed bump-function as a weight-function approximating curves are obtained. These curves are similar in character to B-spline curves but provide C^∞ continuity. The form of the weight-function can be controlled by a shape parameter k. For appropriate choices of k the weight-functions provide a good approximation of uniform B-spline basis functions. Consequently, k is a continuous parameter that mimics the effect of degree in B-spline basis functions, but permits the generation of intermediate forms.

The immediate result of this paper is a simple curve scheme that can generate interpolating and approximating curves with tangent constraints, while supporting the basic properties required for efficient CAD applications. As the curve scheme produces C^∞ curves with compact support, we call them *CINPACT*-splines.

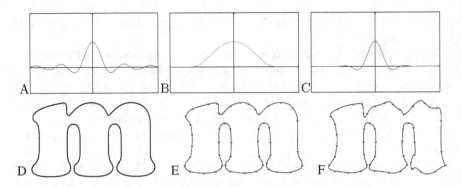

Fig. 1. The PUPs based interpolation scheme proposed by Runions and Samavati [7]. For weight-functions, the method uses a normalized sinc function (A) multiplied by a B-spline basis function (B). Thus, WFs have the form shown in (C). The contour provided in (D) is approximated using the weight-function from panel (C) in (E) and the normalized sinc function in (F).

The remainder of the paper is organized as follows. In Sect. 2, the methods proposed by Runions and Samavati [7] and Zhang and Ma [10] are presented and compared. In Sect. 3, a bump-function based WF is proposed and its application to the generation of interpolating curves is considered. Following this, Sect. 4 outlines the method for interpolating tangents and higher-order derivatives. Approximating CINPACT curves are discussed in Sect. 5, along with their relation to B-splines. Finally, the paper concludes with a summary of the presented results in Sect. 6.

2   Previous Work

2.1   Partition of Unity Parametrics

As proposed by Runions and Samavati [7], a Partition of Unity Parametric (PUP) curve $Q(u)$ takes the form:

$$Q(u) = \sum_{i=0}^{n} R_i(u)C_i \ , \tag{1}$$

where C_i are control points and each $R_i(u)$ takes the form

$$R_i(u) = \frac{W_i(u)}{\sum_{j=0}^{n} W_j(u)} \ , \tag{2}$$

where $W_i(u)$ is a Weight-Function (WF) that controls the contribution of C_i to the curve. Dividing by the sum of the WFs normalizes the values of the functions R_i, guaranteeing that they sum to 1. Thus, the R_i are called *normalized weight functions.* As the R_i provide a partition of unity, the resulting curve must be affine invariant. The final component of the definition of PUPs is the following condition:

$$\sum_{j=0}^{n} W_j(u) \neq 0, \tag{3}$$

which guards against indeterminate forms.

To interpolate, for each P_i an interpolation site i is chosen in the parameter domain. Then, $Q(i) = P_i$ is guaranteed through the appropriate choice of W_i. For this purpose, Runions and Samavati [7] proposed the following weight function:

$$W_i(u) = I_i(u)A_i(u), \tag{4}$$

where $I_i(u)$ can be any continuous function where $I_i(j) = \delta_{ij}$, and $A_i(u)$ is a tempering function which must be non-zero at i (i.e. $A(i) \neq 0$). There are many possible choices for I_i, however they chose

$$I_i(u) = \frac{sin(\pi(u - i))}{\pi(u - i)}, \tag{5}$$

the normalized sinc function (Fig. 1A), which has C^∞ continuity. If this function is supported on the entire domain (i.e. $A_i(u) = 1$), then the typical artifacts afflicting interpolating curves appear (ringing and overshooting of control points, Fig. 1F). To temper these effects they used a compactly-supported C^k function for $A_i(u)$. When I_i and A_i are multiplied together, the curve is still interpolated, but the support of I_i is smoothly truncated to that of A_i. For $A_i(u)$, they employed uniform B-spline basis functions with variable support (Fig. 1B). The form of $I_i(u)A_i(u)$ is shown in Fig. 1C, and a PUPs curve employing this weight-function is shown in Fig. 1E.

The curve produced by the method has a number of properties which are beneficial for geometric modeling:

1. affine invariance, resulting from partition of unity normalization,
2. C^k smoothness, dependent on the choice of $A(u)$,
3. compact support, dependent on the support of $A(u)$,
4. interpolation, resulting from $I(u)$,
5. and a controllable tradeoff between smoothness and overshooting, by tuning the support of $A(u)$.

2.2   Zhang and Ma's Method

A class of interpolating curves, related to the preceding PUPs based method, was proposed by Zhang and Ma [10]. Their method employed curves of the form

$$Q(u) = \sum_{i=0}^{n} R_i(u)C_i \tag{6}$$

with

$$R_i(u) = I_i(u)A_i(u) \tag{7}$$

where, again, the normalized sinc function was proposed for $I_i(u)$. In their case, however, the choice of approximating function was

$$A_i(u) = e^{-\alpha(u-i)^2}. \tag{8}$$

It is important to note that in this case normalization by the sum of the weight-functions does not occur (compare Eqs. 2 and 7), and thus affine invariance is not guaranteed. Nevertheless, for appropriate choices of α, the method produces similar curves to the PUPs based method.

When compared to the interpolation scheme proposed by Runions and Samavati, the method proposed by Zhang and Ma has different properties:

1. It offers a tradeoff between approximate affine invariance, and a more focused influence for R_i. Without the normalization step employed above, the R_i's do not sum to 1 unless $\alpha = 0$. At the same time, Zhang and Ma show that the difference from partition of unity is small even when $\alpha = 1/3$.
2. The curves have C^∞ smoothness.
3. Compact support can only be introduced by forcing $W_i(u) = 0$ outside of some interval. Thus, it comes at the cost of smoothness. As with affine invariance, the magnitude of the discontinuity introduced by truncating W_i decreases quickly as α is decreased, due to the exponential form of R_i.
4. Interpolation, from the choice of $I(u)$.
5. A controllable tradeoff between smoothness and overshooting, by tuning the coefficient α.

Later, Zhang proposed an extension to their method which improved the approximation of partition of unity by introducing an additional term into their basis functions [11]. Aside from this noteworthy difference, the improved method exhibits the same properties as the original method.

3 Weight Functions with C^∞ and Local Support

Examining the interpolation methods described above we notice the following. Runions and Samavati's method provides unconditional affine invariance, compact support, and C^k continuity (for a given finite k). In contrast, Zhang and Ma's method provides conditional (or approximate) affine invariance, compact support and C^∞ continuity. This raises the following question: can we obtain a method with the positive characteristics of both methods? Specifically, can we support unconditional

1. affine invariance,
2. C^∞ continuity,
3. compact support,
4. and interpolation,
5. while providing a controllable tradeoff between smoothness and overshooting?

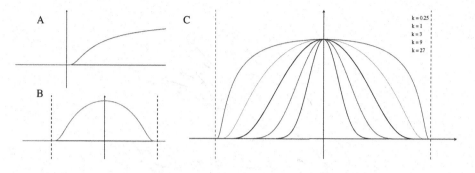

Fig. 2. The form of bump-functions described in the text, dashed vertical lines indicate the radius of support (i.e. $x = \pm c$). (A) The one sided function from Eq. 9. (B) The two-sided bump function from Eq. 10. (C) A family of bump functions created using Eq. 11 by varying k.

To answer this question we use the PUPs framework and follow the methodology employed by Runions and Samavati. Namely, given a set of properties we will seek an appropriate weight-function to satisfy these conditions. The construction of PUPs satisfies condition (1). We thus seek an appropriate weight-function with the same form as Eq. 4. As in the methods presented in Sect. 2, we use the normalized sinc function for $I(u)$, which satisfies condition (4). Thus, it remains to choose our $A(t)$ to satisfy constraints (2), (3) and (5).

These conditions can be satisfied by identifying an $A(t)$ with C^∞, and variable compact support. To motivate our choice of weight-function, we first examine the function

$$f(x) = \begin{cases} e^{-\frac{1}{x}} & x > 0 \\ 0 & otherwise \end{cases}, \tag{9}$$

which is shown in Fig. 2A. This function is zero for negative x-values and smoothly approaches 0 as $x \to 0^+$ (i.e. all derivatives approach zero). Thus, the support of this function is smoothly truncated to positive x values.

To limit the support of our function to an interval $[-c, c]$, while maintaining C^∞ continuity, we modify the denominator of the exponent to introduce singularities at $\pm c$:

$$f(x) = \begin{cases} e^{\frac{-1}{c^2 - x^2}} & x \in (-c, c) \\ 0 & otherwise \end{cases}, \tag{10}$$

which is shown in Fig. 2B. Now our function $f(x)$ has the desired properties: controllable compact support with C^∞ smoothness.

To obtain the $A(u)$ used in our interpolation scheme we modify the above form slightly to obtain:

$$A(u) = \begin{cases} e^{\frac{-ku^2}{c^2 - u^2}} & u \in (-c, c) \\ 0 & otherwise \end{cases}, \tag{11}$$

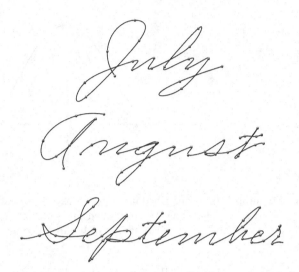

Fig. 3. Cursive writing generated using Interpolating CINPACT curves. The curves spell out the words July, September and August. The bump function used in these curves has radius of support $c = 10$ and shape parameter $k = 10$.

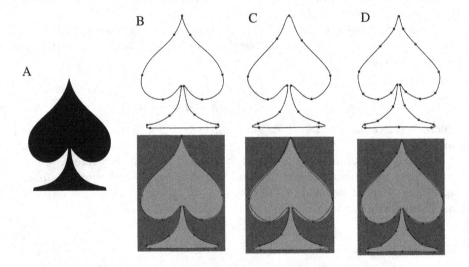

Fig. 4. Comparison of CINPACT curves to those produced using the method of Zhang et al. [10]. A spades symbol (A) is reproduced using a CINPACT curve with 15 points (B) as described in the text, and curves from the method of Zhang et al. with 15 (C) and 24 (D) control points. In (B–D) the top row shows each curve and the bottom row shows the curve overlain on the contour from (A).

where k is a continuous degree-like shape parameter, which can be used in tandem with the radius of support c to obtain a controllable tradeoff between smoothness and overshooting. The form of $A(u)$ differs from the $f(x)$ in Eq. 10, due to the introduction of ku^2 into the numerator of the exponential term. This modification gives $A(u)$ a Gaussian-like form (see Fig. 2C).

The resulting interpolation scheme is demonstrated in Fig. 3, where it is used to reproduce cursive handwriting samples of the words July, August and September. CINPACT splines inherit from PUPs the ability to modify the parameters of weight functions on a per-control-point basis (c.f. Fig. 6 in [7]). This permits refined control over the character of the curve without requiring the introduction of additional control points. In Fig. 4 this permits the *spades* symbol (A)[1] to be produced with a small number of control points (B), which account for the sharp protrusions and clefts along contour as well as the differences in the character of the blade and handle. In comparison, using the scheme of Zhang et al. [10], more points are required to approximate the form from (A) with the same fidelity (see C and D).

4   Specifying Tangents and Higher Derivatives of a Curve

The ability to specify tangents at control points is important for many CAD applications, such as the design of fonts [4], and illustrations in professional software packages such as Adobe Illustrator. Additionally, curves with tangent control are widely employed in computer animation [6] (e.g. for keyframing). This problem was not addressed by the methods proposed by Runions and Samavati [7] or Zhang and Ma [10], which limits their utility. Here, we address this problem for CINPACT and PUP splines by deriving a method for interpolating tangents and higher order derivatives (Figs. 5 and 6).

We first note that, given the formulation of PUPs in Eq. 1, we can add a sum of weighted vectors to the equation without violating partition of unity (i.e. the resulting curve is still affine invariant). This property is exploited to specify the tangents $T_0, T_1, \cdots, T_n$ at the control points of a PUPs curve, yielding the following form

$$Q_T(u) = \sum_{i=0}^{n} R_i(u)P_i + \sum_{i=0}^{n} E_i(u)V_i, \tag{12}$$

where V_i are the vectors we add to the curve to meet our tangent constraints and E_i are weight-functions localizing the contribution of these vectors[2]. For simplicity, let us write

$$Q_T(u) = Q(u) + \sum_{i=0}^{n} E_i(u)V_i, \tag{13}$$

---

[1] The spade symbol in Fig. 4A was obtained from http://commons.wikimedia.org/wiki/File:French_suits.svg.

[2] Note that as V_i are vectors the E_i's do not need to sum to 1.

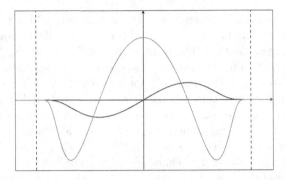

Fig. 5. The weight-function E_i used to interpolate tangents (blue) and its derivative (red). Dashed lines indicate the support of E_i (Color figure online).

where $Q(u)$ is the PUPs curve produced when tangent constraints are ignored. The vector V_i is then chosen so that

$$Q_T(i)' = T_i, \qquad (14)$$

which implies that

$$T_i = Q'(i) + \sum_{i=0}^{n} E_i(i)'V_i. \qquad (15)$$

In general, this provides a system of equations that, for appropriate choices of E_i, we can solve for V_i. When a global system of equations must be solved, however, the choice of tangent at a given point may globally contribute to the resulting V_i and thus the form of the curve. We note that by introducing three constraints on the form of E_i we can compute the V_i's directly (i.e. without solving a system of equations). Additionally, the direct solution we obtain has the added benefit of localizing the impact of tangent constraints on the form of the curve. These three constraints are:

1. The support of E_i is limited to $[i-c, i+c]$ (similar to the CINPACT basis functions).
2. Interpolating tangents should not interfere with the interpolation of positions:

$$E_i(j) = 0 \quad (j \in \mathbb{N}). \qquad (16)$$

3. Interpolation of a tangent at one site shouldn't interfere with interpolation at other sites:

$$E_i(j)' = \delta_{ij} \quad (j \in \mathbb{N}). \qquad (17)$$

Enforcing these constraints simplifies Eq. 15 to

$$T_i = Q'(i) + V_i. \qquad (18)$$

Thus, our V_i become (Fig. 6A)

$$V_i = T_i - Q'(i). \qquad (19)$$

It now remains to find an appropriate choice of E_i, which satisfies our three constraints. This problem is simplified by restricting our search to functions of the following form

$$E_i(u) = \frac{f_i(u)g_i(u)}{(f_i(u)g_i(u))'(i)}, \tag{20}$$

where $(f_i(u)g_i(u))'(i)$ is the derivative of $f_i(u)g_i(u)$ evaluated at i. In this case it then suffices to choose $f_i(u)$ and $g_i(u)$ such that:

1. $f_i(j)g_i(j) = 0$ $(j \in \mathbb{N})$,
2. $(f_i(j)g_i(j))' = f_i'(j)g_i(j) + f_i(j)g_i'(j) = \delta_{ij}$ $(j \in \mathbb{N})$.

while maintaining compact support and C^∞ smoothness. We note that these conditions are met by setting

$$f_i(u) = A_i(u), \tag{21}$$

the bump functions from Eq. 11, and:

$$g_i(u) = u - i. \tag{22}$$

To enforce the second condition $((f_i(j)g_i(j))' = \delta_{ij})$, we use a radius of support $c \leq 1$ (if a $c \geq 1$ is desired it suffices to use $f_i(u) = I_i(u)A_i(u)$, where $I_i(u)$ is the normalized sinc function centered at i).

Thus, our tangent interpolation function is simply (Fig. 5):

$$E_i(u) = \frac{(u - i)A_i(u)}{((u - i)A_i(u))'(i)} . \tag{23}$$

Note that if we do not wish to constrain the tangent at P_i, then $V_i = 0$ and the corresponding term $E_i(u)V_i$ can be omitted from Eq. 12.

With this weight-function, a CINPACT curve that interpolates the specified positions and tangents can be computed in three steps (Fig. 6):

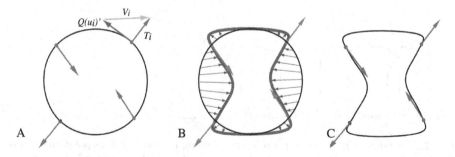

Fig. 6. CINPACT tangent interpolation. (A) The curve $Q(u)$ is shown with desired tangent constraints (orange arrows). The computation of V_i (blue arrow) from $Q(u_i)'$ (green arrow) and the tangent constraint T_i is illustrated. (B) The vectors V_i are multiplied by WFs and added to $Q(u)$ to yield $Q_T(u)$ (red curve). Blue arrows indicate the sum of weighted vectors at each parameter value. (C) The curve $Q_T(u)$ smoothly interpolates control points and tangents (Color figure online).

1. $Q(u)$ is evaluated using Eq. 1 and the interpolating WFs proposed in Sect. 3.
2. V_i is calculated for each T_i (Eq. 19, Fig. 6A).
3. The sum of weighted V_i's is added to $Q(u)$ (Fig. 6B) to obtain $Q_T(U)$ (Fig. 6C).

As the WFs used for the control points and the tangents are C^∞ with compact support (and by definition $\sum_{j=0}^{n} W_j(u) \neq 0$), the resulting curve is as well.

The basic construction employed to specify the first derivative at i (i.e. the tangent) generalizes to higher order derivatives in a straightforward manner. In particular, to specify $D_{i,n}$ the nth derivative at i we simply set $g_i(x) = (x - i)^n$, and $V_{i,n}$ to

$$V_{i,n} = D_{i,n} - Q_{n-1}^{(n)}(i) \tag{24}$$

where Q_{n-1} is the curve resulting from the interpolation of the $n-1^{th}$ derivatives (note that $Q_1 = Q_T$). This generalized construction is analogous to a Taylor series expansion about the parameter value i, with the impact of the series' contribution to the curve localized by the support of $A_i(x)$.

The method for specifying tangents is demonstrated in Fig. 7. In the figure, the contour provided in (A) is reproduced using the control points and tangents shown in (B). Using tangent constraints we are able to approximate the contour using 27 control points (compared to Fig. 1E where 49 control points were used).

In some instances it is cumbersome to specify the tangent at each point (e.g. interactive curve modeling). In such instances, to ease the specification of curves which interpolate a position and tangent, it is useful to generate tangents automatically, from the given set of control points. The Catmull-Rom spline [2], widely employed within computer animation, is a relatively popular method for achieving this goal. These curves are cubic Hermite splines [3, pp. 102–106], where the tangent at each control point P_i is calculated from the neighboring control points

$$T_i = \frac{P_{i+1} - P_{i-1}}{2}. \tag{25}$$

Fig. 7. The contour in panel (A) is reproduced using the tangent interpolation scheme (B,C). Tangents are visualized as orange arrows and control points as orange spheres. The resulting curve is shown with (B) and without (C) visualizing tangents (Color figure online).

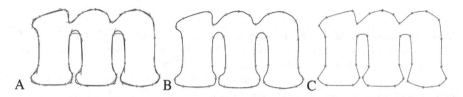

A   B   C

Fig. 8. The contour from Fig. 7(A) is reproduced using a Catmull-Rom spline (red contour) and a CINPACT spline (black contour) with the same tangents specified at each control point. In (A) the curves are overlaid and visualized with the control points and tangent vectors. The curves are shown individually in (B) and (C) (Color figure online).

In Fig. 8(A), an interpolating CINPACT spline with tangents calculated using Eq. 25 is compared to the corresponding Catmull-Rom spline. The curves approximate the contour provided in Fig. 7(A). Note that the CINPACT spline has more evenly distributed curvature, which, in this case, better approximates the original contour (compare panels Fig. 8(B) and (C) with Fig. 7(A)).

5   Approximating Curves

Our function $A(x)$ has a plot similar to that of a B-spline basis function. Given this, it makes sense to examine the curves generated by the weight-functions

$$W_i(u) = A(u - i),\tag{26}$$

which produces curves that approximate control points. The resulting curves are well-behaved, exhibiting affine invariance, compact support and C^∞ continuity. Furthermore, as shown in Fig. 9 the curves are similar in character to B-splines. Here, however, the parameter k serves as a continuous degree-like parameter. To quantify this statement we have numerically fit $W(u)$ to uniform B-spline basis functions of different orders. Measuring the goodness of fit requires a measure of distance between two functions. To this end we employ the commonly used L^2 norm for functions:

$$d(f,g) = \left[\int_{-\infty}^{\infty} (f(u) - g(u))^2 du\right]^{1/2}.\tag{27}$$

Using this distance we can estimate the relative error of the fit of the weight function $W(u)$ to the degree d B-spline basis function $B^d(u)$ with the following equation

$$RelativeError(W, B^d) = \frac{d(W, B^d)}{d(0, B^d)},\tag{28}$$

where the division by the distance between B^d and the constant function 0 normalizes the distance between W and B^d.

Table 1. Parameter values for the best-fit C^∞ weight-functions W for B^d, the B-spline basis of degree d. The second column shows the distance between each B^d and the constant function 0. The distance between B^d and the best-fit W is shown in the third column, with corresponding parameters for W given in the next 3 columns. The final column provides the relative error of the fit (i.e. $d(W, B^d)/d(B^d, 0)$).

Degree	$d(B^d, 0)$	$d(W, B^d)$	k value	Radius of support c	Scalar weight	Relative error
1	0.8165	0.0499	38.94	3.976	0.916264	0.06
2	0.7416	0.0122	25.83	3.898	0.761562	0.016
3	0.6924	0.0025	17.27	3.684	0.662606	0.0036
4	0.6561	0.0021	28.48	5.158	0.599502	0.0032
5	0.6276	0.00055	28.00	5.574	0.549665	0.00088
6	0.6044	0.0018	23.83	5.560	0.509389	0.0029
7	0.5850	0.0033	21.31	5.626	0.476678	0.0056
		0.0020	39.84	7.587	0.477571	0.0034
8	0.5683	0.0038	21.05	5.919	0.450372	0.0067
		0.00097	33.46	7.365	0.452382	0.0017
9	0.5538	0.0015	31.94	7.579	0.429608	0.0027

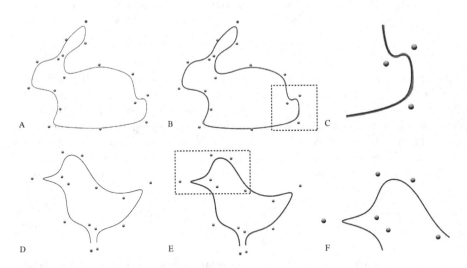

Fig. 9. A comparison of CINPACT and B-spline curves. (A) A CINPACT curve created from WFs with $k = 25.83$ and $c = 3.898$. (B) The curve from (A) is compared to the quadratic B-spline curve generated by the same control points (red curve); the inset is shown in (C). (D) A CINPACT curve created from WFs with $k = 17.27$ and $c = 3.684$. (E) The curve from (D) is compared to the cubic B-spline curve generated by the same control points (red curve); the inset is shown in (F) (Color figure online).

The results of our numerical fitting are reported in Table 1. We observe that by varying k and the support c we can approximate B-spline basis functions of increasing order very well. The particular values of k and c have been determined by numerical optimization, and for B-spline basis functions of degree 2 and higher we obtain small differences between our weight-function and the basis function (less than 0.7 % for $d > 2$).

The close correspondence between the forms of CINPACT and B-spline curves is illustrated in Fig. 9 for quadratic (A–C) and cubic (D–F) B-spline curves. While the overall approximation is very good, the CINPACT curves, being C^∞, exhibit less variation (as seen in inset of the bunny tail shown in C).

These results show that uniform B-spline functions differ only slightly from a C^∞ function with a reasonably simple form. Conversely, this also means that when the parameters k and c are chosen appropriately, the resulting CINPACT curves can be subdivided using B-spline filters with a relatively small error.

More broadly, these results show that CINPACT splines, like PUPs, can be used to generate high quality approximating as well as interpolating curves. This can be contrasted against the methods proposed by Zhang and Ma. [10,11], which only support interpolation.

6   Conclusions

In this paper we have presented CINPACT-splines, which support interpolation and approximation of control points. The resulting curves have a simple, piecewise exponential form. These splines are generated using a class of bump-functions as WFs, and guarantee that the resulting curves are C^∞, with compact support and affine invariance. In this sense, the interpolating CINPACT-splines are an improvement on the previously proposed interpolation methods of Runions and Samavati [7] and Zhang and Ma [10,11]. Additionally, we provide a method for the interpolation of tangents, a common requirement for the curves used in animation and interactive modeling applications. This, in particular, was not considered in the works of Runions and Samavati [7] and Zhang and Ma [10,11].

The approximating CINPACT-splines behave similarly to B-splines, but do not sacrifice continuity in order to achieve compact support. Additionally, the shape parameter k provides a continuous degree like parameter. Consequently, unlike B-splines, the weight-functions are not defined by a recursive relation, but have a simple closed form. Furthermore, when curve fitting is performed the parameter k and radius of support c can be independently optimized as continuous parameters. In contrast, for B-spline curves the degree determines the support, and when optimized during curve fitting introduces a discrete parameter. These advantages highlight the potential of CINPACT-splines as an alternative to B-spline curves. As the curves considered by Runions and Samavti [7] employed B-spline basis functions CINPACT-splines likewise offer a number of advantages over the curves considered in [7]. By employing closed form exponentials (bump-functions) in place of B-spline basis functions our curves have a relatively simple implementation, but can still provide the basis for a comprehensive curve modeling package.

A key goal of future work on CINPACT-splines will be to further examine their relation to B-splines. In this paper we considered the fitting of CINPACT weight-functions to uniform B-spline basis functions, which assume a uniform spacing of knot values. It is an open question, however, what class of bump-function is required to reproduce arbitrary B-spline basis functions (i.e. those generated by non-uniform knot values). A related problem is the local refinement of weight-functions, which for B-splines is accomplished through knot insertion [1]. Pursuing these two directions should allow CINPACT and B-splines to be used interchangeably, opening the door to their widespread use in CAD applications.

References

1. Bartels, R., Beatty, J., Barsky, B.: An Introduction to Splines for Use in Computer Graphics and Geometric Modeling. Morgan Kaufmann, Los Altos (1987)
2. Catmull, E., Rom, R.: A class of local interpolating splines. In: Barnhill, R.E., Riesenfeld, R.F. (eds.) Computer Aided Geometric Design, pp. 317–326. Academic Press, New York (1974)
3. Farin, G.: Curves and Surfaces for CAGD. Morgan Kauffmann, San Francisco (2002)
4. Knuth, D.: The METAFONT Book. Addison-Wesley, Boston (1995)
5. Lee, J.: Introduction to Smooth Manifolds. Springer, New York (2000)
6. Parent, R.: Computer Animation. Morgan Kauffmann, San Francisco (2012)
7. Runions, A., Samavati, F.: Partition of unity parametrics: a framework for meta-modeling. Vis. Comput. **27**, 495–505 (2011)
8. Sederberg, T., Zheng, J., Bakenov, A., Nasri, A.: T-splines and T-NURCCs. ACM Trans. Graph. **22**, 477–484 (2003)
9. Wang, Q., Hua, W., Guiqing, L., Bao, H.: Generalized NURBS curves and surfaces. In: Geometric Modeling and Processing, pp. 365–368 (2004)
10. Zhang, R.-J., Ma, W.: An efficient scheme for curve and surface construction based on a set of interpolatory basis functions. ACM Trans. Graph. **30**, 10:1–10:11 (2011)
11. Zhang, R.-J.: Curve and surface reconstruction based on a set of improved interpolatory basis functions. Comput. Aided Des. **44**, 749–756 (2012)

Error Estimates for Approximate Operator Inversion via Kernel-Based Methods

Kristof Schröder[1,2]($\boxtimes$)

[1] Faculty of Mathematics, Technische Universität München, Boltzmannstraße 3,
85748 Garching, Germany
[2] Helmholtz Centre Munich, Ingolstädter Landstraße 1, 85764 Neuherberg, Germany
kristof.schroeder@helmholtz-muenchen.de

Abstract. In this paper we investigate error estimates for the approximate solution of operator equations $Af = u$, where u needs not to be a function on the same domain as f. We use the well-established theory of *generalized interpolation*, also known as *optimal recovery in reproducing kernel Hilbert spaces*, to generate an approximation to f from finitely many samples $u(x_1), \ldots, u(x_N)$. To derive error estimates for this approximation process we will show *sampling inequalities* on fairly general Riemannian manifolds.

Keywords: Generalized interpolation · Positive definite functions · Reproducing kernel hilbert spaces · Sampling inequalities on manifolds

1 Introduction

The reconstruction of a function f from some given data u is a very common problem in many branches of science and engineering. In many situation the data u are linear measurements of f. This means we consider the solution of operator equations of the form

$$Af = u \tag{1}$$

where $A : H_1 \to H_2$ is a linear continuous operator mapping between two Hilbert spaces. Normally due to experimental restrictions one does not have access to the function u itself but only finitely many samples $u(x_1), \ldots, u(x_N)$ of it. It is therefore necessary to come up with an approximate operator inversion scheme which uses these samples only.

In this paper we consider a kernel-based approach for this approximate inversion. This approach can be sketched as follows. Given samples $u(x_1), \ldots, u(x_N)$ we formulate the *generalized interpolation problem*

$$\lambda_i(f) = (\delta_{x_i} \circ A)f = Af(x_i) = u(x_i), \qquad i = 1, \ldots, N. \tag{2}$$

The functionals λ_i are linear and continuous provided the operator A is a continuous linear operator from H_1 to H_2 and the point evaluation functionals are bounded functionals on H_2.

© Springer International Publishing Switzerland 2015
J.-D. Boissonnat et al. (Eds.): Curves and Surfaces 2014, LNCS 9213, pp. 399–413, 2015.
DOI: 10.1007/978-3-319-22804-4_28

The latter assumption makes it necessary to restrict our considerations with respect to H_2 to *reproducing kernel Hilbert spaces* (RKHS). If we additionally assume that the space H_1 is a RKHS it is well known that a solution of the generalized interpolation problem (2) is given in the form

$$f^+(x) = \sum_{i=1}^{N} a_i \lambda_i^y \Phi(x, y), \qquad (3)$$

where Φ is the uniquely determined positive definite kernel of the RKHS H_1 and the notation λ_i^y indicates that the functional acts on $\Phi(x, y)$ as a function of the second variable y.

In [4] this viewpoint has been called *semi-discrete inverse problem* and error estimates have been derived in the case that H_1 and H_2 are Sobolev spaces defined on the same Lipschitz-domain $\Omega \subset \mathbb{R}^d$.

The motivation for our work on those problems mainly stems from image reconstruction from photoacoustic measurements. The problem here is the reconstruction of a function f from spherical mean measurements. More precisely, let

$$\mathcal{M}f(\xi, t) = \int_{\mathbb{S}^{d-1}} f(\xi + ty) \, d\,\sigma(y) \qquad (4)$$

be the so-called *spherical mean operator*, where $\mathbb{S}^{d-1}$ is the unit sphere in $\mathbb{R}^d$ with normalized surface measure σ. Given to us are samples $\mathcal{M}f(\xi_i, t_j)$, where the points ξ_i are located on a submanifold $\Xi \subset \mathbb{R}^d$ of dimension $d - 1$, the so-called *acquisition surface*.

Having this in mind we cannot assume that the data is given on the same domain as the function which has to be reconstructed. Instead we will assume that the accessible data is given on some manifold. The aim of this article is to derive error estimates in the case that H_2 is a Hilbert space on a fairly general Riemannian manifold and the operator A has only local regularity.

A key ingredient for the analysis of the approximation error is a relation between the norm of the difference $f - f^+$ on the one hand and the discrete norm of the sampling vector $\{A(f - f^+)(x_i)\}_{i=1}^N$ on the other hand. Inequalities of such type are called sampling inequalities. They were studied for bounded and unbounded domains in $\mathbb{R}^d$ by many authors, see for example in chronological order Rieger et. al. [3,13,17], Madych [6] and Arcangéli et. al. [9–12]. See also Fuselier and Wright [1] and Hangelbroek and Narcowich [14] for related considerations on closed manifolds. Due to our studies on the reconstruction problem described above we have been motivated to study sampling inequalities on fairly general Riemannian manifolds. In this paper we present results for this setting.

The paper is organized as follows. In Sect. 2 we will present some background material on reproducing kernel Hilbert spaces and Sobolev spaces on Riemannian manifolds and collect some results on compact operators with certain regularity properties. Section 3 contains the result on sampling inequalities for Sobolev functions and the error analysis for the generalized interpolation problem.

2   Preliminaries

2.1   Reproducing Kernel Hilbert Spaces and Generalized Interpolation

Let X be a topological space and H be a Hilbert space of functions on X such that point evaluations are continuous functionals. Such spaces are called *reproducing kernel Hilbert spaces* (RKHS). The reason for this notation is the fact that by the representer theorem of Riesz there is a unique function $\Phi : X \times X \to \mathbb{R}$ such that

1. $\Phi(\cdot, x) \in H$ for all $x \in X$,
2. $f(x) = \langle f, \Phi(\cdot, x) \rangle_H$ for all $x \in X$ and $f \in H$.

This function is called the *reproducing kernel* of H due to property 2. It is positive definite, i.e.

$$\sum_{i,j=1}^{N} c_i c_j \Phi(x_i, x_j) \geq 0, \tag{5}$$

for all $N \in \mathbb{N}, c_1, \ldots, c_N \in \mathbb{R}, x_1, \ldots, x_N \in X$.

On the other hand for a given positive definite function Φ one can construct a unique Hilbert space $H = \mathcal{N}_\Phi$ such that Φ is the reproducing kernel of $\mathcal{N}_\Phi$. This space is sometimes called *native space* of Φ. This means the connection between positive definite functions and RKHS is one to one.

We will formulate a generalized interpolation problem in a RKHS H. Given a linear independent set of linear continuous functionals $\lambda_1, \cdots, \lambda_N$ on H and a vector $\boldsymbol{u} = (u_i)_{i=1}^{N} \in \mathbb{R}^N$, find a function $f \in H$ such that

$$\lambda_i(f) = u_i, \qquad \text{for } i = 1, \cdots, N. \tag{6}$$

By the results of Golumb and Weinberger, see [5], also known as *optimal recovery* [16, Ch.16], there is a unique solution f^+ of minimal norm to this problem, meaning

$$\|f^+\|_H \leq \|f\|_H, \tag{7}$$

for all $f \in H$ satisfying the generalized interpolation condition (6).

Moreover the solution is a linear combination of the Riesz representers of the functionals λ_i, i.e.

$$f^+ = \sum_{i=1}^{N} a_i v_i, \qquad \lambda_i(f) = \langle f, v_i \rangle_H. \tag{8}$$

Since H has a reproducing kernel Φ these representers are given explicitly by

$$v_i(x) = \lambda_i^y \Phi(x, y). \tag{9}$$

The notation λ_i^y indicates the action of λ_i on $\Phi(x, y)$ as a function in the second variable.

The coefficients $a = (a_i)_{i=1}^N$ of the superposition are given as the solution of the linear system

$$\mathrm{M}a = u, \qquad \mathrm{M}_{i,j} = \langle \lambda_i^y \Phi(\cdot, y), \lambda_j^y \Phi(\cdot, y) \rangle_H. \tag{10}$$

Here again the functionals act on the second variable of the kernel and the inner product of H is taken with respect to the first variable of Φ.

We are interested in functionals of the form

$$\lambda_i = (\delta_{x_i} \circ A), \tag{11}$$

which are continuous if the operator A is mapping continuously into another RKHS. Which spaces and operators are appropriate for this will be the topic of Sects. 2.2 and 2.3.

2.2   Sobolev Spaces on Riemannian Manifolds

Sobolev spaces of a dimension depending order are important examples of RKHS. We introduce Sobolev spaces for more general parameters since the sampling inequality of Theorem 1 holds for a large range of parameters and is interesting on its own. A detailed discussion on the content of this section can be found e.g. in [15].

For $1 < p < \infty$ and $s \geq 0$ set

$$W^{s,p}(\mathbb{R}^d) := \{f \in L^p(\mathbb{R}^d) : \exists v \in L^p(\mathbb{R}^d), f = (I - \Delta)^{-s/2}v\}, \tag{12}$$

equipped with the norm

$$\|f\|_{W^{s,p}(\mathbb{R}^d)} := \|(I - \Delta)^{-s/2}v\|_{L^p(\mathbb{R}^d)}. \tag{13}$$

We have the embedding

$$W^{s,p}(\mathbb{R}^d) \hookrightarrow C_b(\mathbb{R}^d), \qquad s > d/p. \tag{14}$$

This means for $p = 2$ these spaces have a reproducing kernel.

The same is true if we focus on functions defined on a bounded Lipschitz domain $\Omega \subset \mathbb{R}^d$, i.e.

$$W^{s,p}(\Omega) := \{f \in L^p(\Omega), \|f\|_{W^{s,p}(\Omega)} < \infty\}, \tag{15}$$

where

$$\|f\|_{W^{s,p}(\Omega)} := \inf\{\|g\|_{W^{s,p}(\mathbb{R}^d)}, g \in W^{s,p}(\mathbb{R}^d), g_{|\Omega} = f\}. \tag{16}$$

We again have $W^{s,p}(\Omega) \hookrightarrow C_b(\Omega)$ for $s > d/p$.

On a connected smooth d-dimensional Riemannian manifold (X, g) with smooth metric g we will define Sobolev spaces via localization in specific charts. Therefore we assume some restrictions on the manifold.

Denote by $\gamma(p, v, t)$ the unique geodesic with $\gamma(p, v, 0) = p$ and $\dot{\gamma}(p, v, 0) = v$ for $p \in X$ and $v \in T_pX$ where T_pX is the tangent space of X at p.

First of all we assume *geodesic completeness*, i.e. the parametrization in t of each geodesic $\gamma(p, v, t)$ is extendable to the whole of $\mathbb{R}$. By the Theorem of Hopf-Rinow this is equivalent to the completeness of X as a metric space.

The exponential map at p is defined by

$$\exp_p : T_p X \to X, \qquad v \mapsto \gamma(p, v, 1). \tag{17}$$

By identifying the tangent space with $\mathbb{R}^d$ the exponential map yields for sufficiently small $r > 0$ a diffeomorphism from

$$B(r) := \{v \in \mathbb{R}^n : \|v\|_{\mathbb{R}^n} < r\} \quad \text{onto} \quad U_p(r) := \exp_p(B(r)), \tag{18}$$

so $(U_p(r), \phi_p)$ with $\phi_p = exp_p^{-1}$ is a chart for X at p.

These coordinates are called *geodesic normal* coordinates. If r_p denotes the supremum of all $r > 0$ satisfying the diffeomorphism property of the exponential map then

$$r_X = \inf_{p \in X} r_p \tag{19}$$

is called the *injectivity radius* of X.

A connected complete Riemannian manifold is of *bounded geometry* if the injectivity radius is bigger than zero and the change of geodesic normal coordinates is bounded in all derivatives on balls of radius bounded by r_X. This means for $U_r(p) \cap U_r(q) \neq 0$ with $r < r_X$ we have

$$|\mathrm{D}^\alpha(\phi_p \circ \phi_q^{-1})(x)| \le C_k, \qquad x \in B(r), \tag{20}$$

for every multi index α with $|\alpha| \le k$ and all $k \in \mathbb{N}_0$ and all derivatives D^α.

This restriction on the geometry has some important implication, i.e. there are constants $0 < c_1 \le 1 \le c_2 < \infty$ such that for all $p \in X$ and all $x, y \in U_p(r), r < r_X$,

$$c_1\|\phi_p(x) - \phi_p(y)\| \le \mathrm{d}_g(x, y) \le c_2\|\phi_p(x) - \phi_p(y)\|, \tag{21}$$

where $\mathrm{d}_g(p, q)$ denotes the geodesic distance on (X, g). In other words the geodesic distance is globally comparable to the Euclidean distance in the tangent space.

Summarising we will assume that (X, g) is a complete d-dimensional Riemannian manifold of bounded geometry.

Such a manifold give rise to a countable atlas of geodesic normal charts and a subordinate uniform locally finite partition of unity.

Proposition 1. *[15, Proposition 7.2.1] If a connected complete Riemannian manifold (X, g) is of bounded geometry then there exist $\varepsilon > 0$ and a sequence of points $\{p_j\} \subset X$ such that $\{(U_j, \phi_j)\}_{j \in \mathbb{N}}$ with $\phi_j = \exp_{p_j}^{-1}$ is an atlas with a smooth uniform locally finite partition of unity $\{\psi_j\}_{j \in \mathbb{N}}$, i.e. a sequence of functions such that*

1. $\psi_j \in C^\infty(X)$, $\quad 0 \le \psi_j \le 1$, $\quad \sum_{j \in \mathbb{N}} \psi_j(p) = 1$, for all $p \in X$,
2. $\operatorname{supp}(\psi_j) \subset U_j$,
3. there is $L \in \mathbb{N}$, such that for a fixed $i \in \mathbb{N}$ at most L functions ψ_j satisfy $\operatorname{supp}(\psi_i) \cap \operatorname{supp}(\psi_j) \ne \emptyset$.
4. for every multi-index α there is $b_\alpha > 0$ s.t. $|D^\alpha(\psi_j \circ \exp_{p_j})(x)| \le b_\alpha$, for all $|x| \le \varepsilon$.

Given a connected complete d-dimensional Riemannian manifold (X, g) of bounded geometry then by Proposition 1 there is an atlas of geodesic normal charts $\{U_j, \phi_j\}_{j \in \mathbb{N}}$ and a corresponding subordinate partition of unity $\{\psi_j\}_{j \in \mathbb{N}}$. Thus one way to define Sobolev spaces for $s \ge 0$ is to consider those functions $u \in L^q(X)$ for which the norm

$$\|u\|_{W^{s,q}(X)} := \Big(\sum_{j \in \mathbb{N}} \|(\psi_j u) \circ \phi_j^{-1}\|^q \Big)^{1/q} \tag{22}$$

is finite, this means

$$W^{s,q}(X) := \{u \in L^q(X) : \|u\|_{W^{s,q}(X)} < \infty\}. \tag{23}$$

Here the functions $(\psi_j u) \circ \exp_{p_j}$ with support in $\exp_{p_j}^{-1}(U_j)$ are continuously extended to $\mathbb{R}^d$ by zero and this extension is denoted by the same symbol.

Due to the description via localization the embedding theorems for functions on $\mathbb{R}^d$ transfer to the manifold case with the same conditions on the parameters. Especially

$$W^{s,q}(X) \hookrightarrow C_b(X), \qquad \text{for } s > d/q. \tag{24}$$

In particular this means that in the case $q = 2$ the spaces $H^s(X) = W^{s,2}(X)$ are reproducing kernel Hilbert spaces.

2.3 Compact Operators with Local Regularity

In this section we will make some assumptions on the reproducing kernel, the operators and the sampling set that will be needed for the error analysis in Sect. 3.2. Therefore let $\Omega \subset \mathbb{R}^d$ be a simply connected Lipschitz domain and $K \subset \Omega$ compact.

Assumption 1 (Reproducing kernel). *Suppose $\Phi : \Omega \times \Omega \to \mathbb{R}$ is a continuous positive definite function such that*

$$\mathcal{N}_\Phi(\Omega) \cong H_K^s(\Omega), \tag{25}$$

meaning that the native space $\mathcal{N}_\Omega$ is norm-equivalent to the Sobolev space $H_K^s(\Omega)$ of functions with support in the compact set K.

Remark 1. Note that a positive definite function of this kind cannot be the restriction to Ω of a translation invariant function $\Phi(x, y) = \varphi(x - y)$ since there would be for every compact set $K \subset \Omega$ a function $f \in \mathcal{N}_\phi$ with $\operatorname{supp}(f) \nsubseteq K$.

Especially widely used *radial symmetric* positive definite functions also called *radial basis functions* are not adequate for our purpose. Nevertheless the construction of specific kernels having the property (25) is beyond the scope of this article.

For the operators we will assume certain mapping properties in the Sobolev scale. If X is a manifold with properties discussed in the previous section, we will assume that functions having compact support $K \subset \Omega$ will be mapped to functions having compact support $\Lambda \subset X$.

Assumption 2 (Mapping properties). *Suppose A is a linear operator such that*

$$c\|f\|_{H_K^s(\Omega)} \leq \|Af\|_{H_\Lambda^{s+\tau}(X)} \leq C\|f\|_{H_K^s(\Omega)} \qquad (26)$$

for all $f \in H^s(\Omega)$ with $\operatorname{supp} f \subseteq K$ and $K \subset \Omega$, $\Lambda \subset X$ fixed compact sets.

As examples we have certain integral operators in mind that arises in several tomographic problems and are well-known to have this local regularity property.

Example 1 (Radon Transform). Let $K \subset \mathbb{R}^2$ be compact and $f \in C_K(\mathbb{R}^2)$. The *Radon Transform* is defined as

$$\mathcal{R}f(\xi, t) = \int_{\mathbb{R}} f(t\xi + s\xi^{\perp})ds, \qquad \xi \in \mathbb{S}^1, t \in \mathbb{R},$$

which has application e.g. in computerized tomography.

This operator can be extended such that

$$c\|f\|_{H^s(\mathbb{R}^2)} \leq \|\mathcal{R}f\|_{H^{s+1/2}(\mathbb{S}^1 \times \mathbb{R})} \leq C\|f\|_{H^s(\mathbb{R}^2)}$$

for all f with $\operatorname{supp} f \subseteq K$, see e.g. [7]. In this example we would have $\Lambda = \mathbb{S}^1 \times [-T, T]$ with $T = \inf\{T > 0 : K \subseteq \overline{B_T(0)}\}$.

Example 2 (Spherical mean operator). Suppose $\Omega \subset \mathbb{R}^d$ is an open convex connected bounded domain with smooth boundary. For $\xi \in \partial\Omega$, $t \in \mathbb{R}$ and $f \in C_K(\Omega)$ define

$$\mathcal{M}f(\xi, t) = \int_{\mathbb{S}^{d-1}} f(\xi + tu) \, d\,\sigma(u).$$

This operator arises in the application of *Photoacoustic Tomography*. Again $\mathcal{M}$ can be extended such that for $s \geq 0$

$$c\|f\|_{H^s(\Omega)} \leq \|\mathcal{M}f\|_{H^{s+d/2}(\mathbb{S}^{d-1} \times \mathbb{R})} \leq C\|f\|_{H^s(\Omega)},$$

see [8]. Here the compact set Λ is given by $\Lambda = \partial\Omega \times [0, T]$ with $T = \operatorname{diam}(\Omega)$.

Combining this assumptions leads to the fact that the functionals $\delta_x \circ A$ are bounded on the native space of the kernel Φ. To make sure that the generalized interpolant f^+ given by (8) is well-defined, we have to restrict to specific discrete sets that guarantee the unique solvability of the linear system (10).

Assumption 3 (Linear independence). *We restrict to those quasi-uniformly distributed discrete sets $\mathcal{X} \subset X$ such that for a linear operator A satisfying Assumption 2 the functionals*

$$\lambda_x f = (\delta_x \circ A)f, \qquad x \in \mathcal{X}, \tag{27}$$

are linear independent.

3   Approximation Property

3.1   Sampling Inequality

Since we would like to gain knowledge about our approximation from knowledge of our linear measurements of the function we have to make sure that this is possible in a stable way. This means that 'small norm' of our measurements on a sufficiently dense discrete set $\mathcal{X}$ implies 'small norm' of the function itself.

In approximation theory estimates for this purpose originated from the observation that a differentiable function cannot have large values everywhere if its derivative is bounded and it is zero on a dense discrete set, see [2] and [14] for functions on closed manifolds. A *sampling inequality* generalises this observation to arbitrary values of the function which was established in [3,13] and similar in [6].

More concrete one is looking for an inequality of the type

$$\|u\|_{W^{s,q}} \leq C_1(\mathcal{X})\|u\|_{W^{r,p}} + C_2(\mathcal{X})\|u\|_{\ell^k(\mathcal{X})} \tag{28}$$

where the constants are related to the distribution of the discrete set $\mathcal{X}$ and

$$\|u\|_{\ell^k(\mathcal{X})} := \begin{cases} \left(\sum_{x \in \mathcal{X}} |u(x)|^q\right)^{1/q} & , k \neq \infty, \\ \sup_{x \in \mathcal{X}} |u(x)| & , k = \infty, \end{cases} \tag{29}$$

is the ℓ_k-norm on the discrete set $\mathcal{X}$.

For a generalization to a large range of parameters see [9–12].

Suppose $\Omega \subset \mathbb{R}^d$ is a bounded Lipschitz domain with an interior cone condition, which means that there is an angle $\theta \in (0, \pi/2)$ and a radius r and a unit vector $\xi(x)$ such that the cone

$$C(x, \xi(x), \theta, r) = \{x + ty, y \in \mathbb{R}^d, \|y\| = 1, y^T\xi(x) \geq \cos(\theta), t \in [0, r]\} \tag{30}$$

is contained in Ω.

For a discrete set $\mathcal{A} \subset \Omega$ we define its *fill-distance* in Ω by

$$h(\mathcal{A}, \Omega) := \sup_{x \in \Omega} \min_{x_i \in \mathcal{A}} \|x - x_i\| \tag{31}$$

and its *separation distance* by

$$q(\mathcal{A}) = \frac{1}{2} \min_{x_i \neq x_j} \|x_i - x_j\|. \tag{32}$$

These structural constants describe the distribution of the discrete set in Ω. More concrete $h(\mathcal{A}, \Omega)$ represents the radius of the largest ball in Ω that contains no element of $\mathcal{A}$ and $q(\mathcal{A})$ measures the smallest distance between two points of the discrete set. As it is well known these two constants influence the error and stability of scattered data interpolation. For a good reference on this see [16].

The quotient $\frac{h(\mathcal{A},\Omega)}{q(\mathcal{A})}$ of these constants is called *mesh-ratio*. We will say that a family of discrete sets is *quasi-uniformely* distributed if the mesh-ratio is uniformly bounded which means

$$\frac{h(\mathcal{A}, \Omega)}{q(\mathcal{A})} \leq C \tag{33}$$

and the constant does not depend on the specific set $\mathcal{A}$.

In the same way we define those constants for discrete sets on a Riemannian manifold (X, g) by exchanging the Euclidean distance by the geodesic distance on X. More concrete for $K \subset X$ with non-empty interior and a discrete set $\mathcal{X} \subset K$ we set

$$h_g(\mathcal{X}, K) := \sup_{x \in K} \min_{x_i \in \mathcal{X}} \mathrm{d}_g(x, x_i) \quad \text{and} \quad q_g(\mathcal{X}) = \frac{1}{2} \min_{x_i \neq x_j} \mathrm{d}_g(x_i, x_j). \tag{34}$$

To derive a sampling inequality on the manifold (X, g) we will use Euclidean sampling inequalities locally in each chart. A sampling inequality that is appropriate for this is the following.

Proposition 2. *[12, Theorem 3.2 and Remark 3.4] Let $\Omega \subset \mathbb{R}^d$ be a bounded Lipschitz domain satisfying an interior cone condition with angle θ and radius ρ Suppose that $p, k \in [1, \infty]$, $q \in [1, \infty)$ and $r > d/p$ for $1 < p < \infty$, $r \geq d$ for $p = 1$ or $r \in \mathbb{N}^*$ in the case $p = \infty$. Then there exist constants $R > 1$ depending on d, r and $C > 0$ depending only on the domain and the parameters p, q, k, r, d such that for all discrete sets $\mathcal{A} \subset \Omega$ with $h = h(\Omega, \mathcal{A}) \leq \frac{\rho \sin(\theta)}{2R(1+\sin(\theta))}$ and any $u \in W^{r,p}(\Omega)$*

$$\|u\|_{W^{s,q}(\Omega)} \leq C\left(h^{r-s-d(1/p-1/q)_+}\|u\|_{W^{r,p}(\Omega)} + h^{d/\gamma-s}\|u\|_{\ell^k(\mathcal{A})}\right)$$

for all $0 \leq s \leq \lfloor r - d(1/p - 1/q)_+ \rfloor$ and $\gamma = \max\{p, q, k\}$.

Theorem 1. *Let (X, g) be a d-dimensional manifold of bounded geometry, $p, q, k \in [1, \infty]$ and $u \in W^{r,p}(X)$ for $r > d/p$ or $r \geq d$ in the case $p = 1$ with $\mathrm{supp}\, u \subseteq K$, $K \subset X$ compact. Suppose the discrete set $\mathcal{X} = \{x_i\}_{i \in I}$ has a fill-distance $h_g = h_g(\mathcal{X}, K)$ satisfying*

$$h_g \leq \frac{\varepsilon}{2Rc_2}\left(\frac{c_1 \sin(\pi/9)}{(1 + \sin(\pi/9))}\right)^2, \tag{35}$$

and is quasi-uniformely distributed, where $R > 1$ has the same meaning as in Proposition 2. Then there is a constant $C > 0$ such that for $0 \leq s \leq \lfloor r - d(1/p - 1/q)_+ \rfloor$

$$\|u\|_{W^{s,q}(X)} \leq C\left(h_g^{r-s-d(1/p-1/q)_+}\|u\|_{W^{r,p}(X)} + h_g^{d/\gamma-s}\|u\|_{\ell^k(\mathcal{X})}\right), \tag{36}$$

where $\gamma = \max\{p, q, k\}$.

Proof. The idea of the proof is to use locally the result of Proposition 2. Doing this one has to take care of the constants involved.

By definition of the Sobolev norm we can write for $\varepsilon < r_X$

$$\|u\|_{W^{s,q}(X)} = \left(\sum_j \|(\psi_j u) \circ \exp_{p_j}\|^q_{W^{s,q}(\mathbb{R}^n)} \right)^{1/q}. \tag{37}$$

Since the support of the function $u_j := (\psi_j u) \circ \exp_{p_j}$ is contained in $B := \exp_{p_j}^{-1}(U_j) = B_\varepsilon(0)$ we have

$$\|u_j\|_{W^{s,q}(\mathbb{R}^n)} = \|u_j\|_{W^{s,q}(B)}, \qquad \text{for all } j \in J,$$

Moreover there is a finite subset of the U_j with $K \subset \bigcup_{j \in J} U_j$.

Now to employ local sampling inequalities in each U_j that reflects the global behaviour we have to extend the set X. This means for $\bigcup_{j \in J} U_j$ take a point set $\tilde{X}$ with $K \cap \tilde{X} = X$ and fill-distance $h_g(\tilde{X}, \bigcup_j U_j)$ comparable to $h_g(X, K)$. This can be constructed by taking a maximal set of points that fulfills

$$q_g(X) \le q_g(\tilde{X}), \tag{38}$$

which always exists due to Zorn's Lemma. We have

$$h_g(\tilde{X}, \bigcup_{j \in J} U_j) \le h_g(X, K), \tag{39}$$

since otherwise we could add a point to $\tilde{X}$ and still satisfy (38), which contradicts the maximality of the set. On the other hand we also have

$$h_g(X, K) \le C q_g(X) \le C q_g(\tilde{X}) \le C h_g(\tilde{X}, \bigcup_j U_j), \tag{40}$$

where the constants comes from the quasi-uniformety condition on X, which means $\frac{h_g}{q_g} \le C$. This gives the comparability of $h_g(X, K)$ and $h_g(\tilde{X}, \bigcup_j U_j)$.

For abreviation we denote

$$\tilde{h}_g := h_g(\tilde{X}, \bigcup_j U_j). \tag{41}$$

In addition to use locally in the geodesic charts the Euclidean sampling inequality for each u_j we have to make sure that the local Euclidean fill-distances

$$\tilde{h}_j := \sup_{y \in B} \min_{x_i \in \tilde{X} \cap U_j} \|y - \exp_{p_j}^{-1}(x_i)\| \tag{42}$$

are small enough. Moreover the local geodesic distances

$$\tilde{h}_{g,j} := \sup_{x \in U_j} \min_{x_i \in \tilde{X} \cap U_j} d_g(x, x_i), \tag{43}$$

have to be comparable to the global fill-distance $\tilde{h}_g$.

Repeating the same argumentation as in the proof of [1, Theorem 8] one can show using the assumption

$$\tilde{h}_g \leq h_g \leq \frac{\varepsilon}{2Rc_2} \left(\frac{c_1 \sin(\pi/9)}{(1 + \sin(\pi/9))} \right)^2 \leq \frac{c_1 \varepsilon \sin(\pi/9)}{(1 + \sin(\pi/9))}$$

that there are at least two points in $\tilde{X} \cap U_j$ for each $j \in J$ and we can estimate

$$\tilde{h}_{g,j} \leq \frac{c_2}{c_1} \left(\frac{1 + \sin(\pi/9)}{\sin(\pi/9)} \right) \tilde{h}_g.$$

Consequently we have

$$\tilde{h}_j \leq \frac{1}{c_1} \tilde{h}_{g,j} \leq \frac{c_2}{c_1} \left(\frac{1 + \sin(\pi/9)}{\sin(\pi/9)} \right) \tilde{h}_g \leq \frac{\varepsilon \sin(\pi/9)}{2R(1 + \sin(\pi/9))},$$

which means that each $\tilde{h}_j$ fulfill the assumption of Proposition 2 w.r.t to the ball B as this satisfies an interior cone condition with radius ε for each angle $\theta \in (0, \pi/3)$.

On the other hand due to the quasi-uniform distribution of X the extended set $\tilde{X}$ is also quasi-uniformely distributed and therefore

$$\tilde{h}_g \leq C\tilde{q}_g \leq C\tilde{q}_{g,j} \leq C\tilde{h}_{g,j}, \qquad \forall j \in J,$$

where

$$\tilde{q}_{g,j} := \frac{1}{2} \min_{x_m \neq x_n} \mathrm{d}_g(x_m, x_n), \qquad x_n, x_m \in U_j \cap \tilde{X}, \tag{44}$$

denotes the local separation distance, which is well-defined due to the existence of at least two points in each U_j.

This leads to

$$\|u\|_{W^{s,q}(X)} \leq C\Big(\sum_{j \in J} \big(\tilde{h}_j^{r-s-d(1/p-1/q)_+} \|u_j\|_{W^{r,p}(B)} + \tilde{h}_j^{d/\gamma-s} \|\psi_j u\|_{\ell^k(\tilde{X} \cap U_j)} \big)^{1/q} \Big) \tag{45}$$

By the Minkowski inequality we get

$$\|u\|_{W^{s,q}(X)} \leq C\Big[\Big(\sum_{j \in J} \big(\tilde{h}_j^{r-s-d(1/p-1/q)_+} \|u_j\|_{W^{r,p}(B)} \big)^q \Big)^{1/q}$$

$$+ \Big(\sum_{j \in J} \big(\tilde{h}_j^{d/\gamma-s} \|\psi_j u\|_{\ell^k(\tilde{X} \cap U_j)} \big)^q \Big)^{1/q} \Big] \tag{46}$$

and by the metric-equivalence (21)

$$\|u\|_{W^{s,q}(X)} \leq C\Big[\big(\sup_{j \in J} \tilde{h}_{g,j} \big)^{r-s-d(1/p-1/q)_+} \Big(\sum_{j \in J} \|u_j\|_{W^{r,p}(B)}^q \Big)^{1/q}$$

$$+ \big(\inf_{j \in J} \tilde{h}_{g,j} \big)^{d/\gamma-s} \Big(\sum_{j \in J} \|\psi_j u\|_{\ell^k(\tilde{X} \cap U_j)}^q \Big)^{1/q} \Big]. \tag{47}$$

The first part can be estimated by comparing the ℓ^q and ℓ^p norm on $\mathbb{R}^J$. We have

$$\Big(\sum_{j\in J}\|u_j\|^q_{W^{r,p}(B)}\Big)^{1/q} \le |J|^{(1/p-1/q)+}\Big(\sum_{j\in J}\|u_j\|^p_{W^{r,p}(B)}\Big)^{1/p}$$

in the case $p < \infty$ and

$$\Big(\sum_{j\in J}\|u_j\|^q_{W^{r,p}(B)}\Big)^{1/q} \le |J|\sup_{j\in J}\|u_j\|_{W^{r,\infty}(B)}$$

if $p = \infty$.

The second part can be estimated in the same way, i.e.

$$\Big(\sum_{j\in J}\|\psi_j u\|^q_{\ell^k(\tilde{\mathcal{X}})}\Big)^{1/q} \le |J|^{(1/p-1/q)+}\Big(\sum_{j\in J}\|\psi_j u\|^k_{\ell^k(\tilde{\mathcal{X}})}\Big)^{1/k}$$

for $k \ne \infty$ and in the case $k = \infty$

$$\Big(\sum_{j\in J}\|\psi_j u\|^q_{\ell^k(\tilde{\mathcal{X}})}\Big)^{1/q} \le |J|\sup_{j\in J}\|\psi_j u\|_{\ell^\infty(\tilde{\mathcal{X}}\cap U_j)}.$$

Since the covering of the ε-balls is L-uniform locally finite each point $\tilde{x} \in \tilde{\mathcal{X}}$ is contained in at most L different charts U_j. This leads to

$$\Big(\sum_{j\in J}\|\psi_j u\|^k_{\ell^k(\tilde{\mathcal{X}})}\Big)^{1/k} \le L\|u\|_{\ell^k(\tilde{\mathcal{X}})}$$

and an analog estimate holds for $k = \infty$.

Moreover since $\tilde{\mathcal{X}} \cap \operatorname{supp} u = \mathcal{X}$ the norms agree, i.e.

$$\|u\|_{\ell^k(\mathcal{X})} = \|u\|_{\ell^k(\tilde{\mathcal{X}})}.$$

Combining all this facts yields

$$\|u\|_{W^{s,q}(X)} \le C\Big[\big(\sup_{j\in J}\tilde{h}_{g,j}\big)^{r-s-d(1/p-1/q)+}\|u\|_{W^{r,p}(X)} + \big(\inf_{j\in J}\tilde{h}_{g,j}\big)^{d/\gamma-s}\|u\|_{\ell^k(\mathcal{X})}\Big].$$

Using the comparability of the $\tilde{h}_{g,j}$ with $\tilde{h}_g$ and the comparability of $\tilde{h}_g$ with h_g results in

$$\|u\|_{W^{s,q}(X)} \le C\Big(h_g^{r-s-d(1/p-1/q)+}\|u\|_{W^{r,p}(X)} + h_g^{d/\gamma-s}\|u\|_{\ell^k(\mathcal{X})}\Big).$$

Remark 2. Since we restrict ourselves to functions having compact support the assumption of bounded geometry on the manifold seems unnecessary. Indeed for an arbitrary Riemannian manifold (X,g) choosing an arbitrary atlas (U_j,ϕ_j) one could define on a compact subset K local Sobolev spaces in the same way,

$$W^{s,q}(K) := \Big\{u \in L^q(X) : \operatorname{supp}(u) \subseteq K, \Big(\sum_{j\in J}\|(\psi_j u)\circ\phi_j^{-1}\|^q_{W^{s,q}(\mathbb{R}^n)}\Big)^{1/q} < \infty\Big\}, \tag{48}$$

where ψ_j denotes a subordinate partition of unity which is automatically uniform locally finite due to the finiteness of the sub covering $\{U_j\}_{j \in J}$. Now to employ local sampling inequalities one has to make sure that the sets $\phi_j(U_j)$ satisfy the assumption of Proposition 2, i.e. Lipschitz boundary and interior cone condition. To avoid this problem we restrict ourselves to manifolds of bounded geometry. Moreover the restriction on the curvature allows to transfer Theorem 1 to the case of functions with non-compact support, as those functions supported on the whole manifold. In this case the discrete set $\mathcal{X}$ is no longer finite. As long as the discrete norm $\|u\|_{\ell^k(\mathcal{X})}$ is finite the inequality (36) still holds. If the norm is automatically finite is not clear. However, in the case $X = \mathbb{R}^d$ a proof can be found in [9].

3.2 Error Estimate

In this section we will give an error estimate for the generalized interpolant f^+ which is defined for a given discrete set $\mathcal{X} = \{x_1, \ldots, x_N\} \subset X$ and values $\boldsymbol{u} = (u_i)_{i=1}^N \in \mathbb{R}^N$ by

$$f^+ = \sum_{i=1}^N a_i \lambda_i^y \Phi(x, y), \qquad \lambda_i(f) = (\delta_{x_i} \circ A)f, \tag{49}$$

and $\boldsymbol{a} = (a_i)_{i=1}^N$ is the solution of

$$\mathrm{M}\boldsymbol{a} = \boldsymbol{u}, \qquad \mathrm{M}_{i,j} = \langle \lambda_i^y \Phi(\cdot, y), \lambda_j^y \Phi(\cdot, y) \rangle_{\mathcal{N}_\Phi}. \tag{50}$$

Assuming that their is a solution to the continuous problem $Af = u$ and the data is given by $\boldsymbol{u} = (u(x_i))_{i=1}^N$ we have the following error estimate.

Theorem 2. *Suppose $f \in H^s(\Omega), s > d/2$, with $\mathrm{supp}(f) \subset K$, $K \subset \Omega$ compact and $Af = u$, where A is a linear operator fulfilling Assumption 2 for some compact set $\Lambda \subset X$. Furthermore assume that the kernel Φ satisfies Assumption 1 and the discrete set $\mathcal{X} \subset \Lambda$ satisfying Assumption 3 has a fill-distance $h_g = h_g(\mathcal{X}, \Lambda)$ sufficiently small.*
Then the generalized interpolant f^+ for the data $\boldsymbol{u} = (u(x))_{x \in \mathcal{X}}$ fulfils

$$\|f - f^+\|_{H^\alpha(\Omega)} \leq C h_g^{s-\alpha} \|f\|_{H^s(\Omega)} \tag{51}$$

for all $\alpha \geq 0$ such that $\alpha + \tau \leq \lfloor s + \tau \rfloor$ where τ denotes the order of A.

Proof. By *Assumption 2* we can write

$$\|f - f^+\|_{H^\alpha(\Omega)} \leq C \|A(f - f^+)\|_{H^{\alpha+\tau}(X)}. \tag{52}$$

Since $\mathrm{supp}(A(f - f^+)) \subset \Lambda$ we can apply Theorem 1 for a fill-distance sufficiently small to get

$$\|f - f^+\|_{H^s(\Omega)} \leq C\left(h_g^{s-\alpha} \|A(f - f^+)\|_{H^{s+\tau}(X)} + h_g^{d/2-\alpha} \|A(f - f^+)\|_{\ell^2(\mathcal{X})}\right). \tag{53}$$

Since f^+ is a generalized interpolant we have $\|A(f - f^+)\|_{\ell^2(\mathcal{X})} = 0$ and thus again by *Assumption* 2 we have

$$\|f - f^+\|_{H^s(\Omega)} \le Ch_g^{s-\alpha}\|(f - f^+)\|_{H^s(\Omega)}. \tag{54}$$

Since the kernel satisfies *Assumption* 1 and thus the native space $\mathcal{N}_\Phi$ is norm-equivalent to the Sobolev space we can estimate

$$\|f - f^+\|_{H^s(\Omega)} \le Ch_g^{s-\alpha}\|(f - f^+)\|_{\mathcal{N}_\phi}. \tag{55}$$

Since f^+ is the norm-minimal solution and $\mathcal{N}_\Phi$ is norm-equivalent to $H_K^s(\Omega)$ we have

$$\|f - f^+\|_{H^s(\Omega)} \le Ch_g^{s-\alpha}\|f\|_{\mathcal{N}_\Phi} \le Ch_g^{s-\alpha}\|f\|_{H^s(\Omega)}. \tag{56}$$

Remark 3. In application one cannot expect to measure the exact values of the 'ground truth' u but perturbed values $\boldsymbol{u}^\varepsilon$. It is not useful to interpolate these perturbed data so one has to stabilise the reconstruction. One way to do this is to determine the solution of the system

$$(\mathrm{M} + \gamma\,\mathrm{Id})\boldsymbol{a} = \boldsymbol{u}^\varepsilon. \tag{57}$$

This corresponds to the well-known *Thikonov-Phillips regularization* and $\gamma > 0$ is called regularization parameter. For a discussion on error estimates for this regularization method and the choice of optimal parameters see [4].

Acknowledgement. The author would like to thank Grady B. Wright for the possibility to gave a talk in the minisymposium 'Kernel Based Approximation Methods' at the conference 'Curves and Surfaces' 2014. Moreover he would like to thank Frank Filbir for fruitful discussions on the topic and careful proofreading.

References

1. Fuselier, E., Wright, G.B.: Scattered data interpolation on embedded submanifolds with restricted positive definite kernels: Sobolev error estimates. SIAM J. Numer. Anal. **50**(3), 1753–1776 (2012)
2. Narcowich, F.J., Ward, J.D., Wendland, H.: Sobolev bounds on functions with scattered zeros, with applications to radial basis function surface fitting. Math. Comput. **74**, 743–763 (2005)
3. Rieger, C., Wendland, H.: Approximate interpolation with applications to selecting smoothing parameters. Numer. Math. **101**(4), 729–748 (2005)
4. Krebs, J., Louis, A.K., Wendland, H.: Sobolev error estimates and a priori parameter selection for semi-discrete Thikhonov regularization. J. Inverse Ill-Posed Probl. **17**(9), 845–869 (2009)
5. Golomb, M., Weinberger, H.F.: Optimal approximations and error bounds. DTIC Document (1958)
6. Madych, W.: An estimate for multivariate interpolation II. J. Approximation Theor. **142**, 116–128 (2006)

7. Natterer, F.: Mathematics of Computerized Tomography. Teubner, Stuttgart (1986)
8. Palamodov, V.: Remarks on the general funk transform and thermoacoustic tomography. Inverse Probl. Imaging **4**(4), 693–702 (2010)
9. Arcangéli, R., Torrens, J.J.: Sampling inequalities in Sobolev spaces. J. Approximation Theor. **182**, 18–28 (2014)
10. Arcangéli, R., Torrens, J.J., de Silanes, M.C.L.: An extension of a bound for functions in Sobolev spaces with applications to (m, s)-spline interpolation and smoothing. Numer. Math. **107**(2), 181–211 (2007)
11. Arcangéli, R., Torrens, J.J., de Silanes, M.C.L.: Estimates for functions in Sobolev spaces defined on unbounded domains. J. Approximation Theor. **161**, 198–212 (2009)
12. Arcangéli, R., Torrens, J.J., de Silanes, M.C.L.: Extension of sampling inequalities to Sobolev semi-norms of fractional order and derivative data. Numer. Math. **121**(3), 587–608 (2011)
13. Rieger, C.: Sampling inequalities and applications. Ph.D thesis. University Göttingen (2008)
14. Hangelbroek, T., Narcowich, F.J., Ward, J.: Polyharmonic and related kernels on manifolds: interpolation and approximation. Found. Comput. Math. **12**(5), 625–670 (2012)
15. Triebel, H.: Theory of Function Spaces 2. Birkhäuser, Basel (1992)
16. Wendland, H.: Scattered Data Approximation. Cambridge University Press, Cambridge (2005)
17. Rieger, C., Schaback, R., Zwicknagl, B.: Sampling and stability. In: Dæhlen, M., Floater, M., Lyche, T., Merrien, J.-L., Mørken, K., Schumaker, L.L. (eds.) MMCS 2008. LNCS, vol. 5862, pp. 347–369. Springer, Heidelberg (2010)

Boundary Controlled Iterated Function Systems

Dmitry Sokolov[1]($\boxtimes$), Gilles Gouaty[2], Christian Gentil[3], and Anton Mishkinis[3]

[1] LORIA - Alice, Campus Scientifique, BP 239, 54506
Vandoeuvre-les-Nancy Cedex, France
dmitry.sokolov@loria.fr
[2] Laboratoire des Sciences de l'Information et des Systémes – UMR 7296, Polytech
Marseille Campus de Luminy, Case postale 925, 13288 Marseille Cedex, France
gilles_gouaty@yahoo.fr
[3] Laboratoire LE2I – UMR CNRS 6306, Faculté des Sciences Mirande,
Aile de l'Ingénieur, 21078 Dijon Cedex, France
{christian.gentil,anton.mishkinis}@u-bourgogne.fr

Abstract. Boundary Controlled Iterated Function Systems is a new layer of control over traditional (linear) IFS, allowing creation of a wide variety of shapes. In this work, we demonstrate how subdivision schemes may be generated by means of Boundary Controlled Iterated Function Systems, as well as how we may go beyond the traditional subdivision schemes to create free-form fractal shapes. BC-IFS is a powerful tool allowing creation of an object with a prescribed topology (e.g. surface patch) independent of its geometrical texture. We also show how to impose constraints on the IFS transformations to guarantee the production of smooth shapes.

Keywords: Subdivision surfaces · Fractals · Iterated function system · B-splines

1 Motivation

Objects modeled through Computer Aided Geometric Design (CAGD) systems are often inspired by standard machining processes. However, other types of objects, such as objects with a porous structure or with a rough surface, may be interesting to create: porous structures can be used for their lighter weight while maintaining satisfactory mechanical properties, rough surfaces can be used for acoustic absorption.

Fractal geometry is a relatively new branch of mathematics that studies complex objects of non-integer dimensions. Because of their specific physical properties, fractal-like structure is a centre of interest in numerous areas such as architecture [18], jewellery [19], heat and mass transport [14] or antennas [6,17].

The emergence of techniques such as 3D printers allow for new possibilities that are as yet unused and even unexplored. Different mathematical models and algorithms have been developed to generate fractals. We can roughly categorize them into three families. The first gathers algorithms computing the basins of

© Springer International Publishing Switzerland 2015
J.-D. Boissonnat et al. (Eds.): Curves and Surfaces 2014, LNCS 9213, pp. 414–432, 2015.
DOI: 10.1007/978-3-319-22804-4_29

attraction of a given function. Julia sets and the Mandelbrot set [13] or Mandelbulb [1] are some examples. The second is based on simulation of phenomena such as percolation or diffusion [7]. The last one corresponds to deterministic or probabilistic algorithms or models based on the self-similarity property associated fractals like the terrain generator [24], Iterated Function System [3], L-system [16]. Shapes are generated from rewriting rules providing control of the geometry. Nonetheless, most of these models were developed for image synthesis without consideration of fabricability or were developed for very specific applications like Wood modeling [20].

Some studies address this aspect for specific applications for 3D printers [19]. In [5] Barnsley defines fractal homeomorphisms from $[0, 1]^2$ onto the modeling space $[0, 1]^2$. The same approach is used in 3D to build 3D fractals. A 3D standard object is embedded in the domain space $[0, 1]^3$ and then transformed into a 3D fractal object. This approach preserves the topology of the initial object which is an important point for fabricability. The control of the resulting geometry, however, is induced by the definition of the homeomorphism and this is not obvious. And finally, the definition of the initial topology is left up to the user.

We elaborate here a new type of modeling system, using the facilities of existing CAGD software, while extending their capabilities and their application areas. This new type of modeling system will offer designers (engineers in industry) and creators (visual artists, stylists, designers, architects, etc.) new opportunities to design and produce a quick mock-up, a prototype or a single object. Our approach is to expand the possibilities of a standard CAD system by including fractal shapes while preserving ease of use for end users. We enrich Iterated Function Systems by introducing Boundary Representation concepts.

The following section introduces the necessary background information related to Iterated Function Systems and Controlled Iterated Function Systems. Section 3 presents the Boundary Controlled Iterated Function System to control the topology of fractal shapes during the subdivision process.

2   Background

2.1   Iterated Function Systems

Iterated Function Systems (IFS) were introduced by Hutchinson [10] and further developed and popularized by Barnsley [4]. More research has followed on from these seminal studies [3,8,11,22].

IFS are based on the self-similarity property. A modeled object is made up of the union of several copies of itself; each copy is transformed by a function. These functions are usually contractive, that is to say they bring points closer together and make shapes smaller. Hence, the modeled object, called the *attractor*, is made up of several possibly overlapping smaller copies of itself, each copy also made up of copies of itself, ad infinitum.

Definition. Given a complete metric space $(\mathbb{X}, d)$ with the associated metric d, an IFS is defined by a finite set of continuous transformations $T = \{T_i\}_{i=0}^{N-1}$ in the space $\mathbb{X}$. Let $\Sigma = \{0, \ldots, N-1\}$ be the set of IFS transformation indices, thus $|\Sigma| = N$. The IFS is then denoted by $\{\mathbb{X}; T_i \mid i \in \Sigma\}$.

A simple example of an IFS can be constructed for a representation of real numbers in $[0, 1]$. Let

$$T_i : x \rightarrow \frac{x+i}{3} : [0, 1] \rightarrow [0, 1],$$

where $i \in \Sigma = \{0, 1, 2\}$. The IFS $\{[0, 1]; T_0, T_1, T_2\}$ can thus express the representation of any real number in $[0, 1]$.

We are substantially interested in so-called *hyperbolic* IFS, defined as those whose transformations T_i are all contractive.

Definition. A transformation $T : \mathbb{X} \rightarrow \mathbb{X}$ is called *contractive* if and only if there exists a real s, $0 \leqslant s < 1$ such that $d(T(x), T(y)) < s \cdot d(x, y)$ for all $x, y \in \mathbb{X}$.

Definition. For the set of non-empty compacts of $\mathbb{X}$, denoted $\mathcal{H}(\mathbb{X})$, we define the Hausdorff distance $d_{\mathbb{X}}$ induced by the metric d:

$$d_{\mathbb{X}}(A, B) = \max\{\sup_{a \in A} \inf_{b \in B} d(a, b), \sup_{b \in B} \inf_{a \in A} d(a, b)\}.$$

Since the metric space $(\mathbb{X}, d)$ is complete, the set $(\mathcal{H}(\mathbb{X}), d_{\mathbb{X}})$ is also complete.

In the early 80's, Hutchinson [10] used the Banach fixed point theorem to deduce the existence and the uniqueness of an attractor for a hyperbolic IFS, i.e. the fixed point of the associated contractive map. He defined an operator $\mathbb{T} : \mathcal{H}(\mathbb{X}) \rightarrow \mathcal{H}(\mathbb{X})$, now called Hutchinson operator [4], as the union of the IFS transformations T_i:

$$\mathbb{T}(K) = \bigcup_{i=0}^{N-1} T_i(K).$$

If IFS is hyperbolic then $\mathbb{T}$ is also contractive in the complete metric space $(\mathcal{H}(\mathbb{X}), d_{\mathbb{X}})$. According to the Banach fixed point theorem [4], $\mathbb{T}$ has a unique fixed point $\mathcal{A}$. This fixed point is named the attractor of the IFS:

$$\mathcal{A} = \mathbb{T}(\mathcal{A}) = \bigcup_{i=0}^{N-1} T_i(\mathcal{A}). \tag{1}$$

Since the Hutchinson operator is contractive, the attractor of an IFS can be evaluated recursively. That is, the attractor can be approximated by a sequence $\{K_n\}_{n \in \mathbb{N}}$ converging to $\mathcal{A}$. The initial element in the sequence is defined by means of a primitive $K \in \mathcal{H}(\mathbb{X})$. The following elements are defined recursively:

$$K_0 = K$$
$$K_{n+1} = \bigcup_{i \in \Sigma} T_i(K_n).$$

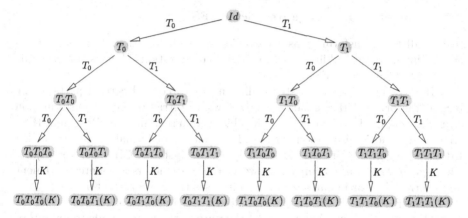

Fig. 1. The IFS evaluation tree calculated to the third level. Internal nodes correspond to the calculation of a composite function. Leaves correspond to subsets of K_3^a to construct or to visualize.

The elements K_n are images of composite functions applied to K.

Each element in the sequence represents an approximation of the IFS attractor. Each term K_n is composed of N^n images of K by a composite of n functions. For example, a sequence of the attractor approximations for an IFS $\{\mathbb{X}; T_0, T_1\}$ is presented here:

$$
\begin{aligned}
K_0 &= K, \\
K_1 &= \mathbb{T}(K_0) & &= T_0(K) \cup T_1(K), \\
K_2 &= \mathbb{T}(K_1) & &= T_0 T_0(K) \cup T_0 T_1(K) \cup T_1 T_0(K) \cup T_1 T_1(K), \\
K_3 &= \mathbb{T}(K_2) & &= T_0 T_0 T_0(K) \cup T_0 T_0 T_1(K) \cup T_0 T_1 T_0(K) \cup T_0 T_1 T_1(K) \cup \\
& & & \quad T_1 T_0 T_0(K) \cup T_1 T_0 T_1(K) \cup T_1 T_1 T_0(K) \cup T_1 T_1 T_1(K), \\
&\;\vdots & &\;\;\vdots \\
K_n &= \mathbb{T}(K_{n-1}) = \bigcup_{\alpha_i \in \{0,1\}} T_{\alpha_1} \ldots T_{\alpha_n}(K_n).
\end{aligned}
$$

In this iterative algorithm a set of transformed primitives K is constructed recursively and calculations can be represented by an *evaluation tree*. Each node on the i-th level of the tree corresponds to the image of a composite of i IFS transformations. This tree is traversed up to a given depth n, where we display the image of K by the composite function associated with the current node, as shown in Fig. 1.

Note that these composite functions are calculated from left to right. The primitive K is transformed finally by a constructed composite function. In practice, the IFS transformations T_i are affine operators and can therefore be represented by matrices. A composite affine transformation can thus be represented by a product of transformation matrices.

2.2   Controlled/Language Restricted IFS

In IFS all the transformations are applied on each iteration. It is possible to enrich this model by adding rules to control the iterations. This is the principle of a *CIFS* (Controlled IFS).

CIFS are more general systems allowing us to control certain parts of the IFS attractor. A CIFS denotes an IFS with restrictions on transformation sequences imposed by a control graph. This system is similar to "Recurrent IFS" (RIFS) [4], and is also described [15,21] by means of formal languages, called LRIFS (Language-Restricted Iterated Function System). CIFS defines objects whose geometry can be complex. However CIFS attractors are more convenient and controllable for manufacturability purposes than IFS attractors.

The attractor of a CIFS can be evaluated by an automaton [12] defined on the control graph. Each validated word of the automaton corresponds to an authorized composition of transformations. Each state of the automaton corresponds to different parts of the modeled object. States are associated with construction spaces. Transitions between states indicate that one sub-part is contained in another one. It is then possible to control the attractor more precisely.

Definition. A CIFS is given by an automaton, where each state q is associated with an attractor $\mathcal{A}^q \in \mathbb{X}^q$, and each transition from q to w is associated with an operator $\mathbb{X}^w \to \mathbb{X}^q$. The following is a list of parameters describing the CIFS:

- An automaton (Σ, Q, δ), where Σ is an alphabet, Q is a set of states and δ is a transition function $\delta : Q \times \Sigma \to Q$;
- A set of complete metric spaces associated with the automaton states $\{\mathbb{X}^q\}_{q \in Q}$;
- An operator associated with each transition $T_i^q : \mathbb{X}^{\delta(q,i)} \to \mathbb{X}^q$;
- A compact set $K^q \in \mathcal{H}(\mathbb{X}^q)$, called a *primitive*, associated with each state $q \in Q$. Primitives are not used to define the attractor, but only to approximate it;
- Finally, the automaton is provided by an initial state, noted by ♮, and all states are final states.

In the following, we denote by Σ^q the restriction of Σ by outgoing transitions from the state q, i.e.:

$$\Sigma^q = \{i \in \Sigma,\ \delta(q,i) \in Q\}$$

CIFS defines a family of attractors associated with the states: $\{\mathcal{A}^q\}_{q \in Q}$, where $\mathcal{A}^q \in \mathcal{H}(\mathbb{X}^q)$. The attractors $\mathcal{A}^q$ are mutually defined recursively:

$$\mathcal{A}^q = \bigcup_{i \in \Sigma^q} T_i^q(\mathcal{A}^{\delta(q,i)}) \tag{2}$$

As for IFS, each CIFS attractor can be approximated by a sequence $\{K_n^q\}_{n \in \mathbb{N}}$ converging to $\mathcal{A}^q$. Each state $q \in Q$ is associated with a primitive $K^q \in \mathcal{H}(\mathbb{X}^q)$,

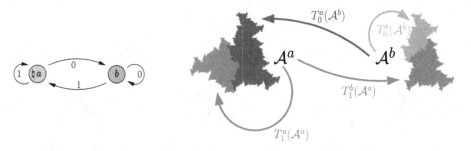

Fig. 2. Fractal kite and dart for Penrose tilings. The CIFS automaton is shown on the left, the attractors with the transformations are shown on the right.

which defines the initial element in the sequence. The following elements are mutually defined recursively:

$$K_0^q = K^q$$
$$K_{n+1}^q = \bigcup_{i \in \Sigma^q} T_i^q(K_n^{\delta(q,i)})$$

In this iterative algorithm, a set of transformed primitives K^q is recursively constructed and the calculations can also be represented by an *evaluation tree*. Each node on the i-th level of the tree corresponds to the image of a composition of i CIFS transformations, i.e. a path of length i in the automaton. This tree is traversed up to a given depth n, where we display the image of K^q by the composite function associated with the current node.

Example 1. Consider an example of the CIFS attractor, illustrated in the right-hand image in Fig. 2. This was introduced by Bandt and Gummelt [2]. The system is described by an automaton with two states: a and b. There are two subdividing operators from state a as well as from b. The left panel in Fig. 2 shows the automaton of this CIFS.

The automaton transition functions are the following:

$$\delta(a,0) = b \qquad \delta(b,0) = b$$
$$\delta(a,1) = a \qquad \delta(b,1) = a$$

Each transition $\delta(q,i) = w$ of the automaton is associated with an operator $T_i^q : \mathbb{X}^w \to \mathbb{X}^q$. **N.B.:** *T and δ act in opposite directions!*

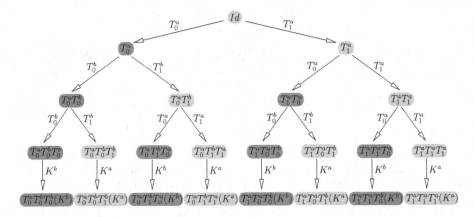

Fig. 3. CIFS evaluation tree calculated to the third level. Internal nodes correspond to the calculation of a composite function. Leaves correspond to subsets of K_3 to construct or to visualize (Color figure online).

In this example, $\mathbb{X}^a$ and $\mathbb{X}^b$ are both in the same Euclidean affine plane. Let us define the mappings as follows:

$$T_0^a\left(\begin{bmatrix} x \\ y \end{bmatrix}\right) = \begin{bmatrix} x \\ y \end{bmatrix}$$

$$T_1^a\left(\begin{bmatrix} x \\ y \end{bmatrix}\right) = \frac{2}{1+\sqrt{5}}\begin{bmatrix} \cos(3/5\pi) & -\sin(3/5\pi) \\ \sin(3/5\pi) & \cos(3/5\pi) \end{bmatrix}\begin{bmatrix} x \\ y \end{bmatrix} + \begin{bmatrix} 0 \\ 1 \end{bmatrix}$$

$$T_0^b\left(\begin{bmatrix} x \\ y \end{bmatrix}\right) = \frac{2}{1+\sqrt{5}}\begin{bmatrix} \cos(4/5\pi) & -\sin(4/5\pi) \\ \sin(4/5\pi) & \cos(4/5\pi) \end{bmatrix}\begin{bmatrix} x \\ y \end{bmatrix} + \begin{bmatrix} \frac{1+\sqrt{5}}{2}\sin(4/5\pi) \\ 1 + \frac{1+\sqrt{5}}{2}\cos(4/5\pi) \end{bmatrix}$$

$$T_1^b\left(\begin{bmatrix} x \\ y \end{bmatrix}\right) = \frac{2}{1+\sqrt{5}}\begin{bmatrix} \cos(7/5\pi) & -\sin(7/5\pi) \\ \sin(7/5\pi) & \cos(7/5\pi) \end{bmatrix}\begin{bmatrix} x \\ y \end{bmatrix} + \begin{bmatrix} 0 \\ 1 \end{bmatrix}$$

Thus, the attractors $\mathcal{A}^a$ and $\mathcal{A}^b$ satisfy the following equations:

$$\mathcal{A}^a = \bigcup_{i \in \Sigma^a} T_i^a(\mathcal{A}^{\delta(a,i)}) = T_1^a(\mathcal{A}^a) \cup T_0^a(\mathcal{A}^b)$$

$$\mathcal{A}^b = \bigcup_{i \in \Sigma^b} T_i^b(\mathcal{A}^{\delta(b,i)}) = T_0^b(\mathcal{A}^b) \cup T_1^b(\mathcal{A}^a)$$

Figure 3 shows the CIFS evaluation tree calculated to the third level. Internal nodes correspond to the calculation of a composite function, while the leaves correspond to subsets of K_3^a to construct or to visualize. The pink and blue highlights help distinguish between current state (space) a and b, respectively.

Example 2. Our second CIFS example is a simple uniform quadratic B-spline curve with 4 control points. Figure 4 gives the automaton (left) and the attractors with corresponding transformations (right). This is a special kind of CIFS, sometimes called Projected IFS in the literature [9,23].

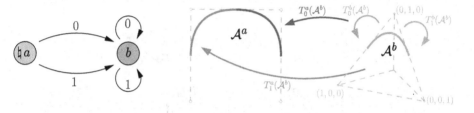

Fig. 4. The $2D$ uniform quadratic B-spline curve can be defined as a projection of an attractor from the three-dimensional barycentric space.

The automaton transition functions are the following:

$$\delta(a,0) = b \qquad\qquad \delta(b,0) = b$$
$$\delta(a,1) = b \qquad\qquad \delta(b,1) = b$$

Now the space associated with state a is still the Euclidean plane $\mathbb{R}^2$, while a three-dimensional barycentric space is associated with state b. Given the coordinates of the 4 control points P_0, P_1, P_2 and P_3, we can express the transformations as follows:

$$T_0^a\left(\begin{bmatrix} x \\ y \\ z \end{bmatrix}\right) = \begin{bmatrix} P_0^x & P_1^x & P_2^x \\ P_0^y & P_1^y & P_2^y \end{bmatrix}\begin{bmatrix} x \\ y \\ z \end{bmatrix} \qquad T_1^a\left(\begin{bmatrix} x \\ y \\ z \end{bmatrix}\right) = \begin{bmatrix} P_1^x & P_2^x & P_3^x \\ P_1^y & P_2^y & P_3^y \end{bmatrix}\begin{bmatrix} x \\ y \\ z \end{bmatrix}$$

$$T_0^b\left(\begin{bmatrix} x \\ y \\ z \end{bmatrix}\right) = \begin{bmatrix} 3/4 & 1/4 & 0 \\ 1/4 & 3/4 & 3/4 \\ 0 & 0 & 1/4 \end{bmatrix}\begin{bmatrix} x \\ y \\ z \end{bmatrix} \qquad T_1^b\left(\begin{bmatrix} x \\ y \\ z \end{bmatrix}\right) = \begin{bmatrix} 1/4 & 0 & 0 \\ 3/4 & 3/4 & 1/4 \\ 0 & 1/4 & 3/4 \end{bmatrix}\begin{bmatrix} x \\ y \\ z \end{bmatrix}$$

The transformations T_0^a and T_1^b are two projections of the same attractor representing the basis functions of the uniform quadratic B-spline illustrated in Fig. 4 (on the left side). Figure 5 gives the evaluation tree for the approximation K_3^a. Note that besides the 1st level of subdivision this is an ordinary IFS (all nodes are blue). The attractors of a CIFS are uniquely defined if operators associated with each cycle in the control graph are contractive. Hence, we do not have any constraints on the coordinates of the control points, as T_i^a do not appear in any cycle.

3   Boundary Controlled IFS

IFS and CIFS can model complex shapes; it is, however, difficult to control their topological properties. These shapes are determined by a set of geometry operators. Modifying these operators leads to both global and local changes in the shape and affects not only geometry but also topology. In order to control the topological structure of the modeled shape, we enrich CIFS by integrating a topological model to obtain *BCIFS* (Boundary Controlled Iterated Function System).

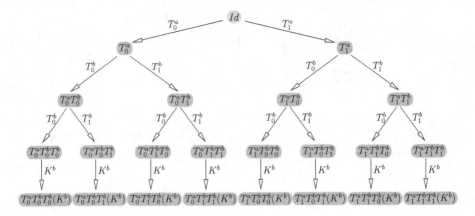

Fig. 5. CIFS evaluation tree calculated to the third level. Note that the pattern of pink/blue nodes has changed completely from the previous example (Color figure online).

In standard CAD systems, topology and geometric properties of shapes are separated. The topological structure is encoded by a set of topological cells (faces, edges, vertices) interconnected by a set of incidence and adjacency relations. The incidence relations are based on the nesting of cells: each face is bounded by a set of edges, and each edge is bounded by two vertices. The adjacency relations are based on sharing cells: two adjacent faces share a common edge, and two adjacent edges are bounded by a common vertex.

Inspired by this approach, we propose to extend the CIFS model by integrating B-rep relations. BCIFS is thus an extension of a CIFS enriched by a description of topology. Sub-parts of the attractor are identified as topological cells by specifying incidence constraints. These cells are assembled during the subdivision process by adjacency constraints. These constraints induce constraints on the subdivision operators of the CIFS.

Our B-rep structure is more general than the standard one. A topological cell may be fractal. For example, a face can be the Sierpinski triangle or an edge can be the Cantor set, but the topological structure remains consistent. Each topological cell corresponds to an attractor in a certain space.

3.1 Specifying the Topology

There are two types of transitions in the BCIFS automaton:

- transitions subdividing a topological cell;
- transitions embedding a topological cell in another one.

The alphabet Σ is also divided into:

- symbols of subdivision $\Sigma_{\div} = \{\div_i \mid i = 0, \cdots, n_{\div}\}$;
- symbols of incidence $\Sigma_{\partial} = \{\partial_i \mid i = 0, \cdots, n_{\partial}\}$.

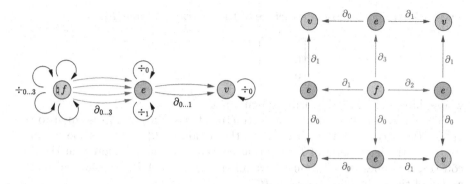

Fig. 6. Left image: automaton representing a quad patch defined by its boundaries. Right image: expanded incidence relations of the automaton.

Each subdividing transition $\delta(q, \div_i) = w$ is associated with a subdividing operator $T_i^q : \mathbb{X}^w \to \mathbb{X}^q$, where $q, w \in Q$ and $\div_i \in \Sigma_\div$. Similarly, each embedding transition $\delta(q, \partial_i) = w$ is associated with an embedding operator $B_i^q : \mathbb{X}^w \to \mathbb{X}^q$, where $q, w \in Q$ and $\partial_i \in \Sigma_\partial$.

Example. Let us illustrate the idea with an example. In this section we generate a continuous patch of a freeform surface with 9 control points. This patch will be defined as the attractor $\mathcal{A}^f$ of the IFS $(\mathbb{X}^f; T_0^f, T_1^f, T_2^f, T_3^f)$, let us call it "facet". We construct it as a B-rep structure with "edges" corresponding to the attractor $\mathcal{A}^e$ of the IFS $(\mathbb{X}^e; T_0^e, T_1^e)$ and "vertices" corresponding to the attractor $\mathcal{A}^v$ of the IFS $(\mathbb{X}^v; T_0^v)$.

Figure 6 gives the corresponding automaton. Note that the automaton has a hierarchical structure: there are three separate IFS linked by incidence operators. We omit the evident step of projecting the attractors into the modeling space. Refer to Fig. 4 to see how the projection is carried out in general.

Incidence Constraints. We start with the definition of the vertex attractor $\mathcal{A}^v$. To keep the example simple, we choose $\mathbb{X}^v$ to be a 1-dimensional barycentric space and T_0^v is simply a 1×1 identity matrix. The edge attractor will be defined in a 3-dimensional barycentric space.

We choose the inclusion of $\mathcal{A}^v$ (recall that it is just a point) inside the attractor $\mathcal{A}^e$, it defines boundaries of the edge $\mathcal{A}^e$. Let us say we want the vertex to be included twice at the coordinates $(1, 0, 0)$ and $(0, 0, 1)$. That is to say, we need certain constraints on the matrices T_0^e and T_1^e to force points $(1, 0, 0)$ and $(0, 0, 1)$ to belong to the attractor $\mathcal{A}^e$.

Let us assume $B_0^e = \begin{bmatrix} 1 & 0 & 0 \end{bmatrix}^\top$ and $B_1^e = \begin{bmatrix} 0 & 0 & 1 \end{bmatrix}^\top$. Now we can express first incidence constraints as

$$B_0^e T_0^v = T_0^e B_0^e$$
$$B_1^e T_0^v = T_1^e B_0^e$$

These constraints impose a structure of the matrices T_0^e and T_1^e:

$$T_0^e = \begin{bmatrix} 1 & \cdot & \cdot \\ 0 & \cdot & \cdot \\ 0 & \cdot & \cdot \end{bmatrix} \quad T_1^e = \begin{bmatrix} \cdot & \cdot & 0 \\ \cdot & \cdot & 0 \\ \cdot & \cdot & 1 \end{bmatrix},$$

where dots stand for arbitrarily chosen reals.

The subset of $\mathcal{A}^e$ defined as an attractor of the IFS $(\mathbb{X}^e; T_0^e)$ is equal to the $B_0^e \mathcal{A}^v$, the attractor $\mathcal{A}^v$ embedded by the action of B_0^e. In the same manner, $(0, 0, 1)$ is the fixed point of T_1^e and thus contained in the $\mathcal{A}^e$. Note that the first constraint implies that T_0^e must have all eigenvectors of T_0^v transformed by the action of the embedding operator B_0^e.

Property. More generally, let us show that the incidence constraints force the inclusion of the boundary CIFS attractors into the corresponding cell. If a cell $\mathcal{A}^Y$ has a number of boundaries defined by attractors $\mathcal{A}^{X_i}$, then we want to show that each $\mathcal{A}^{X_i}$ (when embedded in the space associated with state Y) is a sub part of $\mathcal{A}^Y$: in other words, we want to show that the inclusion $B_i^Y \mathcal{A}^{X_i} \subset \mathcal{A}^Y$ holds.

The incidence constraints have the following expression:

$$B_i^Y T_j^{X_i} = T_{f(i,j)}^Y B_{g(i,j)}^Y,$$

with $i \in \Sigma_\partial^Y$ and $j \in \Sigma_{\div}^{X_i}$. The functions f and g are simply the corresponding ordering of the boundaries and subdivisions. For example, for a square patch with four boundaries, each subdivided by two operators, we have $|\Sigma_\partial^Y| \times |\Sigma_{\div}^{X_i}| = 4 \times 2$ constraints.

For each boundary embedding we can write the following:

$$B_i^Y \mathcal{A}^{X_i} = B_i^Y \bigcup_{j \in \Sigma_{\div}^{X_i}} T_j^{X_i} \mathcal{A}^{X_i} = \bigcup_{j \in \Sigma_{\div}^{X_i}} B_i^Y T_j^{X_i} \mathcal{A}^{X_i} = \bigcup_{j \in \Sigma_{\div}^{X_i}} T_{f(i,j)}^Y B_{g(i,j)}^Y \mathcal{A}^{X_i}.$$

This means that the boundary $B_i^Y \mathcal{A}^{X_i}$ can be obtained as a union of other boundaries $B_{g(i,j)}^Y \mathcal{A}^{X_i}$ under the action of the subdivision operators. We can repeat this process ad infinitum. This means that by restricting the generated language, every boundary $B_i^Y \mathcal{A}^{X_i}$ can be generated solely by operators T^Y and therefore the inclusion $B_i^Y \mathcal{A}^{X_i} \subset \mathcal{A}^Y$ holds.

We can choose any real values for the dots in the expression of T_0^e and T_1^e, the attractor $\mathcal{A}^e$ will include two points $(1, 0, 0)$ and $(0, 0, 1)$. Note that at this point the attractor $\mathcal{A}^e$ can be a disjoint set of points. In the following subsection we add an adjacency constraint that will enable the attractor $\mathcal{A}^e$ to be a continuous curve. Figure 7 provides an example of the attractor $\mathcal{A}^e$ with the subdivision operators chosen as follows:

$$T_0^e = \begin{bmatrix} 1 & 1/2 & 1/4 \\ 0 & 1/2 & 1/2 \\ 0 & 0 & 1/4 \end{bmatrix} \quad T_1^e = \begin{bmatrix} 1/2 & 0 & 0 \\ 1/4 & 1/2 & 0 \\ 1/4 & 1/2 & 1 \end{bmatrix},$$

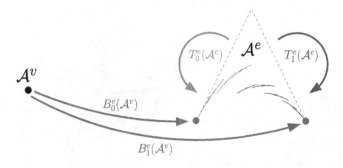

Fig. 7. Incidence constraints ensure the inclusion $B_0^e \mathcal{A}^v \subset \mathcal{A}^e$ and $B_1^e \mathcal{A}^v \subset \mathcal{A}^e$, however they do not guarantee the connectivity of the attractor $\mathcal{A}^e$.

Let us define edge-to-facet embedding operators:

$$B_0^f = \begin{bmatrix} 1\,0\,0\,0\,0\,0\,0\,0\,0 \\ 0\,1\,0\,0\,0\,0\,0\,0\,0 \\ 0\,0\,1\,0\,0\,0\,0\,0\,0 \end{bmatrix}^\top \qquad B_1^f = \begin{bmatrix} 1\,0\,0\,0\,0\,0\,0\,0\,0 \\ 0\,0\,0\,1\,0\,0\,0\,0\,0 \\ 0\,0\,0\,0\,0\,0\,1\,0\,0 \end{bmatrix}^\top$$

$$B_2^f = \begin{bmatrix} 0\,0\,1\,0\,0\,0\,0\,0\,0 \\ 0\,0\,0\,0\,0\,1\,0\,0\,0 \\ 0\,0\,0\,0\,0\,0\,0\,0\,1 \end{bmatrix}^\top \qquad B_3^f = \begin{bmatrix} 0\,0\,0\,0\,0\,0\,1\,0\,0 \\ 0\,0\,0\,0\,0\,0\,0\,1\,0 \\ 0\,0\,0\,0\,0\,0\,0\,0\,1 \end{bmatrix}^\top$$

and the corresponding incidence constraints:

$$B_0^f T_0^e = T_0^f B_0^f \qquad\qquad B_0^f T_1^e = T_1^f B_0^f$$
$$B_3^f T_0^e = T_2^f B_3^f \qquad\qquad B_3^f T_1^e = T_3^f B_3^f$$
$$B_1^f T_0^e = T_0^f B_1^f \qquad\qquad B_1^f T_1^e = T_2^f B_1^f$$
$$B_2^f T_0^e = T_1^f B_2^f \qquad\qquad B_2^f T_1^e = T_3^f B_2^f.$$

This particular form of B_0^f simply signifies that the corresponding edge depends on the first three control points (out of nine total). If we take the first pair of constraints only, it ensures that the attractor of the IFS $(\mathbb{X}^f; T_0^f, T_1^f)$ (recall that it is a sub-attractor of $\mathcal{A}^f$) is an image of the edge $\mathcal{A}^e$ embedded by the action of B_0^f. In the same manner, three other pairs of constraints ensure that $\mathcal{A}^f$ contains edges $B_1^f \mathcal{A}^e$, $B_2^f \mathcal{A}^e$ and $B_3^f \mathcal{A}^e$. Figure 8 shows an example of $\mathcal{A}^f$ with randomly fixed degrees of freedom.

Adjacency Constraints. Here we add the adjacency constraints that enforce connection of corresponding attractors. Recall that our attractors are self-similar, so after a subdivision one smaller copy of the attractor will be adjacent to another smaller copy.

Figure 9 illustrates the idea. Attractor $\mathcal{A}^e$ is defined as a union of its subdivisions $\mathcal{A}^e = T_0^e(\mathcal{A}^e) \cup T_1^e(\mathcal{A}^e)$. Let us apply the following constraint:

$$T_0^e B_1^e = T_1^e B_0^e.$$

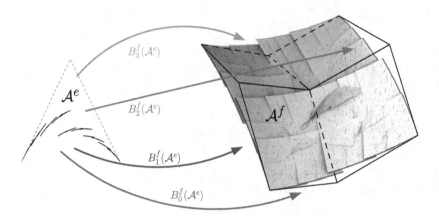

Fig. 8. As for the edge, facet incidence constraints ensure the inclusion $B_0^f \mathcal{A}^e \subset \mathcal{A}^f$, $B_1^f \mathcal{A}^e \subset \mathcal{A}^f$, $B_2^f \mathcal{A}^e \subset \mathcal{A}^f$ and $B_3^f \mathcal{A}^e \subset \mathcal{A}^f$.

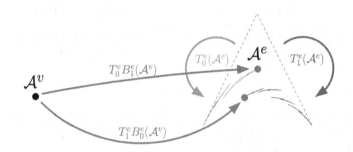

Fig. 9. In order to obtain a continuous curve, it suffices to apply the constraint $T_0^e B_1^e = T_1^e B_0^e$.

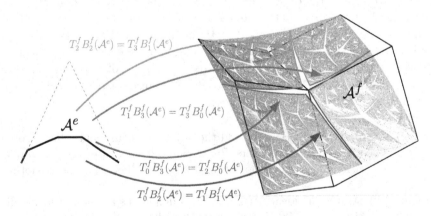

Fig. 10. Non-respect of the adjacency constraints leads to a disconnected patch.

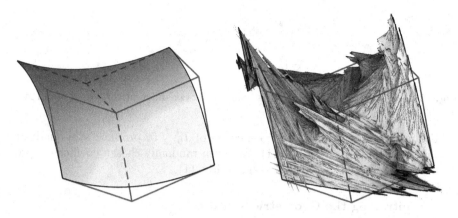

Fig. 11. After applying incidence and adjacency constraints, the attractor of the IFS $(\mathbb{X}^f; T_0^f, T_1^f, T_2^f, T_3^f)$ is guaranteed to have the topology of a quad patch. The degrees of freedom left in the operators can only affect the geometric texture. Left image: the degrees of freedom were fixed to produce a bi-quadratic Bézier patch; right image: even randomly chosen coefficients produce a continuous patch.

This signifies that $T_0^e(\mathcal{A}^e)$ and $T_1^e(\mathcal{A}^e)$ must share a common vertex thus producing a connected attractor $\mathcal{A}^e$. Let us express the corresponding matrices explicitly:

$$T_0^e = \begin{bmatrix} 1 & a_0 & b_0 \\ 0 & a_1 & b_1 \\ 0 & 1 - a_0 - a_1 & 1 - b_0 - b_1 \end{bmatrix} \qquad T_1^e = \begin{bmatrix} b_0 & c_0 & 0 \\ b_1 & c_1 & 0 \\ 1 - b_0 - b_1 & 1 - c_0 - c_1 & 1 \end{bmatrix},$$

We have 6 degrees of freedom left in the matrices; any choice of the coefficients ensures the connectivity of the attractor of the IFS $(\mathbb{X}^e; T_0^e, T_1^e)$.

In exactly the same manner, we apply the adjacency constraints for the facet subdivision operators:

$$T_0^f B_2^f = T_1^f B_1^f$$
$$T_2^f B_2^f = T_3^f B_1^f$$
$$T_0^f B_3^f = T_2^f B_0^f$$
$$T_1^f B_3^f = T_3^f B_0^f$$

Figure 10 provides an illustration. We omit an explicit expression of $T_{0...3}^f$ here, since it is cumbersome but straightforward to obtain.

At this point (after applying incidence and adjacency constraints) the attractor of the IFS $(\mathbb{X}^f; T_0^f, T_1^f, T_2^f, T_3^f)$ is guaranteed to have the topology of a quad patch. The degrees of freedom left in the operators can only affect the geometric texture.

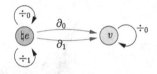

Fig. 12. In this example the edge $\mathcal{A}^e$ is bounded by two different vertices $\mathcal{A}^{v_0}$ and $\mathcal{A}^{v_1}$.

For instance, we can fix the coefficients of $T_{0\ldots3}^f$ to produce a bi-quadratic Bézier patch (left image in Fig. 11). But even randomly chosen coefficients produce a continuous surface (right image in Fig. 11).

3.2   Controlling the Geometric Texture

Full analysis of the differential behaviour of produced shapes is beyond the scope of this article, but in this section we try to illustrate how the BCIFS model can be used to control geometry (in addition to topology).

Let us construct an edge bounded by two vertices, each depending on one control point. Each vertex can be represented by the same state v, the only one possible for a one dimensional space with its trivial subdivision operator : $T_0^v = \begin{bmatrix} 1 \end{bmatrix}$. Figure 12 provides the automaton.

Let us assume $B_0^e = \begin{bmatrix} 1 & 0 & 0 \end{bmatrix}^\top$ and $B_1^e = \begin{bmatrix} 0 & 0 & 1 \end{bmatrix}^\top$ and the usual incidence and adjacency constraints:

$$B_0^e T_0^{v_0} = T_0^e B_0^e$$
$$B_1^e T_0^{v_1} = T_1^e B_1^e$$
$$T_0^e B_1^e = T_1^e B_0^e.$$

Solving for the incidence and adjacency constraints we obtain:

$$T_0^e = \begin{bmatrix} 1 \cdot a \\ 0 \cdot b \\ 0 \cdot c \end{bmatrix} \quad T_1^e = \begin{bmatrix} a \cdot 0 \\ b \cdot 0 \\ c \cdot 1 \end{bmatrix} \text{ with } a + b + c = 1$$

In order to control the differential behaviour at each vertex, we define two additional states, denoted by d_0 and d_1, with a two dimensional barycentric space associated with each one, and two embedding operators, $B_0^{e\prime} = \begin{bmatrix} 1 & 0 & 0 \\ 0 & 1 & 0 \end{bmatrix}^\top$ and

$B_1^{e\prime} = \begin{bmatrix} 0 & 1 & 0 \\ 0 & 0 & 1 \end{bmatrix}^\top$, specifying which control points are implied for each differential behaviour. Each state has its own subdivision operator, respectively $T_0^{d_0}$ and $T_0^{d_1}$.

As for C_0 continuity, we use incidence constraints for the differential continuity.

$$B_0^{e\prime} T_0^{d_0} = T_0^e B_0^{e\prime}$$
$$B_1^{e\prime} T_0^{d_1} = T_1^e B_1^{e\prime}.$$

The consequence is:

$$T_0^{do} = \begin{bmatrix} 1 & 1-\lambda \\ 0 & \lambda \end{bmatrix} \qquad T_0^{d_1} = \begin{bmatrix} \mu & 0 \\ 1-\mu & 1 \end{bmatrix}$$

$$T_0^e = \begin{bmatrix} 1 & 1-\lambda & a \\ 0 & \lambda & b \\ 0 & 0 & c \end{bmatrix} \qquad T_1^e = \begin{bmatrix} a & 0 & 0 \\ b & 1-\mu & 0 \\ c & \mu & 1 \end{bmatrix}$$

Attractors of $\mathcal{A}^{do}$ and $\mathcal{A}^{d_1}$ are points with coordinates given by the dominant eigenvectors of T_0^{do} and $T_0^{d_1}$. Hence we know that $\mathcal{A}^{do}$ is a point with coordinates $(1,0)$ and $\mathcal{A}^{d_1}$ is a point with coordinates $(0,1)$ in corresponding barycentric spaces.

Recall that incidence constraints embed all eigenvectors (and eigenvalues) of T_0^{do} and $T_0^{d_1}$ into T_0^e and T_1^e. Therefore, if λ is a sub-dominant eigenvalue of T_0^e, then the half-tangent at the "left" endpoint of the curve $\mathcal{A}^e$ is the vector $(-1,1,0)$ (the first edge of the control polygon). In the same manner, if μ is subdominant in T_1^e, then the "right" half-tangent is the vector $(0,-1,1)$, the 2nd edge of the control polygon.

As for the C^0 adjacency constraint, we can apply the same constraint for the half-tangents:[1]

$$T_0^e B_1^{e\prime} \begin{bmatrix} -1 \\ 1 \end{bmatrix} = T_1^e B_0^{e\prime} \begin{bmatrix} -1 \\ 1 \end{bmatrix} \Rightarrow T_0^e \begin{bmatrix} 0 \\ -1 \\ 1 \end{bmatrix} = T_1^e \begin{bmatrix} -1 \\ 1 \\ 0 \end{bmatrix}.$$

Solving the incidence and adjacency constraints we obtain the following expression for the subdivision operators:

$$T_0^e = \begin{bmatrix} 1 & 1-\lambda & \frac{1-\lambda}{2} \\ 0 & \lambda & \frac{\lambda+\mu}{2} \\ 0 & 0 & \frac{1-\mu}{2} \end{bmatrix} \qquad T_1^e = \begin{bmatrix} \frac{1-\lambda}{2} & 0 & 0 \\ \frac{\lambda+\mu}{2} & \mu & 0 \\ \frac{1-\mu}{2} & 1-\mu & 1 \end{bmatrix}.$$

Provided that λ and μ are positive subdominant eigenvalues ($\lambda \geq \frac{1-\mu}{2}$ and $\mu \geq \frac{1-\lambda}{2}$), the attractor $\mathcal{A}^e$ is guaranteed to be a C^1 curve.

3.3 Example of Application

Given a set of control points and a subdivision method, construction of a CIFS whose attractor is exactly the limit subdivision surface is straightforward. In this example, we push the concept a bit further. We want to construct a solid arborescent structure, whose boundary is a subdivision surface.

---

[1] Strictly speaking, we do not need the equality of the halftangents, collinearity suffice, so the adjacency constraint can be written as $T_0^e B_1^{e\prime} \begin{bmatrix} -\alpha \\ \alpha \end{bmatrix} = T_1^e B_0^{e\prime} \begin{bmatrix} -1 \\ 1 \end{bmatrix}$ for some $\alpha \neq 0$.

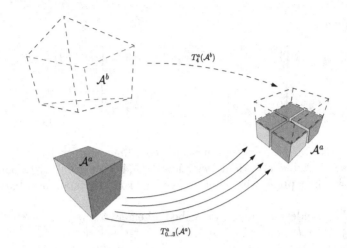

Fig. 13. Arborescent structure of $\mathcal{A}^a$ is defined by iterative condensation of the attractor $\mathcal{A}^b$.

Fig. 14. Left: unfolding of the six facets of $\mathcal{A}^b$. We are free to choose two of them; the other four are fixed automatically by our choice. Right: our choice.

Figure 13 shows the core of the subdivision process. The final arborescent structure $\mathcal{A}^a$ is defined as an iterative condensation of the attractor $\mathcal{A}^b$. Here $\mathcal{A}^b$ is a limit subdivision surface for a given mesh. The idea is simple: $\mathcal{A}^a$ is defined layer-by-layer. $\mathcal{A}^b$ covers the top of the shape, then the second layer is defined by four smaller copies of $\mathcal{A}^b$, the third layer has 16 copies of $\mathcal{A}^b$ and so forth.

Both $\mathcal{A}^a$ and $\mathcal{A}^b$ have six facets; Fig. 14 shows the constraints we obtain on facets from the nature of the subdivision process. We are free to choose two facets (one upper and one lateral) for the attractor $\mathcal{A}^b$, the other four are fixed automatically by our choice.

Finally, Fig. 15 gives the final shape of the attractor $\mathcal{A}^a$.

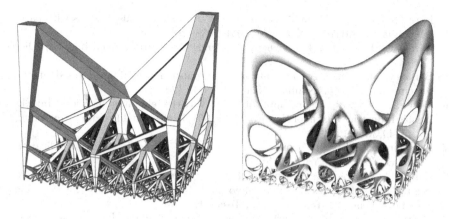

Fig. 15. Left: mesh of control points for the $\mathcal{A}^a$, right: the final shape of $\mathcal{A}^a$.

4   Conclusion

Our goal is to provide a computer aided geometry design software to model fractal shapes with the facilities of standard systems. The model proposed here is based on Iterated Function Systems (IFS) and enriched with Boundary Representation concepts to describe fractal topology. The fractal subdivision process is controlled by introducing incidence and adjacency constraints on topological cells. These constraints induce constraints on transformations composing the IFS. The local aspect of the shapes (rough or smooth) is controlled by the remaining degrees of freedom. The global geometry of the shape is controlled by a set of control points. This model can produce curves, surfaces, volumes, trees, or any complex fractal topology. The main important characteristic of our model is the control of the topological subdivision. According to this topological description, the fractal can be easily approximated by a coherent topological standard structure such as a mesh in order to fabricate it by 3D printing.

Descriptions are quite easy to specify for curves, surfaces or wireframe structures. But for volume subdivisions, the number of incidence and adjacency constraints increases and the description could become tedious. Furthermore, adjacency constraints have to verify orientation conditions (two adjacent cells must share common borders dispatched in a compatible way) to avoid degenerated solutions (attractor reduced to a point). These conditions increase the complexity of the description. However, automatic construction can be provided using topological operators as topological products.

References

1. Aron, J.: The mandelbulb: first 'true' 3D image of famous fractal. New Sci. **204**(2736), 54 (2009)
2. Bandt, C., Gummelt, P.: Fractal Penrose tilings I. Construction and matching rules. Aequationes Math. **53**(1–2), 295–307 (1997)

3. Barnsley, M., Hutchinson, J., Stenflo, O.: V-variable fractals: fractals with partial self similarity. Adv. Math. **218**(6), 2051–2088 (2008)
4. Barnsley, M.: Fractals Everywhere. Academic Press Professional Inc., San Diego (1988)
5. Barnsley, M., Vince, A.: Fractal homeomorphism for bi-affine iterated function systems. Int. J. Appl. Nonlinear Sci. **1**(1), 3–19 (2013)
6. Cohen, N.: Fractal antenna applications in wireless telecommunications. In: Electronics Industries Forum of New England, 1997, Professional Program Proceedings, pp. 43–49, May 1997
7. Falconer, H.J.: Fractal Geometry: Mathematical Foundations and Applications, 2nd edn. Wiley, New York (1990)
8. Gentil, C.: Les fractales en synthése d'images: le modèle IFS. Ph.D. thesis, Université LYON I, March 1992. Jury: Vandorpe, D., Chenin, P., Mazoyer, J., Reveilles, J.P., Levy Vehel, J., Terrenoire, M., Tosan, E
9. Guerin, E., Tosan, E., Baskurt, A.: Fractal coding of shapes based on a projected IFS model. In: Proceedings of 2000 International Conference on Image Processing, 2000, vol. 2, pp. 203–206, September 2000
10. Hutchinson, J.: Fractals and self-similarity. Indiana Univ. J. Math. **30**, 713–747 (1981)
11. Massopust, P.R.: Fractal functions and their applications. Chaos Solitons Fractals **8**(2), 171–190 (1997)
12. Mauldin, R.D., Williams, S.C.: Hausdorff dimension in graph directed constructions. Trans. Am. Math. Soc. **309**(2), 811–829 (1988)
13. Peitgen, H.-O., Richter, P.: The Beauty of Fractals: Images of Complex Dynamical Systems. Springer, Heidelberg (1986)
14. Pence, D.: The simplicity of fractal-like flow networks for effective heat and mass transport. Exp. Therm. Fluid Sci. **34**(4), 474–486 (2010). ECI International Conference on Heat Transfer and Fluid Flow in Microscale
15. Prusinkiewicz, P., Hammel, M.: Language-restricted iterated function systems, koch constructions, and l-systems. In: SIGGRAPH 1994 Course Notes (1994)
16. Prusinkiewicz, P., Lindenmayer, A.: The Algorithmic Beauty of Plants. Springer-Verlag New York Inc., New York (1990)
17. Puente, C., Romeu, J., Pous, R., Garcia, X., Benitez, F.: Fractal multiband antenna based on the sierpinski gasket. Electron. Lett. **32**(1), 1–2 (1996)
18. Rian, I.M., Sassone, M.: Tree-inspired dendriforms and fractal-like branching structures in architecture: a brief historical overview. Front. Archit. Res. **3**(3), 298–323 (2014)
19. Soo, S.C., Yu, K.M., Chiu, W.K.: Modeling and fabrication of artistic products based on IFS fractal representation. Comput. Aided Des. **38**(7), 755–769 (2006)
20. Terraz, O., Guimberteau, G., Mérillou, S., Plemenos, D., Ghazanfarpour, D.: 3Gmap L-systems: an application to the modelling of wood. Vis. Comput. **25**(2), 165–180 (2009)
21. Thollot, J., Tosan, E.: Construction of fractales using formal languages and matrices of attractors. In: Santos, H.P. (ed.) Conference on Computational Graphics and Visualization Techniques, Compugraphics, Alvor, Portugal, pp. 74–78 (1993)
22. Tosan, E.: Surfaces fractales définies par leurs bords. Grenoble (2006). Journées Courbes, surfaces et algorithmes
23. Zair, C.E., Tosan, E.: Fractal modeling using free form techniques. Comput. Graph. Forum **15**(3), 269–278 (1996)
24. Zhou, H., Sun, J., Turk, G., Rehg, J.M.: Terrain synthesis from digital elevation models. IEEE Trans. Vis. Comput. Graph. **13**(4), 834–848 (2007)

Construction of Smooth Isogeometric Function Spaces on Singularly Parameterized Domains

Thomas Takacs$^{(\boxtimes)}$

Department of Mathematics, University of Pavia, Pavia, Italy
thomas.takacs@unipv.it

Abstract. We aim at constructing a smooth basis for isogeometric function spaces on domains of reduced geometric regularity. In this context an isogeometric function is the composition of a piecewise rational function with the inverse of a piecewise rational geometry parameterization. We consider two types of singular parameterizations, domains where a part of the boundary is mapped onto one point and domains where the two parameter lines at a corner of the parameter domain are collinear in the physical domain. We locally map a singular tensor-product patch of arbitrary degree onto a triangular patch, thus splitting the parameterization into a singular bilinear mapping and a regular mapping on a triangular domain. This construction yields an isogeometric function space of prescribed smoothness. Generalizations to higher dimensions are also possible and are briefly discussed in the final section.

1 Introduction

In this paper we consider two different configurations of singular planar NURBS geometry parameterizations, leading to two different types of triangular domains. The goal of our construction is the definition of arbitrarily smooth isogeometric function spaces defined on these domains as well as the construction of a suitable basis. The approach presented here can be generalized to other types of domains and to higher dimensions.

The ability to construct test/trial functions of high smoothness, suitable for numerical simulations, is one of the main features of isogeometric analysis, as introduced in [6]. B-spline and NURBS function spaces on standard tensor-product domains possess the possibility of k-refinement, creating a sequence of non-nested spaces of increasing degree and increasing smoothness. Hence, increasing degree and smoothness may lead to improved convergence [1]. Several applications in isogeometric analysis rely on function spaces of smoothness of higher order, like differential equations of higher order [3], or the analysis of shells [2,7,8], just to name a few examples. In all these applications, the results may deteriorate if singular parameterizations are present. To overcome this deficiency, we present constructions leading to isogeometric functions spaces of arbitrary smoothness on singularly parameterized domains.

We start with some preliminary definitions and notation on B-splines and NURBS in Sect. 2.1 and on isogeometric functions in Sect. 2.2. The smoothness conditions of interest are presented in Sect. 2.3. In Sect. 3 we develop the

© Springer International Publishing Switzerland 2015
J.-D. Boissonnat et al. (Eds.): Curves and Surfaces 2014, LNCS 9213, pp. 433–451, 2015.
DOI: 10.1007/978-3-319-22804-4_30

construction of smooth spaces over singular domains where one edge of the parameter domain is mapped onto one point in the physical domain. In Sect. 4 we present a similar construction for domains where two parameter directions are collinear at the boundary of the physical domain. Both constructions can be used to obtain circular domains, see also [9]. We briefly discuss generalizations to higher dimension in Sect. 5 and conclude the presented results in Sect. 6.

2   Preliminaries

Isogeometric function spaces $\mathcal{V}$, as they are present in isogeometric analysis, are built from an underlying B-spline or NURBS space. Hence, to introduce the notation needed, we start this preliminary section with recalling the notion of B-splines and NURBS. We do not give detailed definitions here and refer to standard literature for further reading [4, 10, 11].

2.1   B-Splines and NURBS

Univariate B-splines are piecewise polynomial functions. Given a degree $p \in \mathbb{Z}^+$ and a knot vector $[S] = (s_{-p}, \ldots, s_{N+p+1})$ of length $N + 2p + 2$, the i-th B-spline, for $i = 0, \ldots, N + p$, is denoted by $B_i^p[S](s)$. We assume that the parameter domain is the unit interval and that the knot vector is open, i.e.

$$0 = s_{-p} = \ldots = s_0 < s_1 \leq \ldots \leq s_N < s_{N+1} = \ldots = s_{N+p+1} = 1. \quad (1)$$

Note that any B-spline basis function can be represented via its local knot vector, which we denote by

$$b[s_{i-p}, \ldots, s_i, s_{i+1}](s) = B_i^p[S](s),$$

where the knot vector S contains the knot sequence $s_{i-p}, \ldots, s_i, s_{i+1}$ for the indices $i - p$ to $i + 1$. Using this notation, the degree p of the B-spline $b[s_{i-p}, \ldots, s_i, s_{i+1}] (= B_i^p[S])$ is implicitly given by the length $p + 2$ of the local knot vector.

The concept of univariate B-splines can easily be generalized to two dimensions via a tensor-product construction. Let $p, q \in \mathbb{Z}^+$ and let S and T be open knot vectors fulfilling Eq. (1). The parameter domain is set to be the box $\mathbf{B} = [0, 1]^2$, leading to the *tensor-product B-spline space*

$$\mathscr{S} = \mathrm{span}\left(\left\{B_i^p[S]\, B_j^q[T] : \mathbf{B} \to \mathbb{R} \mid \text{ for } (0, 0) \leq (i, j) \leq (N_1 + p, N_2 + q)\right\}\right).$$

The B-splines span a piecewise polynomial function space on a grid given by the knot vectors S and T. Given a weight function $g_0 \in \mathscr{S}$, with $g_0(\mathbf{s}) > 0$ for all $\mathbf{s} \in \mathbf{B}$, we can define a *NURBS space* via

$$\mathscr{N} = \left\{\frac{f}{g_0} : \mathbf{B} \to \mathbb{R} \mid f \in \mathscr{S}\right\}.$$

We can now define isogeometric function spaces.

2.2 Isogeometric Functions

We use the following standard definition of isogeometric functions over a physical domain Ω, where we follow the notation in [16]. This definition is based on the concept of *isogeometric analysis* introduced in [6]. For a given NURBS geometry parameterization

$$\mathbf{G} = (G_1, G_2)^T = \left(\frac{g_1}{g_0}, \frac{g_2}{g_0} \right)^T : \mathbf{B} \to \overline{\Omega} \subset \mathbb{R}^2,$$

with $G_1, G_2 \in \mathcal{N}$, the space of *isogeometric functions* defined on the open domain $\Omega = \mathbf{G}(\mathbf{B}°)$ is denoted by

$$\mathcal{V} = \left\{ \varphi : \Omega \to \mathbb{R} \mid \varphi = F \circ \mathbf{G}^{-1}, \text{ with } F = \frac{f}{g_0} \in \mathcal{N} \right\}.$$

We assume that $\mathbf{G}$ is invertible in the interior $\mathbf{B}° =]0,1[^2$ of the box $\mathbf{B}$, hence the functions φ are well-defined. Note that an isogeometric function φ can be defined via its graph surface in homogeneous coordinates

$$\mathbf{f} = (g_0, g_1, g_2, f)^T : \mathbf{B} \to \tilde{\Omega} \times \mathbb{R}, \tag{2}$$

with $r_j \in \mathcal{S}$ for $j = 0, 1, 2, 3$. Here $\tilde{\Omega}$ is given such that $\Pi(\tilde{\Omega}) = \Omega$, where the mapping $\Pi : (x_0, x_1, x_2) \mapsto (x_1/x_0, x_2/x_0)$ is the central projection onto the plane $x_0 = 1$.

In the following subsection we present smoothness conditions which are of interest in isogeometric analysis.

2.3 Smoothness Conditions

We consider a notion of continuity, which may be of interest for any numerical application where a high order of smoothness is necessary.

Definition 1. *The space $\mathscr{C}^k(\overline{\Omega})$ of $\mathscr{C}^k$-continuous functions on the closure of Ω is defined as the space of functions $\varphi : \Omega \to \mathbb{R}$ with $\varphi \in C^k(\Omega)$ such that there exists a unique limit*

$$\lim_{\mathbf{y} \to \mathbf{x}, \mathbf{y} \in \Omega} \frac{\partial^{|\alpha|} \varphi(\mathbf{y})}{\partial x_1^{\alpha_1} \partial x_2^{\alpha_2}} = \frac{\partial^{|\alpha|} \varphi(\mathbf{x})}{\partial x_1^{\alpha_1} \partial x_2^{\alpha_2}}$$

for all $\mathbf{x} \in \partial\Omega = \overline{\Omega}\backslash\Omega$ and for all $|\alpha| = \alpha_1 + \alpha_2 \leq k$. Here $C^k(\Omega)$ is the traditional space of k-times continuously differentiable functions on the open domain Ω.

Since we are considering spline spaces $\mathscr{S}$ of degree p and smoothness $p-1$, the highest reasonable smoothness to consider here is of order $k = p - 1$. However, this may not be feasible for arbitrary domains.

Another way to prescribe smoothness is by regularity in the sense of Sobolev spaces, i.e. $\mathcal{V} \subset H^k(\Omega)$. Note that $\mathscr{C}^k(\overline{\Omega}) \subset H^k(\Omega)$. For many isogeometric function spaces, the condition $\mathscr{V} \subset \mathscr{C}^k(\overline{\Omega})$ is equivalent to $\mathscr{V} \subset H^{k+1}(\Omega)$. If the domain parameterization $\mathbf{G}$ is singular somewhere at the boundary $\partial\mathbf{B}$, then the function space $\mathscr{V}$ may not be regular. For many settings $\mathscr{V} \subset \mathscr{C}^0(\overline{\Omega})$ as well as $\mathscr{V} \subset H^1(\Omega)$ are not fulfilled (e.g. for patches of type A, see Sect. 3). For a detailed study concerning Sobolev regularity on singular parameterizations in isogeometric analysis we refer to [14,15]. The papers present construction schemes for H^1- and H^2-smooth isogeometric function spaces and corresponding bases.

In this paper we generalize the presented approach to $\mathscr{C}^k$-smoothness for arbitrary k. We consider two types of singular parameterizations. The first type A is a class of singular parameterizations where a part of the boundary of $\mathbf{B}$ is mapped onto one point. The second type B covers singular parameterizations where the parameter lines in the physical domain are collinear at the boundary.

3 Singular Tensor-Product Patches of Type A

In this section we construct smooth isogeometric function spaces on singular patches of type A. A parameterization $\mathbf{G} = (g_1/g_0, g_2/g_0)^T$, represented via

$$(g_0, g_1, g_2)^T = \sum_{i=0}^{N_1+p} \sum_{j=0}^{N_2+q} \mathbf{C}_{i,j}\, B_i^p[\mathrm{S}]\, B_j^q[\mathrm{T}]$$

in homogeneous coordinates, is called a *singular mapping of type A* if the control points fulfill $\mathbf{C}_{0,j} = \mathbf{C}_{0,k}$ for all $0 \le j, k \le N_2 + q$, and

$$\det \nabla\mathbf{G}(s,t) = s\, g(s,t) \quad \text{for all } (s,t) \in [0,1]^2,$$

where $g(s,t) > \underline{g} > 0$ for all $(s,t) \in [0,1]^2$. Hence we have

$$\det \nabla\mathbf{G}(\mathbf{s}) = 0 \quad \text{for all } \mathbf{s} \in \{0\} \times [0,1],$$

where the part of the boundary $\{0\} \times [0,1] \subset \partial\mathbf{B}$ of the parameter domain box $\mathbf{B}$ is mapped onto one point in the physical domain. The class of singular patches we consider is derived from triangular Bézier patches. We start with a construction for Bézier patches which we then generalize to B-spline patches.

3.1 Triangular Bézier Patches as Singular Tensor-Product Bézier Patches

As presented by Hu in [5], a triangular Bézier patch

$$\boldsymbol{\rho}\ :\ \Delta \to \mathbb{R}^d\ :\ (u,v) \mapsto \sum_{i+j+k=p} \beta^p_{(i,j,k)}(u,v)\, \boldsymbol{\rho}_{i,j,k},$$

with control points $\boldsymbol{\rho}_{i,j,k} \in \mathbb{R}^d$, parameter domain

$$\Delta = \{(u,v) : 0 \le u \le 1,\ 0 \le v \le u\}$$

and basis functions

$$\beta^p_{(i,j,k)} \; : \; \Delta \to \mathbb{R} \; : \; (u,v) \mapsto \frac{p!}{i!j!k!}(1-u)^i v^j (u-v)^k,$$

can be represented as a tensor-product Bézier patch

$$\mathbf{f} \; : \; [0,1]^2 \to \mathbb{R}^d \; : \; \mathbf{s} = (s,t) \mapsto \sum_{i=0}^{p} \sum_{j=0}^{p} b_i^p(s) \, b_j^p(t) \, \mathbf{f}_{i,j}, \qquad (3)$$

with Bernstein polynomials b_i^p of degree p, where

$$\mathbf{f}_{i,j} = \sum_{\ell=0}^{i} \binom{i}{\ell} \frac{\binom{p-i}{j-\ell}}{\binom{p}{j}} \boldsymbol{\rho}_{p-i,\ell,i-\ell}$$

for $0 \leq i,j \leq p$. The control points for each row are computed via degree elevation. For $i = 0$ all control points are the result of degree elevation of a constant "curve", i.e. all points are equal

$$\mathbf{f}_{0,j} = \boldsymbol{\rho}_{p,0,0},$$

hence the tensor-product Bézier patch $\mathbf{f}(s,t)$ is singular at $s = 0$. For $i = 1$ the control points result from degree elevating a linear, for $i = 2$ from degree elevating a quadratic curve, and so on.

The transformation leading to Eq. (3) can also be interpreted as a change of parameters

$$\mathbf{f} = \boldsymbol{\rho} \circ \mathbf{u} \qquad (4)$$

with the reparameterization

$$(s,t) \mapsto (u,v), \;\; \text{with}$$
$$\mathbf{u}(s,t) = (u(s,t),v(s,t)) = (s, s\,t)$$

with $\mathbf{u} \in (\mathbb{Q}_1(\mathbb{R}^2))^2$, $\boldsymbol{\rho} \in (\mathbb{P}_k(\mathbb{R}^2))^d$ and $\mathbf{f} \in (\mathbb{Q}_k(\mathbb{R}^2))^d$. Note that the bilinear mapping $\mathbf{u}(s,t)$ is singular for s=0. Here $\mathbb{Q}_k(\mathbb{R}^\ell)$ is the space of ℓ-variate polynomials of maximal degree $\leq k$ and $\mathbb{P}_k(\mathbb{R}^\ell)$ is the space of ℓ-variate polynomials with total degree $\leq k$.

Selecting $\boldsymbol{\rho}_{p-i,\ell,i-\ell} \in \mathbb{R}^4$, we get $\mathbf{f} = \boldsymbol{\rho} \circ \mathbf{u} : \mathbf{B} \to \mathbb{R}^4$ which may serve as a homogeneous graph surface of an isogeometric function as in (2). In this case we conclude $\mathscr{V} \subset \mathscr{C}^\infty(\overline{\Omega})$ if the rational triangular Bézier patch $(\rho_1/\rho_0, \rho_2/\rho_0)^T$ is regular. Generalizing this construction we can define a smooth isogeometric function space on a certain class of domains containing a singularity. We will also give a detailed proof of the smoothness result in the following section.

3.2   Smooth Function Spaces over a Singular B-Spline Patch

Given a tensor-product B-spline function space $\mathscr{S}$ of degree (p,q) and prescribed order of smoothness $k \leq \min(p,q)$ we want to construct a function space $\mathscr{S}^k \subset \mathscr{S}$, as well as $\mathscr{V}^k \subset \mathscr{V}$ derived from $\mathscr{S}^k$, such that $\mathscr{V}^k \subset \mathscr{C}^k(\overline{\Omega})$.

In the previous section we constructed a polynomial patch $\mathbf{f} = \boldsymbol{\rho} \circ \mathbf{u}$ that can be split into a singular bilinear part $\mathbf{u}$ and a regular polynomial part $\boldsymbol{\rho}$ defined on a triangular domain. The core idea of the generalized approach is the following. Given a B-spline surface $\mathbf{f} \in (\mathscr{S})^4$, we assume that $\mathbf{f}$ is equivalent to a triangular patch up to order k at the singularity. Hence, we introduce the function space $\mathscr{S}^k(\mathbf{u}, \mathbf{S}) \subset \mathscr{S}$.

Definition 2. *The function space* $\mathscr{S}^k(\mathbf{u}, \mathbf{S}) \subset \mathscr{S}$ *is defined as the space of splines* $f \in \mathscr{S}$*, such that there exists a polynomial* $\rho \in \mathbb{P}_k$ *fulfilling*

$$\frac{\partial^{|\alpha|} f}{\partial s^{\alpha_1} \partial t^{\alpha_2}}(\mathbf{s}) = \frac{\partial^{|\alpha|} (\rho \circ \mathbf{u})}{\partial s^{\alpha_1} \partial t^{\alpha_2}}(\mathbf{s}) \ \text{ for all } \mathbf{s} \in \mathbf{S}, \tag{5}$$

for $0 \le \alpha_1, \alpha_2 \le k$*,* $|\alpha| = \alpha_1 + \alpha_2$*, where* $\mathbf{u}$ *is the mapping*

$$\mathbf{u}: \ [0,1]^2 \ \to \ \Delta = \{(u,v) : 0 \le u \le 1, \ 0 \le v \le u\}$$
$$(s,t)^T \ \mapsto \ (s, s\,t)^T$$

and $\mathbf{S} = \{0\} \times [0,1]$*.*

Note that if $k = q = p$, then $\mathscr{S}^p(\mathbf{u}, \mathbf{S})\, |_{[0,s_1] \times [0,1]} \circ \mathbf{u}^{-1} = \mathbb{P}_p$, where $\mathbb{P}_p$ is the space of polynomials of total degree $\le p$ and s_1 is the first interior knot of the knot vector S.

Using this approach, we obtain linear conditions on the coefficients of the B-spline. Certain linear combinations of B-spline basis functions will correspond to the basis functions on the triangular patch. Just as presented in [5], Definition 2 is equivalent to the first row of control points being constant, the second row forming a linear curve, the third row a quadratic, and so on. This leads to the following definition.

Definition 3. *Let* $k \le \min(p, q)$*. The basis* $\mathbb{S}^k$ *is defined via*

$$\mathbb{S}^k = \big\{ B_i^p[\mathrm{S}](s) b_j^i(t) : 0 \le j \le i \ \text{and} \ 0 \le i \le k \big\}$$
$$\cup \big\{ B_i^p[\mathrm{S}](s) B_j^q[\mathrm{T}](t) : k + 1 \le i \le N_1 + p \ \text{and} \ 0 \le j \le N_2 + q \big\},$$

where $B_i^p[\mathrm{S}] B_j^q[\mathrm{T}]$*, with* $(0,0) \le (i,j) \le (N_1 + p, N_2 + q)$*, is the standard basis of* $\mathscr{S}$ *and* $b_j^i(t)$ *is the* j*-th Bernstein polynomial of degree* i*.*

Fig. 1 gives a schematic depiction of the index set corresponding to $\mathbb{S}^k$ for $k = 3$.

Lemma 1. *Let* $k \le \min(p, q)$*. The set* $\mathbb{S}^k$ *given in Definition 3 is a basis for the space* $\mathscr{S}^k(\mathbf{u}, \mathbf{S})$ *given in Definition 2.*

Proof. Obviously, we have $\operatorname{span}(\mathbb{S}^k) \subset \mathscr{S}$ since

$$b_j^i(t) \in \operatorname{span}\big(\{ B_j^q[\mathrm{T}](t) : 0 \le j \le N_2 + q \} \big)$$

for all i, j with $0 \le j \le i \le q$.

We first show that $\operatorname{span}(\mathbb{S}^k) \subseteq \mathscr{S}^k(\mathbf{u}, \mathbf{S})$, i.e. all functions $f \in \mathbb{S}^k$ fulfill Eq. (5) for some polynomial $\rho \in \mathbb{P}_{\min(p,q)}$. Since the condition (5) needs to be fulfilled for

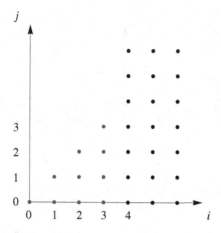

Fig. 1. Index set corresponding to $\mathbb{S}^k$ for $k = 3$

$s = 0$, we assume that $s < s_1$, which is the first knot of the knot vector S. For $i > k + 1$, the functions $B_i^p[\mathrm{S}](s)B_j^q[\mathrm{T}](t)$ fulfill $\frac{\partial^\alpha}{\partial s^\alpha}B_i^p[\mathrm{S}](s)B_j^q[\mathrm{T}](t) = 0$ for all $\alpha \le k$. Hence (5) is fulfilled with $\rho \equiv 0$. For $B_i^p[\mathrm{S}](s)b_j^i(t)$ with $j \le i \le k$ we have that $B_i^p[\mathrm{S}](s) = s^i\, r(s)$, where $r(s)$ is some polynomial in s of degree $\le k - i$. Moreover, $b_j^i(t)$ is a polynomial in t of degree i. Hence, $B_i^p[\mathrm{S}](s)b_j^i(t) = r(s)\, s^i\, b_j^i(t)$ can be represented as a polynomial ρ in $u = s$ and $v = s\,t$ with total degree $\le k$. This is exactly the form $B_i^p[\mathrm{S}](s)b_j^i(t) = \rho \circ \mathbf{u}$ required in Eq. (5).

What remains to be shown is that $\mathscr{S}^k(\mathbf{u}, \mathbf{S}) \subseteq \mathrm{span}(\mathbb{S}^k)$. Assume that $f \in \mathscr{S}$ is equivalent to a monomial $u^i v^j$, $i + j \le k$, with respect to condition (5). In that case we conclude $f(s,t) = s^{i+j}\, t^j + s^{k+1}\, r(s,t)$ for $s < s_1$, where r is some polynomial of degree $p - k - 1$ in s and of degree q in t. One can show easily that $f \in \mathrm{span}(\mathbb{S}^k)$ in that case. Finally, if $f \in \mathscr{S}$ is equivalent to $\rho \equiv 0$ with respect to (5), then $f(s,t) = s^{k+1}\, r(s,t)$ for $s < s_1$, and again $f \in \mathrm{span}(\mathbb{S}^k)$, which concludes the proof. □

One can show that if ρ is regular and $\mathbf{G}$ is C^k-smooth, then the mapping

$$\mathbf{F} : \Delta \to \mathbb{R} \quad \text{with} \quad \mathbf{F} = \mathbf{G} \circ \mathbf{u}^{-1}$$

is a C^k-smooth mapping from the triangle Δ to the domain $\overline{\Omega}$. Here, the inverse of the singular mapping $\mathbf{u}$ is equal to

$$\mathbf{u}^{-1}(u, v) = \left(u, \frac{v}{u}\right)^T.$$

This leads to a split of the mapping $\mathbf{G}$ into a bilinear singular transformation $\mathbf{u}$ and a regular mapping $\mathbf{F}$, via

$$\mathbf{G} = \mathbf{F} \circ \mathbf{u}.$$

The various introduced mappings and domains are depicted in Fig. 2.

Using this definition we can construct an isogeometric function space fulfilling $\mathscr{V}^k \subseteq \mathscr{V} \cap \mathscr{C}^k(\overline{\Omega})$, for $k \le \min(p,q)$.

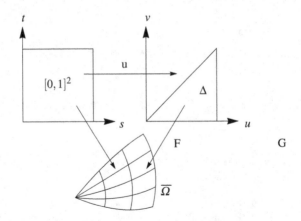

Fig. 2. Mappings **F**, **u**, **G** for an example domain of type A

Theorem 1. *Let $k \leq \min(p, q)$, let $\mathscr{S} \subset C^k(\mathbf{B})$ and let $\mathscr{V}^k$ be the isogeometric function space derived from $\mathscr{S}^k(\mathbf{u}, \mathbf{S})$ with $\mathbf{G} = (g_1/g_0, g_2/g_0)^T$ with $g_0, g_1, g_2 \in \mathscr{S}^k(\mathbf{u}, \mathbf{S})$. Moreover, $\mathbf{G}(s, t)$ is regular for all $s > 0$ and $t \in [0, 1]$.*

Then $\mathscr{V}^k \subset \mathscr{C}^k(\overline{\Omega})$ if there exist polynomials ρ_0, ρ_1 and ρ_2 such that (5) is fulfilled for $f = g_0$, g_1 and g_2, respectively, and $\left(\frac{\rho_1}{\rho_0}, \frac{\rho_2}{\rho_0}\right)^T$ is regular in Δ.

Proof. Given an isogeometric function $\varphi = f \circ \mathbf{G}^{-1} \in \mathscr{V}^k$. Due to Lemma 1 the homogeneous graph surface $\mathbf{f} = (g_0, g_1, g_2, f)^T$ fulfills (5) for some $(\rho_0, \rho_1, \rho_2, \rho_3)^T$. The condition $\varphi \in \mathscr{C}^k(\overline{\Omega})$ is given by

$$\frac{f}{g_0} \circ \left(\frac{g_1}{g_0}, \frac{g_2}{g_0}\right)^{-1} \in \mathscr{C}^k(\overline{\Omega}).$$

Since $\mathbf{G}$ is regular for $s > 0$ and $\mathbf{G} \in C^k(\mathbf{B})$, we conclude that $\mathbf{G}^{-1} \in \mathscr{C}^k(\mathbf{G}([\epsilon, 1] \times [0, 1]))$ by definition. Hence, it remains to be shown that

$$\frac{f}{g_0} \circ \left(\frac{g_1}{g_0}, \frac{g_2}{g_0}\right)^{-1} \in \mathscr{C}^k(\mathbf{G}([0, \epsilon] \times [0, 1])).$$

Due to (5) this is equivalent to

$$\frac{\rho_3}{\rho_0} \circ \mathbf{u}^{-1} \circ \mathbf{u} \circ \left(\frac{\rho_1}{\rho_0}, \frac{\rho_2}{\rho_0}\right)^{-1} \in \mathscr{C}^k(\mathbf{G}([0, \epsilon] \times [0, 1])).$$

Since $\rho_i \in C^\infty$, this condition is equivalent to $\left(\frac{\rho_1}{\rho_0}, \frac{\rho_2}{\rho_0}\right)$ being invertible which concludes the proof.   □

In Definition 3 we have already given a basis $\mathbb{S}^k$ for the function space $\mathscr{S}^k(\mathbf{u}, \mathbf{S})$. In the next section we propose an algorithm to determine the coefficients of the linear conditions with respect to the standard basis of $\mathscr{S}$ yielding the new basis functions in $\mathbb{S}^k$.

3.3   Algorithm to Construct the New Basis Functions

In this section we describe an algorithm to find a representation for the new basis functions

$$B_k^p[S](s)\, b[0,1,\ldots,1](t)$$

$$B_2^p[S](s)\, b[0,1,1,1](t) \qquad \vdots$$
$$B_1^p[S](s)\, b[0,1,1](t)\ \ B_2^p[S](s)\, b[0,0,1,1](t)$$
$$B_0^p[S](s)\, b[0,1](t)\ \ B_1^p[S](s)\, b[0,0,1](t)\ \ B_2^p[S](s)\, b[0,0,0,1](t)\ \ldots\ B_k^p[S](s)\, b[0,\ldots,0,1](t)$$

or equivalently

$$\left\{ b[s_{i-p},\ldots,s_{i+1}](s)b_j^i(t) : 0 \le j \le i \text{ and } 0 \le i \le k \right\},$$

in terms of the basis $B_i^p[S](s)B_j^q[T](t)$ of the space $\mathscr{S}$. Here $b_j^i(t)$ is the j-th Bernstein polynomial of degree i. The algorithm is composed of three steps: degree elevation for t, knot insertion for t and tensor-product multiplication with basis functions in s-direction. The presented algorithms are taken from standard literature [10, 11].

1. Perform degree elevation in t-direction. Let E_i^q be the matrix corresponding to the degree elevation of a Bernstein polynomial of degree i represented in terms of Bernstein polynomials of degree $q > i$, i.e.

$$(b_j^i(t))_{j=0,\ldots,i}^T = \mathrm{E}_i^q (b_j^q(t))_{j=0,\ldots,q}^T.$$

The matrix E_i^q is of dimension $(i+1) \times (q+1)$ and has the form

$$\mathrm{E}_i^q = \left(\frac{\binom{i}{\ell}\binom{q-i}{j-\ell}}{\binom{q}{j}} \right)_{\ell=0,\ldots,i \,\times\, j=0,\ldots,q},$$

see e.g. [11]. In Sect. 3.4 we list some of the matrices E_i^q for example configurations.

2. Perform knot insertion for t-direction. Let K_τ be the matrix corresponding to the knot insertion of interior knots $\tau = (t_1, t_2, \ldots t_N)$ of the knot vector T, leading to

$$(b_j^q(t))_{j=0,\ldots,q}^T = \mathrm{K}_\tau (b[t_{j-q},\ldots,t_j,t_{j+1}](t))_{j=0,\ldots,N+q}^T.$$

The knot insertion matrix K_τ can be defined via iteratively inserting the knots into the knot vector. The following algorithm gives the resulting matrix for insertion of the single knot t_i into the given knot vector with interior knots up to t_{i-1}. A B-spline of degree q with knot vector $(0,\ldots,0,t_1,\ldots,t_{i-1},1,\ldots,1)$ can be represented as a B-spline of degree q with knot vector $(0,\ldots,0,t_1,\ldots,t_{i-1},t_i,1,\ldots,1)$ using knot insertion. The corresponding transformation matrix K_{t_i} is given by

$$
\begin{pmatrix}
\vdots \\
b[t_{i-q-2},\dots,t_{i-1}](t) \\
b[t_{i-q-1},\dots,t_{i-1},1](t) \\
b[t_{i-q},\dots,1,1](t) \\
\vdots \\
b[t_{i-2},t_{i-1},1,\dots,1](t) \\
b[t_{i-1},1,\dots,1](t)
\end{pmatrix}
=
\begin{pmatrix}
\ddots & & & & & \vdots \\
& 1\ 0 & 0 & & 0 & 0 \\
& 0\ 1 & 1-\lambda_{i-q-1} & & 0 & 0 \\
& 0\ 0 & \lambda_{i-q-1} & & 0 & 0 \\
& & & \ddots & & \\
& 0\ 0 & 0 & & 1-\lambda_{i-1} & 0 \\
\dots & 0\ 0 & 0 & & \lambda_{i-1} & 1
\end{pmatrix}
\begin{pmatrix}
\vdots \\
b[t_{i-q-2},\dots,t_{i-1}](t) \\
b[t_{i-q-1},\dots,t_i](t) \\
b[t_{i-q},\dots,t_i,1](t) \\
\vdots \\
b[t_{i-1},t_i,1,\dots,1](t) \\
b[t_i,1,\dots,1](t)
\end{pmatrix},
$$

where

$$
\lambda_j = \frac{t_i - t_j}{1 - t_j}.
$$

Using this construction, the knot insertion matrix K_τ is given via

$$
K_\tau = K_{t_1} K_{t_2} \dots K_{t_N}.
$$

In the last step we multiply with the corresponding s-dependent functions.

3. Compute tensor-product basis. Combining steps 1 and 2 with the tensor product representation leads to

$$
(B_i^p[S](s) b_j^i(t))_{j=0,\dots,i}^T = E_i^q (B_i^p[S](s) b_j^q(t))_{j=0,\dots,q}^T
$$
$$
= E_i^q K_\tau (B_i^p[S](s)\ B_j^q[T](t))_{j=0,\dots,N+q}^T,
$$

for $0 \le i \le p$.

In the following we compute the coefficients for some example configurations.

3.4   Some Example Configurations

We start with an example of a patch of degree $p = q = 2$.

Example 1. Let $S = (0,0,0,h,2h,3h,\dots)$ and $T = (0,0,0,1/4,1/2,3/4,1,1,1)$. We want to find a representation of the basis $\mathbb{S}^2$ for $\mathscr{S}^2(\mathbf{u},\mathbf{S})$ with respect to the standard basis $B_i^2[S] B_j^2[T]$ of $\mathscr{S}$. We denote the new basis functions by $\tilde{B}_{(i,j)}^p =$

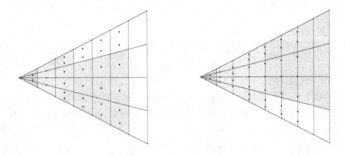

Fig. 3. Quadratic (left) and cubic (right) singular B-spline patch of type A (Color figure online)

$B_i^p[S](s)b_j^i(t)$. The basis functions we need to construct are $\tilde{B}_{(i,j)}^2$, with $0 \le i \le 2$ and $0 \le j \le i$.

The left hand side of Fig. 3 depicts a schematic overview of the bi-quadratic patch. The degree elevation matrices E_i^q for $q = 2$ are given by

$$E_0^2 = \begin{pmatrix} 1 & 1 & 1 \end{pmatrix}, E_1^2 = \begin{pmatrix} 1 & \frac{1}{2} & 0 \\ 0 & \frac{1}{2} & 1 \end{pmatrix}, E_2^2 = \begin{pmatrix} 1 & 0 & 0 \\ 0 & 1 & 0 \\ 0 & 0 & 1 \end{pmatrix}$$

and the knot insertion matrix K_t, with $(b_i^2)_{i=0,1,2}^T = K_t(B_j^2[T])_{j=0,\dots,5}^T$, fulfills

$$K_t = \begin{pmatrix} 1 & \frac{3}{4} & \frac{3}{8} & \frac{1}{8} & 0 & 0 \\ 0 & \frac{1}{4} & \frac{1}{2} & \frac{1}{2} & \frac{1}{4} & 0 \\ 0 & 0 & \frac{1}{8} & \frac{3}{8} & \frac{3}{4} & 1 \end{pmatrix}.$$

Hence we conclude

$$\tilde{B}_{(0,0)}^2 = E_0^2 K_t(B_0^2[S]B_j^2[T])_{j=0,\dots,5}^T = \begin{pmatrix} 1 & 1 & 1 & 1 & 1 & 1 \end{pmatrix} (B_0^2[S]B_j^2[T])_{j=0,\dots,5}^T,$$

$$\begin{pmatrix} \tilde{B}_{(1,0)}^2 \\ \tilde{B}_{(1,1)}^2 \end{pmatrix} = E_1^2 K_t(B_1^2[S]B_j^2[T])_{j=0,\dots,5}^T = \begin{pmatrix} 1 & \frac{7}{8} & \frac{5}{8} & \frac{3}{8} & \frac{1}{8} & 0 \\ 0 & \frac{1}{8} & \frac{3}{8} & \frac{5}{8} & \frac{7}{8} & 1 \end{pmatrix} (B_1^2[S]B_j^2[T])_{j=0,\dots,5}^T$$

and

$$\begin{pmatrix} \tilde{B}_{(2,0)}^2 \\ \tilde{B}_{(2,1)}^2 \\ \tilde{B}_{(2,2)}^2 \end{pmatrix} = E_2^2 K_t(B_2^2[S]B_j^2[T])_{j=0,\dots,5}^T = \begin{pmatrix} 1 & \frac{3}{4} & \frac{3}{8} & \frac{1}{8} & 0 & 0 \\ 0 & \frac{1}{4} & \frac{1}{2} & \frac{1}{2} & \frac{1}{4} & 0 \\ 0 & 0 & \frac{1}{8} & \frac{3}{8} & \frac{3}{4} & 1 \end{pmatrix} (B_2^2[S]B_j^2[T])_{j=0,\dots,5}^T.$$

The newly defined basis functions are visualized via their Greville ascissae (red dots). The part of the domain containing the singularity is the red triangle to the left. The part colored in light red is the support of the newly defined basis functions. The remaining part of the patch is not influenced by the modification of the function space. One standard basis function is visualized via its Greville abscissa and support (colored in blue and light blue, respectively).

Figure 4 depicts the basis functions $\tilde{B}_{(0,0)}^2$, $\tilde{B}_{(1,0)}^2$, $\tilde{B}_{(1,1)}^2$, $\tilde{B}_{(2,0)}^2$, $\tilde{B}_{(2,1)}^2$ and $\tilde{B}_{(2,2)}^2$. The parametrization is the same as the one shown in the left hand side of Fig. 3. Here, only the support of the new basis functions (the region colored in light red) is shown.

The second example is a patch of degree $p = q = 3$.

Example 2. Let S and T be the same knot vectors as in the previous example. Again, we represent the new basis functions $\tilde{B}_{(i,j)}^3$, with $0 \le i \le 3$ and $0 \le j \le i$, of $\mathscr{S}^k$ with respect to the standard basis $B_i^3[S]B_j^3[T]$ of $\mathscr{S}$.

Note that in general the degree elevation matrix E_0^q is a row vector of length $q + 1$ with entry 1 in each column. This derives from the fact that E_0^q arises from degree elevation of a constant function. Obviously, the matrix E_q^q is the unit matrix of size $(q+1) \times (q+1)$. The remaining degree elevation matrices and the knot insertion matrix K_t fulfill

$$
E_1^3 = \begin{pmatrix} 1 & \frac{2}{3} & \frac{1}{3} & 0 \\ 0 & \frac{1}{3} & \frac{2}{3} & 1 \end{pmatrix}, \quad
E_2^3 = \begin{pmatrix} 1 & \frac{1}{3} & 0 & 0 \\ 0 & \frac{2}{3} & \frac{2}{3} & 0 \\ 0 & 0 & \frac{1}{3} & 1 \end{pmatrix}, \quad
K_t = \begin{pmatrix} 1 & \frac{3}{4} & \frac{3}{8} & \frac{3}{32} & 0 & 0 & 0 \\ 0 & \frac{1}{4} & \frac{1}{2} & \frac{13}{32} & \frac{1}{8} & 0 & 0 \\ 0 & 0 & \frac{1}{8} & \frac{13}{32} & \frac{1}{2} & \frac{1}{4} & 0 \\ 0 & 0 & 0 & \frac{3}{32} & \frac{3}{8} & \frac{3}{4} & 1 \end{pmatrix}.
$$

Hence we conclude

$$
\tilde{B}_{(0,0)}^3 = \begin{pmatrix} 1 & 1 & 1 & 1 & 1 & 1 & 1 \end{pmatrix} (B_0^3[S]B_j^3[T])_{j=0,\dots,6}^T,
$$

$$
\begin{pmatrix} \tilde{B}_{(1,0)}^3 \\ \tilde{B}_{(1,1)}^3 \end{pmatrix} = \begin{pmatrix} 1 & \frac{11}{12} & \frac{3}{4} & \frac{1}{2} & \frac{1}{4} & \frac{1}{12} & 0 \\ 0 & \frac{1}{12} & \frac{1}{4} & \frac{1}{2} & \frac{3}{4} & \frac{11}{12} & 1 \end{pmatrix} (B_1^3[S]B_j^3[T])_{j=0,\dots,6}^T
$$

and

$$
\begin{pmatrix} \tilde{B}_{(2,0)}^3 \\ \tilde{B}_{(2,1)}^3 \\ \tilde{B}_{(2,2)}^3 \end{pmatrix} = \begin{pmatrix} 1 & \frac{5}{6} & \frac{13}{24} & \frac{11}{48} & \frac{1}{24} & 0 & 0 \\ 0 & \frac{1}{6} & \frac{5}{12} & \frac{13}{24} & \frac{5}{12} & \frac{1}{6} & 0 \\ 0 & 0 & \frac{1}{24} & \frac{11}{48} & \frac{13}{24} & \frac{5}{6} & 1 \end{pmatrix} (B_2^3[S]B_j^3[T])_{j=0,\dots,6}^T
$$

as well as

$$
\begin{pmatrix} \tilde{B}_{(3,0)}^3 \\ \tilde{B}_{(3,1)}^3 \\ \tilde{B}_{(3,2)}^3 \\ \tilde{B}_{(3,3)}^3 \end{pmatrix} = \begin{pmatrix} 1 & \frac{3}{4} & \frac{3}{8} & \frac{3}{32} & 0 & 0 & 0 \\ 0 & \frac{1}{4} & \frac{1}{2} & \frac{13}{32} & \frac{1}{8} & 0 & 0 \\ 0 & 0 & \frac{1}{8} & \frac{13}{32} & \frac{1}{2} & \frac{1}{4} & 0 \\ 0 & 0 & 0 & \frac{3}{32} & \frac{3}{8} & \frac{3}{4} & 1 \end{pmatrix} (B_3^3[S]B_j^3[T])_{j=0,\dots,6}^T
$$

The right hand side of Fig. 3 depicts a schematic overview of the bi-cubic patch. The structure is the same as for the previous example.

A simple consequence of the special structure of the newly defined basis $\mathbb{S}^k$ of the function space $\mathscr{S}^k(\mathbf{u}, \mathbf{S})$ is that we can also define a corresponding dual basis.

3.5 Dual Basis

In this section we present a construction of a dual basis for the basis presented in Definition 3. Recall that the basis $\mathbb{S}^k$ is given by

$$
\left\{ B_i^p[S](s) b_j^i(t) : 0 \le j \le i \text{ and } 0 \le i \le k \right\}
$$
$$
\cup \left\{ B_i^p[S](s) B_j^q[T](t) : 0 \le j \le N+p \text{ and } k+1 \le i \le M+q \right\}.
$$

Let $\{\lambda_\ell^s\}_{\ell=0,\dots,M+p}$ be a dual basis of $\{B_i^p[S](s)\}_{i=0,\dots,M+p}$ and let $\{\lambda_\ell^t\}_{\ell=0,\dots,N+q}$ be a dual basis of $\{B_j^q[T](t)\}_{j=0,\dots,N+q}$, with

$$
\lambda_\ell^s(B_i^p[S](s)) = \delta_i^\ell \text{ and } \lambda_\ell^t(B_j^q[T](t)) = \delta_j^\ell.
$$

One possibility for such a dual basis is presented in [12]. Moreover, let $\{\mu_\ell^i\}_{\ell=0,\ldots,i}$ be a dual basis to the Bernstein polynomials $\{b_j^i(t)\}_{j=0,\ldots,i}$ of degree i, with

$$\mu_\ell^i(b_j^i(t)) = \delta_j^\ell.$$

Then, since the construction of the function space is tensor-product, the functionals

$$\{\lambda_i^s \mu_j^i : 0 \leq j \leq i \text{ and } 0 \leq i \leq k\}$$
$$\cup \{\lambda_i^s \lambda_j^t : 0 \leq j \leq N + p \text{ and } k + 1 \leq i \leq M + p\}$$

form a dual basis for the basis $\mathbb{S}^k$ as given in Definition 3.

4 Singular Tensor-Product Patches of Type B

In this section we discuss the construction of a smooth basis for singular parameterizations of type B, which have collinear parameter directions at a point of the boundary. A parameterization $\mathbf{G}$ is called a *singular mapping of type B* if the partial derivatives are collinear and in opposite direction at $(s, t) = (0, 0)$, i.e. there exists a $\lambda > 0$ such that

$$\frac{\partial \mathbf{G}}{\partial s}(0, 0) = -\lambda \frac{\partial \mathbf{G}}{\partial t}(0, 0),$$

leading to $\det \nabla \mathbf{G}(0, 0) = 0$. In the following we give a construction for B-spline function spaces leading to smooth isogeometric spaces, similar to the construction presented for patches of type A.

Definition 4. *Let $k \leq \min(p, q)$. The function space $\mathscr{S}^k(\mathbf{u}, \mathbf{S}) \subset \mathscr{S}$ is defined as the space of splines $f \in \mathscr{S}$, such that there exists a polynomial $\rho \in \mathbb{P}_k$ fulfilling*

$$\frac{\partial^{|\alpha|} f}{\partial s^{\alpha_1} \partial t^{\alpha_2}}(\mathbf{s}) = \frac{\partial^{|\alpha|}(\rho \circ \mathbf{u})}{\partial s^{\alpha_1} \partial t^{\alpha_2}}(\mathbf{s}) \text{ for all } \mathbf{s} \in \mathbf{S},$$

for $0 \leq \alpha_1, \alpha_2 \leq k$, $|\alpha| = \alpha_1 + \alpha_2$, where $\mathbf{u}$ is the mapping

$$\mathbf{u}: [0,1]^2 \rightarrow \Delta = \{(u, v) : 0 \leq u \leq 1, \, -1 + u \leq v \leq 1 - u\}$$
$$(s, t)^T \mapsto (s\,t, t - s)^T.$$

and $\mathbf{S} = \{(0, 0)\}$.

Moreover, we can define a Bernstein-like basis $\mathbb{S}^k$ via

$$\mathbb{S}^k = \left\{ \tilde{B}_{(i,j)}^k(s, t) : 0 \leq i, j \leq k \text{ and } i + j \leq k \right\}$$
$$\cup \left\{ B_i^p[S](s) B_j^q[\mathrm{T}](t) : 0 \leq i \leq N_1 + p, \, 0 \leq j \leq N_2 + q \text{ and } \max(i, j) > k \right\},$$

where $B_i^p[S] B_j^q[\mathrm{T}]$, with $(0, 0) \leq (i, j) \leq (N_1 + p, N_2 + q)$, is the standard basis of $\mathscr{S}$ and

$$\tilde{B}_{(i,j)}^k \in span\left(\left\{ B_{\ell_1}^p[S](s) B_{\ell_2}^q[\mathrm{T}](t) : 0 \leq \ell_1, \ell_2 \leq k \right\}\right) \tag{6}$$

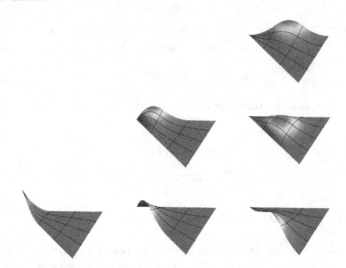

Fig. 4. Isogeometric basis functions $\tilde{B}^2_{(0,0)} \circ \mathbf{G}^{-1}$ to $\tilde{B}^2_{(2,2)} \circ \mathbf{G}^{-1}$ for a quadratic patch of type A (Color figure online)

are defined in such a way that

$$\frac{\partial^{|\alpha|} \tilde{B}^k_{(i,j)}}{\partial s^{\alpha_1} \partial t^{\alpha_2}}(\mathbf{s}) = \frac{\partial^{|\alpha|} (\beta^k_{(i,j)} \circ \mathbf{u})}{\partial s^{\alpha_1} \partial t^{\alpha_2}}(\mathbf{s}) \; \text{ for all } \mathbf{s} \in \mathbf{S},$$

for $0 \leq \alpha_1, \alpha_2 \leq k$, $|\alpha| = \alpha_1 + \alpha_2$, with triangular Bernstein basis functions

$$\beta^k_{(i,j)} : \Delta \to \mathbb{R} : (u,v) \mapsto \frac{k!}{i!j!(k-i-j)!} \left(\frac{1-u-v}{2}\right)^i \left(\frac{1-u+v}{2}\right)^j u^{k-i-j}.$$

Remark 1. One can show easily, that $\mathbb{S}^k$ is in fact a basis for $\mathscr{S}^k(\mathbf{u}, \mathbf{S})$. Moreover, similar to singular mappings of type A, the isogeometric function space $\mathscr{V}^k$ derived from $\mathscr{S}^k(\mathbf{u}, \mathbf{S})$ fulfills $\mathscr{V}^k \subset \mathscr{C}^k(\overline{\Omega})$ if the underlying triangular patch is regular.

We do not go into the details of the construction but present an example configuration allowing for $\mathscr{C}^2$-smooth bi-cubic isogeometric functions.

Example 3. Let $p = q = 3$ and $k = 2$. We construct the geometry mapping from a bi-quadratic rational triangular patch representing a quarter of a circle. Applying degree elevation to a bi-quadratic triangular Bézier parameterization $\mathbf{T}$ given by its homogeneous control points

$\mathbf{t}_{(i,j,2-i-j)}$	$j=0$	$j=1$	$j=2$
$i=0$	$(1,0,0)^T$	$(2,0,1)^T$	$(1,0,1)^T$
$i=1$	$(2,1,0)^T$	$(\sqrt{2},\sqrt{2},\sqrt{2})^T$	
$i=2$	$(1,1,0)^T$		

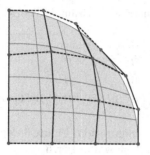

Fig. 5. Parameterization and control points for a singular patch of type B for Example 3

leads to a singular tensor-product patch $\mathbf{G} = \mathbf{T} \circ \mathbf{u}$ with control points as depicted in Fig. 5. The blue control points correspond to standard basis functions and the red control points correspond to basis functions $B_{\ell_1}^p[S]B_{\ell_2}^q[T]$, with $0 \le \ell_1, \ell_2 \le 2$, that span the space containing the new basis functions $\tilde{B}_{(i,j)}^k$, with $0 \le i + j \le 2$, as in Eq. (6).

In this case it is less intuitive to construct a basis as for patches of type A, presented in Subsect. 3.3. We will however give an overview of the basis functions

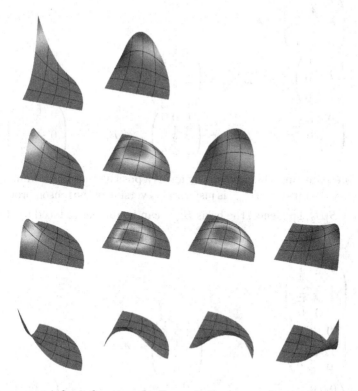

Fig. 6. Isogeometric basis functions for Example 3 corresponding to the functions given in Table 1

Table 1. Basis functions for Example 3 corresponding to Fig. 6

$B_3^3[S]B_0^3[T]$	$\tilde{B}_{(0,2)}^2$		
$B_3^3[S]B_1^3[T]$	$\tilde{B}_{(0,1)}^2$	$\tilde{B}_{(1,1)}^2$	
$B_3^3[S]B_2^3[T]$	$\tilde{B}_{(0,0)}^2$	$\tilde{B}_{(1,0)}^2$	$\tilde{B}_{(2,0)}^2$
$B_3^3[S]B_3^3[T]$	$B_2^3[S]B_3^3[T]$	$B_1^3[S]B_3^3[T]$	$B_0^3[S]B_3^3[T]$

for the presented example configuration. Figure 6 depicts the isogeometric basis functions defined on the patch shown in Fig. 5.

The isogeometric functions as depicted in Fig. 6 correspond to the B-spline functions listed in Table 1.

Given the matrix of tensor-product basis functions

$$\mathscr{B} = \begin{pmatrix} B_2^3[S]B_0^3[T] & B_1^3[S]B_0^3[T] & B_0^3[S]B_0^3[T] \\ B_2^3[S]B_1^3[T] & B_1^3[S]B_1^3[T] & B_0^3[S]B_1^3[T] \\ B_2^3[S]B_2^3[T] & B_1^3[S]B_2^3[T] & B_0^3[S]B_2^3[T] \end{pmatrix},$$

the functions $\tilde{B}_{(i,j)}^2$ are given as

$$\tilde{B}_{(0,2)}^2 = \begin{pmatrix} \frac{2}{3} & \frac{5}{12} & \frac{1}{4} \\ \frac{2}{9} & \frac{5}{36} & \frac{1}{12} \\ 0 & 0 & 0 \end{pmatrix} : \mathscr{B}$$

$$\tilde{B}_{(0,1)}^2 = \begin{pmatrix} 0 & 0 & 0 \\ \frac{1}{3} & \frac{1}{9} & 0 \\ \frac{1}{3} & \frac{1}{9} & 0 \end{pmatrix} : \mathscr{B} \quad \tilde{B}_{(1,1)}^2 = \begin{pmatrix} \frac{1}{3} & \frac{1}{2} & \frac{1}{2} \\ \frac{1}{3} & \frac{1}{2} & \frac{1}{2} \\ \frac{2}{9} & \frac{1}{3} & \frac{1}{3} \end{pmatrix} : \mathscr{B}$$

$$\tilde{B}_{(0,0)}^2 = \begin{pmatrix} 0 & 0 & 0 \\ 0 & 0 & 0 \\ \frac{1}{9} & 0 & 0 \end{pmatrix} : \mathscr{B} \quad \tilde{B}_{(1,0)}^2 = \begin{pmatrix} 0 & 0 & 0 \\ \frac{1}{9} & \frac{1}{9} & 0 \\ \frac{1}{3} & \frac{1}{3} & 0 \end{pmatrix} : \mathscr{B} \quad \tilde{B}_{(2,0)}^2 = \begin{pmatrix} 0 & \frac{1}{12} & \frac{1}{4} \\ 0 & \frac{5}{36} & \frac{5}{12} \\ 0 & \frac{2}{9} & \frac{2}{3} \end{pmatrix} : \mathscr{B}.$$

The basis presented in Table 1 obviously forms a positive partition of unity. As one can easily see, the function $\tilde{B}_{(0,0)}^2$ is just a scaled version of the tensor-product basis function $B_2^3[S]B_2^3[T]$. Hence, the basis $\tilde{B}_{(i,j)}^2$ can be further reduced to obtain the basis $\widehat{B}_{(i,j)}^2$,

$$\widehat{B}_{(0,2)}^2 = \begin{pmatrix} \frac{2}{3} & \frac{5}{12} & \frac{1}{4} \\ \frac{2}{9} & \frac{5}{36} & \frac{1}{12} \\ 0 & 0 & 0 \end{pmatrix} : \mathscr{B}$$

$$\widehat{B}_{(0,1)}^2 = \begin{pmatrix} 0 & 0 & 0 \\ \frac{7}{12} & \frac{7}{36} & 0 \\ 0 & \frac{7}{36} & 0 \end{pmatrix} : \mathscr{B} \quad \widehat{B}_{(1,1)}^2 = \begin{pmatrix} \frac{1}{3} & \frac{1}{2} & \frac{1}{2} \\ 0 & \frac{1}{3} & \frac{1}{2} \\ 0 & 0 & \frac{1}{3} \end{pmatrix} : \mathscr{B}$$

$$\widehat{B}_{(0,0)}^2 = \begin{pmatrix} 0 & 0 & 0 \\ 0 & 0 & 0 \\ 1 & 0 & 0 \end{pmatrix} : \mathscr{B} \quad \widehat{B}_{(1,0)}^2 = \begin{pmatrix} 0 & 0 & 0 \\ \frac{7}{36} & \frac{7}{36} & 0 \\ 0 & \frac{7}{12} & 0 \end{pmatrix} : \mathscr{B} \quad \widehat{B}_{(2,0)}^2 = \begin{pmatrix} 0 & \frac{1}{12} & \frac{1}{4} \\ 0 & \frac{5}{36} & \frac{5}{12} \\ 0 & \frac{2}{9} & \frac{2}{3} \end{pmatrix} : \mathscr{B}$$

as depicted in Fig. 7.

In the following we briefly discuss a way to generalize to higher dimensions.

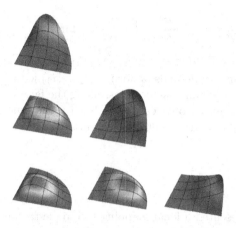

Fig. 7. Isogeometric basis functions $\widehat{B}^2_{(i,j)} \circ \mathbf{G}^{-1}$ for Example 3

5   Constructions for Higher Dimension

The approach presented here can also be generalized to higher dimensions. On the one hand one can generate smooth isogeometric function spaces on surfaces embedded in $\mathbb{R}^3$ directly by substituting the planar triangular patch with a triangular surface patch. This may be of interest when dealing with partial differential equations on surfaces or for implementations of a boundary element method (e.g. [13]).

The basic idea behind this generalization is to consider $\mathbf{f} = (g_0, g_1, g_2, g_3, f)^T$, with $g_i, f \in \mathscr{S}^k(\mathbf{u}, \mathbf{S})$ for either type A or type B. Then, the isogeometric function

$$\varphi: \quad \Omega \to \mathbb{R}$$

$$\mathbf{x} \mapsto \frac{f}{g_0} \circ \left(\frac{g_1}{g_0}, \frac{g_2}{g_0}, \frac{g_3}{g_0} \right)^{-1} (\mathbf{x})$$

defined on the surface $\Omega = \mathbf{G}(\mathbf{B}^\circ) \subset \mathbb{R}^3$ is smooth of order k if the underlying triangular surface patch is regular.

On the other hand, one can define smooth isogeometric spaces on singularly parameterized volumetric domains. Similar to the bivariate case, one can again define an isogeometric function represented via its graph in homogeneous coordinates

$$\mathbf{f} = (g_0, g_1, g_2, g_3, f)^T : [0,1]^3 \to \tilde{\Omega} \times \mathbb{R}$$

with $\tilde{\Omega}$ being the homogeneous representation of the physical domain Ω. On a tetrahedral domain, given by the tri-linear singular mapping $\mathbf{u}(r,s,t) = (r, r\,s, r\,s\,t)$, we can define a basis according to

$$\frac{\partial^{|\alpha|}\tilde{B}_{\mathbf{i}}^k}{\partial r^{\alpha_1}\partial s^{\alpha_2}\partial t^{\alpha_3}}(\mathbf{s}) = \frac{\partial^{|\alpha|}(\beta_{\mathbf{i}}^k \circ \mathbf{u})}{\partial r^{\alpha_1}\partial s^{\alpha_2}\partial t^{\alpha_3}}(\mathbf{s}) \quad \text{for all } \mathbf{s} = (r,s,t) \in \mathbf{S},$$

for $\mathbf{S} = \{0\} \times [0,1] \times [0,1]$, $0 \le \alpha_1, \alpha_2, \alpha_3 \le k$, $|\alpha| = \alpha_1 + \alpha_2 + \alpha_3$, and tri-variate tetrahedral Bernstein polynomials $\beta_{\mathbf{i}}^k$ for $\mathbf{i} = (i_1, i_2, i_3)$ with $0 \le i_1 + i_2 + i_3 \le k$. Moreover, one needs to enforce smoothness along the face $s = 0$ of the unit box, which collapses to a line in physical space. Such a construction corresponds to the findings in [16] about the smoothness conditions of isogeometric functions on volumetric patches.

6   Conclusion

In this paper we presented a local mapping technique to construct isogeometric functions of arbitrary smoothness over singularly parameterized domains. The construction works for domains of arbitrary dimension. We focused on two dimensional patches containing exactly one point of singularity in physical space. However, the concept can be generalized to embedded surfaces and volumes as well as to structurally more complex domains.

One direction of future research is the study and development of a refinement scheme maintaining the smoothness without enforcing additional smoothness conditions after each refinement step. Another area of interest, which arises for the presented isogeometric function spaces, is the question of approximation properties and convergence behavior. It is not clear, following the presented construction, whether or not the approximation properties of the function space are optimal.

Acknowledgments. The work presented here is partially supported by the Italian MIUR through the FIRB "Futuro in Ricerca" Grant RBFR08CZ0S and by the European Research Council through the FP7 Ideas Consolidator Grant *HIgeoM*. This support is gratefully acknowledged.

References

1. Beirão da Veiga, L., Buffa, A., Rivas, J., Sangalli, G.: Some estimates for h-p-k-refinement in isogeometric analysis. Numer. Math. **118**, 271–305 (2011)
2. Benson, D.J., Bazilevs, Y., Hsu, M.C., Hughes, T.J.R.: Isogeometric shell analysis: the Reissner-Mindlin shell. Comput. Meth. Appl. Mech. Eng. **199**(5–8), 276–289 (2010)
3. Cottrell, J.A., Reali, A., Bazilevs, Y., Hughes, T.J.R.: Isogeometric analysis of structural vibrations. Comput. Meth. Appl. Mech. Eng. **195**(41–43), 5257–5296 (2006)
4. Farin, G.E.: NURBS: From Projective Geometry to Practical Use. Ak Peters Series. A.K Peters, Natick (1999)

5. Hu, S.-M.: Conversion between triangular and rectangular Bézier patches. Comput. Aided Geom. Des. **18**(7), 667–671 (2001)
6. Hughes, T.J.R., Cottrell, J.A., Bazilevs, Y.: Isogeometric analysis: CAD, finite elements, NURBS, exact geometry and mesh refinement. Comput. Meth. Appl. Mech. Eng. **194**(39–41), 4135–4195 (2005)
7. Kiendl, J., Bazilevs, Y., Hsu, M.-C., Wüchner, R., Bletzinger, K.-U.: The bending strip method for isogeometric analysis of Kirchhoff-Love shell structures comprised of multiple patches. Comput. Meth. Appl. Mech. Eng. **199**(37–40), 2403–2416 (2010)
8. Kiendl, J., Bletzinger, K.-U., Linhard, J., Wüchner, R.: Isogeometric shell analysis with Kirchhoff-Love elements. Comput. Meth. Appl. Mech. Eng. **198**(49–52), 3902–3914 (2009)
9. Lu, J.: Circular element: isogeometric elements of smooth boundary. Comput. Meth. Appl. Mech. Eng. **198**(30–32), 2391–2402 (2009)
10. Piegl, L., Tiller, W.: The NURBS book. Springer, London (1995)
11. Prautzsch, H., Boehm, W., Paluszny, M.: Bézier and B-Spline Techniques. Springer, New York (2002)
12. Schumaker, L.L.: Spline Functions: Basic Theory. Cambridge University Press, Cambridge (2007)
13. Simpson, R.N., Scott, M.A., Taus, M., Thomas, D.C., Lian, H.: Acoustic isogeometric boundary element analysis. Comput. Meth. Appl. Mech. Eng. **269**, 265–290 (2014)
14. Takacs, T., Jüttler, B.: Existence of stiffness matrix integrals for singularly parameterized domains in isogeometric analysis. Comput. Meth. Appl. Mech. Eng. **200**(49–52), 3568–3582 (2011)
15. Takacs, T., Jüttler, B.: H^2 regularity properties of singular parameterizations in isogeometric analysis. Graph. Model. **74**(6), 361–372 (2012)
16. Takacs, T., Jüttler, B., Scherzer, O.: Derivatives of isogeometric functions on n-dimensional rational patches in $\mathbb{R}^d$. Comput. Aided Geom. Des. **31**(78), 567–581 (2014). Recent Trends in Theoretical and Applied Geometry

Reflexive Symmetry Detection in Single Image

Zhongwei Tang[1], Pascal Monasse[2]([✉]), and Jean-Michel Morel[1]

[1] CMLA-ENS Cachan, 61 Avenue du Président Wilson, 94235 Cachan, France
{zhongwei.tang,jean-michel.morel}@cmla.ens-cachan.fr
[2] LIGM (UMR CNRS 8049), ENPC,
Université Paris-Est, 77455 Marne-la-Vallée, France
monasse@imagine.enpc.fr

Abstract. Reflective symmetry can be used as a strong prior for many computer vision tasks. We interpret the planar reflective symmetry detection by using the property of an involution, which implies that two pairs of matched points are enough to define a planar reflective symmetry observed from a non-frontal viewpoint. This makes the reflective symmetry estimation as efficient as the classical homography estimation in binocular stereovision. This simple reflective symmetry computation can be plugged into any multiple model estimation to detect multiple symmetries at different scales and locations in images. The experimental results show that the proposed method is able to detect single and multiple reflective symmetries both in frontal and non fronto-parallel viewpoints.

Keywords: Reflective symmetry · Perspective distortion · Involution · Multiple model estimation

1 Introduction

Reflective symmetry is very common in artificial environments, and also in the natural world, especially with living forms. For the semantic interpretation of a photograph, an accurate detection of the symmetries is a valuable mid-level input. Leveraging on reliable feature point detection and matching, such as SIFT [7], a few methods were proposed to detect reflective symmetries in images. For example, Loy and Eklundh [8] use a Hough transform approach to symmetry detection, where each pair of matching points contributes with a weight to a parameterized symmetry axis set. They show that SIFT descriptors can be directly manipulated to provide descriptors of the virtual reflected image. Though it may be faster than applying the full SIFT pipeline to the mirror image, we find it simpler to use the latter. A main limitation of the method is the restriction to fronto-parallel symmetry, since the symmetry axis is supposed to be the median line of the segment joining matching points. When the symmetric structure is viewed from a titled angle, this assumption does not hold and the algorithm is not applicable. A related approach is the one of Liu et al. [6], that tries a large set of possible symmetry lines and computes for each a matching score based on edge directions.

© Springer International Publishing Switzerland 2015
J.-D. Boissonnat et al. (Eds.): Curves and Surfaces 2014, LNCS 9213, pp. 452–460, 2015.
DOI: 10.1007/978-3-319-22804-4_31

One of the best performing methods is the one of Cho and Lee [1]. It finds regions of symmetry by growing from a matched pair based on photometric similarity and geometric consistency. Still, it maintains the assumption of fronto-parallel view of the symmetric object.

Being able to detect symmetries without relying on the fronto-parallel view assumption is still a challenge. Though this assumption was lifted in some algorithms for detection of repetitions, such as the one of Tuytelaars et al. [11], to the best of our knowledge it is the first time a fully developped algorithm not relying on this assumption is presented.

From a single pair of matching points, the symmetry axis cannot be detected when the fronto-parallel hypothesis is not satisfied. However, we show that two matching pairs are enough to estimate it. This suggests an algorithm of type RANSAC [4] to recover a symmetry. In Sect. 2, we formulate the geometric problem, propose a parameterized representation of the symmetry and a simple method to estimate it, much akin to homography estimation from four point matches. An algorithm detecting the symmetry is then proposed. Extending it to the detection of multiple symmetries in the image is not straightforward. Section 3 shows that relying on the J-linkage algorithm [10] provides an efficient generalization to the detection of multiple symmetries in skewed views. Experiments on synthetic and real images and qualitatively and quantitatively evaluated in Sect. 4, showing that the proposed method establishes the state of the art in detection accuracy. Finally, some extensions are suggested in Sect. 5.

2 Single Symmetry Detection

2.1 Problem Statement

Let us consider 2D points P_i and P_i' that are image of each other under a planar symmetry. Assuming the points are represented by 3-vectors in homogeneous coordinates p_i and p_i', we can write

$$p_i' = Tp_i \text{ with } T = \begin{pmatrix} -1 & 0 & 0 \\ 0 & 1 & 0 \\ 0 & 0 & 1 \end{pmatrix}. \tag{1}$$

The equalities have to be understood up to a scalar factor. If these points lie on a 2D plane viewed by a pinhole camera, the points undergo a projective transformation, represented by homography matrix H [5]. Let $q_i = Hp_i$ and $q_i' = Hp_i'$ be the projections of P_i and P_i' into the image plane. From (1), we deduce

$$q_i' = \hat{H}q_i \text{ with } \hat{H} = HTH^{-1}. \tag{2}$$

From the obvious equality $T^2 = I_3$, the 3×3 identity matrix, we deduce the analogous equation for $\hat{H}$

$$\hat{H}^2 = I_3, \tag{3}$$

showing that $\hat{H}$, the H-conjugate of T, is still an involution.

To estimate such an involution $\hat{H}$ from point correspondences (q_i, q_i'), we propose two methods, exposed in the following sections. The first one is based on a parameterization of an involution matrix and requires algebraic geometry techniques. The second one is simpler and direct, related to the standard 4-point homography estimation.

2.2 Parameterized Involution Computation

Notice that

$$\begin{pmatrix} a & 0 \\ 0 & B \end{pmatrix} T \begin{pmatrix} a & 0 \\ 0 & B \end{pmatrix}^{-1} = T, \tag{4}$$

for any real $a \neq 0$ and B a 2×2 invertible matrix. We can always choose a and B such that

$$H \begin{pmatrix} a & 0 \\ 0 & B_{2\times 2} \end{pmatrix} = \begin{pmatrix} 1 & x & y \\ z & 1 & 0 \\ t & 0 & 1 \end{pmatrix} := \tilde{H}(x, y, z, t) \text{ with } xz + yt \neq 1. \tag{5}$$

In effect, the equation above parameterizes an involution by four parameters x, y, z, and t. This yields

$$\hat{H} = \tilde{H} T \tilde{H}^{-1} = \frac{1}{1 - xz - yt} \begin{pmatrix} 1 - xz - yt & 2x & 2y \\ -2z & 1 + xz - yt & 2yz \\ -2t & 2xt & 1 - xz + yt \end{pmatrix}. \tag{6}$$

We see that from two pairs of corresponding points (q_i, q_i'), $i = 1, 2$, (2) results in four algebraic equations of degree 2 in the unknowns (x, y, z, t). This can yield up to $2^4 = 16$ real solutions. These solutions can be obtained through Gröbner bases extraction for example [2]. To disambiguate these solutions, a third pair should be used. This method has the advantage that it parameterizes explicitly the unknown matrix $\hat{H}$, but the solution presented in the next section, while providing less insight, has the advantage of being much simpler to compute.

2.3 Direct Involution Computation

Instead of parameterizing explicitly an involution matrix $\hat{H}$, we can just write

$$\begin{cases} q_i' = \hat{H} q_i \\ q_i = \hat{H} q_i' \end{cases} \tag{7}$$

Equalities being up to a scalar factor, a pair of matching points yields 4 independent equations. Considering that $\hat{H}$ has 8 variables (a 3×3 matrix defined up to scale), two distinct pairs provide exactly one linear system for the coefficients of $\hat{H}$. This is very similar to the standard homography estimation from 4 point matches, namely here (q_1, q_1'), (q_1', q_1), (q_2, q_2'), and (q_2', q_2).

Notice that a geometric interpretation of the involution is quite simple. The line of symmetry L can be parameterized by two points:

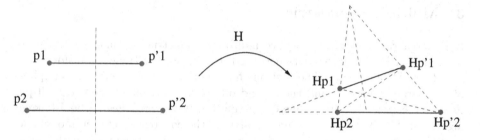

Fig. 1. Reflective symmetry through a homography. See text for the geometric construction illustrated in right figure.

1. The intersection of the lines $p_1 \times p_2$ and $p'_1 \times p'_2$.
2. The intersection of the diagonal $p_1 \times p'_2$ and $p_2 \times p'_1$.

The notation $\times$ corresponds to the cross-product of homogeneous vectors representing the points and gives the three coefficients of the line equation. Now, let C be the intersection of the lines $p_1 \times p'_1$ and $p_2 \times p'_2$. This point is at infinity in the frontal view. The image p'_3 of a point p_3 through the symmetry is obtained through the following process:

1. Let D be the intersection of lines L and $p_1 \times p_3$.
2. The point p'_3 is at the intersection of $C \times p_3$ and $D \times p'_1$.

Of course, the above construction can be obtained also with p_2 and p'_2.

The geometric construction described above is based only on line intersections. The consequence is that it can be translated directly in the non-frontal view through the homography H. This is illustrated in Fig. 1.

2.4 Algorithm

The algorithm to detect a symmetry in the image is to use the RANSAC procedure [4] with the 2-correspondence algorithm presented in previous section. More precisely:

1. Compute reflected image $\tilde{I}(x, y) = I(w - x, y)$ of original image I of width w.
2. Match SIFT keypoints in I and $\tilde{I}$: $(q_i, q'_i)_{i \in \{1, \cdots, n\}}$.
3. Run RANSAC using 2-correspondence solver:
   - Take random pair $(i, j) \in \{1, \cdots, n\}$, and compute $H_{i,j}$.
   - If $\#\{k : d(H_{i,j} q_i, q'_i) < \sigma\}$ exceeds its current maximal value, update the best homography $H \leftarrow H_{i,j}$.
   - If planned number of iterations is not yet reached, iterate.

To make the process more robust, we can use the scale and orientation of SIFT descriptors to check a homography $H_{i,j}$: after rectification, we must have for $k = i$ and $k = j$

$$|s(p_k) - s(p'_k)| < \delta \max(s(p_k), s(p'_k)) \tag{8}$$
$$\cos(\theta_k + \theta'_k) < -1 + \varepsilon, \tag{9}$$

where s is the scale associated to keypoint and θ the orientation.

3   Multiple Symmetries

It is not a rare occurence to have multiple symmetries in images. The standard approaches to detect parameterized models are the Hough transform and the RANSAC algorithm. The first one has the problem of an efficient sampling of the parameter space and the second one is not naturally tuned to multiple model detection. An alternative combining the two approaches is the J-linkage algorithm [10]. It stores in a binary matrix C the coherence between model and data. More precisely, $C(i, j) = 1$ iff data index i is compatible with model number j. A row $C(i, :)$ is the preference set (PS) of sample i and a column $C(:, j)$ is the consensus set (CS) of model j. The J-linkage algorithm clusters the set of data points in a bottom-up procedure. Each point i is initially in its own cluster $C_i = \{i\}$. The two clusters (i, i') with minimal Jaccard distance are merged into cluster $C_i \cup C_{i'}$ whose PS is the intersection of the two PSs of i and i'. This is iterated until the minimal Jaccard distance becomes 1. The Jaccard distance is:

$$d_J(A, B) = \frac{|A \cup B| - |A \cap B|}{|A \cup B|}. \tag{10}$$

A model is adjusted to each final cluster by least square fitting to the data points.

The success of J-linkage relies crucially on a non-uniform sampling of the models, contrary to RANSAC. The reason is that if true models are oversampled, we get stable row features (PS). We use the following sampling strategy to select pairs of matching points that generate a model:

$$\mathbb{P}(m_i) = \frac{1}{Z_1} \exp\left(-\frac{1}{\sigma_d^2}(\|SIFT(q_i) - SIFT(q_i')\| - d_0)^2 - \frac{1}{\sigma_l^2}(\|q_i - q_i'\| - l_0)^2 \right) \tag{11}$$

$$\mathbb{P}(m_j|m_i) = \frac{1}{Z_2} \exp\left(-\frac{1}{\sigma_c^2}(\|c_i - c_j\| - c_0)^2 \right), \tag{12}$$

where $m_i = (q_i, q_i')$. Z_1 and Z_2 are normalization factors. The parameters are estimated based on statistics of distances:

$$d_0 = \frac{1}{2N} \sum_{m_i} \|SIFT(q_i) - SIFT(q_i')\| \tag{13}$$

$$\sigma_d^2 = \frac{1}{10} \max_{m_i} \|SIFT(q_i) - SIFT(q_i')\|^2 \tag{14}$$

$$l_o = \sqrt{w^2 + h^2} \tag{15}$$

$$\sigma_l^2 = \frac{1}{10} \max_{m_i} \|q_i - q_i'\|^2 \tag{16}$$

$$c_0 = \frac{1}{20} \sqrt{w^2 + h^2} \tag{17}$$

$$\sigma_c^2 = \frac{1}{10} \max_{m_i, m_j} \|c_i - c_j\|^2. \tag{18}$$

The rationale is that we want high quality matches to be sampled with high probability and that the second match is neither too close nor too far away from the first one.

Table 1. Performance comparison on the PSU dataset. TP=true positive, FP=false positive and GT=ground truth symmetries.

	Synthetic single symmetry				Synthetic multiple symmetries			
	LE [8]	LHXS [6]	CL [1]	Proposed	LE [8]	LHXS [6]	CL [1]	Proposed
TP/GT	92 %	62 %	100 %	100 %	35 %	28 %	77 %	67 %
FP/GT	12 %	0 %	15 %	0 %	4 %	8 %	33 %	10 %
	Real single symmetry				Real multiple symmetries			
	LE [8]	LHXS [6]	CL [1]	Proposed	LE [8]	LHXS [6]	CL [1]	Proposed
TP/GT	84 %	29 %	94 %	97 %	43 %	18 %	68 %	65 %
FP/GT	68 %	3 %	69 %	39 %	44 %	0 %	17 %	16 %

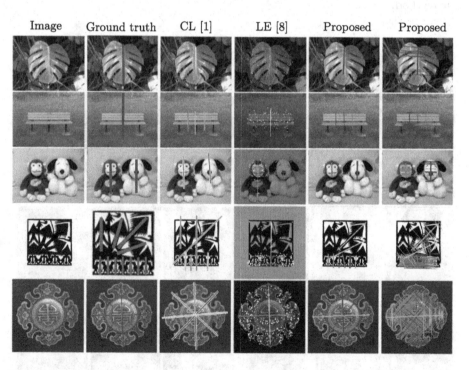

Fig. 2. Some examples of results from the PSU dataset. The last column shows the consensus set of matches for the detected symmetries, on top of the symmetry axes found by our method.

4   Experiments

In all our experiments we fixed parameters $\delta = 0.2$ and $\varepsilon = 0.25$. We first tested the proposed method on the PSU dataset[1]. Performance is reported in Table 1.

On synthetic datasets, the best performing method in the literature, due to Cho and Lee [1], detects all single symmetries, but also some erroneous ones (15 % of false positives), while the proposed algorithm gives a perfect score. The result changes when multiple symmetries are present in the image. More of them are detected by the former method while our algorithm has a significantly lower false positive rate. On real images, the true positives are roughly the same for both methods, but the false positives are smaller with our method when there is a single symmetry per image. In all cases, the method of Liu et al. [6] has a low false positive rate, but with a true positive rate that is well below the best performing methods. Figure 2 shows the results on some images of the dataset.

Additional results on other real images are shown in Fig. 3. Notice that even with significant projective distortion, the symmetries are correctly detected by our method.

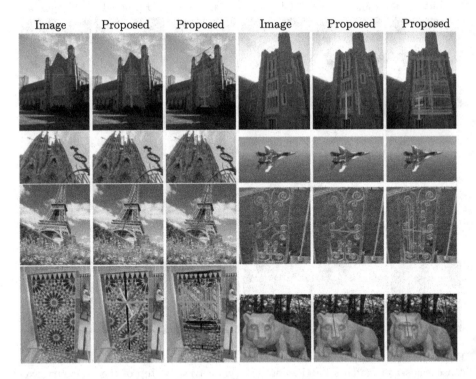

Image   Proposed   Proposed   Image   Proposed   Proposed

Fig. 3. Additional examples of our proposed method, with and without display of the supporting matches in addition of the symmetry axes.

---

[1] http://vision.cse.psu.edu/research/symmComp/.

Fig. 4. Application of the TILT algorithm [13] to the zones of symmetry detected by our algorithm (in green boxes). The rectified zones are delimited by the red boxes (Color figure online).

We applied the TILT algorithm of Zhang et al. [13] to some of the detected symmetry zones. This algorithm rectifies the zone by decomposing the intensity matrix into low-rank factors. Notice that this does not require the homography matrix H, which is not uniquely determined by the computation of $\hat{H}$ from (6) since the latest depends on only four parameters. Therefore a direct rectification of the image is not possible (Fig. 4).

5   Conclusion

We have presented a new method for multiple symmetry detection in images. It relies on multiple model fitting through the J-linkage algorithm. We have demonstrated on experiments that the algorithm is able to detect symmetries with a moderate amount of perspective distortion. This ability could be extended with a more robust invariant feature matching algorithm, such as ASIFT [12]. The main drawback is the somewhat high false positive rate of detection, even if it is not worse than comparable detectors with the highest true positive rate. A way to control that would be to introduce an *a contrario* criterion [3] for validating a model into J-linkage. An alternative would be to use MAC-RANSAC [9]. Extensions to other Euclidean repetitions, like translation and rotation, are possible future work. Finally, an implementation in IPOL journal[2] would be an interesting development.

References

1. Cho, M., Lee, K.M.: Bilateral symmetry detection via symmetry-growing. In: Proceedings of BMVC (2009)
2. Cox, D.A., Little, J.B., O'shea, D.: Using Algebraic Geometry. Graduate Texts in Mathematics, vol. 185. Springer, New York (2005)

---

[2] http://www.ipol.im/.

3. Desolneux, A., Moisan, L., Morel, J.M.: From Gestalt Theory to Image Analysis: A Probabilistic Approach, vol. 34. Springer, Heidelberg (2007)
4. Fischler, M.A., Bolles, R.C.: Random sample consensus: a paradigm for model fitting with applications to image analysis and automated cartography. Commun. ACM **24**(6), 381–395 (1981)
5. Hartley, R., Zisserman, A.: Multiple View Geometry in Computer Vision. Cambridge University Press, Cambridge (2003)
6. Liu, Y., Hays, J., Xu, Y.Q., Shum, H.Y.: Digital papercutting. In: ACM SIGGRAPH 2005 Sketches. p. 99. ACM (2005)
7. Lowe, D.G.: Object recognition from local scale-invariant features. In: Computer Vision, 1999. The Proceedings of the seventh IEEE International Conference on. vol. 2, pp. 1150–1157. IEEE (1999)
8. Loy, G., Eklundh, J.-O.: Detecting Symmetry and Symmetric Constellations of Features. In: Leonardis, A., Bischof, H., Pinz, A. (eds.) ECCV 2006. LNCS, vol. 3952, pp. 508–521. Springer, Heidelberg (2006)
9. Rabin, J., Delon, J., Gousseau, Y., Moisan, L., et al.: Mac-ransac: a robust algorithm for the recognition of multiple objects. In: Proceedings of 3DPTV 2010 (2010)
10. Toldo, R., Fusiello, A.: Robust Multiple Structures Estimation with J-Linkage. In: Forsyth, D., Torr, P., Zisserman, A. (eds.) ECCV 2008, Part I. LNCS, vol. 5302, pp. 537–547. Springer, Heidelberg (2008)
11. Tuytelaars, T., Turina, A., Van Gool, L.: Noncombinatorial detection of regular repetitions under perspective skew. Pattern Anal. Mach. Intell. IEEE Trans. **25**(4), 418–432 (2003)
12. Yu, G., Morel, J.M.: ASIFT: An algorithm for fully affine invariant comparison. Image Processing On Line 1 (2011) http://dx.doi.org/10.5201/ipol.2011.my-asift
13. Zhang, Z., Ganesh, A., Liang, X., Ma, Y.: Tilt: transform invariant low-rank textures. Int. J. Comput. Vis. **99**(1), 1–24 (2012)

The Sylvester Resultant Matrix and Image Deblurring

Joab R. Winkler$^{(\boxtimes)}$

Department of Computer Science, The University of Sheffield, Regent Court,
211 Portobello, Sheffield S1 4DP, UK
j.r.winkler@sheffield.ac.uk

Abstract. This paper describes the application of the Sylvester resultant matrix to image deblurring. In particular, an image is represented as a bivariate polynomial and it is shown that operations on polynomials, specifically greatest common divisor (GCD) computations and polynomial divisions, enable the point spread function to be calculated and an image to be deblurred. The GCD computations are performed using the Sylvester resultant matrix, which is a structured matrix, and thus a structure-preserving matrix method is used to obtain a deblurred image. Examples of blurred and deblurred images are presented, and the results are compared with the deblurred images obtained from other methods.

Keywords: Image deblurring · Sylvester resultant matrix

1 Introduction

The removal of blur from an image is one of the most important problems in image processing, and it is motivated by the many applications, which include medicine, astronomy and microscopy, in which its need arises. If the function that represents the blur, called the point spread function (PSF), is known, then a deblurred image can be computed from the PSF and a blurred form of the exact image. If, however, the PSF is known partially or not at all, additional information, for example, prior information on the image to be deblurred, must be specified. Even if the PSF is known, the computation of a deblurred image is ill-conditioned, and thus regularisation must be applied in order to obtain a stable deblurred image.

It is assumed in this paper that the PSF $\mathcal{H}$ is spatially invariant, in which case a blurred image $\mathcal{G}$ is formed by the convolution of $\mathcal{H}$ and the exact image $\mathcal{F}$, and the addition of noise $\mathcal{N}$,

$$\mathcal{G} = \mathcal{H} \otimes \mathcal{F} + \mathcal{N}, \tag{1}$$

where the PSF is assumed to be known exactly, and thus additive noise is the only source of error (apart from roundoff errors due to floating point arithmetic). The exact image $\mathcal{F}$ and PSF $\mathcal{H}$ are of orders $M \times N$ pixels and $(p+1) \times (r+1)$

© Springer International Publishing Switzerland 2015
J.-D. Boissonnat et al. (Eds.): Curves and Surfaces 2014, LNCS 9213, pp. 461–490, 2015.
DOI: 10.1007/978-3-319-22804-4_32

pixels respectively, and the blurred image is therefore $(M + p) \times (N + r)$ pixels. Equation (1) can be written as a linear algebraic equation,

$$\mathbf{g} = \mathbf{H}\mathbf{f} + \mathbf{n}, \qquad (2)$$

where $\mathbf{g}, \mathbf{f}, \mathbf{n} \in \mathbb{R}^m, \mathbf{H} \in \mathbb{R}^{m \times m}$, $m = MN$, and the vectors $\mathbf{g}, \mathbf{f}$ and $\mathbf{n}$ store the blurred image, the exact image and the added noise, respectively, and the entries of $\mathbf{H}$ are functions of the PSF [10]. The matrix $\mathbf{H}$ is ill-conditioned, and thus a simple solution of (2) will have a large error, and an additional difficulty is introduced by the large size of $\mathbf{H}$, even for small values of M and N. This large size would have considerable implications on the execution time of the computation of the solution $\mathbf{f}$ of (2), but it is shown in [10] that the spatial invariance of the PSF implies that $\mathbf{H}$ is a structured matrix, such that only a small part of $\mathbf{H}$ need be stored and computationally efficient algorithms that exploit its structure can be used. Furthermore, it is shown in [10] that regularisation procedures can be included in these structured matrix algorithms, and thus a stable solution of (2) can be computed rapidly.

It follows from (1) that the computation of $\mathcal{F}$ reduces to the deconvolution of the PSF from the blurred image $\mathcal{G}$ in the presence of noise. This computation is an example of linear deconvolution because the PSF is known, and other methods for linear deconvolution include the Wiener filter, which assumes the PSF, and power spectral densities of the noise and true image, are known, and the Lucy-Richardson algorithm. This algorithm arises from the method of maximum likelihood in which the pixel values of the exact image are assumed to have a Poisson distribution, which is appropriate for photon noise in the image [9].

Methods for linear deconvolution are not appropriate when the PSF is not known, in which case the computation of a deblurred image is called blind image deconvolution (BID). It follows from (1) and (2) that BID is substantially more difficult than linear deconvolution because it reduces to the separation of two convolved signals that are either unknown or partially known. Some methods for BID use a statistical approach, based on the method of maximum likelihood, but unlike linear deconvolution, the PSF is also estimated. Other methods for BID include constrained optimisation [4], autoregressive moving average parameter estimation and deterministic image constraints restoration techniques [13], and zero sheet separation [20, 23].

This paper considers the application of the Sylvester resultant matrix to the calculation of the PSF and its deconvolution from a blurred image.[1] This matrix, which is used extensively for polynomial computations, has been used for BID [5, 16, 19], and the Bézout matrix, which is closely related to the Sylvester matrix, has also been used for BID [15]. The work described in this paper differs, however, from the works in these references in several ways. Specifically, the z and Fourier transforms are used to calculate the PSF in [15, 16, 19], but it is shown in this paper that these transforms are not required because the pixel values of the exact, blurred and deblurred images, and the PSF, are the coefficients of polynomials, and polynomial computations, using the Sylvester matrix, are used

---

[1] This matrix will, for brevity, henceforth be called the Sylvester matrix.

to deblur an image. This matrix is structured, and a structure-preserving matrix method [22] is therefore used in this work to obtain a deblurred image.

The signal-to-noise ratio (SNR) is required in [5,15] for the calculation of the degrees in x (column) and y (row) of the bivariate polynomial $H(x, y)$ that represents the PSF because this ratio is a threshold in a stopping criterion in an algorithm for deblurring an image. A different approach is used in [16,19] because a few trial experiments, with PSFs of various sizes, are performed and visual inspection of the deblurred images is used to determine estimates of the horizontal and vertical extents of the PSF. In this paper, the given blurred image is preprocessed and it is shown this allows the degrees in x and y of $H(x, y)$ to be calculated, such that knowledge of the noise level is not required. This is an important feature of the work described in this paper because the noise level, or the SNR, may not be known, or they may only be known approximately. Furthermore, even if the noise level or the SNR are known, it cannot be assumed they are uniformly distributed across the blurred image. This non-uniform distribution implies that the assumptions of a constant noise level or constant SNR may yield poor results when one or both of these constants are used in normwise termination criteria in an algorithm for deblurring an image.

It is shown in Sect. 2 that the convolution operation defines the formation of a blurred image and the multiplication of two polynomials. This common feature allows polynomial operations, in particular greatest common divisor (GCD) computations and polynomial divisions, to be used for image deblurring. Properties of the PSF that must be considered for polynomial computations are described in Sect. 3, and it is shown in Sect. 4 that the PSF is equal to an approximate GCD (AGCD) of two polynomials. This leads to Sect. 5, in which the computation of an AGCD of two polynomials using their Sylvester matrix is described.

It is assumed the PSF is separable, and the extension of the method discussed in this paper from a separable PSF to a non-separable PSF is considered in Sect. 6. It is shown it is necessary to perform more computations of the same kind, and additional computations, when a non-separable PSF is used. Section 7 contains examples in which the deblurred images obtained using the method discussed in this paper are compared with the deblurred images obtained using other methods. The paper is summarised in Sect. 8.

2   The Convolution Operation

The multiplication of two polynomials reduces to the convolution of their coefficients, and it follows from (1) that, in the absence of noise, a blurred image is formed by the convolution of the exact image and a spatially invariant PSF. This equivalence between bivariate polynomial multiplication and the formation of a blurred image by this class of PSF is quantified by considering these two operations separately, and then showing that they yield the same result. It follows that if the blurred image $\mathcal{G}$ and the PSF $\mathcal{H}$ are represented by bivariate polynomials, then the deblurred image is formed by the division of the polynomial form of $\mathcal{G}$ by the polynomial form of $\mathcal{H}$.

Consider initially bivariate polynomial multiplication. In particular, let the pixel values $f(i,j)$ of $\mathcal{F}$ be the coefficients of a bivariate polynomial $F(x,y)$ that is of degrees $M-1$ and $N-1$ in x and y respectively,

$$F(x,y) = \sum_{i=0}^{M-1} \sum_{j=0}^{N-1} f(i,j) x^{M-1-i} y^{N-1-j},$$

and let the pixel values $h(k,l)$ of the PSF be the coefficients of a bivariate polynomial $H(x,y)$ that is of degrees p and r in x and y respectively,

$$H(x,y) = \sum_{k=0}^{p} \sum_{l=0}^{r} h(k,l) x^{p-k} y^{r-l}. \tag{3}$$

The product of these polynomials is $G_1(x,y) = F(x,y)H(x,y)$,

$$G_1(x,y) = \sum_{i=0}^{M-1} \sum_{j=0}^{N-1} \sum_{k=0}^{p} \sum_{l=0}^{r} f(i,j)h(k,l) x^{M+p-1-(i+k)} y^{N+r-1-(j+l)},$$

and the substitutions $s = i+k$ and $t = j+l$ yield

$$G_1(x,y) = \sum_{i=0}^{M-1} \sum_{j=0}^{N-1} \sum_{s=i}^{p+i} \sum_{t=j}^{r+j} f(i,j)h(s-i,t-j) x^{M+p-1-s} y^{N+r-1-t}.$$

It follows that the coefficient of $x^{M+p-1-s} y^{N+r-1-t}$ in $G_1(x,y)$ is

$$g_1(s,t) = \sum_{i=0}^{M-1} \sum_{j=0}^{N-1} f(i,j)h(s-i,t-j), \tag{4}$$

where $h(k,l) = 0$ if $k < 0$ or $l < 0$.

Consider now the formation of a blurred image by the convolution of $\mathcal{F}$ and $\mathcal{H}$, as shown in (1). A PSF $p(i,s,j,t)$ quantifies the extent to which the pixel value $f(i,j)$ at position (i,j) in $\mathcal{F}$ influences the pixel value $g_2(s,t)$ at position (s,t) in $\mathcal{G}$. The pixel value $g_2(s,t)$ is therefore given by

$$g_2(s,t) = \sum_{i=0}^{M-1} \sum_{j=0}^{N-1} f(i,j)p(i,s,j,t), \tag{5}$$

and if the PSF depends on the relative positions of the pixels in $\mathcal{F}$ and $\mathcal{G}$, and not on their absolute positions, then the PSF is spatially invariant. It follows that a spatially invariant PSF is a function of $s-i$ and $t-j$,

$$p(i,s,j,t) = h(s-i,t-j), \tag{6}$$

and thus (5) becomes

$$g_2(s,t) = \sum_{i=0}^{M-1} \sum_{j=0}^{N-1} f(i,j)h(s-i,t-j). \tag{7}$$

It follows from (4) and (7) that $g_1(s,t) = g_2(s,t)$, and thus the blurring of an image $\mathcal{F}$ by a spatially invariant PSF $\mathcal{H}$ is equivalent to the multiplication of the polynomial forms of $\mathcal{F}$ and $\mathcal{H}$.

The equivalence of polynomial multiplication and the formation of a blurred image $\mathcal{G}$ shows that $\mathcal{G}$ is larger than the exact image $\mathcal{F}$. Specifically, $\mathcal{F}$ is represented by a polynomial of degrees $M-1$ in x (column) and $N-1$ in y (row), and the PSF is represented by a polynomial of degrees p in x and r in y. It therefore follows that $\mathcal{G}$ is represented by a polynomial of degrees $M+p-1$ and $N+r-1$ in x and y respectively, and thus $\mathcal{G}$ has p and r more pixels than $\mathcal{F}$ along the columns and rows, respectively. These extra pixels are removed after the PSF is computed, when it is deconvolved from $\mathcal{G}$, thereby yielding a deblurred image that is the same size as $\mathcal{F}$.

The polynomial computations that allow a PSF to be calculated require that the different forms of a PSF be considered because they affect these computations. This issue is addressed in the next section.

3   The Point Spread Function

The applicability of polynomial computations to blind image deconvolution assumes the PSF is spatially invariant (6), as shown by the equivalence of (4) and (7). A spatially invariant PSF may be separable or non-separable, and it follows from (3) that a PSF is separable if

$$H(x,y) = H_c(x)H_r(y), \tag{8}$$

where $H_c(x)$ and $H_r(y)$ are the column and row blurring functions respectively. The assumption of separability is more restrictive than the assumption of spatial invariance, but it is included in the work described in this paper because, as discussed in Sect. 6, the removal of the separability condition introduces additional computations.

Spatial invariance of the PSF is satisfied in many problems, but there also exist problems in which it is not satisfied, and thus a spatially variant PSF must sometimes be considered. The mathematical implications of a spatially variant PSF are considerable because (5) does not reduce to (7) in this circumstance since (6) is not satisfied by a spatially variant PSF, and thus the convolution operation is not appropriate. Furthermore, the matrix $\mathbf{H}$ in (2) is not structured if the PSF is spatially variant, which has obvious implications for the computational cost, with respect to complexity and memory requirements, of the algorithm used for the solution of (2). It is therefore desirable to retain in a deblurring algorithm that uses a spatially variant PSF the computational properties associated with a spatially invariant PSF. Nagy et $al.$ [17] address this issue by considering a class of spatially variant PSFs that are formed by the addition of several spatially invariant PSFs, where each spatially invariant PSF is restricted to a small subregion of the blurred image. Piecewise constant interpolation is used to join these spatially invariant PSFs in each subregion in order to form a spatially variant PSF.

4 Polynomial Computations for Image Deblurring

It was shown in Sect. 2 that the formation of a blurred image by a spatially invariant PSF can be represented by the multiplication of the bivariate polynomials that represent the exact image and the PSF. This polynomial multiplication is considered in detail in this section and it is shown that the PSF is equal to an AGCD of two polynomials.

It follows from Sect. 2 that (1) can be written as

$$G(x, y) = H(x, y)F(x, y) + N(x, y), \tag{9}$$

where the PSF satisfies (8), and thus if it is assumed $E(x, y)$ is the uncertainty in the PSF, then (9) is generalised to

$$G(x, y) = (H_c(x)H_r(y) + E(x, y)) F(x, y) + N(x, y). \tag{10}$$

Consider two rows $x = r_1$ and $x = r_2$, and two columns $y = c_1$ and $y = c_2$, of $\mathcal{G}$,

$$G(r_1, y) = (H_c(r_1)H_r(y) + E(r_1, y)) F(r_1, y) + N(r_1, y),$$
$$G(r_2, y) = (H_c(r_2)H_r(y) + E(r_2, y)) F(r_2, y) + N(r_2, y),$$
$$G(x, c_1) = (H_c(x)H_r(c_1) + E(x, c_1)) F(x, c_1) + N(x, c_1),$$
$$G(x, c_2) = (H_c(x)H_r(c_2) + E(x, c_2)) F(x, c_2) + N(x, c_2),$$

where $H_c(r_1), H_c(r_2), H_r(c_1)$ and $H_r(c_2)$ are constants. It is adequate to consider either the equations for the rows r_1 and r_2, or the equations for the columns c_1 and c_2, because they have the same form. Consider, therefore, the equations for r_1 and r_2, which are equations in the independent variable y,

$$G(r_1, y) = (H_c(r_1)H_r(y) + E(r_1, y)) F(r_1, y) + N(r_1, y),$$
$$G(r_2, y) = (H_c(r_2)H_r(y) + E(r_2, y)) F(r_2, y) + N(r_2, y). \tag{11}$$

The polynomials $E(r_1, y)$ and $E(r_2, y)$ represent uncertainties in the PSF, and if their magnitudes are sufficiently small, then (11) can be written as

$$G(r_1, y) \approx H_c(r_1)H_r(y)F(r_1, y) + N(r_1, y),$$
$$G(r_2, y) \approx H_c(r_2)H_r(y)F(r_2, y) + N(r_2, y). \tag{12}$$

Also, if the magnitudes of the polynomials $N(r_1, y)$ and $N(r_2, y)$ that represent the noise are sufficiently small, then the approximations (12) simplify to

$$G(r_1, y) \approx H_c(r_1)H_r(y)F(r_1, y),$$
$$G(r_2, y) \approx H_c(r_2)H_r(y)F(r_2, y).$$

It follows that if the polynomials $F(r_1, y)$ and $F(r_2, y)$, that is, the polynomial forms of the rows r_1 and r_2 of $\mathcal{F}$, are coprime, then

$$H_r(y) = \text{AGCD}\,(G(r_1, y), G(r_2, y)). \tag{13}$$

The arguments r_1 and r_2 appear on the right hand side of this equation, and not on the left hand side, because it is assumed the PSF is separable, and thus r_1 and r_2 are the indices of any two rows of $\mathcal{G}$.

Equation (13) shows that the row component of a separable and spatially invariant PSF is equal to an AGCD of any two rows of a blurred image if the pixel values of these rows are the coefficients of two polynomials. It is clear that the column component $H_c(x)$ of a separable and spatially invariant PSF can be computed identically, and the PSF can then be calculated from (8). After the PSF components $H_c(x)$ and $H_r(y)$ have been calculated, the deblurred image $\tilde{\mathcal{F}}$ is calculated from an approximate form of (10),

$$\tilde{F}(x,y) \approx \frac{G(x,y)}{H_c(x)H_r(y)},\qquad(14)$$

which involves two approximate polynomial divisions.

Consider initially the approximate division for the calculation of $Q(x,y)$, which is the polynomial representation of the partially deblurred image $\mathcal{Q}$ formed after the row component of the PSF has been deconvolved from $\mathcal{G}$,

$$Q(x,y) \approx \frac{G(x,y)}{H_r(y)}.\qquad(15)$$

It follows that $H_r(y)Q(x,y) \approx G(x,y)$, and this approximate polynomial equation is applied to all the rows of $\mathcal{G}$. This approximate equation is written in matrix form,

$$\mathbf{T}_r\mathbf{q}_i \approx \mathbf{g}_i,\qquad i = 1,\ldots,M+p,\qquad(16)$$

where $\mathbf{T}_r \in \mathbb{R}^{(N+r)\times N}$ is a lower triangular Tœplitz matrix whose entries on and below the leading diagonal are the coefficients of $H_r(y)$, $\mathbf{q}_i \in \mathbb{R}^N$ is the transpose of the ith row of the partially deblurred image $\mathcal{Q}$, $\mathbf{g}_i \in \mathbb{R}^{N+r}$ is the transpose of the ith row of $\mathcal{G}$, and p and r are defined in (3). The approximations (16) are solved, in the least squares sense, for each vector $\mathbf{q}_i$ and thus the partially deblurred image $\mathcal{Q}$, which is of order $(M+p) \times N$ pixels, is obtained.

The column component of the PSF must be deconvolved from $\mathcal{Q}$ in order to obtain the deblurred image $\tilde{\mathcal{F}}$, and it follows from (14) and (15) that $\tilde{\mathcal{F}}$ is computed from the approximation

$$H_c(x)\tilde{F}(x,y) \approx Q(x,y),$$

which is applied to all the columns of $\mathcal{Q}$. These N approximations can be written in a form that is identical to (16),

$$\mathbf{T}_c\tilde{\mathbf{f}}_j \approx \mathbf{q}_j,\qquad j = 1,\ldots,N,\qquad(17)$$

where $\mathbf{T}_c \in \mathbb{R}^{(M+p)\times M}$ is a Tœplitz matrix that is similar to $\mathbf{T}_r$ except that its entries are the coefficients of the column component $H_c(x)$ of the PSF, $\tilde{\mathbf{f}}_j \in \mathbb{R}^M$ is the jth column of $\tilde{\mathcal{F}}$ and $\mathbf{q}_j \in \mathbb{R}^{M+p}$ is the jth column of the partially deblurred image $\mathcal{Q}$. It follows that the image $\tilde{\mathcal{F}}$ that results from the approximate polynomial divisions (14) is formed from the vectors $\tilde{\mathbf{f}}_j, j = 1, \ldots, N$,

$$\tilde{\mathbf{F}} = \begin{bmatrix} \tilde{\mathbf{f}}_1 \ \tilde{\mathbf{f}}_2 \cdots \tilde{\mathbf{f}}_{N-1} \ \tilde{\mathbf{f}}_N \end{bmatrix} \in \mathbb{R}^{M \times N}.$$

The analysis in this section shows that the computation of a separable and spatially invariant PSF involves two AGCD computations. The computation of an AGCD of two polynomials must be done with care because it is an ill-posed operation, and it is considered in the next section.

5   Approximate Greatest Common Divisors

This section considers the computation of an AGCD of two polynomials from their Sylvester matrix and subresultant matrices. This computation is required for the determination of the column and row components of the PSF, as shown in (13) for the row component. It is appropriate, however, to consider initially exact polynomials that have a non-constant GCD, and then describe the modifications required when inexact forms of these polynomials, which have an AGCD, are considered.

Let $\hat{p}(y)$ and $\hat{q}(y)$ be exact polynomials of degrees r and s respectively,[2] and let $\hat{d}(y) = \mathrm{GCD}(\hat{p}, \hat{q})$, where $\hat{t} = \deg \hat{d}(y) > 0$,

$$\hat{p}(y) = \sum_{i=0}^{r} \hat{p}_i y^{r-i} \quad \text{and} \quad \hat{q}(y) = \sum_{i=0}^{s} \hat{q}_i y^{s-i}. \tag{18}$$

It is shown in [1] that $\hat{d}(y)$ can be computed from the Sylvester matrix $\mathbf{S}(\hat{p}, \hat{q})$ of $\hat{p}(y)$ and $\hat{q}(y)$, and its subresultant matrices $\mathbf{S}_k(\hat{p}, \hat{q}), k = 2, \ldots, \min(r, s)$, where $\mathbf{S}_1(\hat{p}, \hat{q}) = \mathbf{S}(\hat{p}, \hat{q})$ and $\mathbf{S}_k(\hat{p}, \hat{q}) \in \mathbb{R}^{(r+s-k+1) \times (r+s-2k+2)}, k = 1, \ldots, \min(r, s)$. The Sylvester matrix $\mathbf{S}(\hat{p}, \hat{q})$ is

$$\mathbf{S}(\hat{p}, \hat{q}) = \begin{bmatrix}
\hat{p}_0 & & & & \hat{q}_0 & & & \\
\hat{p}_1 & \hat{p}_0 & & & \hat{q}_1 & \hat{q}_0 & & \\
\vdots & \hat{p}_1 & \ddots & & \vdots & \hat{q}_1 & \ddots & \\
\hat{p}_{r-1} & \vdots & \ddots & \hat{p}_0 & \hat{q}_{s-1} & \vdots & \ddots & \hat{q}_0 \\
\hat{p}_r & \hat{p}_{r-1} & \ddots & \hat{p}_1 & \hat{q}_s & \hat{q}_{s-1} & \ddots & \hat{q}_1 \\
& \hat{p}_r & \ddots & \vdots & & \hat{q}_s & \ddots & \vdots \\
& & \ddots & \hat{p}_{r-1} & & & \ddots & \hat{q}_{s-1} \\
& & & \hat{p}_r & & & & \hat{q}_s
\end{bmatrix},$$

where the coefficients of $\hat{p}(y)$ and $\hat{q}(y)$ occupy the first s columns and last r columns, respectively. It is seen that $\mathbf{S}(\hat{p}, \hat{q})$ has a partitioned structure, and it

---

[2] The degree r of $\hat{p}(y)$ is not related to the degree r in y of $H(x, y)$, which is defined in (3).

is shown in the sequel that this property may cause numerical problems. The subresultant matrices $\mathbf{S}_k(\hat{p}, \hat{q})$, $k = 2, \ldots, \min(r, s)$, are formed by deleting rows and columns from $\mathbf{S}(\hat{p}, \hat{q})$, and they also have a partitioned structure that must be considered when computational issues are addressed. The application of these matrices to the calculation of the GCD of $\hat{p}(y)$ and $\hat{q}(y)$ is based on the following theorem.

Theorem 1. *Let $\hat{p}(y)$ and $\hat{q}(y)$ be defined in (18).*

1. The degree $\hat{t}$ of the GCD of $\hat{p}(y)$ and $\hat{q}(y)$ is equal to the rank loss of $\mathbf{S}(\hat{p}, \hat{q})$,

$$\hat{t} = r + s - \operatorname{rank} \mathbf{S}(\hat{p}, \hat{q}).$$

2. The coefficients of $\hat{d}(y)$ are contained in the last non-zero rows of the upper triangular matrices U and R obtained from, respectively, the LU and QR decompositions of $\mathbf{S}(\hat{p}, \hat{q})^T$.

3. The value of $\hat{t}$ is equal to the largest integer k such that the kth subresultant matrix $\mathbf{S}_k(\hat{p}, \hat{q})$ is singular,

$$\operatorname{rank} \mathbf{S}_k(\hat{p}, \hat{q}) < r + s - 2k + 2, \quad k = 1, \ldots, \hat{t},$$
$$\operatorname{rank} \mathbf{S}_k(\hat{p}, \hat{q}) = r + s - 2k + 2, \quad k = \hat{t} + 1, \ldots, \min(r, s),$$

where

$$\mathbf{S}_k(\hat{p}, \hat{q}) = [\, C_k(\hat{p}) \ \ D_k(\hat{q}) \,], \tag{19}$$

and $C_k(\hat{p}) \in \mathbb{R}^{(r+s-k+1) \times (s-k+1)}$ and $D_k(\hat{q}) \in \mathbb{R}^{(r+s-k+1) \times (r-k+1)}$ are Tœplitz matrices.

It follows from the definition of $\hat{d}(y)$ that there exist coprime polynomials $\hat{u}(y)$ and $\hat{v}(y)$, of degrees $r - \hat{t}$ and $s - \hat{t}$ respectively, that satisfy

$$\hat{p}(y) = \hat{u}(y)\hat{d}(y) \qquad \text{and} \qquad \hat{q}(y) = \hat{v}(y)\hat{d}(y). \tag{20}$$

The next lemma considers the rank of the $\hat{t}$th subresultant matrix $\mathbf{S}_{\hat{t}}(\hat{p}, \hat{q})$.

Lemma 1. *The rank of $\mathbf{S}_{\hat{t}}(\hat{p}, \hat{q})$ is equal to $r + s - 2\hat{t} + 1$.*

Proof. Since the degree of the GCD of $\hat{p}(y)$ and $\hat{q}(y)$ is $\hat{t}$, there exists exactly one set of polynomials $\hat{u}(y)$ and $\hat{v}(y)$, defined up to an arbitrary non-zero scalar multiplier, that satisfies (20). The elimination of $\hat{d}(y)$ between the equations in (20) yields an equation, the matrix form of which is

$$\mathbf{S}_{\hat{t}}(\hat{p}, \hat{q}) \begin{bmatrix} \hat{\mathbf{v}} \\ -\hat{\mathbf{u}} \end{bmatrix} = \mathbf{0},$$

where $\mathbf{S}_k(\hat{p}, \hat{q})$ is defined in (19), and $\hat{\mathbf{u}}$ and $\hat{\mathbf{v}}$ are, respectively, vectors of the coefficients of $\hat{u}(y)$ and $\hat{v}(y)$. Since $\hat{u}(y)$ and $\hat{v}(y)$ are unique up to an arbitrary non-zero scalar multiplier, the dimension of the null space of $\mathbf{S}_{\hat{t}}(\hat{p}, \hat{q})$ is one, and it therefore follows that the rank of $\mathbf{S}_{\hat{t}}(\hat{p}, \hat{q})$ is $r + s - 2\hat{t} + 1$.

The exact polynomials $\hat{p}(y)$ and $\hat{q}(y)$ are subject to errors in practical problems and thus inexact forms of these polynomials, $p(y)$ and $q(y)$ respectively, must be considered. These polynomials are, with high probability, coprime, but an algorithm that returns $\deg \mathrm{GCD}\,(p, q) = 0$ is unsatisfactory because this result is governed entirely by the errors in the coefficients of $p(y)$ and $g(y)$, and it does not consider the proximity of $(p(y), q(y))$ to $(\hat{p}(y), \hat{q}(y))$. An effective algorithm for the computation of an AGCD of $p(y)$ and $q(y)$ should return

$$\mathrm{AGCD}\,(p, q) \approx \mathrm{GCD}\,(\hat{p}, \hat{q}),$$

such that the error in this approximation is small.

The computation of an AGCD is performed in two stages:

Stage 1: Compute the degree t of an AGCD of $p(y)$ and $q(y)$.
Stage 2: Compute the coefficients of an AGCD of degree t.

It is shown in [28] that $p(y)$ and $q(y)$ must be processed by three operations before their Sylvester matrix $\mathbf{S}(p, q)$ is used to compute an AGCD. The first preprocessing operation arises because $\mathbf{S}(p, q)$ has, as noted above, a partitioned structure, which may cause numerical problems if the coefficients of $p(y)$ are much smaller or larger in magnitude than the coefficients of $q(y)$ since $\mathbf{S}(p, q)$ is not balanced if this condition is satisfied. It is therefore necessary to normalise $p(y)$ and $q(y)$, and Theorem 2 shows that the normalisation of an arbitrary polynomial $s(y)$,

$$s(y) = \sum_{i=0}^{m} s_i y^{m-i}, \tag{21}$$

by the geometric mean of its coefficients is better than the normalisation by the 2-norm of the vector $\mathbf{s}$ of its coefficients.

Theorem 2. *Consider the polynomial $s(y)$, which is defined in (21), and the polynomials $s_1(y)$ and $s_2(y)$ that are formed from $s(y)$ by two different normalisations,*

$$s_1(y) = \frac{s(y)}{\|s\|_2} \quad and \quad s_2(y) = \frac{s(y)}{\left(\prod_{i=0}^{m} s_i\right)^{\frac{1}{m+1}}},$$

where it is assumed, for simplicity, $s_i > 0$. The ith coefficients of $s_1(y)$ and $s_2(y)$ are therefore

$$s_{1,i} = \frac{s_i}{\|s\|_2} \quad and \quad s_{2,i} = \frac{s_i}{\left(\prod_{i=0}^{m} s_i\right)^{\frac{1}{m+1}}},$$

and the relative errors of the ith coefficients of $s(y), s_1(y)$ and $s_2(y)$ when the coefficients of $s(y)$ are perturbed are, respectively,

$$\Delta s_i = \frac{|\delta s_i|}{s_i}, \qquad \Delta s_{1,i} = \frac{|\delta s_{1,i}|}{s_{1,i}} \quad and \quad \Delta s_{2,i} = \frac{|\delta s_{2,i}|}{s_{2,i}}.$$

The ratio of the relative error of $s_{1,i}$ to the relative error of s_i is

$$r_1(s_i) = \frac{\Delta s_{1,i}}{\Delta s_i} = 1 - \frac{s_i^2}{\|s\|_2^2},$$

and the ratio of the relative error of $s_{2,i}$ to the relative error of s_i is

$$r_2 = \frac{\Delta s_{2,i}}{\Delta s_i} = \frac{m}{m+1}.$$

It follows that $r_1(s_i)$ is a function of the coefficients s_i, and thus this form of normalisation may change the relative errors of the coefficients on which computations are performed. This is different from r_2, which is constant and independent of these coefficients, and thus normalisation by the geometric mean of the coefficients of $s(y)$ is preferred. This form of normalisation is therefore used in the work described in this paper.

It follows that the normalised forms of $p(y)$ and $q(y)$, which are inexact forms of the exact polynomials $\hat{p}(y)$ and $\hat{q}(y)$ that are defined in (18), are

$$\dot{p}(y) = \sum_{i=0}^{r} \bar{p}_i y^{r-i}, \qquad \bar{p}_i = \frac{p_i}{\left(\prod_{i=0}^{r} |p_i|\right)^{\frac{1}{r+1}}}, \qquad p_i \neq 0, \qquad (22)$$

and

$$\dot{q}(y) = \sum_{i=0}^{s} \bar{q}_i y^{s-i}, \qquad \bar{q}_i = \frac{q_i}{\left(\prod_{i=0}^{s} |q_i|\right)^{\frac{1}{s+1}}}, \qquad q_i \neq 0, \qquad (23)$$

and it is clear that this computation must be changed if the non-zero condition on the coefficients of $p(y)$ or $q(y)$ is not satisfied. It was noted above that this normalisation defines the first of three operations that must be implemented before computations are performed on $\mathbf{S}(p,q)$, and the second and third preprocessing operations are now considered.

The second preprocessing operation arises because an AGCD of $\dot{p}(y)$ and $\dot{q}(y)$ is a function of their coefficients, and it is independent of the magnitudes of their coefficient vectors. It therefore follows that if α is an arbitrary non-zero constant, then

$$\mathrm{AGCD}(\dot{p}, \dot{q}) \sim \mathrm{AGCD}(\dot{p}, \alpha\dot{q}), \qquad (24)$$

where $\sim$ denotes equivalence to within an arbitrary non-zero scalar multiplier. The optimal value of the constant α must be calculated, and this issue is addressed below.

It is shown in [6] that the ratio of the entry of maximum magnitude of an arbitrary matrix X to the entry of minimum magnitude of X is believed to be a useful condition number of X, and it is therefore desirable to minimise this ratio. The objective of the third preprocessing operation is therefore the minimisation of the ratio $\mathcal{R}$ of the entry of maximum magnitude, to the entry of minimum magnitude, of the Sylvester matrix. A change in the independent variable from y to w is made,

$$y = \theta w, \tag{25}$$

and α and θ are constants whose optimal values minimise $\mathcal{R}$. It follows from (22)-(25) that an AGCD of the polynomials

$$\bar{p}(w, \theta) = \sum_{i=0}^{r} \left(\bar{p}_i \theta^{r-i} \right) w^{r-i} \quad \text{and} \quad \alpha \bar{q}(w, \theta) = \alpha \sum_{i=0}^{s} \left(\bar{q}_i \theta^{s-i} \right) w^{s-i},$$

is computed, and thus the optimal values α_0 and θ_0 of α and θ, respectively, minimise $\mathcal{R}$,

$$\alpha_0, \theta_0 = \arg\min_{\alpha, \theta} \left\{ \frac{\max\left\{ \max_{i=0,\ldots,r} \left| \bar{p}_i \theta^{r-i} \right|, \max_{j=0,\ldots,s} \left| \alpha \bar{q}_j \theta^{s-j} \right| \right\}}{\min\left\{ \min_{i=0,\ldots,r} \left| \bar{p}_i \theta^{r-i} \right|, \min_{j=0,\ldots,s} \left| \alpha \bar{q}_j \theta^{s-j} \right| \right\}} \right\}.$$

It is shown in [26] that this minimisation problem can be transformed to a linear programming (LP) problem from which α_0 and θ_0 are easily computed. The arguments of this problem are the coefficients $\bar{p}_i$ and $\bar{q}_j$ of, respectively, the noisy polynomials $\bar{p}(w, \theta)$ and $\bar{q}(w, \theta)$, and it is therefore necessary that α_0 and θ_0 be insensitive to noise. The sensitivity of the LP problem to perturbations in its arguments is addressed in [18], and some guidelines for performing sensitivity analysis are stated.

The LP problem is solved by an iterative procedure, and its stability must therefore be considered. In particular, if the iterations fail to converge or the maximum number of iterations is exceeded, then the values $\alpha_0 = 1$ and $\theta_0 = 1$ are used, that is, the only preprocessing operation is the normalisation of the coefficients of $p(y)$ and $q(y)$. Numerous computational experiments on blurred images showed, however, that the LP problem always converged, even with images of about 500×500 pixels and PSFs whose degrees in x and y are large.

It follows that an AGCD of the given inexact polynomials $p(y)$ and $q(y)$ is computed from the Sylvester matrix $\mathbf{S}(\bar{p}, \alpha_0 \bar{q}) = \mathbf{S}(\bar{p}(w, \theta_0), \alpha_0 \bar{q}(w, \theta_0))$. Computational experiments in [28] show the importance of the inclusion of α_0 and θ_0 for AGCD computations, and it is also shown that incorrect values of α_0 or θ_0 may lead to an incorrect result.

The next section considers the computation of the degree and coefficients of an AGCD of $\bar{p}(w, \theta_0)$ and $\alpha_0 \bar{q}(w, \theta_0)$ from their Sylvester matrix.

5.1   The Sylvester Resultant Matrix

This section describes the use of the Sylvester matrix for the computation of an AGCD of $\bar{p}(w, \theta_0)$ and $\alpha_0 \bar{q}(w, \theta_0)$. The simplest method of performing this calculation is by using the LU or QR decompositions of $\mathbf{S}(\bar{p}, \alpha_0 \bar{q})^T$, as stated in Theorem 1, but these decompositions yield poor results, and a structure-preserving matrix method [22] yields much better results.

Consider Stage 1 of the calculation of an AGCD of $\bar{p}(w, \theta_0)$ and $\alpha_0 \bar{q}(w, \theta_0)$, which is, as noted in Sect. 5, the determination of its degree t. Two methods for the calculation of t that use the singular value decomposition (SVD) of the Sylvester matrix of $\bar{p}(w, \theta_0)$ and $\alpha_0 \bar{q}(w, \theta_0)$, and its subresultant matrices, are described in [28], and it is shown in more recent work [25] that t can also be computed from the QR decomposition of these matrices. Since the subresultant matrix $\mathbf{S}_{k+1}(\bar{p}, \alpha_0 \bar{q})$ is formed by the deletion of two columns and one row from the subresultant matrix $\mathbf{S}_k(\bar{p}, \alpha_0 \bar{q})$, it follows that the update formula of the QR decomposition allows efficient computation, and it is therefore preferred to the SVD, whose update formula is complicated, for the calculation of t.

It follows from Lemma 1 that, using the calculated value of t, the numerical rank loss of $\mathbf{S}_t(\bar{p}, \alpha_0 \bar{q})$ is one, and there therefore exists a vector $\tilde{\mathbf{x}}$ such that

$$\mathbf{S}_t(\bar{p}, \alpha_0 \bar{q})\tilde{\mathbf{x}} \approx \mathbf{0}, \qquad \frac{\|\mathbf{S}_t(\bar{p}, \alpha_0 \bar{q})\tilde{\mathbf{x}}\|}{\|\mathbf{S}_t(\bar{p}, \alpha_0 \bar{q})\| \, \|\tilde{\mathbf{x}}\|} \ll 1, \qquad (26)$$

where the ith column of $\mathbf{S}_t(\bar{p}, \alpha_0 \bar{q})$ is $\mathbf{c}_i \in \mathbb{R}^{r+s-t+1}$,

$$\mathbf{S}_t(\bar{p}, \alpha_0 \bar{q}) = \begin{bmatrix} \mathbf{c}_1 \ \mathbf{c}_2 \ldots \mathbf{c}_{q-1} \ \mathbf{c}_q \ \mathbf{c}_{q+1} \ldots \mathbf{c}_{r+s-2t+1} \ \mathbf{c}_{r+s-2t+2} \end{bmatrix}.$$

It follows from (26) that one column $\mathbf{c}_q$ of $\mathbf{S}_t(\bar{p}, \alpha_0 \bar{q})$ is almost linearly dependent on the other columns, and the methods for the calculation of t described in [28] also return the index q of the column $\mathbf{c}_q$. It follows that the matrix $\mathbf{A}_q$ formed by the removal of $\mathbf{c}_q$ from $\mathbf{S}_t(\bar{p}, \alpha_0 \bar{q})$,

$$\mathbf{A}_q = \begin{bmatrix} \mathbf{c}_1 \ \mathbf{c}_2 \ldots \mathbf{c}_{q-1} \ \mathbf{c}_{q+1} \ldots \mathbf{c}_{r+s-2t+1} \ \mathbf{c}_{r+s-2t+2} \end{bmatrix} \in \mathbb{R}^{(r+s-t+1)\times(r+s-2t+1)},$$

has full rank, and thus (26) can be written as

$$\mathbf{A}_q \mathbf{x} \approx \mathbf{c}_q, \qquad (27)$$

from which it follows that the qth entry of $\tilde{\mathbf{x}}$, which is defined in (26), is equal to -1. It is shown in [26,27] that the entries of $\mathbf{x}$ are the coefficients of the coprime polynomials $\bar{u}(w, \theta_0)$ and $\bar{v}(w, \theta_0)$, which are of degrees $r - t$ and $s - t$ respectively, and that they satisfy

$$\bar{p}(w, \theta_0) \approx \bar{d}(w, \theta_0)\bar{u}(w, \theta_0) \qquad \text{and} \qquad \alpha_0 \bar{q}(w, \theta_0) \approx \bar{d}(w, \theta_0)\bar{v}(w, \theta_0),$$

where $\bar{d}(w, \theta_0)$ is an AGCD of $\bar{p}(w, \theta_0)$ and $\alpha_0 \bar{q}(w, \theta_0)$.

It follows from Lemma 1 that $\bar{p}(w, \theta_0)$ and $\alpha_0 \bar{q}(w, \theta_0)$ have a GCD of degree t if (26) is cast into an exact equation, that is, the coefficient matrix of a modified form of this equation has unit rank loss exactly and not approximately. This modification is achieved by adding the Sylvester subresultant matrix,

$$\mathbf{T}_t = \mathbf{T}_t(\ddot{p}(w, \theta_0), \alpha_0 \ddot{q}(w, \theta_0)),$$

of the polynomials $\ddot{p}(w, \theta_0)$ and $\alpha_0 \ddot{q}(w, \theta_0)$, which are defined as

$$\ddot{p}(w, \theta_0) = \sum_{i=0}^{r} \left(\ddot{p}_i \theta_0^{r-i} \right) w^{r-i} \qquad \text{and} \qquad \alpha_0 \ddot{q}(w, \theta_0) = \alpha_0 \sum_{i=0}^{s} \left(\ddot{q}_i \theta_0^{s-i} \right) w^{s-i},$$

to $\mathbf{S}_t(\bar{p}, \alpha_0 \bar{q})$, such that (26) becomes

$$(\mathbf{S}_t + \mathbf{T}_t)\,\bar{\mathbf{x}} = \mathbf{0}, \tag{28}$$

where $\mathbf{S}_t + \mathbf{T}_t$ is the tth Sylvester subresultant matrix of $\bar{p}(w, \theta_0) + \ddot{p}(w, \theta_0)$ and $\alpha_0\,(\bar{q}(w, \theta_0) + \ddot{q}(w, \theta_0))$, and the rank of $\mathbf{S}_t + \mathbf{T}_t$ is $r + s - 2t + 1$.

The homogeneous equation (28) is transformed to a linear algebraic equation by the removal of the qth column of the coefficient matrix, where the index q is defined in (27), to the right hand side. In particular, if $\mathbf{e}_q$ is the qth column of $\mathbf{T}_t$, and $\mathbf{E}_q$ is formed from the remaining $r + s - 2t + 1$ columns of $\mathbf{T}_t$, then it follows from (27) that (28) becomes

$$(\mathbf{A}_q + \mathbf{E}_q)\,\bar{\mathbf{x}} = \mathbf{c}_q + \mathbf{e}_q, \tag{29}$$

where $\mathbf{A}_q$ and $\mathbf{E}_q$ have the same structure, and $\mathbf{c}_q$ and $\mathbf{e}_q$ have the same structure. The matrix $\mathbf{T}_t$ is not unique because there exists more than one set of polynomials $(\ddot{p}(w, \theta_0), \alpha_0 \ddot{q}(w, \theta_0))$ such that $\bar{p}(w, \theta_0) + \ddot{p}(w, \theta_0)$ and $\alpha_0\,(\bar{q}(w, \theta_0) + \ddot{q}(w, \theta_0))$ have a GCD of degree t. Uniqueness is imposed by requiring that, of all the polynomials $\ddot{p}(w, \theta_0)$ and $\alpha_0 \ddot{q}(w, \theta_0)$ that can be added to, respectively, $\bar{p}(w, \theta_0)$ and $\alpha_0 \bar{q}(w, \theta_0)$, the polynomials $\ddot{p}(w, \theta_0)$ and $\alpha_0 \ddot{q}(w, \theta_0)$ of minimum magnitude be sought, that is, the perturbed polynomials must be as near as possible to the transformed forms $\bar{p}(w, \theta_0)$ and $\alpha_0 \bar{q}(w, \theta_0)$ of the given inexact polynomials. It is shown in [26] that the imposition of this constraint on (29) yields a non-linear equation, which is solved iteratively. The first order approximation of this equation generates a least squares minimisation subject to an equality constraint, the LSE problem, at each iteration $j = 1, 2, \dots$,

$$\min_{\delta \mathbf{y}^{(j)}} \left\| \delta \mathbf{y}^{(j)} - \mathbf{h}^{(j)} \right\|_2 \qquad \text{subject to} \qquad \mathbf{P}^{(j)} \delta \mathbf{y}^{(j)} = \mathbf{r}^{(j)}, \tag{30}$$

where

$$\mathbf{y}^{(j)} = \mathbf{y}^{(j-1)} + \delta \mathbf{y}^{(j)}, \tag{31}$$

$$\mathbf{h}^{(j)} = \mathbf{y}^{(0)} - \mathbf{y}^{(j-1)}, \tag{32}$$

$$\mathbf{y}^{(j)} = \mathbf{y}\left(\ddot{p}^{(j)}, \ddot{q}^{(j)}, \alpha_0^{(j)}, \theta_0^{(j)}, \bar{\mathbf{x}}^{(j)} \right),$$

$$\mathbf{P}^{(j)} = \mathbf{P}\left(\bar{p}, \bar{q}, \ddot{p}^{(j)}, \ddot{q}^{(j)}, \alpha_0^{(j)}, \theta_0^{(j)}, \bar{\mathbf{x}}^{(j)} \right),$$

$$\mathbf{r}^{(j)} = \mathbf{r}\left(\bar{p}, \bar{q}, \ddot{p}^{(j)}, \ddot{q}^{(j)}, \alpha_0^{(j)}, \theta_0^{(j)}, \bar{\mathbf{x}}^{(j)} \right),$$

$\alpha_0^{(0)} = \alpha_0$ and $\theta_0^{(0)} = \theta_0$. The LSE problem (30) is solved by the QR decomposition and its convergence is considered in Sect. 5.2.

The blurred image is of order $(M + p) \times (N + r)$ pixels, and it is therefore represented by a polynomial of degrees $M_1 = M + p - 1$ and $N_1 = N + r - 1$ in x and y respectively. The computation of the degree of the row component of the polynomial form of the PSF is therefore calculated from a Sylvester matrix of order $2N_1 \times 2N_1$, and thus $8O(N_1^3)$ flops are required for its QR decomposition. The computation of the QR decomposition of each subresultant matrix requires $O(N_1^2)$ flops, and since $N_1 - 1$ subresultant matrices are required for the computation of the degree of the row component of the polynomial form of the PSF, it follows that this computation requires a total of $(N_1 - 1)O(N_1^2) = O(N_1^3)$ flops. The repetition of this computation for the column component of the polynomial form of the PSF shows that $9O(N_1^3) + 9O(M_1^3)$ flops are required for the computation of the degrees of the row and column components of the polynomial form of the PSF. Each iteration for the solution of the LSE problem (30) by the QR decomposition requires approximately $(M_1 + N_1)^3$ flops, and numerous computational experiments [27] showed that, even for high levels of noise, convergence of this iterative procedure is achieved in a few iterations. It therefore follows that the AGCD computations required for the solution of the BID problem are cubic in complexity.

It was shown in Sect. 4 that this AGCD computation is applied to the rows r_1 and r_2 of the blurred image $\mathcal{G}$, thereby yielding a partially deblurred image $\mathcal{Q}$, and it is then repeated for the columns c_1 and c_2 of $\mathcal{Q}$. These computations enable the PSF to be computed, and the deblurred image is then obtained by two deconvolutions, as shown in (16) and (17). The coefficient matrices in these approximate equations are Tœplitz and thus a structure-preserving matrix method [21] can be used to obtain an improved deblurred image. In this work, however, the least squares solutions of (16) and (17) are used because the examples in Sect. 7 show that these simple solutions yield deblurred images of high quality, and that they are better than the deblurred images obtained from four other methods of image deblurring.

There is an extensive literature on AGCD computations and many methods have been developed, including methods based on the QR decomposition [3,24,29,30] and optimisation [2,12]. The AGCD in (13) and its column equivalent are computed from a structured low rank approximation of $\mathbf{S}(p,q)$ for the work described in this paper because this approximation exploits the structure of $\mathbf{S}(p,q)$ [26,27]. A structure-preserving matrix method is also used for the AGCD computation in [11,14], and it is similar to the method used in this paper. There are, however, three important differences between the method used in [11,14] and the method used in this paper:

1. The preprocessing operations discussed in Sect. 5 form an important part of the method for the AGCD computations used in this paper, but preprocessing operations are not used in [11,14]. Two of these processing operations introduce the parameters α_0 and θ_0, and their inclusion in the problem formulation implies that (29) is a non-linear equation. A non-linear structure-preserving matrix method [22] is therefore used, and it must be compared with the linear structure-preserving matrix method [21] used in [11,14] because the parameters α_0 and θ_0 are not included in the AGCD computation in these references, which is equivalent to the specification $\alpha_0 = \theta_0 = 1$.

2. The penalty method is used in [11,14] to solve the LSE problem (30), but the QR decomposition is used in this paper to solve this problem. The penalty method requires a parameter $\eta \gg 1$, but a parameter is not required when the QR decomposition is used.

3. It is shown in Sect. 5.1 that it is necessary to determine the optimal column $\mathbf{c}_q$ of $\mathbf{S}_t(\bar{p}, \alpha_0 \bar{q})$, such that the residual of the approximate linear algebraic equation (27) assumes its minimum value with respect to the residuals when the other $r + s - 2t + 1$ columns of $\mathbf{S}_t(\bar{p}, \alpha_0 \bar{q})$ are moved to the right hand side. The optimal column of $\mathbf{S}_t(p, q)$ is defined as its first column in [11, 14], but it is shown in [27] that this choice may yield bad results, and that the optimal column of $\mathbf{S}_t(\bar{p}, \alpha_0 \bar{q})$ must be determined for each problem.

5.2   Convergence of the LSE Problem

This section considers the convergence of the LSE problem (30), which is a problem of the form

$$\min_{\mathbf{y}} \|\mathbf{y} - \mathbf{p}\|_2 \quad \text{subject to} \quad \mathbf{Dy} = \mathbf{q},$$

where $\mathbf{D} \in \mathbb{R}^{r \times s}, \mathbf{y}, \mathbf{p} \in \mathbb{R}^s, \mathbf{q} \in \mathbb{R}^r$ and $r < s^3$. This problem can be solved by the QR decomposition, as shown in Algorithm 1 [7].

Algorithm 1: The solution of the LSE problem by the QR decomposition

(a) Compute the QR decomposition of $\mathbf{D}^T$,

$$\mathbf{D}^T = \mathbf{QR} = \mathbf{Q} \begin{bmatrix} \mathbf{R}_1 \\ \mathbf{0} \end{bmatrix}.$$

(b) Set $\mathbf{w}_1 = \mathbf{R}_1^{-T} \mathbf{q}$.
(c) Partition $\mathbf{Q}$ into

$$\mathbf{Q} = \begin{bmatrix} \mathbf{Q}_1 \ \mathbf{Q}_2 \end{bmatrix}, \tag{33}$$

where $\mathbf{Q}_1 \in \mathbb{R}^{s \times r}$ and $\mathbf{Q}_2 \in \mathbb{R}^{s \times (s-r)}$.
(d) Compute $\mathbf{w}_2 = \mathbf{Q}_2^T \mathbf{p}$.
(e) Compute the solution

$$\mathbf{y} = \mathbf{Q} \begin{bmatrix} \mathbf{w}_1 \\ \mathbf{w}_2 \end{bmatrix} = \mathbf{Q} \begin{bmatrix} \mathbf{R}_1^{-T} \\ \mathbf{0} \end{bmatrix} \mathbf{q} + \mathbf{Q}_2 \mathbf{Q}_2^T \mathbf{p}. \tag{34}$$

Consider the application of this algorithm to the solution of (30). In particular, it follows from (32) and (34) that

$$\delta \mathbf{y}^{(j)} = \mathbf{Q}^{(j)} \begin{bmatrix} \mathbf{R}_1^{-T} \\ \mathbf{0} \end{bmatrix}^{(j)} \mathbf{r}^{(j)} + \left(\mathbf{Q}_2 \mathbf{Q}_2^T \right)^{(j)} \left(\mathbf{y}^{(0)} - \mathbf{y}^{(j-1)} \right),$$

---

[3] The integers r and s are not related to the degrees of the polynomials $\hat{p}(y)$ and $\hat{q}(y)$, and polynomials derived from them, that are introduced in Sect. 5.

at the jth iteration. If $\bar{\mathbf{y}}$ is the solution of (30), and $\mathbf{e}^{(j)}$ and $\mathbf{e}^{(j-1)}$ are the errors at the jth and $(j-1)$th iterations, then it follows from (31) that

$$\mathbf{e}^{(j)} = \mathbf{y}^{(j)} - \bar{\mathbf{y}}$$

$$= \mathbf{e}^{(j-1)} + \left(\mathbf{Q}_2\mathbf{Q}_2^T\right)^{(j)} \left(\mathbf{y}^{(0)} - \mathbf{y}^{(j-1)}\right) + \mathbf{Q}^{(j)} \begin{bmatrix} \mathbf{R}_1^{-T} \\ \mathbf{0} \end{bmatrix}^{(j)} \mathbf{r}^{(j)},$$

and thus

$$\mathbf{e}^{(j)} - \mathbf{e}^{(j-1)} = \left(\mathbf{Q}_2\mathbf{Q}_2^T\right)^{(j)} \left(\mathbf{y}^{(0)} - \mathbf{y}^{(j-1)}\right) + \mathbf{Q}^{(j)} \begin{bmatrix} \mathbf{R}_1^{-T} \\ \mathbf{0} \end{bmatrix}^{(j)} \mathbf{r}^{(j)},$$

$$= \mathbf{Q}^{(j)} \left(\left(\mathbf{Q}^T\right)^{(j)} \left(\mathbf{Q}_2\mathbf{Q}_2^T\right)^{(j)} \left(\mathbf{y}^{(0)} - \mathbf{y}^{(j-1)}\right) + \begin{bmatrix} \mathbf{R}_1^{-T} \\ \mathbf{0} \end{bmatrix}^{(j)} \mathbf{r}^{(j)} \right),$$

$$= \mathbf{Q}^{(j)} \left(\begin{bmatrix} \mathbf{0} \\ \mathbf{Q}_2^T \end{bmatrix}^{(j)} \left(\mathbf{y}^{(0)} - \mathbf{y}^{(j-1)}\right) + \begin{bmatrix} \mathbf{R}_1^{-T} \\ \mathbf{0} \end{bmatrix}^{(j)} \mathbf{r}^{(j)} \right),$$

from (33) because

$$\mathbf{Q}_1^T\mathbf{Q}_1 = \mathbf{I}_r, \qquad \mathbf{Q}_2^T\mathbf{Q}_2 = \mathbf{I}_{s-r} \qquad \text{and} \qquad \mathbf{Q}_1^T\mathbf{Q}_2 = \mathbf{0}.$$

It follows that the iterative scheme converges if

$$\lim_{j\to\infty} \left\| \mathbf{e}^{(j)} - \mathbf{e}^{(j-1)} \right\|_2 = \lim_{j\to\infty} \left\| \mathbf{y}^{(j)} - \mathbf{y}^{(j-1)} \right\|_2$$

$$= \lim_{j\to\infty} \left\| \begin{bmatrix} \left(\mathbf{R}_1^{-T}\mathbf{r}\right)^{(j)} \\ \left(\mathbf{Q}_2^T\right)^{(j)} \left(\mathbf{y}^{(0)} - \mathbf{y}^{(j-1)}\right) \end{bmatrix} \right\|_2 = 0,$$

and thus two conditions must be satisfied for its convergence:

1. It is necessary that $\mathbf{r}^{(j)} \to \mathbf{0}$ as $j \to \infty$.

2. The vector $\mathbf{y}^{(0)} - \mathbf{y}^{(j-1)}$ must lie in the nullspace of $\left(\mathbf{Q}_2^T\right)^{(j)}$ as $j \to \infty$.

Many computational experiments showed that (30) converges in fewer than five iterations, even when a large amount of noise is added to the exact image to form a highly blurred image.

6 Extension to a Non-separable PSF

It has been shown that a separable PSF can be calculated from one blurred image, and it can then be deconvolved from the blurred image, thereby yielding a deblurred form of the blurred image. It is shown in this section that if the PSF is non-separable, then two blurred images are required for its calculation, and that computations that are not required for a separable PSF must be included for the computation of a deblurred image that is formed by a non-separable PSF. It is therefore appropriate to consider changes to the method, such that its modified form can be used for the solution of the BID problem when a non-separable PSF is used.

It was shown in Sect. 4 that the solution of the BID problem requires one blurred image if the PSF is separable, and that the computation of the PSF requires the selection of two rows and two columns from the blurred image $\mathcal{G}$. If, however, the PSF is non-separable, two blurred images $\mathcal{G}_1$ and $\mathcal{G}_2$ are required, and it is assumed, for simplicity, they are the same size, $(M + p) \times (N + r)$ pixels, where the exact images are $M \times N$ pixels and the PSF is $(p+1) \times (r+1)$ pixels. If $r_1(i)$ and $r_2(i)$ are the ith rows of $\mathcal{G}_1$ and $\mathcal{G}_2$ respectively, then the algorithm discussed in Sect. 5 is applied to every pair of rows $(r_1(i), r_2(i)), i = 1, \ldots, M + p$, of $\mathcal{G}_1$ and $\mathcal{G}_2$ when a non-separable PSF is used. The LSE problem (30) is therefore solved for each of these pairs of rows, and thus a deblurred form of each row of $\mathcal{G}_1$ and each row of $\mathcal{G}_2$, and the PSF of each row, are obtained.

This procedure is repeated for the columns $c_1(i)$ and $c_2(i)$, $i = 1, \ldots, N + r$, of, respectively, $\mathcal{G}_1$ and $\mathcal{G}_2$, and thus a deblurred form of each column of $\mathcal{G}_1$, and each column of $\mathcal{G}_2$, and the PSF of each column, are obtained. It follows, therefore, that there are two sets of results, one set obtained by considering the rows of $\mathcal{G}_1$ and $\mathcal{G}_2$, and one set obtained by considering the columns of $\mathcal{G}_1$ and $\mathcal{G}_2$. The deblurred forms of each row and column of $\mathcal{G}_1$ and $\mathcal{G}_2$, and the components of the PSF along each row and column, are obtained by a sequence of independent AGCD computations, and they therefore have scale factors associated with them. These scale factors must be removed in order to compute the deblurred images and the PSF, and a method for their removal is described in [15,16,19].

The method discussed above differs from the method used by Pillai and Liang [19] because they assume that only one blurred image $\mathcal{G}$ is available and that the degrees in x and y of the PSF are small. They propose that $\mathcal{G}$ be partitioned into two regions $\mathcal{R}_1$ and $\mathcal{R}_2$, such that each region contains the entire PSF. The assumption of small supports of the PSF is necessary in order to localise the effects of the partition to the region $\mathcal{L}$ of the common edge of $\mathcal{R}_1$ and $\mathcal{R}_2$, such that the region $\mathcal{L}$ is small and the effects of the partition do not propagate into the interiors of $\mathcal{R}_1$ and $\mathcal{R}_2$. There are, however, differences between the work described in this paper and the work in [19], such that the method described in [19] may be limited. As noted above, it is assumed that the supports of the PSF are small, and it is stated in [19] that the method produces reliable results when the SNR is greater than 40 dB, which is much larger than the SNRs of 15.62 dB and 13.14 dB of the blurred images in Examples 1 and 2 in Sect. 7. Also, it is assumed in [19] that only additive noise $N(x, y)$ is present, and that the uncertainty $E(x, y)$ in the PSF is zero. By contrast, it is assumed in this paper that $N(x, y) \neq 0$ and $E(x, y) \neq 0$, and Example 3 in Sect. 7 shows that the uncertainty in the PSF is a significantly greater cause of blur than is additive noise.

7   Examples

This section contains examples that show the deblurred images that result from the AGCD computations discussed in Sects. 4 and 5, and they are compared with the deblurred images from four other methods. The examples show that the deblurred images obtained from the AGCD computations and polynomial divisions are significantly better than the deblurred images that result from the other methods, even though these other methods require that the PSF be known. This must be compared with the method described in this paper, which allows the PSF to be calculated from the blurred image.

A blurred image is formed by perturbing the PSF and adding random noise, and thus the blurring model used is obtained by including in (1) the uncertainty $\mathcal{E}$ in the PSF,

$$\mathcal{G} = (\mathcal{H} + \mathcal{E}) \otimes \mathcal{F} + \mathcal{N}, \tag{35}$$

which can also be expressed as a generalised form of (2),

$$\mathbf{g} = (\mathbf{H} + \mathbf{E})\mathbf{f} + \mathbf{n}. \tag{36}$$

The PSF that is applied to the exact images in Examples 1, 2 and 3 is shown in Fig. 1. It is seen that the row and column components of the PSF are represented by polynomials of degrees 8 and 10 respectively, that the decays of the PSF in the directions of increasing and decreasing row index are not equal, and that the decays of the PSF in the directions of increasing and decreasing column index are not equal.

The deblurred images obtained from the AGCD computations and polynomial divisions discussed in this paper are compared with the deblurred images obtained from the Lucy-Richardson algorithm, blind image deconvolution, a regularised filter and the Wiener filter [9]. The deblurred images from these methods are obtained using the image processing toolbox in MATLAB. The function calls for these methods are now considered.

Lucy-Richardson algorithm. The function call is

$$[\mathtt{f}] = \mathtt{deconvlucy(g,psf,numit,dampar,weight)}$$

where f is the deblurred image, g is the blurred image, psf is the PSF, numit is the number of iterations, dampar is a scalar threshold that defines the deviation of the pixel values of the deblurred image from the pixel values of the blurred image, such that iterations are suppressed for pixels that deviate less than dampar from their original values, and weight is an array, of the same size as g, that assigns a weight to each pixel that reflects its quality. The default values numit=10 and dampar=0, which corresponds to no damping, are used, and weight is equal to the matrix, all of whose entries are equal to one. This specification implies that all pixels in g have the same error, and all pixels are therefore treated equally in the algorithm.

Point spread function

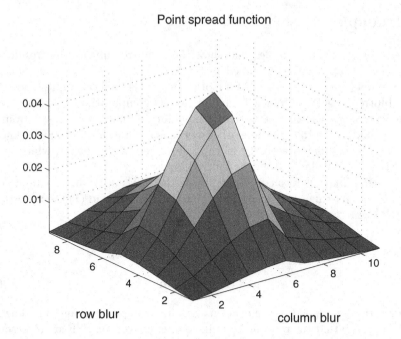

Fig. 1. The PSF that is applied to the exact images in Examples 1, 2 and 3.

Blind image deconvolution. The function call is

$$[\mathtt{f,psf}] = \mathtt{deconvblind(g,initpsf,numit,dampar,weight)}$$

where f,g,numit,dampar and weight are defined in the specification of the function call for the Lucy-Richardson algorithm, initpsf is an initial estimate of the PSF, and psf is an improved estimate of the PSF. The default values of numit,dampar and weight, which are defined in the specification of the Lucy-Richardson algorithm, are used in Examples 1, 2 and 3, and the parameter initpsf is set equal to the exact PSF.

The effect on the deblurred image f of the number of iterations numit in the Lucy-Richardson algorithm and blind image deconvolution is considered in [8].

Regularised filter. The function call is

$$[\mathtt{f}] = \mathtt{deconvreg(g,psf,noisepower,range)}$$

where f,g and psf are defined in the specification of the function call for the Lucy-Richardson algorithm, noisepower is equal to the noise power and range is the range within which the optimal value of a parameter for the satisfaction of a constraint is sought. The default value of range, which is equal to $\left[10^{-9},10^{9}\right]$, is used.

Wiener filter. The function call is

$$[\mathtt{f}] = \mathtt{deconvwnr(g,psf,r)}$$

where f, g and psf are defined in the specification of the function call for the Lucy-Richardson algorithm. The argument r is the ratio of the noise power to the signal power, and it follows from (36) that, assuming $\|E\| \approx 0$,

$$r \approx \frac{\|n\|_2^2}{\|f\|_2^2}. \tag{37}$$

If the argument r is omitted, the function returns an ideal inverse filter.

The default values of the arguments of these functions are used in Examples 1 and 2.

Example 1. Figure 2(a) shows an exact image that is blurred by adding uncertainty $\mathcal{E}$ to the PSF and random noise $\mathcal{N}$, as shown in (35). In particular, let $((\mathcal{H} + \mathcal{E}) \otimes \mathcal{F})_{i,j}$, $\mathcal{E}_{i,j}$, $\mathcal{H}_{i,j}$ and $\mathcal{N}_{i,j}$ denote entries (i,j) of $(\mathcal{H} + \mathcal{E}) \otimes \mathcal{F}$, $\mathcal{E}$, $\mathcal{H}$ and $\mathcal{N}$ respectively. These entries satisfy

$$0 < \frac{\mathcal{E}_{i,j}}{\mathcal{H}_{i,j}} \leq 10^{-5}, \qquad i = 0, \ldots, p; j = 0, \ldots, r, \tag{38}$$

and

$$0 < \frac{\mathcal{N}_{i,j}}{((\mathcal{H} + \mathcal{E}) \otimes \mathcal{F})_{i,j}} \leq 10^{-5}, \qquad i = 0, \ldots, M + p - 1; j = 0, \ldots, N + r - 1, \tag{39}$$

Original image   Blurred image

(a)   (b)

Fig. 2. (a) An original (exact) image and (b) a blurred image obtained after the addition of uncertainty to the PSF and random noise, for Example 1.

and the values of $\mathcal{E}_{i,j}$ and $\mathcal{N}_{i,j}$ yield a SNR of

$$20\log_{10}\frac{\|\mathbf{f}\|_2}{\|\mathbf{g}-\mathbf{f}\|_2} = 15.62\text{ dB}, \tag{40}$$

where $\mathbf{f}$ and $\mathbf{g}$ are defined in (36). The uncertainty $\mathcal{E}$ and noise $\mathcal{N}$ were used to blur the image, thereby obtaining the blurred image shown in Fig. 2(b). The AGCD computations and polynomial divisions discussed in Sects. 4 and 5 were then used to deblur this image, and the deblurred image is shown in Fig. 3. The relative errors in (38) and (39) were not used in the algorithm for the computation of the deblurred image in Fig. 3, as discussed in Sect. 1.

This deblurred image was compared with the deblurred images computed by the Lucy-Richardson algorithm, blind image deconvolution, a regularised filter and the Wiener filter. The results are shown in Figs. 4 and 5, and it is clear that the best deblurred image is obtained using AGCD computations and polynomial divisions. It is important to note that the deblurred images in Figs. 4 and 5 were obtained by specifying the exact PSF, but the PSF was calculated in order to obtain the deblurred image in Fig. 3.

Quantitative comparison of the exact image and the deblurred images in Figs. 3, 4 and 5 requires care because the deblurred image in Fig. 3 is obtained by AGCD computations and polynomial divisions. In particular, it follows from (24) that an AGCD is defined to within an arbitrary non-zero scalar multiplier, and since the computed AGCD is equal to the PSF, which is deconvolved from the blurred image, it follows that the deblurred image from AGCD computations and polynomial divisions is also defined to within an arbitrary non-zero scalar multiplier. Comparison of the exact image with the deblurred images in Figs. 3, 4 and 5 requires, therefore, that all the images be normalised.

Restored image

Fig. 3. The deblurred image obtained by AGCD computations and polynomial divisions for Example 1.

If $\hat{\mathbf{F}}$ and $\bar{\mathbf{F}}$ are, respectively, the normalised forms of the matrix representations of the exact image $\mathcal{F}$ and a deblurred image $\ddot{\mathcal{F}}$ whose matrix representations are $\mathbf{F}$ and $\ddot{\mathbf{F}}$ respectively, then

$$\hat{\mathbf{F}} = \frac{\mathbf{F}}{\|\mathbf{F}\|_2} \quad \text{and} \quad \bar{\mathbf{F}} = \frac{\ddot{\mathbf{F}}}{\left\|\ddot{\mathbf{F}}\right\|_2}.$$

The SNR between the exact image and a deblurred image is therefore

$$\mu = 20 \log_{10} \frac{\left\|\hat{\mathbf{F}}\right\|_F}{\left\|\hat{\mathbf{F}} - \bar{\mathbf{F}}\right\|_F} \text{ dB,} \tag{41}$$

where the subscript F denotes the Frobenius norm. Table 1 shows the values of μ for the deblurred images in Figs. 3, 4 and 5, and it is seen that the maximum value of μ occurs for the deblurred image obtained by AGCD computations and polynomial divisions, which is evident from the deblurred images in these figures.

It follows from Table 1 and (40) that μ is approximately equal to the SNR of the given blurred image $\mathcal{G}$ for blind image deconvolution, the Wiener filter and the Lucy-Richardson algorithm, which shows their regularisation property. The deblurred image from the regularised filter is obtained by the least squares minimisation of the error between the estimated image and the true image, with a constraint on the preservation of the smoothness of the image. The effect of different values of the arguments noisepower and range on the deblurred image obtained from the function deconvreg.m is considered in [8] and it is noted that experiments may be needed to determine the values of these parameters that yield the best deblurred image. Specifically, an example in [8] shows that the best deblurred image is obtained, by experiment, when the noise power is equal to 10 % of its initial estimate and range is equal to $\left[10^{-7}, 10^7\right]$, which is tighter than the default range.

The deblurred image from the Wiener filter was obtained by including the value of r, which is defined in (37), in the arguments of the function deconvwnr.m. The value of r may not, however, be known in practical problems, but Table 1 shows that even when it is known, the value of μ is approximately equal to the values of μ for the deblurred images from blind image deconvolution and the Lucy-Richardson algorithm. The function deconvblind.m that implements blind image deconvolution requires an initial estimate of the PSF, and the function returns a better estimate. In this example, the function deconvblind.m was called with the exact PSF, which is known, as shown by the classification of the PSF in Table 1.

An expression for the error in the computed PSF requires that it be normalised so that the sum of the elements of its matrix representation is one. In particular, if $\hat{\mathbf{H}}$ and $\bar{\mathbf{H}}$ are the matrices of the exact and computed PSFs that are normalised such that their elements, $\hat{h}_{i,j}$ and $\bar{h}_{i,j}$ respectively, satisfy

Lucy–Richardson   Blind deconvolution

(a)   (b)

Fig. 4. Deblurred images of the image in Fig. 2(b) obtained by (a) the Lucy-Richardson algorithm and (b) blind image deconvolution, for Example 1.

$$\sum_{i,j} \hat{h}_{i,j} = 1 \qquad \text{and} \qquad \sum_{i,j} \bar{h}_{i,j} = 1,$$

then the error between the computed and exact PSFs is

$$\lambda = \sum_{i,j} \left| \hat{h}_{i,j} - \bar{h}_{i,j} \right|. \tag{42}$$

The error λ between the computed and exact PSFs is $1.42\exp-05$. This is approximately equal to the lower bound of the componentwise error in the exact PSF, which is specified in (38).   □

Example 2. The procedure described in Example 1 was repeated for the exact image shown in Fig. 6(a), and the blurred image shown in Fig. 6(b) was obtained

Table 1. The SNRs of the deblurred images for Example 1.

Method	PSF known/not known	μ (dB)
Blind image deconvolution	Known	15.26
Regularised filter	Known	9.19
Wiener filter	Known	15.67
Lucy-Richardson	Known	15.16
AGCDs and poly. divisions	Not known (calculated)	62.63

Regularised filter   Wiener filter

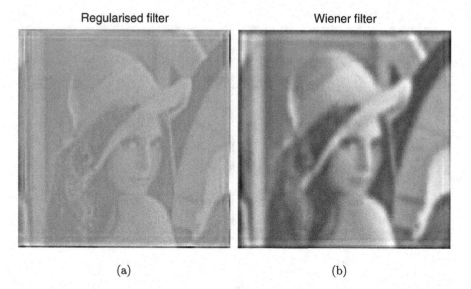

(a)   (b)

Fig. 5. Deblurred images of the image in Fig. 2(b) obtained by (a) a regularised filter and (b) the Wiener filter, for Example 1.

by perturbing the PSF and adding random noise, as shown in (38) and (39) respectively. The SNR of the blurred image is 13.14 dB.

The deblurred image obtained by AGCD computations and polynomial divisions is shown in Fig. 7, and the deblurred images obtained using the Lucy-Richardson algorithm, blind image deconvolution, a regularised filter and the Wiener filter are shown in Figs. 8 and 9. Table 2 shows the value of the SNR, which is defined in (41), for each deblurred image in Figs. 7, 8 and 9, and it is seen that the results of Example 2 are very similar to the results of Example 1 because the deblurred image obtained from the AGCD computations and polynomial divisions has the largest value of μ.

The error between the computed and exact PSFs, using the error measure (42), is 1.41 exp −05. This is approximately equal to the lower bound of the componentwise error in the exact PSF, which is defined in (38).   □

Table 2. The SNRs of the deblurred images for Example 2.

Method	PSF known/not known	μ (dB)
Blind image deconvolution	Known	13.12
Regularised filter	Known	7.94
Wiener filter	Known	13.72
Lucy-Richardson	Known	12.96
AGCDs and poly. divisions	Not known (calculated)	62.64

Original image

Blurred image

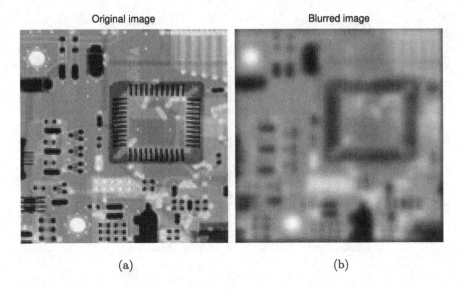

(a)   (b)

Fig. 6. (a) An original (exact) image and (b) a blurred image obtained after the addition of uncertainty to the PSF and random noise, for Example 2.

Restored image

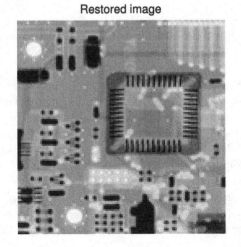

Fig. 7. The deblurred image obtained by AGCD computations and polynomial divisions for Example 2.

Example 3. The blur in Examples 1 and 2 is caused by the uncertainty $\mathcal{E}$ in the PSF $\mathcal{H}$, and additive noise $\mathcal{N}$. The relative importance of these sources of blur was investigated by performing three experiments on the exact image shown in Fig. 6(a):

Lucy–Richardson   Blind deconvolution

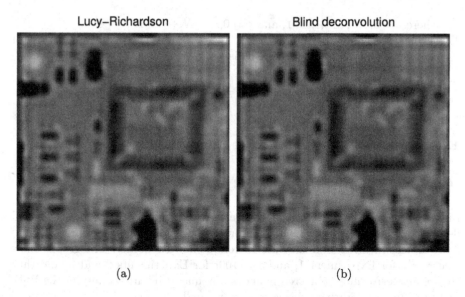

(a)   (b)

Fig. 8. Deblurred images of the image in Fig. 6(b) obtained by (a) the Lucy-Richardson algorithm and (b) blind image deconvolution, for Example 2.

Regularised filter   Wiener filter

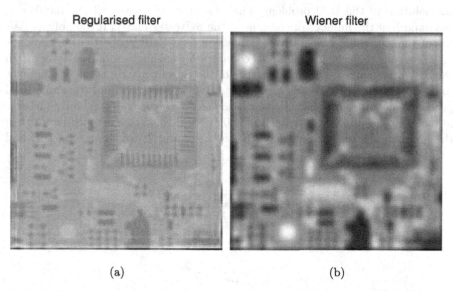

(a)   (b)

Fig. 9. Deblurred images of the image in Fig. 6(b) obtained by (a) a regularised filter and (b) the Wiener filter, for Example 2.

Experiment 1: The uncertainty in the PSF is zero, $\mathcal{E} = 0$, and the pixel value $\mathcal{N}(i,j)$ of the noise $\mathcal{N}$ satisfies

$$0 < \frac{\mathcal{N}(i,j)}{\mathcal{S}(i,j)} \leq \epsilon, \qquad \mathcal{S} = (\mathcal{H} + \mathcal{E}) \otimes \mathcal{F}, \qquad \epsilon = 10^{-10}, 10^{-9}, \ldots, 10^{-3},$$

where $i = 0, \ldots, M + p - 1$, and $j = 0, \ldots, N + r - 1$.
Experiment 2: The additive noise is zero, $\mathcal{N} = 0$, and the pixel value $\mathcal{E}(k,l)$ of the uncertainty $\mathcal{E}$ in the PSF $\mathcal{H}$ satisfies

$$0 < \frac{\mathcal{E}(k,l)}{\mathcal{H}(k,l)} \leq \epsilon, \quad \epsilon = 10^{-10}, 10^{-9}, \ldots, 10^{-3},$$

for $k = 0, \ldots, p$, and $l = 0, \ldots, r$.
Experiment 3: The uncertainty $\mathcal{E}$ in the PSF and additive noise $\mathcal{N}$ satisfy

$$0 < \frac{\mathcal{E}(k,l)}{\mathcal{H}(k,l)}, \frac{\mathcal{N}(i,j)}{\mathcal{S}(i,j)} \leq \epsilon, \quad \epsilon = 10^{-10}, 10^{-9}, \ldots, 10^{-3}.$$

Figure 10 shows the variation of the relative error γ between the exact and deblurred images with the relative error ϵ. The results for Experiments 2 and 3 are almost identical and they are therefore shown together. The graphs show that $\gamma = \epsilon$ for Experiment 1, and $\gamma = 100\epsilon$ for Experiments 2 and 3, and thus the relative error in a deblurred image is dominated by uncertainty in the PSF, and the effect of additive noise is relatively small. □

The Bézout matrix, rather than the Sylvester matrix, is used in [15] for the solution of the BID problem. The Bézout matrix has half the number of rows and half the number of columns of the Sylvester matrix if the polynomials are of the same degree, and it is symmetric, and both properties suggest it is advantageous to perform the AGCD computations using this matrix, rather than the Sylvester matrix. The disadvantage of the Bézout matrix is the requirement to evaluate terms of the form $p_i q_j - p_j q_i$ for its formation, and thus numerical problems may arise because of cancellation. Also, the SNR of the blurred images in the examples in [15] is greater than 50 dB, and this value is much larger than the SNRs of the blurred images in Examples 1 and 2, which are 15.62 dB and 13.14 dB respectively.

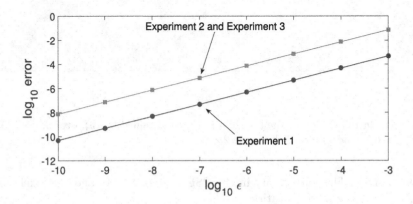

Fig. 10. The relative contributions of the additive noise and uncertainty in the PSF to the error in a deblurred image, for Experiment 1, and Experiments 2 and 3.

Equation (28) arises because the addition of two Sylvester matrices is also a Sylvester matrix, assuming the degrees of the polynomials are consistent. This equation does not apply to the Bézout matrix $\mathbf{B}(f, g)$ because the addition of two Bézout matrices does not yield a Bézout matrix since each entry is a bilinear function of the coefficients of $f = f(y)$ and $g = g(y)$.

8   Summary

This paper has considered the application of AGCD computations and polynomial divisions to the removal of blur from an image. These polynomial operations can be used if a blurred image is formed by the convolution of the exact image and a spatially invariant PSF because the multiplication of two polynomials reduces to the convolution of their coefficients if this condition on the PSF is satisfied. The deblurred image obtained from these polynomial computations was compared with the deblurred images obtained from four other methods, and the deblurred image of highest quality was obtained from the method that uses polynomial computations.

The method of deblurring discussed in this paper can be extended to a nonseparable PSF, which requires that two blurred images be specified for its calculation. More computations of the same type as occur when a separable PSF is used, and additional computations, are required when the PSF is non-separable. It is also necessary to remove arbitrary scale factors that appear in the rows and columns of the deblurred images when a non-separable PSF is used.

References

1. Barnett, S.: Polynomials and Linear Control Systems. Marcel Dekker, New York (1983)
2. Chin, P., Corless, R.M.: Optimization strategies for the approximate GCD problem. In: Proceeding of International Symposium Symbolic and Algebraic Computation, pp. 228–235, Rostock, Germany (1998)
3. Corless, R.M., Watt, S.M., Zhi, L.: QR factoring to compute the GCD of univariate approximate polynomials. IEEE Trans. Signal Process. **52**(12), 3394–3402 (2004)
4. Cornelio, A., Piccolomini, E., Nagy, J.: Constrained numerical optimization methods for blind deconvolution. Numer. Algor. **65**, 23–42 (2014)
5. Danelakis, A., Mitrouli, M., Triantafyllou, D.: Blind image deconvolution using a banded matrix method. Numer. Algor. **64**, 43–72 (2013)
6. Fulkerson, D., Wolfe, P.: An algorithm for scaling matrices. SIAM Rev. **4**, 142–146 (1962)
7. Golub, G.H., Van Loan, C.F.: Matrix Computations. John Hopkins University Press, Baltimore (2013)
8. Gonzalez, R.C., Woods, R.E., Eddins, S.L.: Digital Image Processing Using Matlab. Gatesmark Publishing, Knoxville (2009)
9. Gunturk, B., Li, X.: Image Registration: Fundamentals and Advanves. CRC Press, Florida (2013)
10. Hansen, P.C., Nagy, J.G., O'Leary, D.P.: Deblurring Images: Matrices, Spectra, and Filtering. SIAM, Philadelphia (2006)

11. Kaltofen, E., Yang, Z., Zhi, L.: Structured low rank approximation of a Sylvester matrix. In: Wang, D., Zhi, L. (eds.) Trends in Mathematics, pp. 69–83. Birkhäuser Verlag, Basel (2006)

12. Karmarkar, N.K., Lakshman, Y.N.: On approximate GCDs of univariate polynomials. J. Symb. Comput. **26**, 653–666 (1998)

13. Kundur, D., Hatzinakos, D.: Blind image deconvolution. IEEE Signal Process. Mag. **13**(3), 43–64 (1996)

14. Li, B., Yang, Z., Zhi, L.: Fast low rank approximation of a sylvester matrix by structured total least norm. J. Jpn Soc. Symb. Algebraic Comp. **11**, 165–174 (2005)

15. Li, Z., Yang, Z., Zhi, L.: Blind image deconvolution via fast approximate GCD. In: Proceedings of International Symposium Symbolic and Algebraic Computation, pp. 155–162 (2010)

16. Liang, B., Pillai, S.: Blind image deconvolution using a robust 2-D GCD approach. In: IEEE International Symposium Circuits and Systems, pp. 1185–1188 (1997)

17. Nagy, J., Palmer, K., Perrone, L.: Iterative methods for image deblurring: a Matlab object-oriented approach. Numer. Algor. **36**, 73–93 (2004)

18. Nash, S.G., Sofer, A.: Linear and Nonlinear Programming. McGraw-Hill, New York (1996)

19. Pillai, S., Liang, B.: Blind image deconvolution using a robust GCD approach. IEEE Trans. Image Process. **8**(2), 295–301 (1999)

20. Premaratne, P., Ko, C.: Retrieval of symmetrical image blur using zero sheets. IEE Proc. Vis. Image Signal Process. **148**(1), 65–69 (2001)

21. Rosen, J.B., Park, H., Glick, J.: Total least norm formulation and solution for structured problems. SIAM J. Mat. Anal. Appl. **17**(1), 110–128 (1996)

22. Rosen, J.B., Park, H., Glick, J.: Structured total least norm for nonlinear problems. SIAM J. Mat. Anal. Appl. **20**(1), 14–30 (1998)

23. Satherley, B.L., Parker, C.R.: Two-dimensional image reconstruction from zero sheets. Optics Lett. **18**, 2053–2055 (1993)

24. Triantafyllou, D., Mitrouli, M.: Two resultant based methods computing the greatest common divisor of two polynomials. In: Li, Z., Vulkov, L.G., Waśniewski, J. (eds.) NAA 2004. LNCS, vol. 3401, pp. 519–526. Springer, Heidelberg (2005)

25. Winkler, J.R.: Polynomial computations for blind image deconvolution (2015) Submitted

26. Winkler, J.R., Hasan, M.: A non-linear structure preserving matrix method for the low rank approximation of the Sylvester resultant matrix. J. Comput. Appl. Math. **234**, 3226–3242 (2010)

27. Winkler, J.R., Hasan, M.: An improved non-linear method for the computation of a structured low rank approximation of the Sylvester resultant matrix. J. Comput. Appl. Math. **237**(1), 253–268 (2013)

28. Winkler, J.R., Hasan, M., Lao, X.Y.: Two methods for the calculation of the degree of an approximate greatest common divsior of two inexact polynomials. Calcolo **49**, 241–267 (2012)

29. Zarowski, C.J., Ma, X., Fairman, F.W.: QR-factorization method for computing the greatest common divisor of polynomials with inexact coefficients. IEEE Trans. Signal Process. **48**(11), 3042–3051 (2000)

30. Zeng, Z.: The approximate GCD of inexact polynomials. Part 1: A Univariate Algorithm (2004) (Preprint)

Author Index

Alexandre, Radjesvarane 202
Al-Sahli, Reyouf S. 32
Apprich, Christian 1
Aràndiga, Francesc 16

Bang, Børre 60
Bejancu, Aurelian 32
Bobenko, Alexander I. 47
Bougleux, Sébastien 341
Bratlie, Jostein 60

Cantón, Alicia 70
Caputo, Manuel 80
Cavoretto, Roberto 96
Chkifa, Moulay Abdellah 109
Clarysse, Patrick 216
Cohen, Elaine 129
Cohen, Laurent D. 341
Coulaud, Benjamin 151

Dahl, Heidi E.I. 160
Dalmo, Rune 60
De Rossi, Alessandra 96
Delachartre, Philippe 216
Demers, Éric 169
Denker, Klaus 80, 181
Diop, El Hadji S. 202

Fernández-Jambrina, L. 70
Floater, Michael S. 210
Franz, Mathias O. 80

Gentil, Christian 414
Girard, Patrick R. 216
Gouaty, Gilles 414
Grohs, Philipp 243
Gruberger, Nira 263
Guibault, François 169
Guseinov, Khalik G. 280

Haimes, Robert 129
Hamann, Bernd 181
Hofreither, Clemens 272
Höllig, Klaus 1

Hörner, Jörg 1
Huseyin, Anar 280
Huseyin, Nesir 280

Jüttler, Bert 321

Kagan, Yael 289
Keller, Andreas 1
Khoromskij, B.N. 321

Langer, Ulrich 321
Laube, Pascal 80
Levin, David 263, 289

Mahdavi-Amiri, Ali 304
Mantzaflaris, Angelos 321
Mayr, Martin 96
Mille, Julien 341
Mishkinis, Anton 414
Monasse, Pascal 452
Morel, Jean-Michel 452
Mulet, Pep 16

Nava Yazdani, Esfandiar 1

Obermeier, Axel 243

Peter, Jacques 357
Pujol, Romaric 216

Qiao, Hanli 96
Quatember, Bernhard 96

Recheis, Wolfgang 96
Richard, Frédéric J.P. 151
Riesenfeld, Richard 129
Rosado María, E. 70
Runions, Adam 384

Samavati, Faramarz 304, 384
Schröder, Kristof 399
Sokolov, Dmitry 414

Takacs, Thomas   433
Tang, Zhongwei   452
Tribes, Christophe   169

Umlauf, Georg   80, 181

Vázquez-Gallo, M.J.   70

Wang, Liang   216
Weidner, Martin P.   47
Winkler, Joab R.   461

Yáñez, Dionisio F.   16

Zulehner, Walter   272

Printed in the United States
By Bookmasters